Clean Combustion Technologies

Energy, Combustion and the Environment

A series edited by Scott Samuelsen
Professor of Mechanical, Aerospace, and Environmental Engineering
Director, UCI Combustion Laboratory
University of California, Irvine

The subject of this series is the production of energy and the relationship of this production to environmental impact. The scope focuses on: (1) fossil fuels, since these are projected to serve as the principal energy sources for more than 100 years; (2) three levels of environmental air quality impact, including urban air quality, tropospheric global warming and acid rain, and stratospheric ozone degradation; and (3) both stationary and aeroengine sources of energy production.

Volume 1
Combustion Technologies for a Clean Environment
Edited by Maria da Graça Carvalho, Woodrow A. Fiveland, F. C. Lockwood, and Christos Papadopoulos

Volume 2
Clean Combustion Technologies — Parts A and B
Edited by Maria da Graça Carvalho, Woodrow A. Fiveland, F. C. Lockwood, and Christos Papadopoulos

Clean Combustion Technologies

Selected Papers from the Proceedings of the Second International Conference
Lisbon, Portugal, July 19–22, 1993

Part B

Edited by

Maria da Graça Carvalho
Instituto Superior Técnico
Lisbon, Portugal

F. C. Lockwood
Imperial College of
Science, Technology
and Medicine
London, UK

Woodrow A. Fiveland
Babcock & Wilcox
Alliance, Ohio, USA

Christos Papadopoulos
CINAR SA
Athens, Greece

Gordon and Breach Science Publishers
Australia • Canada • China • France • Germany
• India • Japan • Luxembourg • Malaysia • The Netherlands
• Russia • Singapore • Switzerland

Amsteldijk 166
1st Floor
1079 LH Amsterdam
The Netherlands

Some of the articles appearing in this book were originally published in *Combustion Science and Technology*, volume 108, numbers 4–6.

British Library Cataloguing in Publication Data

Clean combustion technologies : proceedings of the Second
  International Conference
  Part B - (Energy, combustion and the environment ; v. 2
  – ISSN 1073-7804)
  1. Combustion engineering - Environmental aspects -
  Congresses
  I. Carvalho, Maria da Graça
  628.5'32

  ISBN   90-5699-608-8 (part A)
         90-5699-621-5 (part B)
         90-5699-622-3 (two-part set)

CONTENTS

SECTION 9 — LOW EMISSIONS INTERNAL COMBUSTION ENGINES

SECTION 10 — NEW POWER GENERATION CYCLES OR CONCEPTS

SECTION 11 — CONTROL TECHNOLOGIES

SECTION 3 — OIL COMBUSTION GENERATED POLLUTANTS

SECTION 4 — GAS COMBUSTION GENERATED POLLUTANTS

INTRODUCTION TO THE SERIES

The world is in transition toward acknowledging, and accepting, that a basic conflict exists between the provision of energy to support the world's economies and standard of living, and the protection of the air resource that sustains the world's life. The *Energy, Combustion and the Environment* book series is designed to place this conflict into perspective, and to provide readers—students, consulting engineers, regulators, legislative staff, and environmental action groups—with the information and tools required to understand the challenges and conduct analyses to address these challenges.

PREFACE

The conference sequence on Combustion Technologies for a Clean Environment, popularly known as the "Clean Air" conferences, was initiated by its organizers because they felt that, despite the many events dealing with environmental issues, this particular topic had rather surprisingly been largely missed. The popularity and overall success of the first conference bore testament to the validity of their perception. The second conference proved even more successful; contained in this volume are 75 of the reviewed, refereed papers presented at the second Clean Air conference. Topics covered range from chemical kinetics fundamentals to applied research in power stations. New subject areas include: pollutant dispersion, hybrid fueled combustors, new power generation cycles and concepts, and catalytic combustion.

The mix of academic and industry contributions that drew praise during the first Clean Air conference has been retained and, we hope, enhanced. Contributions from invited lecturers, who again comprise geographically dispersed and internationally reputed experts, are included. At this second conference, because it is a European-initiated conference series, two of the invited speakers lectured on themes relevant to the European Union.

It is probably useful at a scientific conference to have some comment from government sources since one thing is certain; pollution abatement measures, like aircraft safety, generally cost money; and in a world that is now almost completely based on market economy concepts, disasters are—in the absence of legislation—inevitable. Most of the atmospheric pollution we experience is due to combustion processes, as much from motor vehicles as from industry. Worldwide, this is accelerating rather than diminishing and the same is true even in some developed countries. To an extent, it may be argued that pollution abatement technology has failed to keep pace with market and, therefore, combustion growth. But the failure is also attributable to inability of government to effect more stringent regulations in a world where public material expectations seem invariably to exceed green aspirations, since the latter undoubtedly imply a degree of economic restraint.

It is the hope of the organizers that the Clean Air conferences can continue to provide a forum for combustion technologies that will assist in suppressing the current dichotomy of economic prosperity and environmental protection.

ADVISORY COMMITTEE

SPONSORING ORGANIZATIONS

- ADIST – Associação para o Desenvolvimento do Instituto Superior Técnico
- Babcock & Wilcox
- Banco Comercial Português
- Banco Nacional Ultramarino
- Câmara Municipal de Lisboa
- Direcção-Geral da Quai Idade do Ambiente
- Department of Defense (USA)
- FLAD–Fundação Luso-Americana para o Desenvolvimento
- Fundação Calouste Gulbenkian
- GALP
- GDP Gás de Portugal, SA
- ICEP–Investimentos, Comércio e Turismo de Portugal
- IST–Instituto Superior Técnico
- ITEC–Instituto Tecnológico para a Europa Comunitária
- JNICT–Junta Nacional de Investigação Cientifica e Tecnológica
- Palácio de Queluz
- SECIL–Companhia Geral de Cal e Cimento, SA
- SHELL
- THERMIE

Abatement of NO$_x$ Emission from a Circulating Fluidized Bed Combustor

X. S. WANG[a,*], N. A. AKHTAR[a], B. M. GIBBS[a] and M. J. RHODES[b]

[a]*Department of Fuel and Energy, University of Leeds, LS2 9JT, U.K.;*
[b]*Department of Chemical Engineering, University of Bradford, BD7 1DP, U.K.*

Abstract—An investigation of NO$_x$ and N$_2$O emissions from a circulating fluidized-bed (CFB) combustor is reported. It was found that the NO$_x$ emission increased whilst the N$_2$O emission decreased with increasing temperature. Both NO$_x$ and N$_2$O emissions were found to increase with increasing excess air factor. Compared to unstaged combustion, the staged combustion appeared to be more effective in controlling the NO$_x$ emission. The effect of air staging on the N$_2$O emission was found to be small. In order to minimise NO$_x$ and N$_2$O emissions from a CFB combustor, the emission of NO$_x$ and N$_2$O needs to be controlled in stages.

Key Words: NO$_x$; N$_2$O; circulating fluidized bed; coal combustion

INTRODUCTION

Continuing efforts have been made to optimise the control of NO$_x$ and N$_2$O emissions from coal combustion in fluidized-bed combustors. The work of Roby and Bowman (1987) showed that the reduction of N$_2$O became more effective at higher temperatures. Boyd (1990) and Leckner *et al.* (1992) observed an increase in NO$_x$ but a decrease in N$_2$O emission with increasing temperature. Moritomi *et al.* (1990) suggested that an effective reduction in N$_2$O emission was possible at higher temperatures. Gavin and Dorrington (1993) also observed a higher NO$_x$ and a lower N$_2$O emission with increasing temperature. Lundqvist *et al.* (1991) found that the NO$_x$ emission generally increased with increasing temperature. Gierse (1990) and Basak *et al.* (1991) observed the sensitivity of NO$_x$ emission with both temperature and excess air factor. Kullendorff and Andersson (1985) reported an increase in NO$_x$ emission with increasing excess air factor. Leckner and Gustavsson (1991) reported a reduction in the NO$_x$ emission of upto 80% at 2% O$_2$ in the flue. Mjornell *et al.* (1991) reported N$_2$O emission in the range 40 to 250 ppmv (at 6% O$_2$), whilst Hiltunen *et al.* (1991) reported the N$_2$O emission from 30 to 130 ppmv (3% O$_2$). The staged combustion has been demonstrated to be effective in controlling the emission of NO$_x$ by Suzuki *et al.* (1990), Steinruck (1990) and Asai *et al.* (1990). In this paper we report on results of an experimental study of the effect of operating conditions on NO$_x$ and N$_2$O emissions from a CFB combustor. Strategies for controlling NO$_x$ and N$_2$O emissions from the CFB are discussed based on the experimental results.

*Research Fellow on a joint project between the University of Bradford and the University of Leeds.

EXPERIMENTAL

A schematic of the CFB combustor used in this investigation is shown in Figure 1. It consisted of a 0.161-m id and 6.2-m tall riser, primary and secondary cyclones, an external heat exchanger (EHE) and a solids return valve (i.e., L-valve). Silica sand having a mean diameter of 120 μm a density of 2500 kg/m^3 was used as the main circulating material. A Dawmill bituminous coal was used as the fuel. Ultimate and proximate analyses of the coal are given in Table I. Primary air was supplied at the bottom of the riser through an air distribution plate, and secondary air could be injected at 1.32 or 1.83 m above the air distribution plate. Solids separated by the primary cyclone from gas-solid suspensions were fed to the EHE via a dipleg. The solids from the EHE were returned to the main combustion chamber (riser) through the L-valve.

The CFB system was warmed-up to around 600°C by a gas burner before coal was fed to the riser. Temperature and pressure profiles were measured by thermocouples and pressure transducers, respectively. Concentrations of combustion gases were measured by on-line analysers. Here, O_2 concentration was measured by a paramagnetic analyser, CO concentration was measured by an NDIR analyser and the concentration of NO_x was measured by a chemiluminescent analyser. Dry flue gas was also sampled with Teflon bags which enabled the concentration of N_2O to be measured

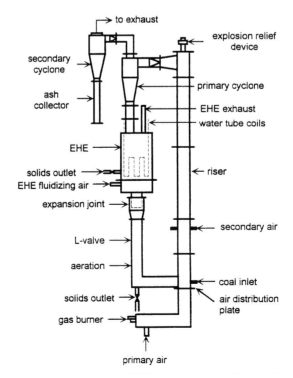

FIGURE 1   A schematic of the CFB combustion system.

TABLE I

Ultimate and proximate analyses of the coal used

Ultimate	(wt.%)	Proximate	(wt.%)
C	74.4	Moisture	6.3
H	4.7	Ash	4.5
O	7.5	Volatile	36.6
N	1.1	Fixed carbon	52.6
S	1.4		

using gas chromatography. The CFB was operated at a solids circulation rate and a superficial gas velocity (above the secondary air jets) of 20 kg/m²s and 5 m/s, respectively.

RESULTS AND DISCUSSION

An example of the concentration profiles of NO_x and N_2O is shown in Figure 2. Here, the CFB was operated at 830°C and at an excess air factor of 20%. Secondary air was injected at 1.83 m above the primary air inlet, and the secondary air/total air ratio was 0.20. The detected high level of NO_x concentration in the bottom of the riser is thought to be due to volatile NO_x formation during devolatilization of coal by oxidation of coal-N. The NO_x formed from coal devolatilization then diffuses into the boundary layer around the burning char particles with CO, where it is either reduced by direct reaction with CO (catalysed by the char surface) or by the reaction between CO and chemisorbed oxygen deposited by NO on the char surface (Suzuki *et al.*, 1990). The N_2O concentration profile in Figure 2 suggests that N_2O is continuously generated

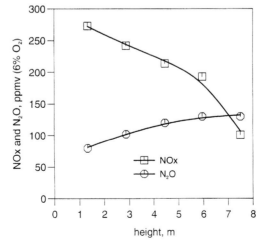

FIGURE 2 NO_x and N_2O profiles in the CFB furnace.

along the riser height, and that the N_2O generation could occur preferentially in the secondary combustion zone where the char combustion is predominant. This result agrees with that of Moritomi *et al.* (1990). It is believed that the N_2O could be generated from HCN during volatiles combustion and from char-N during char combustion.

The effect of temperature on NO_x and N_2O emissions is shown in Figure 3. In this case, the secondary air was injected at 1.32 m above the primary air inlet and other conditions were the same as those for Figure 2. The N_2O emission at the exit of the combustor decreased from 196 to 125 ppmv (3% O_2) when the temperature in the riser was increased from 740° to 860°C. On the other hand, an increase in temperature from 710°C to 850°C caused the NO_x emission to increase from 50 to 160 ppmv (3% O_2). The results shown in Figure 2 agree with those of other researchers (e.g., Hiltunen *et al.*, 1991; Wojtowicz *et al.*, 1991; Gavin and Dorrington, 1993). The sum of N_2O and NO_x emissions also increased with temperature for the operating conditions examined, which is different from what was observed by Gavin and Dorrington (1993). The reason for the increase in the NO_x emission with temperature may be that higher temperatures promoted more rapid devolatilization of coal which resulted in a higher volatile-NO_x emission. Since both NO_x and N_2O are generated from the same precursor (HCN), the lower N_2O emission at a higher temperature may be at the price of a higher NO_x emission.

The effect of excess air factor on NO_x and N_2O emissions is shown in Figure 4. Here, the operating conditions (apart from temperature) were the same as those for Figure 3. It shows that both NO_x and N_2O emissions in the CFB increased with increasing excess air factor. Taking as an example when the excess air factor in the riser was increased from 5% to 65%, the N_2O emission increased from 100 to 210 ppmv (3% O_2), and the NO_x emission increased from 95 to 265 ppmv. This result agrees with that of Hiltunen *et al.* (1991) but differs from what was reported by Moritomi *et al.* (1990), in which the emissions of NO and N_2O were almost invariable with increasing excess air

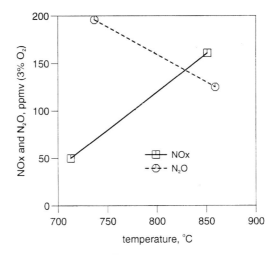

FIGURE 3   Effect of temperature on NO_x and N_2O emissions.

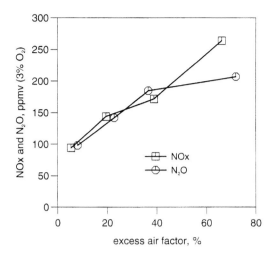

FIGURE 4   Effect of excess air factor on NO_x and N_2O emissions.

factor. The increase in both NO_x and N_2O emissions with increasing excess air factor may be explained in light of the combustion characteristics of coal under similar conditions. During experiments we observed an increase in CO concentration at the base of the riser from 0.1% to 0.14%, and then to 0.3% when the excess air factor was increased from 5% to 20%, and then to 40%, respectively. This indicates that coal devolatilization could be enhanced by an increase in the excess air factor. The subsequent combustion of volatiles therefore produced more NH_3 and HCN, which caused both NO_x and N_2O emissions to increase.

The effects of secondary air ratio (i.e., secondary air/total air ratio) and air staging height (above the primary air inlet) on NO_x and N_2O emissions are shown in Figure 5

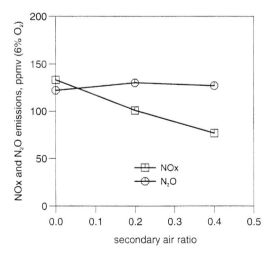

FIGURE 5   Effect of secondary air ratio on NO_x and N_2O emissions.

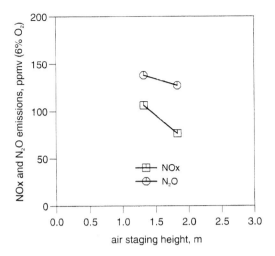

FIGURE 6   Effect of air staging height on NO_x and N_2O emissions.

(with air staging height 1.83 m) and Figure 6 (with secondary air ratio 0.40), respective-ly. It can be noted that the NO_x emission decreased with increasing secondary air ratio and air staging height, whilst the N_2O emission was relatively insensitive to the secondary air ratio and it decreased only slightly with the air staging height. Detailed analyses of the effects of air staging on NO_x and N_2O emissions are described elsewhere (Wang *et al.*, 1994).

In view of above experimental results, the CFB combustion is advantageous in controlling the NO_x emission whilst it is a source of the N_2O emission. It appears, from Figure 4, that both NO_x and N_2O emissions could be controlled by reducing the overall excess air factor. Since a lower excess air factor could result in a lower combustion efficiency (Gulyurtlu and Cabrita, 1985), the strategy of applying a very low excess air factor may only be applicable when techniques are available which could enhance the combustion performance. The results shown in Figures 2 and 3 and in Wang *et al.* (1994) suggest, on the other hand, that the NO_x and N_2O emissions could be controlled in stages. For example, the NO_x could be controlled by adjusting the primary air stoichiometry and temperature in the lower part of the riser. The resulted N_2O emission could then be reduced by increasing the local temperature, or by adding catalytic material, at the exit of cyclones.

CONCLUSIONS

Experimental results show that (1) the NO_x emission increases whereas the N_2O emission decreases with increasing temperature; (2) both NO_x and N_2O emissions increase with increasing excess air factor, and (3) the emission of NO_x could be effectively controlled by air staging. In order to minimise NO_x and N_2O emissions from a CFB, the emission of NO_x and N_2O has to be controlled in stages.

ACKNOWLEDGEMENT

The research work described in this paper was carried out with the financial support of the Science and Engineering Research Council.

REFERENCES

Amand, L. E. and Andersson, S. (1989) Emissions of nitrous oxide from fluidized bed boilers. *Proceedings of the 10th Int. Conf. on FBC*, San Francisco, USA, ASME, 49.

Asai, M., Aoki, K., Shimoda, H., Makino, K., Watanabe, S. and Omata, K. (1990) Optimization of circulating fluidized bed combustion. *Proceedings of the 3rd Int. Conf. on CFB*, Nagoya, Japan, 379.

Basak, A. K., Sitkiewitz, S. D. and Friedman, M. A. (1991) Emission performance summary from the Nucla circulating fluidized bed boiler demonstration project. *Proceedings of the 11th Int. Conf. on FBC*, Montreal, Canada, 211.

Boyd, T. J. and Friedman, M. A. (1990) Operations and test programme summary at the 110 MW Nucla CFB. *Proceedings of the 3rd Int. Conf. on CFB*, Nagoya, Japan, 297.

Gavin, D. G. and Dorrington, M. A. (1993) Factors in the conversion of fuel nitrogen to nitric and nitrous oxides during fluidized bed combustion. *Fuel*, **72**, 381.

Gierse, M. (1990) some aspects of the performance of three different types of industrial circulating fluidized bed boilers. *Proceedings of the 3rd Int. Conf. on CFB*, Nagoya, Japan, 347.

Gulyurtlu, I. and Cabrita, I. (1985) Combustion of woodwastes in a circulating fluidized bed. In P. Basu, (ed.), Circulating Fluidized Bed Technology, Pergammon Press, Toronto, 247.

Hiltunen, M., Kilpinen, P., Hupa, M. and Lee, Y. Y. (1991) N_2O emissions from CFB boilers: experimental results and chemical interpretation. *Proceedings of 11th Int. Conf. on FBC*, Montreal, Canada, *ASME*, 687.

Kullendorff, A. and Andersson, S. (1985) A general review on combustion in circulating fluidized beds. In P. Basu, (ed.), Circulating Fluidized Bed Technology, Pergammon Press, Toronto, 83.

Leckner, B. and Gustavan, L. (1991). Reduction of N_2O by gas injection in CFB boilers. *J. Inst. Energy*, **64**, 176.

Leckner, B., Karlsson, M., MJornell, M. and Hagman, U. (1992) Emissions from a 165 MW circulating fluidized bed boiler. *J. Inst. Energy*, **65**, 122.

Lundqvist, R., Basak, A. K., Smedley, J. and Boyd, T. J. (1991) An evaluation of process performance and scale up effects for Ahlstrom Pyroflow circulating fluidized bed Boilers using results from the 110 MW Nucla boiler and a 0.6 MW pilot plant. *Proceedings of the 11th Int. Conf. on FBC*, Montreal, Canada, ASME, 131.

MJornell, M., Leckner, B., Karlsson, M. and Lyngfelt, A. (1991) Emission control with additives in CFB coal combustion. *Proceedings of the 11th Int. Conf. on FBC*, Montreal, Canada, ASME, 655.

Moritomi, H., Suzuki, Y., Kido, N. and Ogisu, Y. (1990) NO_x emission and reduction from a circulating fluidized bed combustor. *Proceedings of 3rd Int. Conf. on CFB*, Nagoya, Japan, 399.

Roby, R. J. and Bowman, C. T. (1987) Formation of N_2O in laminar, premixed, fuel rich flames. Combustion and Flame, **70**, 119.

Steinruck, P. (1990) The SGP fast internally circulating bed process. Urja/Dec. 1990/28/6/33.

Suzuki, T., Hirose, R., Takemura, M., Moritoma, A., Yano, K. and Hyvarinen, K. (1990) Comparison of NO_x emissions between laboratory modelling and full scale Pyroflow boilers. *Proceedings of the 3rd Int. Conf. on CFB*, Nagoya, Japan, 387.

Wang, X. S., Gibbs, B. M. and Rhodes, M. J. (1994) Impact of air staging on the fate of NO and N_2O in a CFB combustor. Combustion and Flame (in press).

Wojtowicz, M. A., Oude, J. A., Tromp, P. J. J. and Moulijn, J. A. (1991) N_2O formation in fluidized bed combustion of coal. *Proceedings of the 11th Int.Conf. on FBC*, Montreal, Canada, ASME, 1013.

Simulation of a Circulating Atmospheric Fluidized Bed Including Flow, Combustion, Pollutant Formation and Retention

R. LEITHNER, S. VOCKRODT, J. WANG, A. SCHULZ and J. MÜLLER

Institut für Wärme- und Brennstofftechnik (IWBT), Franz-Liszt-Straße 35, 38106 Braunschweig, Federal Republic of Germany

Abstract—For simulation of Circulating Atmospheric Fluidized Bed Combustion (CAFBC) a 1D- and 3D-model have been developed at the IWBT.

The 1-dimensional model can calculate the two-phase flow and coal combustion process including raw coal drying, coal pyrolysis and char combustion within CFB Combustors.

The 3-dimensional model consists of two coupled submodels (flow field- and combustion-calculation model including NO_x- and SO_2-formation and retention) and two postprocessing models (detailed combustion- and pollutant model). All 3-dimensional terms will be calculated by the finite-volume-method.

The theoretical model predictions have been compared with the measurements carried out in a CFBC test rig with a thermal capacity of 1, 2 MW in the power plant of RWE at Niederaussem.

Key Words: Mathematical model; fluidized bed

1. INTRODUCTION

During recent years many facilities using Fluidized Bed Combustion (FBC) went into service or were under construction. In the course of this a shift took place from Stationary FBC to pressurized FBC and Circulating Atmospheric FBC (CAFBC), respectively.

Especially the breakthrough of the CAFBC can be explained by the fact that the CAFBC leads to very low emissions of Sulphur Dioxide (SO_2) and Nitric Oxides (NO_x) without secondary flue gas purification. Furthermore, it is possible to burn the most different and poor quality solid, fluid or gaseous fuels in nearly any combination.

But there are still problems with the relatively high emissions of Carbon Monoxide (CO) and chloridic and fluoridic compounds with the flue gas and fly ash, respectively. The allowed limits of emission of those air pollutants can hardly be kept.

Although the cause for high emission of CO is principally known, the necessary details of the combustion and the burn out behaviour of solid fuels and the influence of SO_2- reducing additives on the formation and retention of air pollutants are still widely unknown.

Until today, most research on combustion modelling and air pollution control inside Fluidized Bed Combustors is based on current work on Stationary of Bubbling FBC's or on experimental results from laboratory or model furnaces. On the other hand there are many results from existing CAFBC commercial plants which can be taken into account.

This paper presents the development of a mathematical model to calculate combustion and burn out behaviour of some different European solid fuels in the CAFBC including terms to calculate the kinetics of air pollutant formation and retention.

Two different model approaches are considered:

- A special 1D-Model is developed based on the classifying effect of fluidized beds.
- Parts of an existing 3D-furnace model are used to calculate flow conditions, heat transfer coefficients and kinetics of burn out and air pollutant formation for the conditions of two phase gas/solid flows with a high particle load.

In order to validate the mathematical models and to find optimal combustion conditions for commercial plants the model calculations are compared with experimental results taken from an experimental CAFBC test rig of 1, 2 MW thermal capacity in Niederaussem.

2. THE ONE-DIMENSIONAL CAFBC-MODEL

2.1   The Flow Model

The polydisperse solid particles inside the fluidized bed show a special particle size distribution. Building up on this fact, the flow model is based on the following approach: if a single particle is streamed upwards by gas, the particle will fall, float or rise, it depends on whether the velocity of gas flow is lower, equal to or higher than the terminal velocity of the corresponding particle. As an averaged flow velocity, the superficial gas velocity divided by the porosity is used. With this model, the continuous distribution of porosity resp. the continuous pressure loss over the height of the fluidized bed dependent on known quantities as the particle size distribution, the gas flow rate etc. can be calculated. The recirculating solids mass flow can be calculated simultaneously. This model is described with further details at Leithner and Wang (1990a, 1990b and 1991) and Wang (1993).

2.2   Coal Particle Distribution inside the Fluidized Bed

To model the combustion of coal in a CAFBC, the coal particle distribution over the height of the fluidized bed has to be determined. For that, the fluidized bed combustion chamber will be subdivided into several layers. Furthermore, the continuous coal combustion is discretised by introducing groups of matter (e.g., 3 groups: raw coal, dried coal and devolatilized coal (char)), each of which has its own density.

Because of their little amount, compared with the amount of inert material, the coal particles do not hardly influence the flow and e.g. the distribution of porosity of the inert material. It can be supposed simplifying, that those coal particles, that are fed into the combustion chamber at certain intervals and those particles, that originate from them by combustion, distribute among their position (layer), in which they float or rise without loss of time. The position can be determined by using the flow model described in Leithner and Wang (1990a, 1990b and 1991), Wang and Leithner (1993) and Wang (1993). Figure 1 shows the schematic display of the process of coal distribution. Every

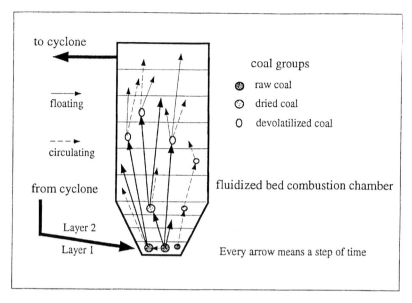

FIGURE 1   Schematic display of the processes of coal distribution.

arrow in the figure means a step of time. If there is no sign at the end of the arrow, the particle is burnt out and the ash is regarded as disappeared.

For every coal particle distribution over the height of the fluidized bed resp. after every step of time, the calculation of coal drying, of devolatilization, of combustion of the volatile matter and of the residual charp resp. the decrease of the residual char diameter is carried out step by step from the bottom to the top. Furthermore, only particular fractions can reach the following steps (e.g., from raw coal to dried coal or directly to the residual char, or from dried coal to the residual char) according to the interval, the particle diameter, the gas temperature, the velocity of gas flow etc. The non-dried part of the raw coal particles will remain in their original layer. It behaves in the same way for those particles, that were devotilized resp. partly burnt out. During large intervals, it may happen, that e.g. the raw coal particles begin to devolatilize or to burn out. The dried part of the coal resp. the devolatilized part of the coal after drying is equal to the ratio of the given interval to the time used by drying resp. the devotilization. The amount of the burnt coal results from the model of char combustion. There are stored as many coal particles, till the supplied energy (coal flow) is equal to the energy laid off by the combustion of the volatile matter and the residual char, so that the balance of energy and mass is solved in combination.

2.3   Combustion of a Single Particle

Raw coal particles, that are burnt out inside the fluidized bed combustion, run through different processes: heating up, drying, devolatilization and combustion of the volatiles and of the char (see Fig. 10 in section 3.2). The drying time consists of the time necessary to heat up the particle and to release its moisture. During the evaporation of the water, there is not only procured the evaporation enthalpy of the water, but also the isoteric desorption heat of the coal moisture. The total drying time is calculated from the energy balance of a single coal particle.

For coal devotilization, the multicomponent-model of Suuberg *et al.* (1978) is used, which is based on the assumption, that the volatiles consist of several (j) components. The formation of each component proceeds independent from the other and each volatile species (or the reaction steps i of the component) reaches a special substance and has a special activation energy and a special frequency factor. So the total devolatilization is modeled by many single step reactions. Each of these reactions is described by the following approach:

$$\frac{dV_i}{d\tau} = -k_i \cdot V_i^n; \qquad k_i = k_{i0} \cdot e \frac{-E_i}{RT}, \tag{1}$$

where V_i is the actual amount of volatile species i at the moment τ, E_i the appearant activation energy, k_{i0} the frequency factor and n the order of reaction.

The reaction rate of the char combustion M_k is calculated according to Turnbull and Davidson (1984):

$$\frac{dM_k}{d\tau} = -\pi d_k^2 \, \phi \, \tilde{M}_c \, k' \, CO_2, \tag{2}$$

where CO_2 is the molar concentration of oxygen, ϕ the coefficient of the reaction mechanism, d_k the particle diameter and $\tilde{M}_c$ the molar of the carbon. The coefficient k' of the combustion reaction rate at the surface of the particles takes into account both, chemical reaction and the influence of the mass transfer. It is defined by:

$$k' = \frac{1}{\dfrac{1}{k'_c} + \dfrac{1}{k'_D}}, \tag{3}$$

where k'_c is the coefficient of superficial reaction and k'_D the diffusion coefficient.

2.4   Heat Transfer and Energy Balance

The single heat transfer coefficients of the particle convection, the gas convection and the radiation of gas and particles can be added to a total heat transfer coefficient. For the radiation heat transfer, the Mie-theory (1908) is used with a two-flux-model.

The combustion chamber is subdivided over the height into a number of vertical layers (lying on top of each other), so that the balancing inside the combustion chamber is only caused by the entering and escaping flue gas mass flow, by the feeded gas flow and possibility also the recirculation gas flow, by the floating inert material, by the recirculation solid mass flow as well as by the water resp. steam temperature at the wall.

2.5   Description of the Test Rig

In order to validate the mathematical models and to investigate the burn out behaviour and pollutant emission of different coals and additives an experimental CAFBC test plant was erected and put into service at the Niederaussem power plant of RWE near

Cologne. The test rig was erected and is owned by the EVT, Stuttgart. The operator is the KF-FS research group of RWE.

The test rig with a thermal capacity of 1, 2 MW consists mainly of the following parts: a rectangular combustion chamber, an ash classifier to discharge the large particles, a cyclone to separate the recirculated solids from the exhaust gas flow, a flue gas cooler, an air preheater and a bag filter to separate the fine fly ash material.

The combustion chamber with a total height of 9, 5m consists of two sections. The lower one is the furnace hopper with a height of 3, 15 m. It is totally refractory lined and the cross section enlarges from 0, 39 × 0, 45 m² at the distributor plate to 0, 54 × 0, 81 m² at the top of the hopper.

The rectangular freeboard with a cross section of 0, 54 × 0, 81 m² and a height of about 6, 35 m is constructed by finned tube walls. Cooling medium is reheater steam from boiler D. Two of the walls with 26 tubes (from a total of 48) are refractory lined. The steam flow through these walls is lowered by throttling. While cooling the furnace chamber, the reheater steam is heated up from about 300°C to 310°C. The total pressure drop of the reheater steam is about 8 bars caused by throttling. The geometry of the combustion chamber is shown in Figure 2.

The combustion air is preheated by the flue gas in a tubular air preheater and fed as primary air through the distributor plate. Additional primary air is fed into the bottom region of the furnace hopper and with the fuel and recirculated bed material.

Secondary air can be fed at five inlets along the combustion chamber (two inlets and air inlet at auxiliary burner in the hopper and two inlets in the freeboard). The first inlet is the cooling air for the propane fired ignition and auxiliary burner. Three other air inlets allow the operation of the test plant with different air staging levels to investigate the NO_x-formation and retention. The possible air staging range differs from a value of 14% secondary Air/86% primary air to a staging of 50% to 50% between secondary and primary air.

To reduce the bed temperature, cold flue gas can be recirculated and mixed with the combustion air. Normally the recirculating gas is only mixed with distributor and the coal feeder air. But some runs took place with feeding of recirculating flue gas to all air inlets. The levels of the air inlets are shown in Figure 2, too.

The solid material, consisting of sand, ash and additives like lime-stone (about 98%) and fuel (about 2%) is fluidized by the primary air. While burning fuels with low ash content or a very fine ash particle size distribution sand material has to be fed into the combustor to keep the mass of the bed material constant. On the other hand the ash classifier allows to separate large ash particles from the bed during operation.

2.6 Examples of Calculation

Together with the partial models of heating up, drying, devolatilization and char combustion, the distribution of temperature, the flue gas concentration, the heat flux density and the coal particles etc. over the height of the fluidized bed can be determined with the energy balance and the model of the distribution of the coal particles based on the two-phase flow model.

Figure 3 shows the concentration profiles O_2, H_2O, CO_2, N_2 and SO_2 (without retention) for the test case A, calculated by the global model. The water vapour reaches

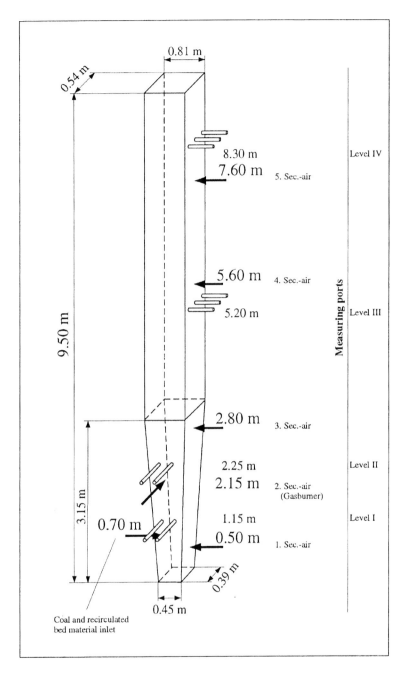

FIGURE 2   Combustion chamber of the Nederaußem test rig.

about 25 vol.-% at the bottom below 0.4 m due to the coal drying below this height. The water vapour is then diluted by the first secondary air inlet reaching water vapour concentration in the flue gas of 18 vol.-%. This high level of water vapour concentration

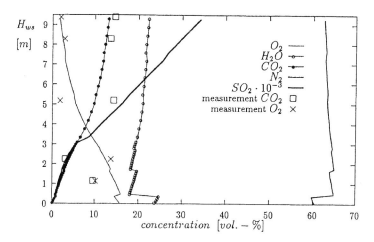

FIGURE 3   Concentration in the CAFB pilot plant at Niederaußem.

(21.5 vol.-% at the top) is related to the high moisture content in the raw coal particle (57 w-%). The CO_2 and SO_2 concentrations increase rather uniformly along the height. The O_2 concentration decreases uniformly along the height. The secondary air at the test case A is relatively uniform and less significant in comparison with primary air.

Figure 4 shows the comparison of temperature profiles between the model predictions and the measurements carried out in the CAFB pilot plant at the test cases A and B.

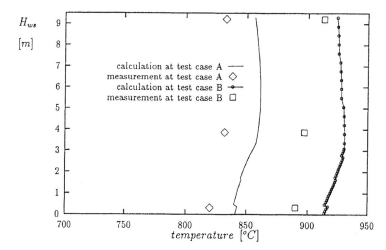

FIGURE 4   Temperature profiles.

3. THE 3-DIMENSIONAL COUPLED CAFBC-MODEL

The 3-dimensional coupled CAFBC-model consists of 6 parts (see Fig. 5):

- Calculation of the two-phase flow
- Calculation of the heat transfer, of the global combustion and of the global energy balance
- Calculation of the detailed combustion as postprocessing
- Calculation of the NO formation
- Calculation of the SO_2 retention
- Graphical postprocessing

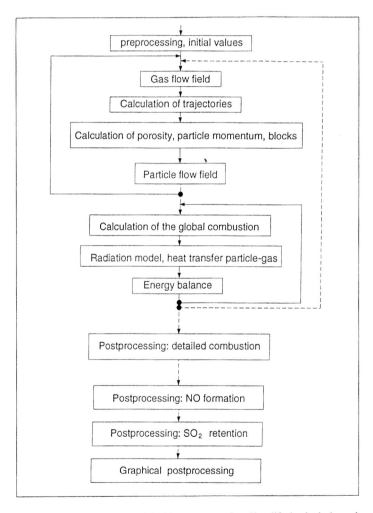

FIGURE 5   Coupled CAFBC-model with postprocessing: Simplified calculation scheme.

After the calculation of the two-phase flow, the heat transfer particle-gas is determined. Afterwards the global combustion is calculated and the energy balance is solved. With the new temperature field, the heat transfer etc. is calculated again, until the temperatures do not change any more. The two-phase flow is determined by the new calculated quantities and the iteration process begins again.

At the end, the detailed combustion and the formation and retention of pollutants is modelled, without coupling and calculating the flow and temperature field once again.

3.1 3-D Flow Model

At the Institut für Wärme- und Brennstofftechnik, Müller (1992) developed a 3-D code, able to model three-dimensional flow including combustion and chemical reactions. The model solves the mass, momentum and energy balance equations using the finite-volume-method. To calculate the turbulence, the k-ε model is used. First, a preprocessing code fixes the geometry (see Fig. 6, see also Fig. 2) of the combustion chamber as well as the initial values (entering velocities, temperatures, dissipations, eddy energies, physical characteristics and wall temperatures). The central processor solves the balance and the calculated quantities for each volume.

The flow of the particle phase is calculated using the Lagrangian approach.

For lots of different particle locations, starting velocities and diameters, trajectories are determined inside the gas flow field and provided with a solid matter load.

Trajectories, returning to the neighbourhood of their starting location, are considered as closed. Only escaping and closed trajectories are used for the simulation. Due to the high concentration of solids inside a circulating fluidized bed, particle–particle-interactions cannot be neglected, therefore trajectory collisions are calculated with exchange of mass and momentum between each other. The calculation bases on the following assumptions:

- The trajectories do not change their directions after a collision.
- All collisions are central impacts without any losses.
- The flying particles are supposed as momentum flows.
- During a collision, it occurs only a loading exchange.

After the calculation of the crossing points of the trajectories the momentum flows at this location are estimated.

This calculation causes, that some trajectory sections take up substance of all the crossing trajectory sections and therefore their mass flow rises continuously, whereas other trajectory sections impoverish within their loading. This causes a partial segregation and areas with many particles and other nearly free of solids. The main problem of this modelling is the high computer time. So for the test cases with a big number of trajectories this model is not used.

In order to consider the turbulent particle transport, the Gaussian plume-model (Menke, 1989) was adapted. With this model the behaviour of many particles can be modelled with only few trajectories.

For the simulation of the distribution the following topics are assumed:

- The calculated trajectory is an averaged curve, where all starting particles have the same properties (density, diameter, etc.).

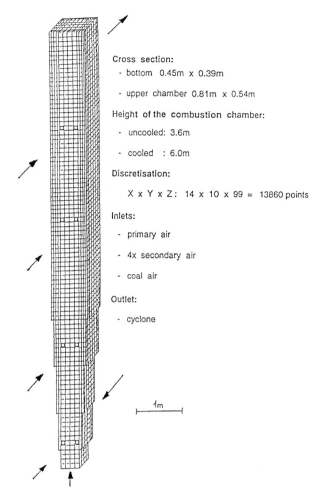

Cross section:
- bottom 0.45m x 0.39m
- upper chamber 0.81m x 0.54m

Height of the combustion chamber:
- uncooled: 3.6m
- cooled   : 6.0m

Discretisation:

    X x Y x Z: 14 x 10 x 99 = 13860 points

Inlets:
- primary air
- 4x secondary air
- coal air

Outlet:
- cyclone

1m

FIGURE 6   Finite volume geometry of the combustion chamber at the CAFBC-test rig at Niederaussem.

- The trajectory is idealized with sections of straight lines.
- For every section the distribution will be calculated and added to the distribution before.

As a result of the calculation, a continual widened trajectory will be modelled (Fig. 7). Some tests show that the distribution-diameters of these widened trajectories do not attain the dimension of the grid geometry, so the calculation has no influence on the calculation of the porosity.

At high mass loads, like in Circulating Atmospheric Fluidized Bed Combustors, the particle-gas interactions are of particular significance.

The influence of the gas phase on the solid phase is modelled with the calculating of the trajectories. The feedback effects between the solids and the gas are realized by two couplings:

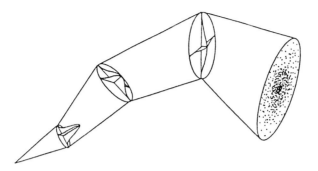

FIGURE 7   The Gaussian plume model [18].

- The momentum coupling is caused by the well-known PSIC-method (**Particle-Source-In-Cell**).
- If a trajectory enters a control volume, the penetration area is found out and afterwards blocked.

Figure 8 shows the coupled gas velocity flow field in different cross sections. It can be seen, that there are several eddies and downward-flow sections near the wall of the

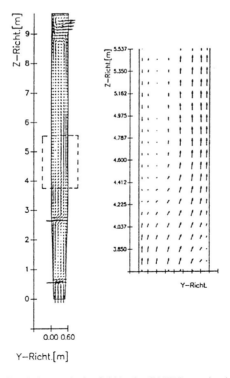

FIGURE 8   Coupled gas velocity field in the CAFBC-test rig niederaussem.

reactor. The highest velocity is reached at the cyclone outlet. Figure 9 shows the corresponding particle velocity field. There is no main direction of velocity visible. Near the wall the particles are moving downwards.

3.2 The Modelling of the Coal Combustion and of the Energy Balance

The combustion model of the CAFBC has to meet two tasks. On the one hand, it has to deliver source terms for the energy balance, concerning the energy release by combustion. On the other hand, it has to calculate the concentrations of radicals, necessary for the calculation of the formation of NO_x within the flue gases. The latter task requires a detailed kinetic reaction scheme, which is too extensive to be solved coupled with the calculation of the flow and temperature field. So, the combustion model is divided into two particular models (see the flow chart, Fig. 5).

- A simple global model to solve the energy balance and to determine the essential flue gas components.
- A more detailed combustion model to calculate the composition of flue gases including the radicals (postprocessing).

Contrasting the inert bed particles, which movement is modelled with the Langragian approach, the coal particles are modelled with a continuous formulation, the

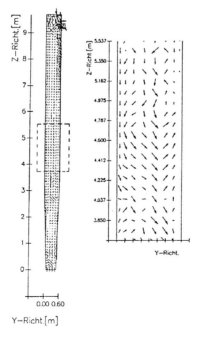

FIGURE 9   Coupled particle velocity field in the CAFBC-test rig Niederaussem

Eulerian approach. Due to the small mass fraction of the fuel phase, this is allowed. So the energy balance can be solved by using the existing combustion model of Müller (1992).

Then the following calculation scheme results to solve the energy balance:

- Calculation of the combustion using the following partial models: drying, devolatilization, combustion of volatiles and char (Fig. 10).
- Gas-particle heat transfer by convection and radiation.
- Energy balance of the gas and particle phase and calculation of the temperature field.

The changed particle size distribution and the combustion of raw lignite are taken into account by modelling of several particle sizes and use of a drying model. The single overall reaction model for devolatilization used by Müller (1992) is used with kinetic parameters adapted to the conditions of the fluidized bed combustion and the modelled coal (Krugmann, 1992).

The 6-flux model, used to calculate the heat transfer by radiation, is extended to the high loading of particles by adapting the Mie-scattering theory (Fischer, 1991 and Mie, 1908). Terms to calculate the particle convective part have still to be implemented.

The concentrations of the inert particles in every single volume cell and the entering and escaping mass flow can be calculated from the calculated trajectories and their loadings. Though the inert particles do not influence the chemical reactions, the energy balance of the particle phase consists only of heat exchange with the surrounding gas phase (conduction, convection and radiation).

With this model, the corresponding temperature field of the calculation velocity distribution is solved. Afterwards the flow field is calculated again, until the criterions of convergence are met. The released and transmitted amounts of heat, the temperatures inside the combustion chamber as well as the composition of the flue gases (O_2, CO_2, CO, H_2O and N_2) are calculated. The distribution of the fuel, consisting of wet coal, dried coal and char is calculated for the different particle sizes.

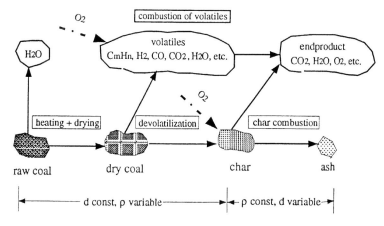

FIGURE 10   Structural change of a coal particle during combustion.

3.3 Postprocessing: A More Detailed Combustion Model

To simulate the formation of the pollutants, especially the formation and retention of NO_x, a good knowledge of the distribution of several intermediate products of the combustion, especially of the O-, H- and OH-radicals, is necessary. For this, the volatile combustion is very important, because it controls the stability of reaction as well as the formation and concentration of the radicals. Therefore an extended combustion model is enclosed as a postprocessor, using as input data the distribution of the dried coal and of the particle temperatures and the composition of the flue gases, calculated by the global model. It consists of three parts:

- A multicomponent devolatilization model.
- A char combustion model.
- A kinetic reaction model for the volatile and CO combustion within the gas phase.

The devolatilization of the volatiles is calculated with the multi-component model of Suuberg *et al.* (1978 a and 1978 b). The total yields correspond very well with the results of the single overall reaction model used to calculate the global combustion.

Combustion of the residual char is modelled with the assumption, that both CO and CO_2 are formed at the surface of the particles. The ratio of both species is a function of the particle temperature and the particle size.

To examine detailed the process of the formation of NO_x at pulverized coal combustion, Lendt (1991) has used a model of volatile combustion at the IWBT, that mainly considers the combustion of CH_4 and the $H_2/O_2/H_2O$ mechanism and was derived from the scheme published by Smoot and Pratt (1979). This model is the basic attempt for the volatile combustion model for the fluidized bed combustion. The calculations of Lendt show, that a neglect of the C_2-hydrocarbons represents a big simplification. This also corresponds with the detailed examinations of the volatile formation of Suuberg *et al.* (1978 a and 1978 b) and Solomon *et al.* (1992), who indicate the following substances as the main parts of the volatiles: CO_2, CO, H_2O, H_2, CH_4, C_2H_6 and C_2H_2 as well as tars and other higher hydrocarbons.

It is supposed within this model, that the latter ones are cracked by secondary reactions inside or at the surface of the particles and breakup into lighter volatiles and semichar (Waldmann, 1991). So the existent model was first extended by the decomposition reactions of C_2H_6, C_2H_4 and C_2H_2. The resulting system of 21 species and 122 (61 forward and 61 back) elementary reactions, proved to be very difficult in handling. The model required very big calculation times. The accuracy and reproducibility of the results show only few satisfactory. To find out the influence of the single reactions on the total model and thus to simplify the model, the system was undertaken a sensitivity analysis (Gaebel, 1992). The influence of the different reactions on the total model was determined first and the reduced system, derived that way, was undergoing a thorough sensitivity analysis again. Figure 11 shows the main reaction chain with cross links of the C_1- and C_2-hydrocarbons.

Figure 12 shows the O_2 and CO_2-content along the combustor height of the CAFBC-test rig Niederaussem for the test case A, calculated by the detailed model and the code FIREZWS (only mean values). The calculated values agree well with the measured ones. The agreement is even a little bit better than obtained from the

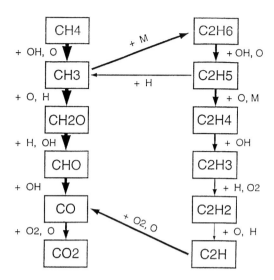

FIGURE 11   Reaction scheme of CH_4- and C_2H_6-combustion.

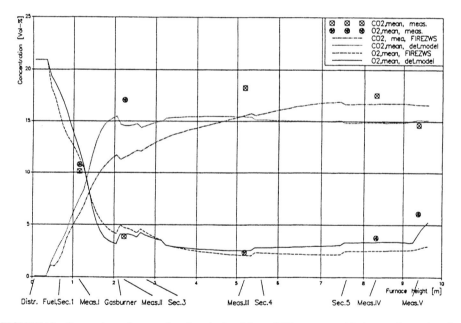

FIGURE 12   Comparison of mean values for calculated O_2 and CO_2-content with the experimental results.

FIREZWS model. The mean values fit well for the higher parts of the furnace, too. The kinetic model agrees a little better with the slightly increase of the O_2-values measured in the higher parts of the furnace and in front of the cyclone.

The detailed model gives a better fit for the CO_2-content along the whole combustor. However, both models underestimate the CO_2-content. But the measured CO_2-values of about 18 Vol-% by a corresponding O_2-content of about 4 Vol-% as measured in level IV (see Fig. 2) seems a bit high. Nevertheless, the slightly decrease in CO_2 and increase in O_2 measured for the test case A is calculated with the detailed model, too.

Considering the CO-predictions of both models, as shown in Figure 13 it can be clearly seen, that the detailed model achieves an excellent agree with the measured values. However, this must be proved by calculating the other test cases with this model, too.

But the CO-content in the higher regions of the furnace is still underestimated by a difference of 28% ÷ 20%. The measured mean CO-content at cyclone inlet was 3 times larger than the mean value at the combustion chamber outlet calculated by the detailed model. If one consider only the calculated values in the outlet cells, a mean CO-content of 2665 mg/m^3 is obtained, lying about 40% lower than the measured value. The results show that the detailed model is above to predict the CO as well as CO_2 and O_2 contents with a better accuracy than the global code used in FIREZWS.

3.4 Postprocessing: NO_x-Formation and Retention

For homogeneous NO-formation the influence of species mixing (micro and macro mixing) competes with the time scale of species conversion.

The time scale for reaching NO reaction equilibria (τ_{reac}, GGW) in CAFBC temperature interval is situated in the range of 10^{-3} to 10^{-2} s.

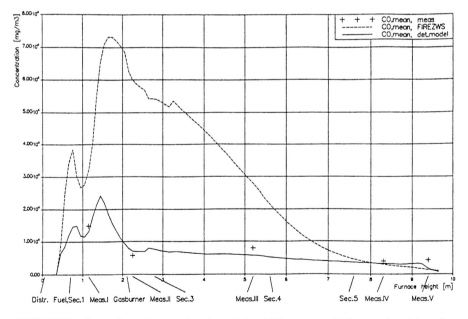

FIGURE 13   Comparison of mean values for calculated CO-content with the experimental results.

Hydrogen and radical concentrations could be obtained from partial equlibria. Sinks and sources of the species CO, CO_2, H_2O, O_2 with higher concentration must be calculated with the global combustion model if the hydrocarbon combustion is not included in the NO reaction system.

The elementary reaction system used in the work was presented by Rohse (1985), Lendt (1991), Ro (1992) and has undergone several parameter studies of species time history and equilibrium states. 35 reactions are involved in the $HCN \rightarrow N_2/NO$ conversion. Intermediate species are NCO, the NH_i-Pool, HNCO, NCO and CN.

The stiff coupled differential equation system for species concentration gradient:

$$\frac{\partial c_i}{\partial \tau} \sum v_i (k_{hin} c_A^{vA} c_B^{vB} - k_{rueck} c_C^{vC} c_D^{vD}) \tag{4}$$

for the common chemical reaction:

$$V_{AB} + V_B B \leftrightarrow V_C C + V_D D \tag{5}$$

was solved by a backward differential algorithm with an integrated NEWTON-algorithm (NAG-FORTRAN-Library).

Comparisons were made for a gas mixture containing 500 ppm HCN and 500 ppm NO (at $t = 0$) in order to simulate HCN devolatilization and NO release in the entire 3D NO-model. It was detected that a HCN/NO mixture can reduce or produce NO depending on O_2/CO and radical concentrations and species time history between 1023 and 1423 K.

Besides the heterogeneous NO formation from residual char after or simultaneously with the devolatilization a reduction of NO at char surface is relevant for CAFBC conditions. In pulverized coal combustion the path contributes only 1% to total conversion because of low char concentrations and burn-out effects (Schultz, 1985).

The volumetric solid phase concentration in CAFBC is theoretical between minimal fluidization point ($c_v \approx 0,57$ for bed material) in the lower parts and values below, 0,001 at the outlet. Fixed by an overall averaged pressure difference of 7000 Pa at the Niederaussem test rig an averaged solid concentration of:

$$c_v = \frac{\Delta p}{(\rho_s \cdot g \cdot H)} \tag{6}$$

approximately 3 Vol.-% is occupied by the solid phase.

Coal concentration in the re-entering solid flow is between 2% and 5%, an exact value could not be obtained because recirculation flow has not been measured. Local coal-concentrations in the combustion chamber reach values up to 1 kg/kg gas. ($\rho_{char} = 900$ kg/m^3, $\rho_{gas} = 0,4$ kg/g^3).

The development of the FORTRAN computer code for calculation of NO profiles in circulating fluidized bed combustors is based on the program architecture of the simulation program CAFIRE (Müller, 1992).

The procedures for NO formation include the following aspects:

- The released nitrogen mass flow is proportional to the total combusted coal mass (Smoot and Smith, 1985).
- Within the given temperature range for CAFBC the nitrogen content in volatile

matter and residual char is distributed in the following way:

$$m_{N,coal} = m_{N,VM} + m_{N,char} \tag{7}$$

$$\frac{m_N}{m_{VM}} = \frac{m_N}{m_{char}} = \frac{m_N}{m_{coal}} \cdot \frac{1}{1 - a_{RK} - \omega_{RK}} \tag{8}$$

what means that ash and moisture is free of nitrogen and N is neither enriched in volatile matter nor in char.

- Volatile nitrogen devolatilizates as hydrogen cyanide. The HCN mass flow per control volume is:

$$\dot{m}_{HCN} = \dot{m}_{VM} \frac{m_N}{m_{coal}} \frac{m_{coal}}{m_{VM}} \frac{M_{HCN}}{M_N} \tag{9}$$

- The remaining nitrogen is released simultaneously to the residual char burn-out as NO in a percentage of 30% (Griwatz, 1993):

$$\dot{m}_{HCN} = -0.3 \, \dot{m}_C \frac{m_N}{m_{coal}} \frac{m_{coal}}{m_{char}} \frac{M_{NO}}{M_N} \tag{10}$$

- The remaining 70% of solid bound nitrogen is released as hydrogen cyanide and must be added to the homogeneous volatile HCN release in order to obtain the total HCN per cell:

$$\dot{m}_{HCN} = \dot{m}_{VM} \frac{m_N}{m_{coal}} \frac{m_{coal}}{m_{VM}} \frac{M_{HCN}}{M_N} - 0.7 \cdot \dot{m}_C \frac{m_N}{m_{coal}} \frac{m_{coal}}{m_{char}} \frac{M_{HCN}}{M_N} \tag{11}$$

- Released HCN and NO react homogeneously in an elementary reaction system forming intermediate species like HNCO, CN, NH_i and producing finally N_2 and NO. An elementary reaction system containing 41 reactions (Song and Bartok, 1982) has been applied. The production or reduction of each species has been used as sources/sinks in each cell.
- Simultaneously to the homogeneous conversion a heterogeneous reduction of produced NO at char particle surface take place.

In order to test the devolatilization and char burn-out model, their influence on N-species distribution and the iteration behavior of the complex post-processing code for the test case A initially NO and HCN profiles were calculated by switching off the kinetic reaction and species sources calculation. The following 2 profiles in Figure 14 for theoretical distributions have been calculated.

The profiles confirm the experiences of a rapid volatile matter release (finished at $h = 2$ m shortly after coal inlet, see Fig. 2) predicted by the implemented one-step pyrolysis and a much slower NO-release simultaneously to the char burn-out model. No influence of secondary air inlets on the NO/HCN-profiles could be remarked in the middle of the chamber.

The presentations in Figure 15 compare the calculated 3D NO-values for minima, maxima and average concentrations with the measured values at 5 levels ($h = 1$, 15; 2, 25; 5, 2; 8, 3; 9, 4 m) (see Fig. 2).

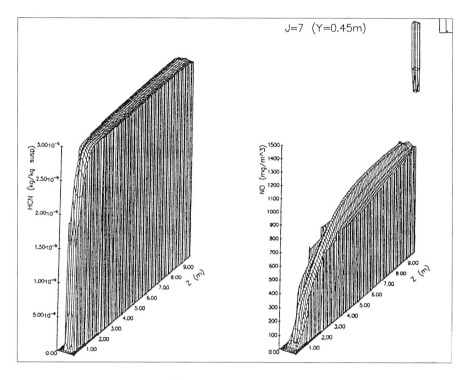

FIGURE 14 Theoretical NO [mg/m^3_N] and HCN [kg/kgsusp] profiles of the released species without reactions for test case A.

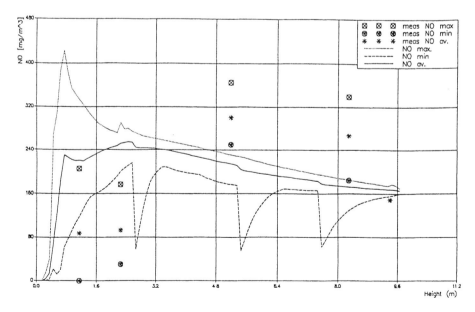

FIGURE 15 Comparison between measured and simulated No-profile [mg/m^3_N] over furnace hight for CAFBC test rig RWE Niederaussem, case A.

The following test case in Figure 16 has a higher percentage of secondary air and thermal capacity than that in Figure 15. One half of the air flow has been used for staging.

The simulated overall conversion rates for the test cases for total fixed nitrogen to NO is about 10%.

For the test case in Figure 15 the averaged measured NO concentration at $h = 9.4$ m (outlet) of 149.5 mg/m_N^3 is well predicted by a value of 168 mg/m_N^3 in the simulation. The averaged temperature is around 831°C. The rise of measured NO concentrations is smaller due to a slower devolatilization. The measured high maximum values at $h = 5.2$ m and $h = 8.3$ m of about 350 mg/m^3 were smaller in the prediction, deviations are caused by very low measured local O_2 concentrations. The low NO concentrations at secondary air inlets are clearly visible. Quenching and reaction effects of secondary air inlets are small due to low secondary air staging. Differences between minima and maxima values are close to the measurements. The low measured NO values around 80 mg/m_N^3 at the lower levels can be interpreted as temperature and mixing effects.

The simulated test case in Fig. 16 shows the effects of air staging. Secondary air inlets show a greater influence on averaged and minima NO-concentrations. The temperatures over is furnace height rise from 890 °C to 913 °C. That's why the higher measured NO values at the upper levels must be caused by local temperature peaks. In the middle of the furnace the NO values were about one third smaller than in Figure 15 despite of the higher temperature.

Absolute deviations over reactor height for both test cases may caused by the simplified combustion and flow model. Despite of the EULER'ian consideration and the rapid one-stage devolatilization the calculations are in a good agreement with the

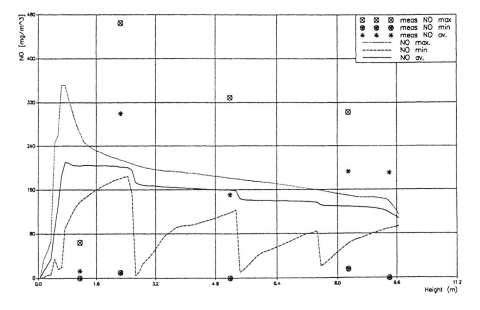

FIGURE 16   Comparison between measured and simulated NO-profile [mg/m^3, N] over furnace height for CAFBC test rig RWE Niederaussem, case B.

measurements. Consequently it can be pointed out that a complicated but chemical proved elementary reaction system is an appropriated tool to predict NO-conversion in circulating stationary fluidized bed combustors fired with coal.

3.5 Postprocessing: SO_2-Formation and Retention

The presented modified shrinking core CO_2—CaO model for sulphur-dioxide retention in CAFBC gives the following formulation per cell

$$\frac{dmSO_2}{dt} = -\frac{n_{part}}{V_{reac}} k_{eff} m_{SO_2}, \quad \text{with} \tag{12}$$

$$k_{eff} = \frac{4\pi R_{core}^2 k_S}{1 + k_s/k_m (R_{core}/R_p)^2 + k_s/D_S' (1 - R_{core}/R_p)}$$

For the mass of SO_2 formed in a sufficient short time interval t_{reac} the following expression can be directly obtained:

$$m_{SO_2}(t_0 + \tau_{reac}) = m_{SO_2}(t_0) \cdot e\left(-\frac{n_{part}}{V_{reac}} k_{eff}\right) t_{reac}, \tag{13}$$

$$\Delta m_{SO_2} = m_{SO_2}(t_0 + t_{reac}) - m_{SO_2}(t_0)$$

Because per 1 Mol SO_2 1 Mol CaO converges to 1 Mol $CaSO_4$:

$$M_{CaO}(t_0 + t_{reac}) = m_{CaO}(t_0) - \Delta m_{SO_2} \frac{M_{CaO}}{M_{SO_2}} \tag{14}$$

and

$$m_{CaSO_4}(t_0 + t_{reac}) = m_{CaSO_4}(t_0) - \Delta m SO_2 \frac{M_{CaSO_4}}{M_{SO_2}} \tag{15}$$

A purely diffusion-controlled reaction in the grain-model formulation reaction rate is defined by Milne et al. (1990):

$$k_D' = \frac{6 \cdot M_{CaO} \cdot D_S \cdot c_{SO_2}}{\rho_{CaO}' \cdot r_{grain}^2} = \frac{6 \cdot M_{CaO} \cdot D_S \cdot c_{SO_2} \cdot A_{BET}^2 \cdot \rho_{CaO}'^2}{\rho_{CaO}' \cdot 9} \tag{16}$$

The reaction rate k_D should by the same for the entire Ca-particle with the core radius R_{core} and the modified diffusion rate D_S':

$$k_D = \frac{6 \cdot M_{CaO} \cdot D_S' \cdot c_{SO_2}}{\rho_{CaO}' \cdot R_{core}^2} \tag{17}$$

With the approach by Milne et al. (1990) for the grain radius the modified diffusion coefficient can be calculated:

$$D_S' = \frac{A_{BET}^2 \cdot \rho_{CaO}'^2 \cdot R_{core}^2}{9} \cdot D_S, \tag{18}$$

If the very small diffusion coefficient ($D_S < 10^{-8} \text{ m}^2/\text{s}$) for a solid product layer ($CaSO_4$) is used for evaluation instead of the CaO particle core surface ($4 \cdot \pi \cdot R_{core}^2$) the effective inner reactive surface must take into account (Vonderbank and Schiewer, 1993).

That inner specific surface of a CaO particle in the range of:

$$A_{\mathrm{BET}} = 5000 \ldots 60000 \ \mathrm{m^2/kg} \tag{19}$$

in diminished by sulfation rate because of greater molar volume of product $CaSO_4$ compared with CaO. By this way the former shrinking unreacted core model with low reactive surfaces gives a formulation for a growing grain model which shows a good agreement with values from literature (Milne *et al.*, 1990 a) and b).

The number of calcium containing particles per control cell in a single phase EULER'ian consideration for CAFBC SO_2-retention can be calculated by making the following assumptions:

- Lignite coal types have a Ca content in their ash, this calcium content is known from coal and ash oxide analysis. For hard coals or anthrazites Ca/S mol ratio is nearly 0.
- Another part of Ca-particles consists of $CaCO_3$ or CaO from additive material. Because calcination is for CAFBC condition a rapid process additive is regarded as CaO for which porosity ϵ_{CaO} and specific reactive surface A_{BET} are known from literature or analysis. Initial coal ash calcium content and additive Ca content are summarized in the external Ca/S-ratio.
- The majority of calcium is part of the recirculating material containing CaO and $CaSO_4$. An iterative process gives the real internal Ca/S-ratio which can be verified by CAFBC bed material analysis at an equilibrium state of the process.

  Calcium containing particles and initially sulphur containing coal flow into the furnace at the same location and will be distributed according to the one-phase EULER'ian approach simultaneously.

If the raw coal content for a cell (V_{reac}) X_{Rk} (kg RK/kg susp) is known the number of Ca-particles is:

$$n_{\mathrm{part}} = \frac{6 \cdot X_{Rk} \cdot \rho_{\mathrm{ges}} \cdot V_{\mathrm{reac}} \cdot n_{\mathrm{Ca/S,int}} \cdot S_{RK}}{\pi \cdot M_S \cdot ((d_{\mathrm{core,0}}^3 \cdot \rho'_{\mathrm{CaO}}/M_{\mathrm{CaO}}) + (d_{\mathrm{p,lime}}^3 - d_{\mathrm{core,0}}^3) \, \rho'_{\mathrm{CaSO^+}}/M_{\mathrm{CaSO_4}})} \tag{20}$$

The internal Ca/S ratio exceeds the external ratio.

Mass flows of calcium containing material (CaO/$CaSO_4$) in the CAFBC furnace are coal ash Ca flow ($\dot{m}_{\mathrm{CaO,Ash}}$). recirculated CaO flow ($\dot{m}_{\mathrm{CaO,rec}}$, $\dot{m}_{\mathrm{CaSO_4,rec}}$) and Ca additive flow ($\dot{m}_{\mathrm{CaO,Add}}$, $\dot{m}_{\mathrm{CaSO_4Add}}$).

The core radius at particle inlet Rcore, in has to be calculated as follows:

1. Mixing of the Ca mass flows:

$\dot{m}_{\mathrm{CaO,new}} = \dot{m}_{\mathrm{CaO,Ash}} + \dot{m}_{\mathrm{CaO,Add}}$.

$\dot{m}_{\mathrm{CaSO_4new}} = \dot{m}_{\mathrm{CaSO_4,Ash}} + \dot{m}_{\mathrm{CaSO_4,Add}}$.

$\dot{m}_{\mathrm{CaO,in}} = \dot{m}_{\mathrm{CaSO_4,rec}} + \dot{m}_{\mathrm{CaSO_4,new}}$.

$$\dot{m}_{CaSO_4,\,in} = \dot{m}_{CaSO_4,\,rec} + \dot{m}_{CaSO_4,\,new}$$

$$R_{core,\,in} = R_{p,\,lime}\left(\frac{1}{1 + (\rho_{CaO}/\rho_{CaSO_4})\,(\dot{m}_{CaSO_4,\,in}\,/\,\dot{m}_{CaO,\,in})}\right)^{1/3} \qquad (21)$$

$$\eta_{Ca/S,\,extern} = \frac{\dot{n}_{Ca,\,Ash} + \dot{n}_{Ca,\,Add} + \dot{n}_{Ca,\,rec}}{\dot{n}_{S,\,in}} \qquad (22)$$

$$\eta_{Ca/S,\,intern} = \frac{\dot{n}_{Ca,\,in}}{\dot{n}_{S,\,in}} \qquad (23)$$

The averaged Ca-particle diameter gives:

$$d_p = \frac{\sum_{i=1}^{NC} m_i}{\sum_{i=1}^{NC} (m_i/d_{p,i})} \qquad (24)$$

because the fractions have to be weighted by the surface area for effective diameter $d_{p,\,lime}$.

Figure 17 shows a comparison between measured and simulated SO_2-profiles for the test case A. The plot shows a good agreement with the measured averaged SO_2 values at the lowest and the upper levels, the low concentration at the second level could not be simulated due to assumed the particle distribution without segregation.

Figure 18 shows a comparison between measured and simulated SO_2-profiles for the test case B. The real profiles show a slower descent of the averaged SO_2 concentration. The values at the second and the outlet level are predicted close to the measurement.

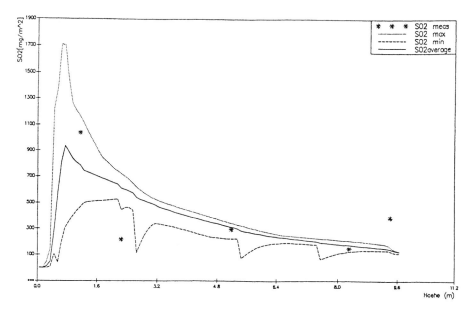

FIGURE 17   Comparison between measured and simulated SO_2-profile [mg/m³. N] over furnace height for CAFBC test rig RWE Niederaussem, case A.

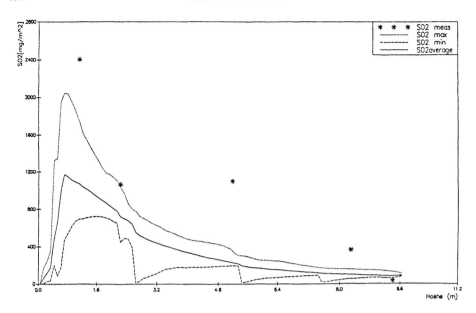

FIGURE 18   Comparison between measured and simulated SO_2-profile [mg/m^3. N] over furnace height for CAFBC test rig RWE Niederaussem, case B.

The simulation of the SO_2 concentrations for Saar hard coal (test case C) shows an underpredicted inlet value (first level) and an over-predicted outlet concentration (see Fig. 19). That relatively constant distribution is due to the very low Ca-content of the coals, the measured decrease may be caused by higher Ca-content in CAFBC ash. Secondary air inlets are visible in minimum and averaged profile.

Summarizing it can be pointed out that the presented suphurdioxide retention model is able to predict the SO_2 concentration profile in CAFBC for lignite and hard coal combustion by means of a modified shrinking core model and a calcium enrichment in the bed ash. The internal Ca/S-ratio could exceed the external value by the approximately 2 scales.

4. CONCLUSIONS

In this paper, total models for the one-dimensional and the three-dimensional case to simulate the CAFBC, are presented. The one-dimensional two-phase flow model, that assumes global particles are distributed over the height corresponding to their size with the smaller particles leaving the fluidized bed and being recirculated. The averaged flow velocity, which is equal to the superficial gas velocity divided by the porosity, is the base of this model. With the partial models for heating up, drying, devolatilization and char combustion, the energy balance and the model of the coal particle distribution, one can find out the distributions of temperature, flue gas concentration, heat flow rate per unit area and the distribution of the coal particles, basing on the two-phase flow model.

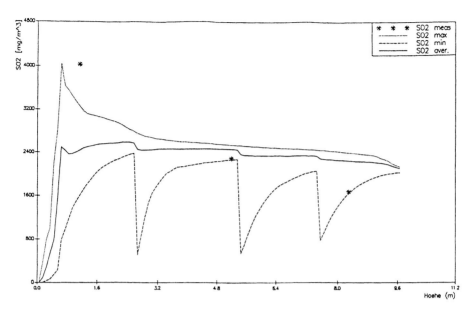

FIGURE 19   Comparison between measured and simulated SO_2-profile [mg/m^2. N] over furnace height for CAFBC test rig TWE Nieddraussem, case C.

The three-dimensional simulation of the gas and solids flow is realized by a coupled Euler-Lagrange model. After the determination of the turbulent gas flow field, the disperse phase was calculated by a trajectory model. Collisions, turbulent fluctuations, momentum exchange and blocks were taken into consideration. The calculation is iterative and averaging.

The models for heat up, drying and devolatilization of the raw coal and the volatile and char combustion, developed for the three-dimensional total models of the circulating fluidized bed, consists of a global model to solve the energy balance and a detailed model to find out the concentrations of the radicals, that are necessary to calculate the NO_x. To reduce the expenditure of calculation, the latter one was simplified by a sensitivity analysis as far as possible.

The model of the calculation of the three-dimensional NO-concentrations inside circulating fluidized beds runs as a postprocessing. It assumes, that 50‡ of the nitrogen inside the fuel is released as HCN homogeneous reaction model by Song and Bartok (1982). The residual nitrogen remaining in the char is released as NO during char combustion. Furthermore the NO is reduced by the char particles. First results without a complete coupling of the models show a reliable agreement with experimental results, using Rheinian lignite.

Within another postprocessing program, the sulphur is released together with the volatiles and is captured by $CaO/CaSO_4$ particles. This process is calculated by a modified shrinking core model (outer shell $CaSO_4$, core CaO), where the recirculation of these particles is taken into consideration. The first results without a complete coupling of the models also show a quite good agreement with measured SO_2 — concentrations and sulphur formation rates, using Rheinian lignite.

ACKNOWLEDGEMENTS

This paper was carried out at the Institut für Wärme-und Brennstofftechnik (IWBT) of the Technische Universität Braunschweig and was partly supported by the Commission of the European Communities within the JOULE-program (contract number 0030).

REFERENCES

Beer, J. M., Jacques, M. T., Farmayan, W. and Talor, R. R. (1981) Fuel-Nitrogen Conversion in Staged Combustion of a High Nitrogen Petroleum Fuel. *18th Symposium (Int.) on Combustion*, 101–110, the Combustion Institute, Pittsburgh.

Borgwardt, R. H. (1970) Kinetics of the Reaction of SO_2 with Calcined Lime-stone. *Environm. Sci. Technol.*, **4**(1), 54–63.

Brummel, H.-G. and Kakaras, E. (1990) Wärmestrahlungsverhalten von Gas/Feststoffgemischen bei niedrigen, mittleren und hohen Staubbeladungen, Wärme- und Stoffbübertragung **25**, 129–140.

Dorendorf, V. (1989) Entwicklung eines Rechenprogramms zur Berechnung des Kohleabbrandes mit Hilfe eines einfachen reaktionskinetischen Modellansatzes, Studienarbeit, IWBT.

Fischer, K. C. (1991) Numerische Berechung des partikelgebundenen Strahlungswärmeaustausches in der zirkulierenden Wirbelschichtfeuerung, Studienarbeit, IWBT.

Gaebel, C. (1992) Erweiterung und Sensitivitätsanalyse eines reaktionskinetischen Modelles zur Berechnung des Flüchtigenabbrandes. Studienarbeit, IWBT.

Griwatz, I. (1993) Kinetik der Stickstoffmonoxid-Bildung bei der Wirbelschichtverbrennung von einzelnen Koksteilchen, Diss., Uni Hannover.

Johnnson, J. E. (1989) A Kinetic Model for NO_x Formation in Fluidized Bed Combustion. Proceedings of the 1989 Int. Conf. on FBC, Ed. Manaker, A. M., **2**, 118–1122.

Krugmann, R. (1992) Vergleich und Anpassung verschiedener Pyrolysemodelle für die Entgasung von Kohlen in einer zirkulierenden Wirbelschichtfeuerung, Diplomarbeit, IWBT.

La Nauze, R. D. (1985) Fundamentals of Coal Combustion in Fluidized Beds. *Chem. Eng. Res. Des.*, **63**(1), 3–33.

Leithner, R. (1989) Einfluß unterschiedlicher Wirbelschichtfeuerungssysteme auf Auslegung, Konstruktion und Betriebsweise der Dampferzeuger, VGB Kraftwerkstechnik **69**(7), 679 ff.

Leithner, R. (1992) Betriebserfahrungen und zükünftige Entwicklung von Dampferzeugern mit Wirbel-schichtfeuerungen, VGB-Konferenz 'Wirbelschichtsysteme 1992', Essen, 2.-3. Sept.

Leithner, R. and Wang, J. (1990 a) Eindimensionales Wirbelschichtmodell, BWK Bd. **42**(7/8), 436 ff.

Leithner, R. and Wang, J. (1990 b) Eindimensionales Wirbelschichtfeuerungsmodell, Internationale VGB-Konferenz 'Wirbelschichtsysteme 1990', 14.-15. Nov., Essen.

Leithner, R. and Wang, J. (1991) One-Dimensional Fluidized Bed Furnace Model. First International Conference on Combustion Technologies for a Clean Environment, Vilamoura (Algarve)-Portugal, 3.-6. Sep.

Lendt, B. (1991) Numerische Berechnung der Stickoxidkonzentration in Kohlestaubflammen, Dissertation am IWBT, TU Braunschweig, 1990, VDI-Fortschrittberichte, **6**(254), Dsseldorf.

Levy, J. M. *et al.* (1981) NO/Char Reactions at Pulverized Coal Flame Conditions. *18th Symp. (Int.) on Combustion*, The Combustion Institute.

Menke, M. C. (1989) Variation der Energieversorgungsstruktur von Braunschweig und deren Einfluß auf Emissionen, Immissionen, Energiebedarf und Kosten, Dissertation am IWBT, TU Braunschweig.

Merrick, D. (1987) The Thermal Decomposition of Coal: Mathematical Models of the Chemical and Physical Changes. Coal Science and Technology 10: Coal Science and Chemistry, A. Volborth (Ed.), Elsevier, Amsterdam.

Mie, G. (1908) Beiträge zur Optik trüber Medien, speziell kolloidaler Metallösungen; Annalen der Physik, 4. Folge, Band **25**, 377–445.

Milne, C. R., Silcox, G. S. and Pershing, D. W. (1990) Calcination and Sintering Models for Application to High-Temperature Short-Time Sulfation of Calcium-Based Sorbents, *Ind. Eng. Chem. Res.*, **29**, 139–149.

Milne, C. R., Geoffrey, D. S., Pershing, D. W. and Kirchgessner, D. A. (1990) High-Temperature, Short-Time Sulfation of Calcium-Based Sorbents, 1. Theoretical Sulfation Modell, *Ind. Eng. Chem. Res.*, **29**, 2192–2201.

Müller, H. (1992) Numerische Berechnung dreidimensionaler turbulenter Strömungen in Dampferzeugern mit Wärmeübergang und chemischen Reaktionen am Beispiel des SNCR-Verfahrens und der Kohleverbrennung, Dissertation am IWBT, TU Braunschweig, 1991, VDI-Fortschrittberichte **6**(268), Düsseldorf.

Ro, S. (1992) Numerische Berechnung der NO_x Bildung in Kohlestaubflammen-Einfluß des Dralles und des Brennstoffstickstoffgehaltes. Dissertation am IWBT, TU Braunschweig, 1992, VDI-Fortschrittberichte **6**(271).

Rohse, H. (1985) Entwicklung eines Rechenprogrammes zur Berechnung des Brennstoff-NO mit Hilfe eines reaktionskinetischen Modellansatzes, Diplomarbeit, TU Braunschweig.

Schnell, U. (1991) Berechnung der Stickoxidemissionen von Kohlestaubfeuerungen. VDI-Verlag, Reihe **6**(250), Düsseldorf.

Schulz, W. (1985) Experimentelle Untersuchung von Stickstoffoxiden bei der Kohlenstaubverbrennung. Dissertation, Bochum.

Sharma, M. P., Cornelius, D. K., Rice, J. G. and Dougan, D. R. (1981) Numerical Computation of Swirling Gas-Particle Flows: Application to Pulverized Coal Classifiers. 80-WA/HT-31 *ASME*.

Smoot, L. D. and Pratt, D. T. (1979) Pulverized Coal Combustion and Gasification. Plenum Press, New York.

Smoot, L. D. and Smith, P. J. (1985) Coal Combustion and Gasification. Plenum Press, New York.

Solomon, P. R., Serio, M. A. and Suuberg, E. M. (1992) Coal Pyrolysis: Experiments Kinetic Rates and Mechanisms. *Prog. Energy Combustion Science*, **18**, 133–220.

Song, Y. H. and Bartok, W. (1982) Rate Controlling Reactions in Fixed Nitrogen Conversion to N_2. *19th Symp. (Int.) on Combustion*, The Combustion Institute, 1291–1299.

Suuberg, E. M., Peters, W. A. and Howard, J. B. (1978 a) Product Composition and Kinetics of Lignite Pyrolysis. *Ind. Eng. Chem. Process Des. Dev.*, **17**(1), 37–46.

Suuberg, E. M., Peters, W. A. and Howard, J. B. (1978 b) Product Compositions and Formation Kinetics in Rapid Pyrolysis of Pulverized Coal-Implications for Combustion. *Proc. 17th Symposium (Int.) on Combustion*, The Combustion Institute.

Turnbull, E. and Davidson, J. F. (1984) Fluidized Combustion of Char and Volatiles. *AIChE Journal*, **30**(6), 881–889.

Vockrodt, S., Wang, J. and Leithner, R. (1992) Combustion Mechanism in Fluidized Beds. CEC-Comett II Programme 1992, Development of a Training Programme in Fluid Flow and Heat Transfer Engineering, 18.–19. June, Athens.

Vonderbank, R. S. and Schiewer, S. (1993) Modellierung paralleler Calcinierung und Sulfatierung von $CaCO_3$-Partikeln, *Chem.-Ing.-Tech.*, **65**(3), 321–323.

Waldmann, E. (1991) Mathematische Modellierung der Pyrolyse großer Kohlestücke. Dissertation an der TU Graz.

Wang, J. and Leithner, R. (1993) Kohleverbrennung in der zirkulierenden Wirbelschichtfeuerung, BWK **45**(7/8).

Wang, J. (1993) Eindimensionale Simulation der zirkulierenden Wirbelschichtfeuerungen. VDI-Fortschritt-berichte **6**(289), Düsseldorf.

Wang, J., Vockrodt, S., Schultz, A. and Müller, J. (1992) Messungen an der ZAWSF-Versuchsanlage im Kraftwerk Niederaussem, Meßserien 1, 2 und 3. Bericht des Instituts für Wärme- und Brennstofftechnik der TU Braunschweig.

Warnatz, J. (1985) Elementarreaktionen in Verbrennungsprozessen, BWK 37, Bd. 1–2, 11–19.

Warnatz, J. (1984) Reaction Rates of the C/H/O System, in: Gardiner, C., Combustion Chemistry, New York.

Wendt, J. O. L. and Schulze, O. E. (1976) On the Fate of Fuel Nitrogen during Coal Char Combustion. *AIChE J.*, **22**(1), 102–110.

Factors Influencing the Combustion Efficiency of Coal in a Circulating Fluidized Bed

X. S. WANG[a,*], N. A. AKHTAR[a], B. M. GIBBS[a] and M. J. RHODES[b]

[a]Department of Fuel and Energy, University of Leeds, LS2 9JT, U.K.;
[b]Department of Chemical Engineering, University of Bradford, BD7 1DP, U.K.

Abstract—This paper presents an experimental study of the combustion efficiency of a bituminous coal in a 0.161 – mid circulating fluidized-bed. The combustion efficiency was found to be in the range 91.62% to 99.99%, and it increased with increasing temperature and excess air factor. The combustion efficiency decreased, however, with increasing secondary air/total air ratio.

Key Words: Circulating fluidized bed; combustion efficiency; coal combustion

INTRODUCTION

Due to rapid depletion of oil and gas reserves, coal has come back into the forefront as a primary source of energy. Coal combustion in the circulating fluidized-bed (CFB) is now on the rise. Recently a number of studies have been carried out on the combustion behaviour of solid fuels in CFB combustors. Lundqvist (1991) reported a combustion efficiency of 98% at a temperature of 850 °C. Park *et al.*, (1990) observed an increase in the combustion efficiency with increasing temperature and excess air factor. Gulyurtlu and Cabrita (1985) reported a combustion efficiency of around 99.5% at an excess air factor above 10%. They found that the combustion efficiency was independent of temperature, coal particle size and gas velocity. Ishizuka *et al.*, (1988) found that the combustion efficiency was in the range from 94% to 99%, and it was insensitive to air staging. Tsuboi and Iwasaki (1988) also found negligible influence of secondary air/total air ratio on the combustion efficiency. In this paper, we present an investigation of the combustion behaviour of a bituminous coal in a CFB combustor. In order to characterise the behaviour of coal combustion in a CFB combustor, we also present results of O_2 and CO measurements in the CFB combustor under similar conditions.

EXPERIMENTAL

A schematic of the CFB combustor used in this investigation is shown in Figure 1. It consisted of a 0.161 – m id and 6.2 – m tall riser, primary and secondary cyclones, an external heat exchanger (EHE) and a solids return valve (i.e. L-valve). Silica sand having a mean diameter of 120 μm and a density of 2500 kg/m^3 was used as the main circulating material. A Dawmill bituminous coal was used as the fuel. Ultimate and

---

*Research Fellow on a joint project between the University of Bradford and the University of Leeds.

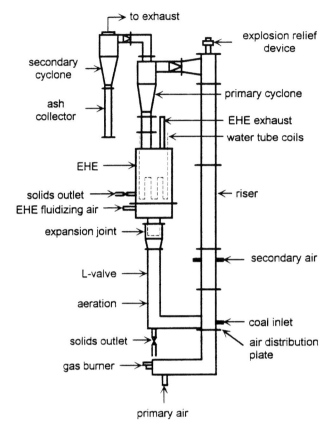

FIGURE 1   A schematic of the CFB combustion system.

proximate analyses of the coal are given in Table I. Primary air was supplied at the bottom of the riser through an air distribution plate, and secondary air could be injected at 1.32 or 1.83 m above the air distribution plate. Solids separated by the primary cyclone from gas-solid suspensions were fed to the EHE via a dipleg. The solids from the EHE were returned to the main combustion chamber (or riser) through the L-valve.

TABLE I

Ultimate and proximate analyses of the coal used

Ultimate (wt.%)		Proximate (wt.%)	
C	74.4	Moisture	6.3
H	4.7	Ash	4.5
O	7.5	Volatile	36.6
N	1.1	Fixed carbon	52.6
S	1.4		

The CFB system was warmed-up to around 600 °C by a gas burner before coal was fed to the riser. Temperature and pressure profiles were measured by thermocouples and pressure transducers, respectively. Concentrations of combustion gases were measured by on-line analysers. Here, the O_2 concentration was measured by a paramagnetic analyser and the CO concentration was measured by an NDIR analyser. The CFB was operated at a solids circulation rate and a superficial gas velocity (above the secondary air jets) of 20 kg/m²s and 5 m/s, respectively.

RESULTS AND DISCUSSION

The effect of temperature and excess air factor on the combustion efficiency is shown in Figures 2 and 3, respectively. Here, the secondary air ratio (i.e. the secondary air/total air ratio) was 0.20 and the secondary air was injected at 1.32 m above the primary air inlet. It can be seen that the combustion efficiency in a CFB combustor is generally very high (normally above 95%). The combustion efficiency increased from 91.62% to 98.74% at an excess air factor of 20% when the temperature was increased from 805 °C to 830 °C. When the excess air factor was increased from 5% to 66% at a temperature of 830 °C, the combustion efficiency increased from 95.42% to 99.99%. It is believed that the increase of temperature and excess air factor can enhance the combustion of char and CO, thus resulting in a higher combustion efficiency.

Figure 4 shows the effect of the secondary air ratio on the combustion efficiency. Here the secondary air was introduced at 1.32 m above the primary air inlet, and the

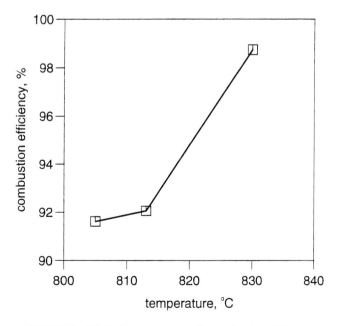

FIGURE 2   Effect of temperature on the combustion efficiency.

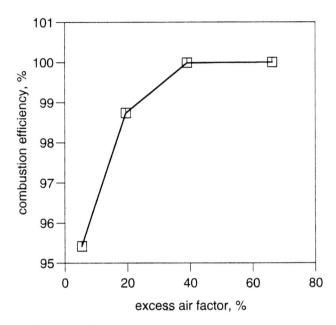

FIGURE 3   Effect of excess air factor on the combustion efficiency.

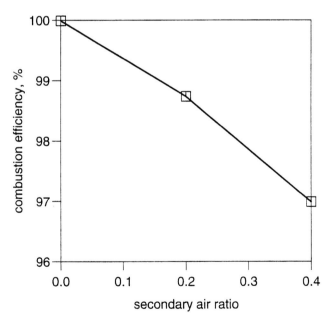

FIGURE 4   Effect of secondary air ratio on the combustion efficiency.

mean temperature in the riser was 830 °C. It shows that the combustion efficiency decreased from 99.99% to 98.74%, and then to 96.99% when the secondary air ratio was increased from 0 to 0.20, and then to 0.40, respectively.

The effect of temperature on the O_2 and CO concentration is shown in Figures 5 and 6, respectively. Here, the experimental conditions were the same as those for Figure 2. As expected, the O_2 concentration in the flue decreased with increasing temperature,

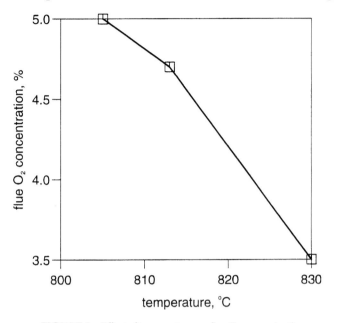

FIGURE 5   Effect of temperature on flue O_2 concentration.

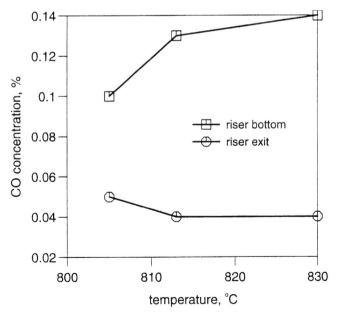

FIGURE 6   Effect of temperature on CO concentration in the riser.

which may be a result of more efficient combustion of char and volatiles. The CO concentration decreased at the riser exit, but increased at the riser bottom, when the temperature was increased. It is believed that higher temperatures promote more rapid devolatilization of coal and therefore result in a higher level of CO emission at the riser bottom. The decrease in the CO concentration at the riser exit with increasing temperature indicates an increase in the rate of CO combustion with increasing temperature.

Variations of O_2 and CO concentrations with excess air factor are shown in Figures 7 and 8. Here, the experimental conditions were the same as those for Figure 3. The O_2 concentration in the flue increased with increasing excess air factor, whereas the CO concentration at the riser exit was insensitive to the excess air factor. The CO concentration at the riser bottom, however, increased with increasing excess air factor. The increase of the CO concentration at the riser bottom with increasing excess air factor suggests that a higher availability of O_2 could enhance coal devolatilization and residual char burning, whilst the small variations in the concentration of CO at the riser exit with increasing excess air factor indicate that the combustion of CO in the upper part of the riser can be enhanced by the higher availability of O_2.

Variations of O_2 and CO concentrations with secondary air ratio are shown in Figures 9 and 10. The operating conditions were the same as those for Figure 6. The O_2 concentration in the flue increased whereas the CO concentration at the riser exit was only slightly affected when the secondary air ratio was increased from 0 to 0.40. The concentration of CO at the riser bottom decreased, however, with increasing secondary

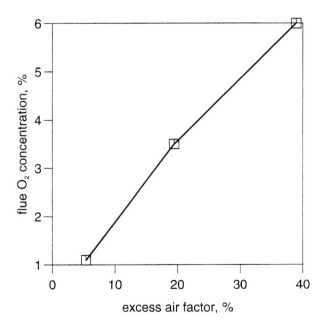

FIGURE 7   Effect of excess air factor on flue O_2 concentration.

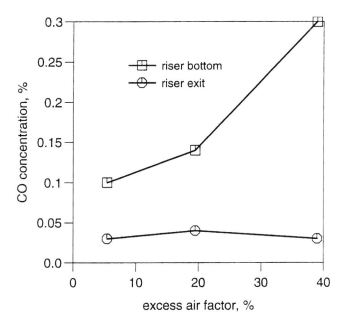

FIGURE 8   Effect of excess air factor on CO concentration in the riser.

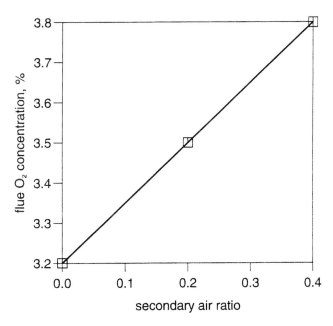

FIGURE 9   Effect of secondary air ratio on flue O_2 concentration.

air ratio. The effect of secondary air ratio on O_2 and CO concentrations is very similar to that of the excess air factor (in the primary combustion zone), and has been discussed earlier in the paper.

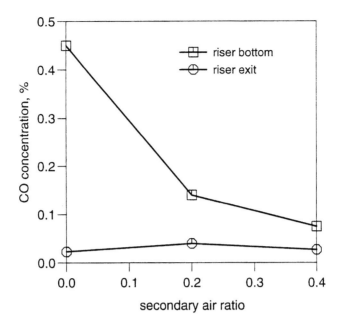

FIGURE 10   Effect of secondary air ratio on CO concentration in the riser.

CONCLUSIONS

The experimental results suggest that the combustion efficiency of the bituminous coal is typically above 95%. The combustion efficiency increased with increasing temperature and excess air factor, but decreased with increasing secondary air ratio. Measurements of O_2 and CO concentrations at selected positions of the combustion chamber (i.e. riser) enable the influences of operating parameters on the combustion efficiency to be explained.

ACKNOWLEDGEMENTS

The research work described in this paper was varied out with the financial support of the Science and Engineering Research Council.

REFERENCES

Gulyurtlu, I. and Cabrita, I. (1985) Combustion of woodwastes in a circulating fluidized bed. In P. Basu (ed.), Circulating Fluidized Bed Technology, Pergammon Press, Toronto, 247.
Ishizuka, H., Hyvarien, K., Morita, A., Suzuki, T., Yano, K. and Hirose, R. (1988) Experimental study on NO_x reduction in CFB coal combustion. In P. Basu and J. F. Large (eds.), Proceedings of the 2nd Int. Conf. on CFB, Compiegne, France, 437.
Lundqvist, R., Basak, A. K. Smedley, J. and Boyd, T. J. (1991) An evaluation of process performance and scale up effects for Ahlstrom Pyroflow circulating fluidized bed boilers using results from the 110 MW Nucla boiler and a 0.6 MW pilot plant. Proceedings of the 11th Int. Conf. on FBC, Montreal, Canada, ASME, 131.

Park, S. S., Choi, Y. T., Lee, G. S. and Kim, S. D. (1990) Coal combustion characteristics in an internal circulating fluidized bed combustor. Proceedings of the 3rd Int. Conf. on CFB, Nagoya, Japan, 497.

Tsuboi, H. and Iwasaki, T (1988) Coal combustion in circulating fluidized bed. In P. Basu (ed.), *Proceedings of the 2nd Int. Conf. on CFB*, France, 327.

Analysis of CAFBC Behaviour Using a Numerical Model

P. C. SARAIVA, J. L. T. AZEVEDO and M.G. CARVALHO   *Instituto Superior Técnico, Mechanical Engineeering Department, Av. Rovisco Pais, 1096 Lisbon Codex, Portugal*

Abstract—Modelling fluidised beds has been in general accomplished using one dimensional numerical models accounting however for non-uniformity's in the cross section. The present paper shows the application of a previously developed model of this category, to experimental conditions corresponding to a hydrodynamic structure of the circulating fluidized bed where the secondary air flow is very high and the flow departs considerably from the one-dimensional approach. The limitations of the model are shown and discussed.

INTRODUCTION

Modelling fluidised beds has been usually performed assuming a given two phase flow structure represented by simplified models where some empirical coefficients are introduced from experimental findings. The solution of the two phase motion equations for fluidised beds has also been performed but due to the unsteady behaviour of the processes in fluidised beds, the simulations have been limited to hydrodynamic studies (see e.g., Tsuo and Gidaspow, 1990) with application on erosion problems.

The alternative one-dimensional approach remains a fair aproximation but depend on the physical understanding that is included in the simplified models. It has been clearly recognized and characterized in the literature the existence of a lower dense region in the circulating fluidized bed (see e.g., Kwauk *et al.*, 1987) while in the upper zone the presence of two zones have been identified: a dilute phase flowing upwards and particles flowing downwards in agglomerates or forming a film close to the walls (see e.g., Rhodes, 1990). The formulation of a one-dimensional approach has to include this basic structure and the capability to recognize their relative importance.

In the lower dense region the gas velocity is usually larger than for conventional bubbling beds leading to the formation of a turbulent regime for which no simple models have been proposed. The behaviour deviates from the two phase flow model for bubbling beds with the formation of alongated voids of gas and clusters and strands of particles. The dense zone can thus be regarded as a two phase region but there is a lack of information concerning the mass transfer between the two phases, the gas and particle mixing and the gas flow division between the phases. The behaviour of the lower dense region will certainly affect the upper zone due to the important particle circulation between the two regions considered.

The model of Wirth (1990) shows the possible presence of two distinct zones, dense and dilute but both uniform. The model is based on the concept of clusters formation with a characteristic dimension but ignores the presence of particles falling close to the walls. This model has been used in a numerical model (see e.g., Hannes *et al.*, 1993) supported by the IEA, to calculate the voidage in the lower region and close to the bed exit (assuming that the flow is totally developed).

Along the freeboard height the particle distribution is usually calculated assuming a mass transfer rate between the core and annulus region (see e.g., Rhodes and Geldart, 1987, Kunnii and Levenspiel, 1990). The mass transfer between the core and annulus region is represented by an empirical constant which represents the result of several complex phenomenae.

The model used in this paper is based on the consideration of a lower dense region described by a two phase bubbling bed theory and an upper region described by a core-annulus structure. The model features a strong coupling between the two zones, through the consideration of the projected particles from the lower region and the particles falling from the upper region close to the walls.

The following section presents a short description of the model (for details see Saraiva *et al.* (1993a,b)). The results obtained for operating conditions with a large fraction of secondary air are presented and discussed in light of the model assumptions.

MODEL DESCRIPTION

Dense Region

The two phase fluidization theory is considered assuming that bubbles rise in plug flow and the emulsion phase is fully mixed. Under this scenario the gas concentration in the bubbles may be expressed along the height as:

$$C_b = C_p + (C_o - C_p)\exp\left(-\frac{K_{bp}\varepsilon_b x}{\beta U_g}\right) \tag{1}$$

with C_b, C_p and C_o represention the gas concentration in the bubbles, emulsion and at the distributor, respectively. Considering heterogeneous first order combustion in the emulsion phase an overall balance to the bottom region can be written as:

$$U_g C_o = \beta U_g C_{bH} + (1 - \beta) U_g C_p + k_r(1 - \varepsilon_b) C_p H C_{\mathrm{char}} \tag{2}$$

where k_r represents the global heterogeneous reaction rate. The particles fed to the dense region are assumed to release moisture and volatile matter at its surface. The solution of equations (1) and (2) together allow the prediction of the gas and char concentration in the dense bed and define the initial concentration for the upper region. The bubble void fraction (ε_b) and the bubble flow rate fraction, (β) are required in the above equations. Assuming that gas is moving in the particulate phase with a velocity larger than the minimum fluidisation velocity by a K_e factor, ε_b may be calculated through:

$$\varepsilon_b = \frac{U_o - K_e U_{mf}}{U_b + U_{mf}(n + 1 - K_e)} \tag{3}$$

Considering the particulate phase voidage equal to the incipient fluidization voidage, the fraction of gas moving within the bubbles is:

$$\beta = 1 - \frac{U_{mf}K_e}{U_o}(1 - \varepsilon_b)$$ (4)

A mass balance is performed for each particle size allowing the determination of the amount of coal particles for each size. This balance accounts for the flux of particles falling in the wall film and for the particle ejection rate evaluated by the correlation of Wen an Chen (1982).

Upper Region

For the core region plug flow is considered. The particle velocities are calculated through the solution of the momentum equation for the different particles (p) inerts (i) and fuel particles (f):

$$u_p \frac{du_p}{dx} = \frac{1}{\tau_p}(u_g - u_p) - g$$ (5)

where τ_p is the particle's relaxation time. In order to calculate the bulk density profiles a mass balance is considered. For the fuel particles it is written as:

$$\frac{d}{dx}(\rho_f u_f) = -\frac{3\dot{m}_{v,c}\rho_f}{\delta_f R_f} - a(\rho_f u_f)$$ (6)

The first term on the right hand side is the mass consumption due to combustion and the last term represents the radial core-annulus particle mass exchange. For the evaluation of the annulus thickness together with the convection heat transfer coefficient the model of Mahalingham and Kolar (1991) was adopted.

The energy balance for the reacting coal particles is written as:

$$C_{pf} m_f \frac{dT_f}{dt} = A_f \dot{m}_{v,c}(H_{v,c} - Q_v) + A_f h(T_g - T_f)$$

$$+ A_f \sigma \varepsilon_f (T_g^4 - T_f^4) + H_{H_2O} \frac{d\dot{m}_{H_2O}}{dt}$$ (7)

where Q_v represents the fraction of heat released by combustion that is transferred to the particle and $\dot{m}_{v,c}$ is the mass release during volatile release (v) and char (c) combustion respectively.

In char combustion both kinetic and diffusion effects are taken into account. It is assumed that CO or CO_2 may be formed depending on the temperature and particle dimensions according to the reaction:

$$C + \frac{1}{\phi}O_2 -> \left(2 - \frac{2}{\phi}\right)CO + \left(\frac{2}{\phi} - 1\right)CO_2$$ (8)

where the reaction mechanism is a function of particle diameter and temperature (see e.g., Arthur (1951)). For small particles CO formed during char combuation diffuses out fast due to the rapid mass transfer and is burned to CO_2 outside the particle, while for the larger particles CO is burned inside the boundary layer and the result of char combustion is CO_2.

A global energy balance is also solved accounting for the energy exchanged with the walls by convection and radiation.

$$\frac{d}{dx}(\rho_g U_g C p_g T_g + \rho_f U_f (C p_f T_f + H_{v,c} - Q_v) + \rho_s U_s C p_s T_s) = \frac{4h}{D}(T_w - T_g) + S_{rad} \qquad (9)$$

In the energy balance the radiation exchange terms (S_{rad}) are evaluated by the discrete heat transfer method assuming that the gas is a mixture of gray gases and particles obey the geomerical optics theory. For further details of the model the reader is referred to Saraiva *et al.* (1993a,b).

DISCUSSION OF RESULTS

The present model was applied to a circulating fluidized bed 1.2 MWt shown schematically in Figure 1. Operating conditions corresponding to a flow division

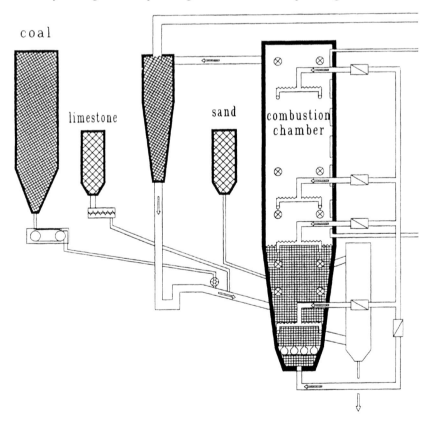

FIGURE 1   Sketch of the simulated circulating bed fluidised bed combustor.

between primary and secondary air of 40/60% were considered in calculations presented by Saraiva *et al.* (1993b,c) while the present paper considers a primary to secondary air division of 20/80%. In this case the influence of the lower region in the overall behaviour is smaller due to the lower velocity through the distributor. As 30% of the total secondary air flow is fed just above the dense region, strong three dimensional effects are generated and the real particle residence time are larger than predicted by the one dimension model.

Figure 2 shows the O_2, CO_2 and CO concentration profiles. The model predicts in the lower dense region and immediately after high CO concentrations a trend that is present in the experimental results. The O_2 and CO_2 concentrations profiles predicted show steep variations and that combustion rate is underpredicted.

The predicted gas temperature along the upper region decreases and is underpredicted as shown in Figure 3 due to the low rates of combustion. Predicted temperature for 1100 µm particles are also shown in the same figure. The fresh particles are predicted to take a large part of their first passage for drying, being completely devolatilized untill the exit. The recirculated particles assumed initially at a lower temperature are heated up overcoming the gas temperature. The projected particles, initially at the dense bed temperature are always hotter than the gas.

CONCLUSIONS

A numerical model to simulate a CAFBC was implemented. The model couples a bottom region with an upper core-annulus structure region. The model allows the

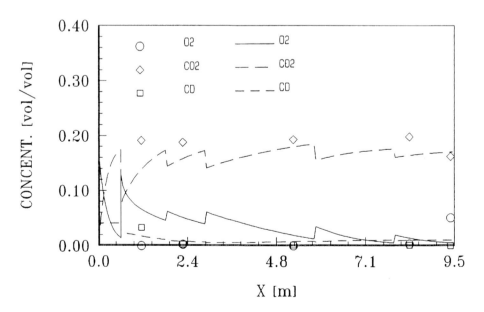

FIGURE 2   Comparison of the predicted main gas species with the measurements.

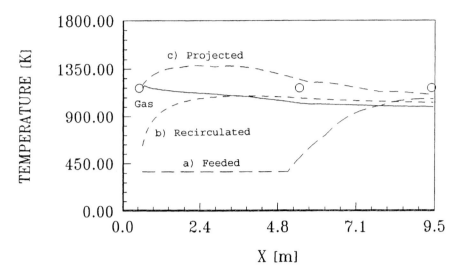

FIGURE 3   Predicted and measured gas temperature (full line). Predicted particle temperature for 1100 μm coal particles a) Feeded, b) Recirculated, c) Projected.

calculation of gas and particle velocities, gas concentration, temperature and heat fluxes, along the reactor. The numerical results for extreme operating conditions as tested in this paper show important deviation from the measurements when compared to other operating conditions with a more uniform heat release. Nevertheless the model predicts the large CO concentration where it is formed.

ACKNOWLEDGEMENTS

This work was partially supported by the Commission of the European Communities under the contract 0030-C (MB) entitled "Minimization of the Formation of Air Pollutants in the CAFBC by using European Fuels and Additives" of the JOULE subprogram of Solid Fuels. The experimental results used in this work were provided by Vockrodt (1992) within the above mentioned project. One of the authors (P. C. Saraiva) also acknowledge financial support of JNICT-CIENCIA BOLSAS program (BM/144/90-IB).

REFERENCES

Arthur, J. R. (1951) Reactions Between Carbon and Oxygen. *Trans. Faraday Soc.*, **47**, 164–178.
Hannes, J. P., Bleek, C. M., Svoboda, K. and Renz, U. (1993) Mathematical Modelling of CFBC: An Overall Modular Programming Frame using 1.5 Dimensional Riser Model. *12th Int. Conf. on Fluidised Bed Combustion, San Diego*, 9–13, May.
Kunni, D. and Levenspiel, O. (1990) Entrainment of Solids from Fluidised Beds. *Powder Technology*, **61**, 193–206.
Kwauk, M., Ningde, W., Youchu, L., Bingyu, C. and Zhiyuan, S. (1987) Fast Fluidization at ICM. *Circulating Fluidised Bed Technology*, 33–45.
Mahalimgam, M. and Kolar, A. K. (1991) Emulsion Layer Model for Wall Heat Transfer in a Circulating Fluidised Bed. *AIChE J.*, **37**, 1139–1150.
Rhodes, M. J., Geldart, D. (1987) A Model for the Circulating Fluidised Bed. *Powder Technology*, **53**, 155–162.
Rhodes, M. J. (1990) Modelling the Flow Structure of Upward-Flowing Gas-Solid Suspensions. *Powder Technology*, **60**, 27–38.
Saraiva, P. C., Azevedo, J. L. T. and Carvalho, M. G. (1993a) Mathematical Simulation of a Circulating Atmospheric Fluidised Bed Combustor. *Combustion Science and Technology*, **91**, 1–21.

Saraiva, P. C., Azevedo, J. L. T. and Carvalho, M. G. (1993b) Modelling Combustion, NO_x Emissions and SO_2 Retention in a Circulating Atmospheric Fluidised Bed Combustor. *12th Int. Conf. on FBC*, San Diego, 375–380.

Saraiva, P. C., Azevedo, J. L. T. and Carvalho, M. G. (1993c) Modelling Flow, Combustion and Pollutants Emission in a Semi-Industrial CAFBC". Preprint volume of the *4th Int. Conf. on Circulating Fluids Beds*, *Sommerset*, 72–79.

Tsuo, Y. P. and Gidapow, D. (1990) Computation of Flow Patterns in Circulating Fluidised Beds. *AIChE J.*, **36**(6), 885–896.

Vockrodt, S. *et al.* (1992) Minimization of the Formation of Air Pollutants in the CAFBC by using European Fuels and Aditives. 4th Progress Report (Contract No. JOUF-0030 C(MB).

Wen, C. Y. and Chen, L. H. (1982) Fluidized Bed Freeboard Phenomena: Entrainment and Elutriation. *AIChE J.*, **28**(1), 117–128.

Wirth, K. E. (1990) Steady-state Diagram for Circulating Fluidized Beds. *3rd Int. Conf. on Circulating Fluidized Bed Technology*, 99–105.

Clean Combustion of Energetic Materials in a Fluidised Bed

A. PIRES and J. CAMPOS   *Laboratory of Energetics and Detonics*
   Mech. Eng. Dep. - Fac. of Sciences and Technology - University
   of Coimbra 3000 COIMBRA - Portugal

Abstract—There are some different ways of eliminating energetic substances, ranking from their storage in silos, their open air combustion or explosion. The present work presents a method for industrial incineration of explosives in a fluidised bed to reduce the amount of the pollutants. The selected industrial explosives are ANFO, emulsion, dynamite. Classical experimental mixtures are generally based on TNT. Gases products composition is predicted using numerical code THOR. It shows more important NO_x concentrations for dynamite than for ANFO and emulsion explosives. For NO_x concentrations increasing and CO concentrations decreasing with low equivalence ratio related to stoichiometry. An original fluidised bed combustion system for energetic materials is presented. It is warmed by the combustion of a propane air mixture as fluidising gas. The energetic materials are pre-mixtured with silica sand particles and injected by a twin screw extruder. A cyclone and a exhaust gas system close the combustion equipment. Measured combustion products concentration of NO_x, CO and CO_2 and measured temperatures, stable near 1200 K, show minimum fluctuations for equivalence ratio to stoichiometry near 2.

Key Words: Incineration; explosives; propellants; fluidised bed; combustion; pollutants; chemical equilibria; chemical composition

INTRODUCTION

The solid energetic substances, explosives and propellants can be classified in two groups related to their normal regime of combustion, respectively detonation and deflagration. The common characteristics of these substances result from their high reaction rates and high pressures. This causes a continuous threat to the environment concerning the explosion risk. This is why explosives and propellants must be disposed of at the end of its lifetime.

There are some different ways of eliminating energetic substances, ranking from their storage in silos, their open air combustion or explosion. The most common way to dispose of obsolete equipment with energetic substances has been to dump them in isolated areas, preferably into deep spots in the sea. Meanwhile, there is a certain inadequacy concerning the laws in the E.E.C. about the emission of polluting gases such as NO_x, SO_x and CO and solid ashes. Aiming to solve this inadequacy the present work will try to develop a method for industrial to reduce the amount of the pollutants, which will benefit both physical and environmental conditions.

The method now proposed, after longer discussion (vd. ICT Conference, 1991, 1992), is the incineration of energetic substances using a fluidised silica sand bed, considered like one of the safest (vd. ICT Conference, 1991; and van Ham, 1991). In a fluidised bed, due to the action of air flow, the particles of the bed float and act as a liquid. It allows the use of different kinds of combustibles particles in the same fluidised bed. Consequently, the fluidised bed combustion (vd. Basu *et al.*, 1991) has some advantages over other processes of combustion:

— security related to a better temperature homogeneity of bed;
— easy control of temperature, pressure and equivalent ratio of mixture;
— better heat transfer which means lower combustion temperatures and consequently a reduction of emission of polluting gases;
— higher density power involved.

The great disadvantage of a fluidised bed combustion equipment comes not only from its ignition and security, during initial working period, but also from the necessity of keeping the fluidised bed warmed before introduction of energetic substances.

The aim of present work is to contribute to a better fluidised bed combustion technology of energetic substances, presenting an original equipment, its performance, tests, and evaluating the obtained results with theoretical predictions using the numerical code THOR.

SELECTED EXPLOSIVES AND EXPERIMENTAL MIXTURES

The most common industrial explosives in Portugal are ammonium nitrate-fuel oil compositions (ANFO), and dynamite explosives, representing an annual production of 8000 tonnes for open air and underground mines applications. Recently ammonium nitrate based emulsion explosives are more and more used in those industrial applications.

In this study three candidate types of explosives were selected:

— one composition of ANFO, an ammonium nitrate-fuel oil composition, with 6 wt-pct of fuel oil, with initial density 870 kg/m^3,
— one composition of dynamite explosive, formed by 30 wt-pct of nitroglycerine, 6 wt-pct of DNT, 60 wt-pct of ammonium nitrate and 4 wt-pct of amidon, with density 1400 kg/m^3,
— one composition of emulsion explosive, formed by an aqueous solution of 10 wt-pct of water, of ammonium and sodium nitrates, respectively 72 and 10 wt-pct, emulsified with oils, wax and emulsifiers, 5.5 wt-pct, with hollow glass spheres as sensitizer, 2.5 wt-pct, with density 1170 kg/m^3.

The standard experimental energetic explosive is TNT, dispersed in an aqueous slurry (vd. van Ham, 1991). It is a very well known chemical defined explosive, with a carbon content larger than the stoichiometry. It is also the main component of classical ammunitions. Consequently the theoretical predictions are based on water slurries of TNT, with 5%, 10%, 15% and 20% of water concentrations.

The calibration and test mixtures are silica sand based mixtures, with successively grass oil, emulsion and ANFO explosives, with a mass content from 5% to 10%.

The incinerator uses inert silica sand as the bed material having a size range of 21.9 μm to 564 μm ($d_{50} = 377.7$ μm) (vd. Fig. 1).

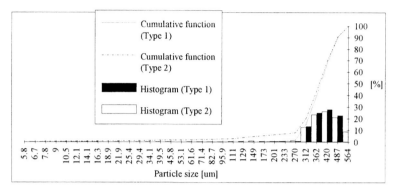

FIGURE 1   Cumulative curve of two types of sand.

THEORETICAL GAS PRODUCTS COMPOSITION AND EVALUATION

The theoretical study starts with the prediction of combustion products using THOR code, based on theoretical work of Heuzé *et al.*, 1985, 1989, and later modified (vd. IEPG Reports, 1989, 1990, Campos, 1991) in order to calculate the composition and thermodynamic properties of explosive compositions, for isobar adiabatic combustion conditions. Several equations of state can be used, namely BKW, Boltzmann, H9, H12 and JCZ3. In previous works (vd. Campos, 1991, 1993) the results have been compared within themselves and with results of different codes in open literature (Quatuor Code from Heuzé *et al.*,1985, 1989, using the BKW EoS the TIGER Code (vd. Chaiken *et al.*,1975) and Mader Code, 1987, and using the KHT EoS the Tanaka Code, 1983).

Chemical Equilibrium Conditions

The classical combustion system is generally a CHNO system. In our computer code it is possible to consider up to m atomic species and to form n chemical components with these atomic species. Among these n chemical components, m are considered "basic" chemical components and n-m "non basic". The selection "ab initio" of the "basic" chemical components depends on the equivalence ratio r of the mixture, related to the stoichiometry ($r = 1$), and they are those which are expected to have significant concentrations in final products composition.

For a CHNO system it has been selected the "basic" combustion products (vd. Manson, 1976; Heuzé, 1989):

— CO_2, H_2O, O_2 and N_2, for poor mixtures ($r < 1$),
— CO_2, H_2, H_2O and N_2, for rich mixtures ($r > 1$) of initial low density, and $C(s)$, CO_2, H_2O and N_2, for rich mixtures of initial high density (initial condensed or solid components).

The mass balance yields a linear system involving m equations. In order to solve the problem it is necessary to add more (n-m) equations. These n-m equilibrium equations are determined by the method of Lagrange multipliers or the equilibrium constants (vd

Brinkley, 1947, White *et al.*,1958). Consequently the system of equations is formed by *m* linear mass balance equations and (*n-m*) non linear equilibrium equations.
The solution of the composition problem involves simultaneously:

— the thermodynamic equilibrium, obtained with the mass and species balance, and the equilibrium condition $G = G_{min}$ (*P,T,*xi), previously described, generally applying to the condensed phase the model proposed by Tanaka, 1983;
— the thermal equation of state (EoS);
— the energetic equation of state, related to the internal energy $E = \Sigma x_i e_i(T) + \Delta e$, $e_i(T)$ calculated from JANAF Thermochemical Tables, 1971, and from polynomial expressions of Gordon and McBride, 1971;
— the combustion regime, being $P_b = P_o$ constant for the isobar adiabatic combustion (equal final and initial total enthalpy $H_b = H_o$).

The selection of components are dependent of atomic initial composition. For a classical CHNO system it is considered the equilibrium compositions of CO_2, CO, H_2O, N_2, O_2, H_2, OH, NO, H, N, O, HCN, NH_3, NO_2, N_2O, CH_4 gases and two kinds of solid carbon (graphite and diamond). It is possible to include more species in final products composition, like S, SO, SO_2, using data from JANAF Thermochemical Tables, 1971, and polynomial expressions of Gordon and McBride, 1971.

In order to evaluate the evolution of gas combustion composition, as a function of final combustion temperature, it is used to dilute gas products with an inert gas, generally N_2, and recalculate the composition for this new equilibrium conditions.

Theoretical Results

Calculated final gas composition of isobaric combustion of selected industrial explosives are shown in Figure 2, using BKW, H9, and H12 equations of state. There are no significant differences from the used equations of state.

Calculated final gas products composition of TNT, as a function of dilution with air and N_2, for different combustion temperatures, are show in Figures 3 and 4.

A parametric study of adiabatic combustion temperatures of water slurries of TNT and its NO and CO concentrations, as function of combustion chamber pressures, are respectively shown in Figures 5 and 6.

Gases products composition is predicted using numerical code THOR. It shows more important NO_x concentrations for dynamite than for ANFO and emulsion explosives.

Obtained results related to NO and CO concentrations, from incineration of TNT-water slurries, proves the low influence of pressure on reduction of pollutants. Consequently our fluidised bed will work at atmospheric pressure.

EXPERIMENTAL EQUIPMENT

Fluidised Bed Combustion Unit

An original fluidised bed combustion system was designed and built to incinerate selected explosives. It has a nominal power of 61 KW and it is composed by (vd. Figs. 7 and 8):

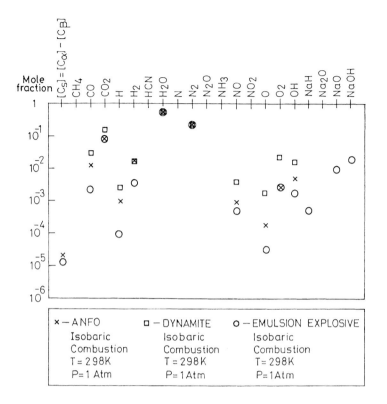

FIGURE 2   Gas and solids products composition of industrial explosives.

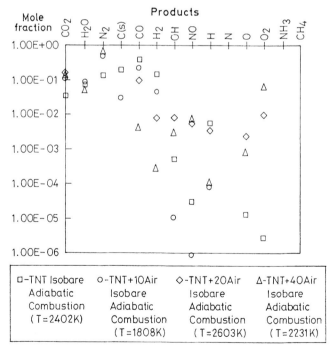

FIGURE 3   Gas and solids products composition of TNT as function of dilution with air.

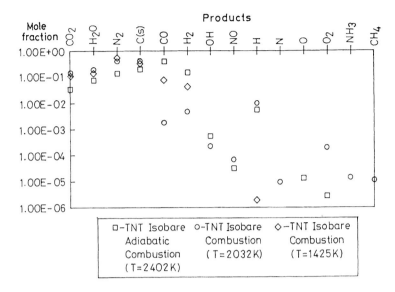

FIGURE 4   Gas composition of TNT as function of dilution with N_2.

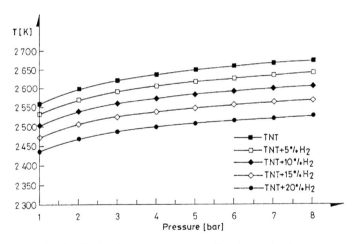

FIGURE 5   Combustion temperatures as a function of chamber pressure.

— a fluidised bed of constant height;
— a slurry feeding twin screw equipment;
— a pressure and temperature control with a real time data acquisition system;
— separation and filtration equipment for the gas combustion products and ash and solids products exhaust conducts;
— gas combustion analyser.

The combustion chamber, executed in stainless steel, is composed of four parts: a fluidised bed incineration chamber, outlet tubes for the solids, a TDH (Transport

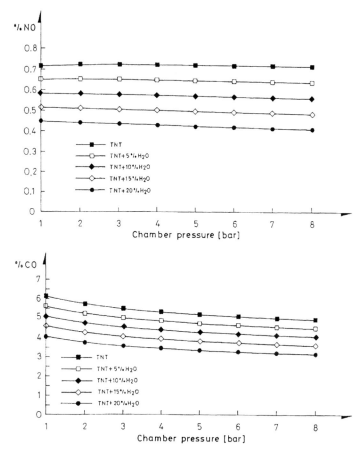

FIGURE 6   NO and CO concentrations as a function of chamber pressure.

disengaging height) chamber and an outlet conduct system for the combustion gases. The incineration chamber consists of an elutriator of internal diameter of $\phi_{int} = 266$ mm and 500 mm in height, which is wrapped by a water cooled liner.

Eight tubes placed on the base of the chamber allow the propane diffusion in air, conducted by an other system of eight tubes peripherally and one in the center. This distributed system ensure a final homogeneous gas mixture (vd. Fig. 9). The fluidised bed combustion system is permanently warmed by the injection of a propane-air mixture as fluidising gas.

Four exhaust tubes, for the sand and solids products, keeps constant the height of fluidised bed. These discharge tubes can be linked to the slurry feed system in a close circuit.

A chimney for the exhaustion of the combustion gases is placed in the upper part of the chamber over the TDH chamber. As the blow up of the bubble in the top of the fluidised bed carries combustible particles in gas stream, the maximum distance reach by these particles above the top of the bed is called Transport Disengaging Height

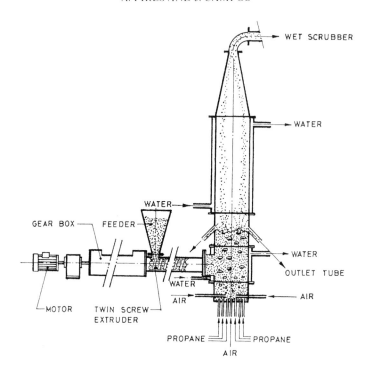

FIGURE 7   Fluidised bed incinerator equipment.

(TDH). The TDH chamber, executed in stainless steel, have an internal diameter of $\phi_{int} = 266$ mm and 735 mm in height, and is also wrapped by a water cooled liner. Finally the exhaust gas system is composed by a cyclone, a wet-scrubber, a mechanical filter and an electrostatic filter. NO, NO_x, and CO, CO_2 gas analysers (types AC31M and CO11M from Emission S.A.) can be connected in future to the end of this exhaust system.

Feeding Systems

The energetic material feeding system is formed by a twin screw extruder for solids and two sonic flow meters for gases. To avoid the formation of slurry dunes in the connexion tube, between the extruder outlet and the incineration chamber inlet, an entry of air was connected to the part of extruder outlet (vd. Fig. 7). Two sonic flow meters ensure the rigorous control of air and propane flows (vd. Fig. 10).

The sonic flow meters used in the experiment are composed of a tranquillisation chamber and an orifice. The nozzles are $\phi = 4.5$ mm and $\phi = 1.5$ mm diameter respectively for the sonic flow meter of air and propane. The mass flow that passes through the orifice is maximum for critical conditions, i.e., when $(P_o/P_1) \geqslant (\gamma + 1/2)^{(\gamma/\gamma - 1)}$ In this case the maximum flow is expressed by $\dot{m}_{max} = (2/\gamma + 1)^{[\gamma + 1/2(\gamma - 1)]}$. $A^* \cdot \rho_o \cdot \sqrt{R \cdot T_o \cdot \gamma}$. Below the critical conditions, $(P_{crit_{AIR}} = 1.89$ bar and $P_{crit_{C_3H_8}} = 1.73$ bar) the mass flow is expressed by $\dot{m} = C_D \cdot A^* \cdot \sqrt{(2/o)} \cdot (P_o - P_1) \cdot \rho$.

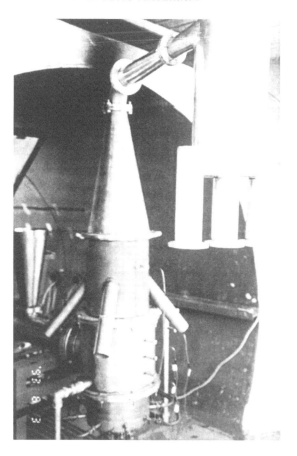

FIGURE 8 Fluidised bed incinerator equipment.

The mass flow, passing through the flow meter, depends not only on the temperature and pression but also on the type of gas. All the calibrations of flow meter have been done with air. Consequently it was necessary to make the correction rate to the flow rate meter for propane. The factor of correction for the critical conditions is $m_{AIR}/m_{C_3H_8} = 0.878$ and under the critical conditions is $m_{AIR}/m_{C_3H_8} = 0.80$. An example of the linear regression, obtained by calibration of the sonics flow meters are showed in Figures 11 and 12.

Control and Safety Systems

The control and safety systems of the incinerator are composed (vd. Fig. 13) by:

— a real time data acquisition temperature system (vd. Fig. 14), composed by a thermocouple amplifier, a analogue digital converter (3D equipment) and a personal computer;
— velocity control of the extruder changing rotating speed of driving motor;
— ignition device (vd. Fig. 15) composed by a vortice propane/air flame;

FIGURE 9   Combustion chamber and feeding system.

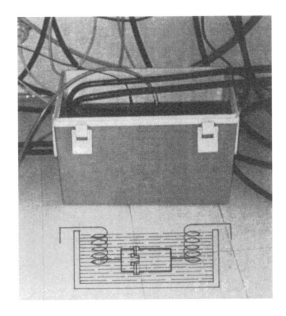

FIGURE 10   Sonic flow rate system.

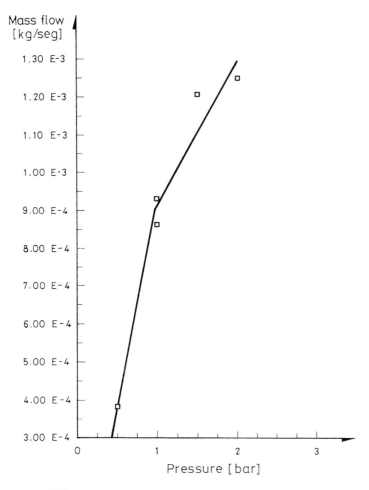

FIGURE 11   Calibration of sonic flow rate meter for gas.

— water cooling system;
— opening outlet tubes.

FLUIDISED BED – COMBUSTION EXPERIMENTS AND RESULTS

Design Characteristics and Optimisation

The bubble velocity inside elutriator can be expressed by (vd. Davidson and Harrison, 1985) , where $U_b = U - U_{mf} + 7.11 \times 10^{-13} \cdot \sqrt{D_b \cdot g}$, where U, U_b U_{mf}, D_b and g are respectively, superficial gas velocity, bubble velocity, minimum fluidization velocity, bubble diameter and gravitational acceleration. The theoretical terminal velocity for average sand particle ($d_{50} \approx 350\,\mu$m) is about 0.4 m/seg and the theoretical superficial

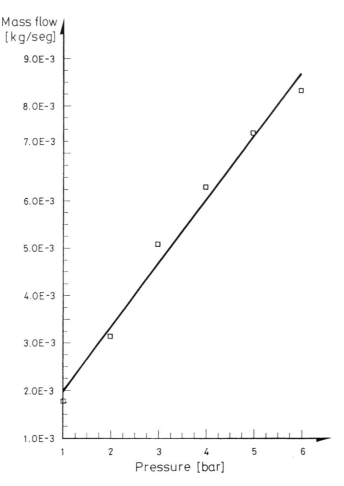

FIGURE 12   Calibration of sonic flow rate meter for air.

gas velocity is about 0.3 m/seg. Consequently the theoretical bubble velocity, $U_b \approx 3$ m/seg, (vd. Fig. 16), is greater than the superficial gas velocity–this fact leads to bubble bursting at the surface of bed and particles returning to bed.

Observed experimental bubble diameter, at the surface of bed (vd. Fig. 17), between 0.12 and 0.15 m, shows a good agreement with the theoretical values ($d \approx 0.18$ m).

The experimental combustion conditions, were shown in following table

Combustion and Incineration Regimes

There are two phases in a normal operation:

— the transient warming regime at the beginning and
— a permanent regime during its effective work.

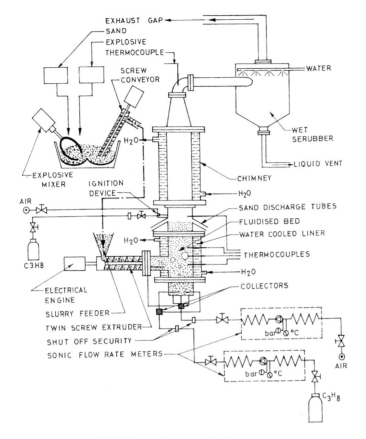

FIGURE 13   Control and safety systems.

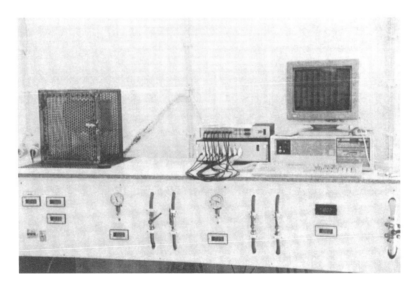

FIGURE 14   Real time data acquisition system and extruder controller.

FIGURE 15   Ignition device.

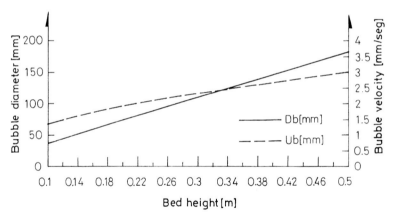

FIGURE 16   Theoretical bubble diameter and velocity as a function of bed height.

In the transient regime the aim is to heat the sand bed. The combustion of the mixture of the fluidising gas (air and C_3H_8) heats the sand bed while the extruder is continuously sending clean sand into the chamber and the refrigeration circuit protects the chamber from any possible overheating.

Optimising conditions leads to pressures in the sonic flow meters approximately at 4.0 bar and 1.0 bar at 20 °C respectively for air and propane. Throughout all tests the sand bed was initially heated above 120 °C. The freeboard temperature was kept between 1000 K and 1200 K (vd. Fig. 18).

When nominal designed temperatures are attained, inside the chamber, the extruder gradually begins to send a mixture of sand and energetic material, while the flowing of

FIGURE 17   Bursting of a bubble.

TABLE I

Design values

Superficial gas velocity, U [m/seg]	0.4
Terminal velocity, U_t [m/seg]	11.8
Minimum fluidization velocity, U_{mf} [m/seg]	0.12
Bed height at incipient fluidization, H_{mf} [m]	0.3
Bubble diameter, D_b [mm]	~ 180
Bubble velocity, U_b [m/seg]	~ 3
Residence time [min]	~ 15

TABLE II

Experimental combustion conditions

Static bed height, [mm]	500
Diameter of the bed [mm]	266
TDH chamber [mm]	735
Average silica sand size, d_{50} [μm]	377.7

C_3H_8 is being gradually reduced. It is necessary to take a special care with slurry mass flow feed in order to avoid the formation of big aggregates, inside the fluidised bed. During the experiment the sand fills the chamber and are expelled through the discharge tubes (vd. Fig. 19). Then the energetic material feeds the combustion chamber and the permanent regime is attained.

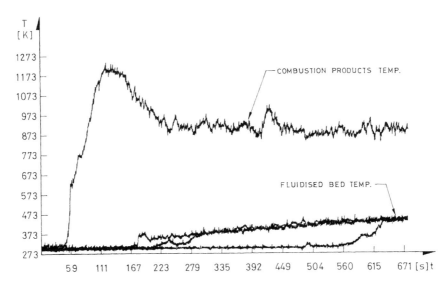

FIGURE 18   Fluidised bed combustion temperature as a function of time during pre-heating regime.

FIGURE 19   Fluidised bed combustion.

When the energetic material begins to burn the temperature inside the chamber rises (vd. Figs. 20 and 21). It shows that the combustion is taking place. When the incineration is finished it is necessary to feed the combustion chamber with clean sand for preventing pack aggregated slurry (vd. Fig. 22).

The experiments were done with air/propane mixtures of equivalence ratio to stoichiometry from 0.701 to 2.099. The fluidization/temperature ratio of combustion

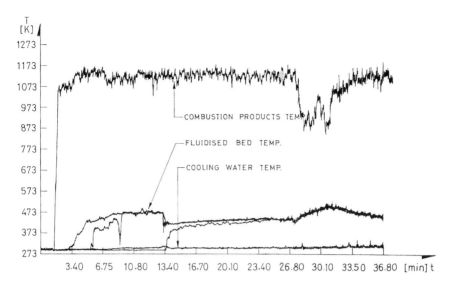

FIGURE 20   Combustion temperature as a function of time - incinerating wax slurry (5%).

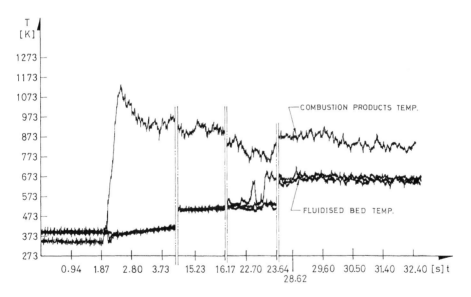

FIGURE 21   Combustion temperature as a function of time - incinerating emulsion slurry (5%).

regime was considered best for the equivalence ratio 2.074 (3.9 bar $\leqslant P_{AIR} \leqslant$ 4.5 bar; 0.5 bar $\leqslant P_{PROPANE} \leqslant$ 1.0 bar), when the temperature fluctuation reaches a minimum.

In the chimney the emission of nitrogen oxides is about 20 ppm for explosive emulsion (3% in sand mixture). The reason of this relatively high concentration of nitrogen oxides, during the incineration, can be explained by the dominant source of

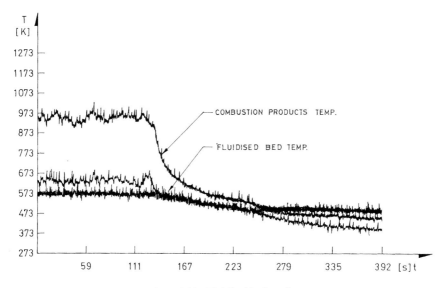

FIGURE 22   Fluidised bed cooling.

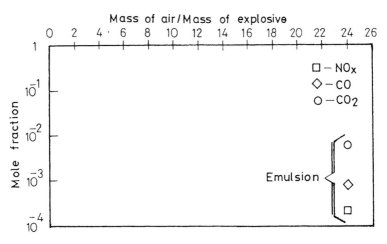

FIGURE 23   Measured concentrations of NO_x, CO and CO_2.

NO_x the "fuel" nitrogen. The CO emissions are about 75 ppm for explosive emulsion (3%). The emissions can be reduced with the perfect control of the air excess. Measured products concentration of NO_x, CO and CO_2, of some tests are presented in Figure 23.

CONCLUSIONS

The selected industrial explosives for incineration are ANFO, emulsion, dynamite. Classical experimental mixture is based on TNT explosives.

Gases products composition is predicted using numerical code THOR. It shows more important NO_x concentrations for dynamite than for ANFO and emulsion explosives.

An original fluidised bed combustion system for energetic materials is presented. It is composed by a fluidised bed, warmed by the combustion of a propane air mixture as fluidising gas. The energetic materials are pre-mixtured with silica sand particles and injected by a twin screw extruder. A cyclone and a exhaust gas system close the combustion equipment. An acquisition and safety data system controls all the incinerating system.

There are two regimes of incineration–the warming transient regime and constant incineration regime. Obtain real temperatures are stable near 1200 K and its minimum fluctuations are detected for equivalence ratio to stoichiometry near 2.

Measured of NO_x, CO, and CO_2 in incinerating products proves the validity of this fluidised bed incinerating system.

ACKNOWLEDGEMENT

The authors gratefully acknowledge to João Moreira for his invaluable work in the constructions of technical devices.

NOMENCLATURE

A^*	Area of the nozzle
C_D	Coefficient
D_b	Bubble diameter
E	Internal energy
G	Global Gibbs free energy
H_{mf}	Static bed height
P_1	Pressure after the nozzle
P_o	Pressure before the nozzle
R	Particular gas constant
T	Temperature
U	Superficial gas velocity
U_b	Bubble velocity
U_{mf}	Minimum fluidization velocity
e_i	Internal energy of component i
g	Gravitational acceleration
m_i	Mass flow of component i
x_i	Mass fraction of component i
γ	C_p/C_v
ρ	Density
μ_i	Gibbs free energy of component i

REFERENCES

Basu, P. and Fraser, S. A. (1991) *Circulating Fludized Bed Boilers – Design and Operations*, Butterworth-Heinemann.
Brinkley, Jr. S. R. (1947) Calculation of the Equilibrium Composition of Systems of many Constituents. *J. Chem. Phys.*, n° 15 107.
Campos, J. (1991) Thermodynamic Calculation of Solid and Gas Combustion Pollutants using Different Equations of State, *First International Conference on Combustion Technologies for a Clean Environment*, Vilamoura, Portugal.
Campos, J., Pires, A., Gois, J. C. and Portugal, A. (1993) Gas Pollutants from Detonation and Combustion of Industrial Explosives, *Second International Conference on Combustion Technologies for a Clean Environment*, Lisboa, Portugal.
Chaiken, R. F., Cook, E. B. and Ruhe, T. C. (1975) *Toxic Fumes from Explosives: Ammonium Nitrate-Fuel Oil Mixtures.* Report of Investigation n° 7867 - Pittsburgh Mining and Safety Research Center, Pittsburgh, Pa., USA.
Davidson, J. F., Clift, R. and Harrison, D. (1985) *Fluidization*, Cambridge, Academic Press Inc., 2nd Edition.
Gordon, S. and McBride, B. J. (1971) Computer Program for Calculation of Complex Chemical Equilibrium Compositions, Rocket Perfomance Incident and Reflected Shocks and Chapman-Jouguet Detonations. *Report NASA SP* 273, NASA Lewis Research Center.
Heuzé, O., Presles, H. N. and Bauer, P. (1985) Computation of Chemical Equilibria. *J. Chem. Phys.*, n° **83** (9), 4734–4735.
Heuzé, O. (1989) *Cálculo Numérico das Propriedades das Misturas Gasosas em Equilbrio Termodinâmico.* Universidade de Coimbra, Portugal.
ICT Conference, 1991. *22nd International Annual Conference. Combustion and Reaction Kinetics.* Fraunhofer Institut fur Chemische Technologie, Karlsruhe, Federal Republic of Germany
ICT Conference, 1992. *23rd International Annual Conference. Waste Management of Energetic Materials and Polymers.* Fraunhofer Institut fur Chemische Technologie, Karslruhe, Federal Republic of Germany.
IEPG - Portuguese Research Group - Campos, J., Luz, P. and Marques, C. (1989, 1990) *BKW- KHT- H9 Equation of State Calculations*, Report of Technological Area 25, Collaborative Technical Programme 1, Portugal.
JANAF, 1971 *Thermochemical Tables* - 2nd Edition. National Bureau of Standards, Washington DC.
Mader, C. L. , (1987) *Numerical modeling of detonations.* University of Poitiers, France; (vd. *Numerical Modelling of Detonations*, 1979. University of California Press, Berkeley, U. S. A.).
Manson, N. (1976) Cours de Thermodynamique des Hautes Temperatures. Université de Poitiers. France.
Tanaka, K. (1983) *Detonations Properties of Condensed Explosives Computed Using the Kihara-Hikita-Tanaka Equation of State.* Report from National Chemical Laboratory for Industry, Ibaraki, Japan.
Van Ham, N. H. A. (1991) Environmentally Acceptable Disposal of Ammunition and Explosives, *22nd International Annual Conference of ICT 1991 – Combustion and Reaction Kinetics*, Karlsruhe, Federal Republic of Germany.
White, W. B., Johnson, S. M. and Dantzig, G.B. (1958) Chemical Equilibrium in Complex Mixtures. *J. Chem. Phys.*, **28**, 751.

Control of Emissions of Mercuric Chloride by Adsorption on Sorbalit™

A. LANCIA[a], D. MUSMARRA[b], F. PEPE[a] and
G. VOLPICELLI[a]   [a]*Dipartimento di Ingegneria Chimica,
Università di Napoli "Federico II", P.le Tecchio 80, 80125 Napoli, Italy;*
[b]*Istituto di Ricerche sulla Combustione, C.N.R., P.le Tecchio
80, 80125 Napoli, Italy*

Abstract—Emissions of mercuric chloride to the atmosphere from municipal solid waste incineration plants have been identified as the cause of severe environmental damages. A growing demand exists for technologies capable of controlling, together with acid gases, also mercury emissions. Sorbalit™, a mixture of impregnated activated carbon and $Ca(OH)_2$, has been tested for such purpose. A laboratory scale apparatus was used to study $HgCl_2$ adsorption on Sorbalit at different temperatures and $HgCl_2$ gas phase concentrations. The adsorption isotherms were found to obey the Langmuir's equation, but an unusual dependence of the adsorption capacity on temperature was found in the range of $200°$–$250°C$.

Key Words: Adsorption; mercuric chloride; MSW incineration

INTRODUCTION

The increasing awareness of the environmental damages caused by the anthropogenic emissions of mercury is pushing toward the identification of suitable technologies for the reduction of such emissions. The combustion of Municipal Solid Waste (MSW) has been identified as one of the main sources of atmospheric mercury, second only to the combustion of coal (Pacyna and Munch, 1991). The toxicity and the relatively high concentration of mercury in the flue-gas can result in severe environmental damages. In the lakes surrounding the emission sources, elevated levels of this pollutant have been found, probably as a consequence of deposition from air. High concentrations of methylmercury have been found in tissues of some kinds of comestible lake fishes, making them unsuitable for consumption, since methylmercury is highly poisonous for man and higher animals (Roberts, 1981).

Recently the problem of controlling mercury emissions from MSW was addressed by the EU. The new directives set limits which will be particularly difficult to meet for small to medium sized industrial plants such as urban wastes incinerators and hospital incinerators, which presently are only equipped with *dry or semi-dry* scrubbing processes for acid gas control (Lancia *et al.*, 1992). A great need exists for technologies capable of controlling the mercury emissions together with acid gases, but at the same time relatively few works are available in the public literature dealing with the fundamental aspects of the capture of mercury compounds on solid sorbents.

The thermodynamic analysis (Lancia *et al.*, 1991) and the experimental results (Hall *et al.*, 1990) show that, at the temperature typical of the combustion of refuses, all the mercury compounds are transformed into elemental mercury. Moreover, during the cooling of the gas, down to 200 °C, a high fraction of the mercury is oxidized by alogenated compounds, by NO_2, and by H_2S. Tseng *et al.* (1990) indicated that $Ca(OH)_2$, which is used in most plants as sorbent for acid gas control, is capable of adsorbing $HgCl_2$, therefore reducing the mercury emissions. The authors indicated that two elements are critical for such reduction: that bag filters, instead than ESP, are installed for particulate removal, and that the temperature at which such filters are operated is lower than 200 °C. Such observations could be interpreted by considering that the layer of adsorbent present on the filter tissue behaves an adsorbing fixed bed, and that the adsorption phenomena are strongly sensitive to temperature. On the other hand Schager (1990) observed that CaO and $CaSO_4$ have negligible retention capacity for mercury compounds. In his work he pointed out to the fact that the steel in the flue-gas ducts may adsorb and reduce mercuric chloride, making the interpretation of the experimental results much more complex, and even misleading. A similar effect of steel, which is particularly strong when its surface has been "activated" by corrosion, was also observed by Yan (1991).

More recently, with the aim of realizing high removal efficiencies in a broad temperature range, some impregnated activated carbons have been considered for mercury capture. Such materials could be injected into the spray chamber together with $Ca(OH)_2$, and then captured on the bag filter, where they could react with mercury. Since stronger (chemical) bonds are formed between mercury and the impregnants, mercury is irreversibly fixed, therefore enhancing the removal efficiency. In particular Otani and coworkers (1986, 1988) compared fly ash and activated carbons with sulfur-impregnated zeolites and activated carbons, finding that sulfur-impregnated sorbents are capable of adsorbing much higher quantities of mercury than "raw" sorbents, reaching 1/1 molar ratios between mercury and sulfur.

In this work, in the context of a broader project aimed at the study of a variety of adsorbing solids, a reagent expressly designed to control the emissions of both acid gas and mercury (Sorbalit^{TM}, made of 95% of $Ca(OH)_2$ and 5% of impregnated activated carbon) is considered. Some tests are carried out into a laboratory scale apparatus to measure the adsorption capacity of Sorbalit using a gas stream containing $HgCl_2$ vapours in order to simulate MSW incinerators flue-gas. In particular the equilibrium adsorbate loading is measured as a function of both gas phase concentration and temperature, and the shape of the breakthrough curves in different experimental conditions is evaluated and discussed.

EXPERIMENTAL APPARATUS AND TECHNIQUE

The apparatus used for the experiments of $HgCl_2$ vapours adsorption on Sorbalit is sketched in Figure 1. In this apparatus the simulated flue-gas, at the required temperature and $HgCl_2$ concentration, was produced, and the mercuric chloride vapours adsorption on the Sorbalit fixed bed was carries out. The simulated flue-gas was obtained by sublimating reagent grade solid $HgCl_2$ contained into a stainless steel (AISI 316) cylindrical saturator (S) into a stream of pure nitrogen. The saturator was kept at a fixed temperature by a heating tape driven by a P.I.D. temperature controller. The required $HgCl_2$ concentration in the gas stream fed to the reactor was obtained by

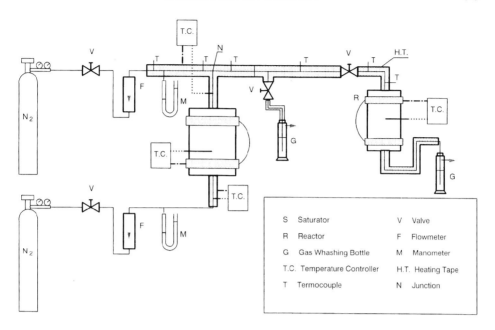

FIGURE 1   Sketch of the experimental apparatus.

varying the temperature of the saturator and by diluting the saturated stream with a stream of pure nitrogen in the junction N. The diluted stream was then fed to a stainless steel (AISI 316) reactor (R), 26 mm I.D. and 250 mm high, the temperature of which was kept constant using heating tapes driven by a P.I.D. temperature controller. Into the reactor the mercuric chloride vapours contacted a bed containing 12.1 mg of Sorbalit particles (Tab. I), sustained by a AISI 316 sintered plate. In order to avoid losses of powder in the gas stream a downward flow was used in the reactor. Furthermore, with the aim of operating with a bed deep enough to avoid channelling and with a small reactive surface, the active particles were mixed with 3.0 g of glass beads of the same size, so that the bed thickness was 4 mm: such material was used since runs performed using a bed made only of glass showed no adsorption. Before each run the steel surfaces of the reactor were saturated for about two hours with a gas containing $HgCl_2$ at the same concentration and temperature as those to be used in the experiment. This procedure ensured that no $HgCl_2$ were adsorbed on the steel surfaces

TABLE I

Characteristics of Sorbalit™

Composition (w/w):	$Ca(OH)_2$	95%
	Brown coke	5%
Mean particle diameter (μm)		90
Bulk density (kg/m^3)		390
BET specific surface area (m^2/g)		20

during the experiments, so that the inlet $HgCl_2$ concentration to the fixed bed was equal to the one measured upstream the reactor.

The experiments were conducted varying the $HgCl_2$ concentration in the inlet stream to the bed (c_0) and the temperature of the adsorbing bed (T), while the gas flow rate was fixed in all runs to keep constant the effect of the mass transfer external resistance. In particular c_0 ranged from 1000–6000 mg/Nm^3, while T was fixed at three different values of 150°, 200°, and 250 °C and the gas flow rate was kept at about 0.17 Nm^3/h.

The $HgCl_2$ concentration in the gas streams entering and exiting the reactor was determined by absorbing the gas into a 12% HNO_3 aqueous solution (Feldman, 1974) by means of gas washing bottles (G), and then analyzing the samples by means of cold vapor atomic absorption (CVAA), using $NaBH_4$ as reducing agent. Such procedure was considered accurate, since no mercury could be found in the gas exiting the gas washing bottle when another bottle was placed in series to the first. Besides, the quantity of mercuric chloride adsorbed on the bed was measured by leaching after each run the bed material with aqua regia ($HNO_3 + 3HCl$) and then analyzing the solution by means of CVAA, and a difference not larger than 8% was found in the mercury material balance.

RESULTS

The experimental runs of $HgCl_2$ adsorption on Sorbalit particles were performed by determining the breakthrough curves of the fixed bed. For each experimental run the breakthrough curves were obtained by measuring the $HgCl_2$ concentration in the inlet stream at different times, and the experiment was considered concluded when the outlet concentration became equal to the inlet concentration.

Preliminary runs were performed to test the reliability of the apparatus varying the amount of adsorbent in the bed. As expected such runs showed that the adsorbate loading does not depend on the quantity of sorbent. The breakthrough curves obtained using the lowest and the highest values of c_0 were reported in Figures 2 and 3 respectively. In such figures, for the three different temperatures of the fixed bed considered, the ratio of the outlet to the inlet concentration c/c_0 was reported as a function of time. Figures 2 and 3 show that, as the solid becomes saturated, the outlet concentration tends to become equal to the inlet concentration. The time necessary to reach the saturation of the adsorbing solids increased as the temperature of the bed increased and the inlet $HgCl_2$ concentration decreased, and for the experimental conditions considered it was in the range of 3–5 h. In particular, even though the results taken at 150° and 200 °C did not evidence a clear trend of the adsorption capacity versus temperature, the results taken at 250 °C showed a definitely higher adsorption capacity.

The equilibrium values of the adsorbate loading, expressed as mass of $HgCl_2$ adsorbed per unit mass of Sorbalit, are reported in Figure 4 versus $HgCl_2$ gas phase concentration. Such figure shows the same trend observed in Figures 2–3 with regard to the effect of the bed temperature. Indeed, while the scatter of the values taken at 150°

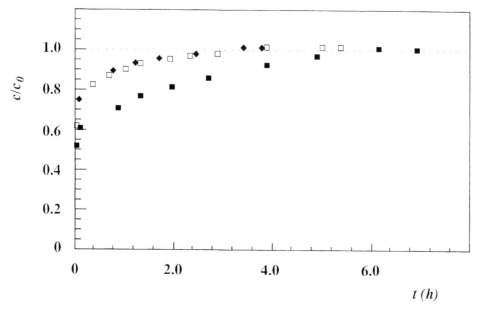

FIGURE 2   Breakthrough curves. ◆ $T = 150°C$; $c_0 = 1.062$ mg/Nm³. □ $T = 200°C$; $c_0 = 0.927$ mg/Nm³.
■ $T = 250°C$; $c_0 = 0.832$ mg/Nm³.

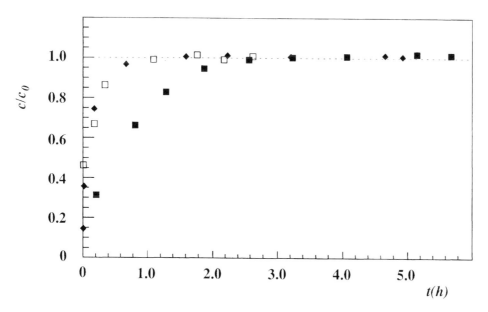

FIGURE 3   Breakthrough curves. ◆ $T = 150°C$; $c_0 = 6.057$ mg/Nm³. □ $T = 200°C$; $c_0 = 6.138$ mg/Nm³.
■ $T = 250°C$; $c_0 = 4.945$ mg/Nm³.

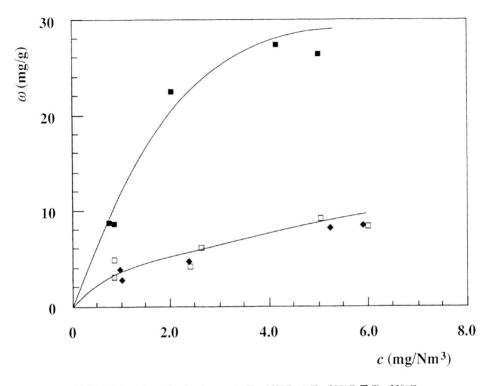

FIGURE 4   Adsorption isotherms. ◆ $T = 150\,^{\circ}C$. □ $T = 200\,^{\circ}C$. ■ $T = 250\,^{\circ}C$.

and 200 °C does not allow a sharp separation between the two adsorption isotherms, the isotherm relative to 250 °C is clearly above the others. While the isotherms have a characteristic "Langmuir" shape, the effect of temperature is somewhat unusual: in fact the exothermic nature of the adsorption phenomena should cause a decrease of the adsorption capacity when the temperature increases. A possible explanation for this behaviour was proposed by Do *et al.* (1990), who developed a theory for adsorption in sorbents with bimodal pore size distribution. According to their theory, for such solids as activated carbons and molecular sieves, a transition temperature may exist at which the increase in the micropore diffusivity determines a large increase in the surface available for the adsorption and therefore in the adsorption capacity.

CONCLUSIONS

Mercury emissions from MSW incinerators are causing a considerable environmental problem, which was addressed by the EU with more stringent emission limits. Mercury is present in MSW incinerators flue-gases mostly as gaseous $HgCl_2$. $Ca(OH)_2$, which is used as acid gases sorbent in dry and semi-dry scrubbers, has a limited retention capability for $HgCl_2$.

Sorbalit, a mixture of $Ca(OH)_2$ and impregnated activated carbon, is suitable for $HgCl_2$ emissions control. The gas-solid equilibrium isotherms for $HgCl_2$ adsorption on such sorbent can be described by Langmuirs equation. However, a strong favourable effect of temperature on adsorption is observed in the interval 200–250 °C. This has been tentatively related to the fact that activated carbon has a bimodal pore size distribution. According to such explanation, at a temperature between 200 °C and 250 °C an increase in the $HgCl_2$ diffusivity makes the micropores accessible, greatly enhancing the surface available for adsorption. Further work is required in order to establish the parameters crucially influencing the adsorption process and its applicability to industrial conditions.

SYMBOLS

c $HgCl_2$ outlet concentration (mg/Nm^3)
c_0 $HgCl_2$ inlet concentration (mg/Nm^3)
T temperature (°C)
t time (h)
ω equilibrium adsorbate loading (mg/g)

REFERENCES

Do, D. D., Pham, T. V. and Do, H. D. (1990) "The role of temperature and length scale on the sorption of gases in microporous solids. II. Nonlinear isotherms", *Chem. Eng. Comm.*, **89**, 1.
Feldman, C. (1974) "Preservation of diluted mercury solutions", *Anal. Chem.*, **46**, 99.
Hall, B., Lindqvist, O. and Ljungström, E. (1990) "Mercury chemistry in simulated flue-gases related to incineration conditions", *Env. Sci. Techn.*, **24**, 108.
Lancia, A., Musmarra D., Pepe F. and Volpicelli, G. (1992) "Analysis of relevant parameters of the process of mercuric chloride vapours removal from municipal incinerators", *Proceedings of 9th World Clean Air Congress & Exhibition*, Montreal (Canada), August 30–September 4, 1992.
Lancia, A., Musmarra, D., Pepe, F. and Volpicelli, G. (1993) "Adsorption of mercuric chloride vapours from incinerator flue-gases on calcium hydroxide", *Comb. Sci. Techn.*, **93**, 277.
Otani, Y., Kanaoka, C., Usui, C., Matsui, S. and Emi, H. (1986) "Adsorption of mercury vapor on particles", *Env. Sci. Techn.*, **20**, 735.
Otani, Y., Emi, H., Kanaoka, C., Uchijima, I. and Nishino, H. (1988) "Removal of mercury vapor from air with sulfur impregnated adsorbents", *Env. Sci. Techn.*, **22**, 708.
Pacyna, J. M. and Munch, J. (1991) "Anthropogenic mercury emission in Europe", *Water, Air, and Soil Pollution*, **56**, 51.
Roberts, R. H. (1981) *Food Safety*, Wiley and Sons, New York, p. 148–169.
Schager, P. (1990) *The Behaviour of Mercury in Flue-Gases*, Ph. D. Dissertation, University of Göteborg, Göteborg (Sweden), November 1990.
Tseng, S. C., Chang, J. C. S. and Sedman, C. B. (1990) "Mercury emissions controls for municipal waste combustors", *Proceedings of AIChE Summer Meeting*, San Diego, (CA), August 20, 1990.
Volland, C. S. (1991) "Mercury emissions from municipal solid waste combustion", *Proceedings of 84th Annual Meeting & Exhibition*, Vancouver (Canada), June 16–21, 1991.
Yan, T. Y. (1991) "Reaction of trace mercury in natural-gas with dilute polysulfide solutions in a packed column", *Ind. Eng. Chem. Res.*, **30**, 2592.

Evaluation of Long-Term Full-Scale Incineration Tests with Differenct Types of Refuse Derived Fuel

M. SCHIPPER-ZABLOTSKAJA   *The National Institute of Public Health and Environmental Protection, A. van Leeuwenhoeklaan 9, P.O. Box 1, 3720 BA Bilthoven, The Netherlands*

Abstract—A Refuse Derived Fuel (RDF) incineration concept as an option in municipal solid waste treatment in the Netherlands is dealt with in the framework of an evaluation study on long-term incineration trials with different types of RDF. In 1988–90 a number of long-term full-scale combustion trials were carried out with different types of RDF, prepared on basis of curbside separation of kitchen/garden waste or mechanical processing of integrally collected MSW. Assessing the effects of a waste-mix composition on the incineration process, i.e. on the energy aspects and emissions to air, water and soil, was the objective of a comparative evaluation study based on the results of the incineration tests.

Results indicated that both curbside separation and mechanical processing lead to a considerable decrease in kitchen/garden waste in the mix and weight reduction. Combustion of RDF with a higher calorific value influences, positively and unequivocally, the burning process itself but does not lead to a considerable reduction of noxious emissions (except CO). Moreover, the total waste treatment costs might be higher due to the necessity of extra processing facilities when compared to the maintenance costs of an incineration plant for treatment of the same waste.

Key Words: Waste incineration; RDF; curbside separation; mechanical processing.

INTRODUCTION: AN OVERVIEW OF WASTE DISPOSAL POLICY IN THE NETHERLANDS

Nowadays about 7.7 million tons of waste a year are produced by Dutch households, offices and services; included are bulky, street and market waste. About 5 million tons of MSW come from households, more than 400 kg per capita a year (at the beginning of the century it was about 150 kg). The amount and composition of MSW in the Netherlands is shown in Table I. Conflicting aims of best environmental option and lowest cost, which must be reconciled, make the best alternative in waste disposal methods a complex issue. About 35% of MSW in the Netherlands is incinerated, the rest is landfilled. According to the principle "energy from waste", the Dutch incinerating capacity by the year 2000 has to be raised to approximately 5 or even 7.5 million tons per year, depending on the development of other processing alternatives like digesting or composting (Tab. II).

There are eight large-scale waste incineration facilities (MWI) operating at the moment in the Netherlands and three under construction. All of them are waste-to-energy plants equipped with a heat recovery boiler (except MWI Alkmaar which is soon going to be replaced by a new plant). Together they produce 1% of the electrical energy distributed in the Netherlands. In addition, low-grade heat can be used for greenhouse- and district heating or for special purposes such as sewage sludge drying or distilled water production. All the installations are for years equipped with an

TABLE I

Total amount of MSW in the Netherlands in 1990 and the distribution
per component (here MSW = household, bulky and trade waste)

Component	kt/a	% of total MSW
Putrescibles/fines	2930	38
Paper & Cardboard	2700	35
Plastic	540	7
Glass	540	7
Metals	390	5
Textiles	150	2
Miscellaneous:		
combustibles	310	4
non-combustibles	150	2
Total	7700	100

TABLE II

MSW treatment in the Netherlands (1990) and expected by the
year 2000

Treatment	M S W			
	1990		2000	
	kt/a	%	kt/a	%
Recycling[1]	1550	20	3850	51
Landfill	3450	45	0	0
Combustion	2700	35	3650	49
Total	7700	100	7500	100

[1] Composting included

electrostatic precipitator for particulate control, a dry sorbent injection for acid
emission control or wet scrubbers. Tightened environmental standards need modern
engineering systems for air pollution control. At the moment de-NO_x equipment is
undergoing practical tests at four incinerators.

In the context of government-supported programmes much research has been done
in the field of waste disposal. One of the most important programmes is the National
Research Programme on Reuse (NOH), which is supervised jointly by NOVEM (The
Netherlands Agency for Energy and Environment) en RIVM (The National Institute of
Public Health and Environmental Protection) on behalf of the Ministries of Economic
Affairs and of Housing, Physical Planning and Environment (VROM). At the moment
the aspects of thermal treatment of waste are covered by another programme, "Energy
Recovery from Waste and Biomass" (EWAB), funded by the same ministries. Though
prevention and recycling of waste are generally considered the most preferable options
(recycling rates of glass and paper are very high, about 60–70%), an integral waste-to-
energy combustion remained the basic option for a long period of time. However, still

more municipalities are introducing separate collection (curbside separation) of putrescibles, or kitchen/yard waste, for reuse as compost or for biogas extraction through fermentation. A separation can also be realized through mechanical processing of waste at special installations to form an RDF with a high calorific value.

LARGE-SCALE INCINERATION TESTS

The Aim of Incineration Tests

Both developments – the separate collection and the mechanical processing prior to incineration – could lead to a change in the composition and calorific value of waste offered to an incineration plant. This would surely influence the incineration process, the composition and the quality of ash and the incinerator's emissions. A number of different waste-to-RDF schemes have been assessed and tested on various scales, including combustion trials at practical level. A number of large-scale long-term experiments were carried out in 1988–90. The concrete aim of the tests was to establish the influence of a changed waste composition on the process management as a result of different processing schemes. Different aspects, such as the incinerator's capacity (throughput), the incineration process itself, energy aspects and the emissions to air, water and soil (flue-gas, slag, flue ash en waste water), were assessed.

Evaluation of the Incineration Tests

In 1991–1992 a study to evaluate these large-scale incineration tests was carried out by the KEMA and TNO research institutes under the direction of RIVM. The aim was to summarize and compare the results of these combustion trials and if possible to draw some general conclusions. It was thought possible to predict the consequences of different schemes of waste processing for the incineration branch and the waste branch in general.

Five series of full-scale combustion trials (in total 13 test runs) were carried out at four MWI plants (Tab. III). The mass of burnt waste in each test ranged from 67 to 1100 ton, and the duration of a test from 12 to 69 hours. Three of the four incinerators were supplied with gas-cleaning equipment (Tab. IV). Four main mix types of waste were

TABLE III

Comparative inventory of long-term full-scale incineration tests (1988–1990)

MSW plant, place	date tests	weight mix, t	duration h	number fractions
VVI Leeuwarden	October 1988	67–125	12–24	2
Philips, Eindhoven	February 1989	160	38	1
ARN, Weurt	July 1990	165	18	1
VVI The Hague	October 1990	520–1110	51–69	3
ARN, Weurt	January, May, October 1990	283–549	40–56	3

TABLE IV

Comparative inventory of the MSW plants. 1 – VVI Leeuwarden, 2 – AVI
Philips (Eindhoven), 3 – ARN Weurt, 4 – VVI The Hague

Plant →	1	2	3	4
Capacity t/h	6	3.5	9	13
LHV MJ/kg[1]	5.9–8.4	9.8–14.7	9.2–15.0	6.0–10.0
Type waste	MSW	RDF	RDF	MSW
Flue-gas cleaning	ESP	ESP, wet scrubbers	ESP, wet scrubbers	lime injection, ESP
Heat recovery	–	+	+	+
Air preheating	–	–	–	+
Moving grate type	reverse, Martin	forward, K + K	forward, K + K	forward, von Roll

[1] the range of calorific values according to the design.

artificially prepared in the way as it was supposed to happen in the future when curbside separation and mechanical processing are assumed to be more common. A reference mix was the integral MSW just as it was usually collected (containing putrescibles and a ferrous fraction, but without admixture of bulky waste). The following main types of mix to incinerate were:

a) the remnant fraction after a separate collection of putrescibles (curbside separation of kitchen/yard waste);
b) the fraction after mechanical processing of integral MSW: putrescibles, ferrous and inert (non-combustible) fractions were mostly excluded here;
c) the combustible fraction after separate collection of putrescibles, followed by mechanical processing; in some cases construction and industrial waste were admixed.

The fractions b) and c) can really be called Refuse Derived Fuel, because they contain less uncombustible matter and have a higher (in extreme cases more than double) calorific value, than integral waste.

Each test had its specific aim, therefore the set-up of the tests and the parameters measured were not entirely identical. For instance, in some tests (Eindhoven) more detailed composition analyses have been done (sorting in components and elemental analyses). The optimization of mechanical processing was the most important objective of another series of tests. In order to compare the results of the tests properly, the system boundaries were defined as closely as possible to the combustion furnace (Fig. 1). Only the measurements at the boundaries or related to the boundaries were taken into consideration.

The compositions of similar waste fractions used in different series of experiments was slightly different because of the seasonal influence and the geografical and social

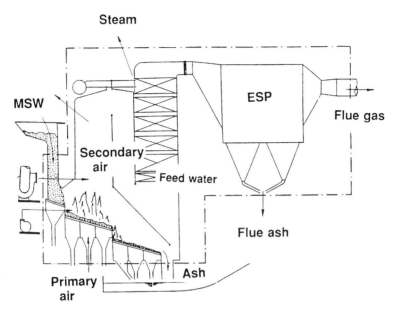

FIGURE 1   System boundaries in the comparative evaluation study on the incineration tests.

INPUT:
– waste, feed chute entrance;
– combustion air (entrance primary,
   secondary and wall panel air);
– feed water, entrance economizer.

OUTPUT:
– bottom ash, end grate;
– flue ash, after ESP;
– flue gas, after ESP;
– steam, end superheater.

factors (village/town). The incineration facilities were also fairly different (throughput and thermal capacity). The results were sometimes unexpected and contradictory for the same parameters from different tests. Still, it is possible to derive some general important conclusions for the waste incineration as a whole.

Four different processing plants were used for RDF production. In Figures 2 and 3 the processing schemes shown are of the VAM waste processing plant in Wijster and the ARN plant in Weurt, integrated with an incineration facility for RDF. The general aim of MSW mechanical processing is to separate an undersize fraction < 20 mm (mostly inert), a very fine fraction < 32 mm (mostly organic, or putrescible) and a ferrous fraction. The largest difference between the installations is the particle size of the resulting RDF (from 32 to 55 mm).

In all the tests mass- and energy balances have been made. The last two series (chronologically) - MWI The Hague (dHg 1, 2, 3) and at ARN Weurt (ARN 1, 2, 3) were the most durable (thus more representative) and provided the most complete information. Therefore these tests were considered as the basis of the evaluation. A conventional scheme at ARN plant (test ARN 1) and "a scenario" for waste processing in the future (test ARN 3) are given as examples of waste processing to RDF in Figures 4 and 5 in Sankey diagrams for mass streams.

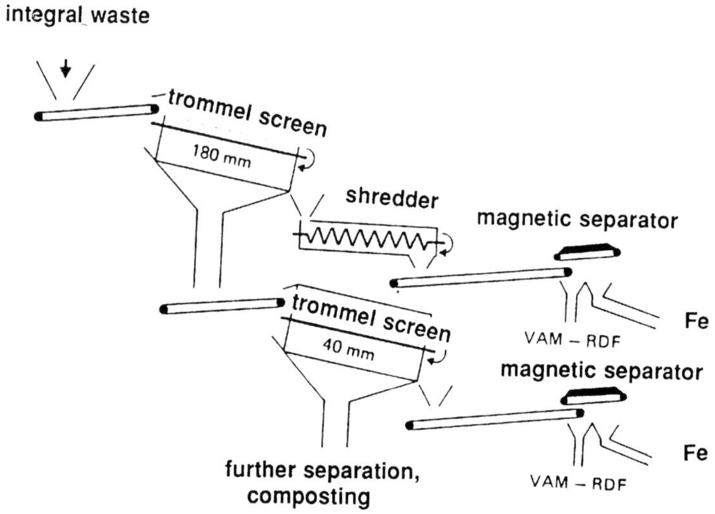

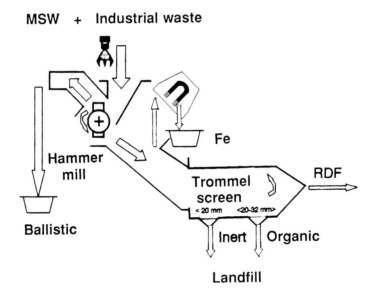

FIGURE 2   Production scheme Waste-to-RDF at the VAM processing plant, Wijster.

FIGURE 3   Production scheme Waste-to-RDF at the ARN processing plant, Weurt.

THE RESULTS OF EVALUATION

Weight Reduction through Processing

Different ways of waste treatment before combustion lead to a certain weight reduction and an increase of calorific value (Tab. 5). The latter was derived from the energy

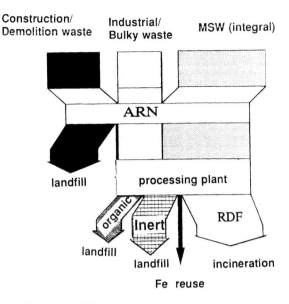

FIGURE 4   Conventional RDF production diagram, ARN, Weurt (test ARN 1).

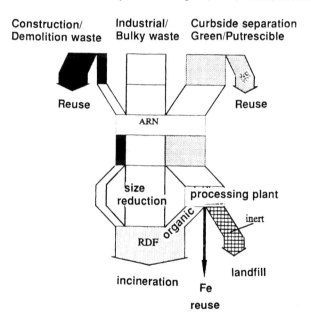

FIGURE 5   Production diagram for an "Ideal Mix" (test ARN 3).

balance rather than from sorting analyses. A curbside separation of kitchen/yard waste leads to a weight reduction between 30 and 45%, depending on the response and the initial content of this ingredient (average 50%) which is influenced by the season and geographical origin. Curbside separation or mechanical processing of integral waste

TABLE V

Weight reduction (%) and calorific value of the MSW fractions and RDF mix

Incinerator plant, test run No	weight decrease, %	L HV, MJ/kg
Integral collection MSW and no pretreatment		
Leeuwarden 4	0	7.7
Leeuwarden 2	0	6.7
Philips 2	0	–
The Hague 1(dHg1)*	0	6.7
Rest fraction of curbside separation of kitchen/yard waste		
Leeuwarden 3	40	9.1
Leeuwarden 1[1]	45	9.0
The Hague 2 (dHg2)*	35	7.7
RDF after mechanical processing of MSW		
ARN/Grontmij**	60	12.1
Philips 1	53	11.2
ARN 1	61	10.7
RDF after combined treatment: curbside/mechanical		
The Hague 3 (dHg3)*	61	10.2
ARN 2	30	14.5
ARN 3[2]	51	13.3

1) MSW with admixture of bulky and industrial waste;
2) idem + construction waste;
*) abbreviation MWI plant;
**) different from ARN 1, 2, 3 and carried out at ARN by Grontmij engineering company.

lower this percentage twice, and when combined reduce the content of kitchen/yard waste to ca. 12%. Mechanical processing of integral waste leads to a weight reduction of 50%–60%, the same as combined processing (curbside + mechanical).

Combustion Emissions and Residues

In the diagrams (Figs. 6–11) some quantitative results, based on measurements during the tests in The Hague (dH1, 2, 3,) and Weurt (ARN1, 2, 3), are presented; mainly emissions to air (before filters) and residues are referred to.

It is rather surprising that, with a few exceptions, the moisture content in the processed RDF remains approximately the same as in the integral waste, ca.35 wt.%. The percentage of solid residues (total slag + flue ash + flue dust) varies between 18 and 38 wt.% (Fig. 6). The quantity of bottom ash (slag) varies between 154 and 346 kg/ton waste. Mechanical processing, lowering the content of inert materials in the waste, leads to a lower slag production. Curbside separation does not seem to influence the production of slag. Flue ash production (particulates) in the tests was between 10 and 44 kg/ton waste. It seems to be influenced by the construction of the incinerator (Fig. 7). The flue gas volume, water vapour and CO_2 emission (per ton waste) increases with the increase in calorific value.

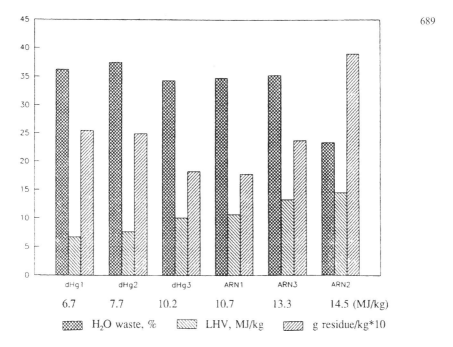

FIGURE 6   Moisture content of waste mix (%), calorific value LHV (MJ/kg waste) and solid residues (g/kg waste). Tests The Hague, ARN.

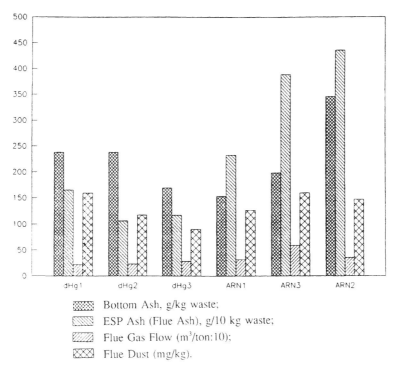

FIGURE 7   Production of solid residues per type (bottom ash of slag, ESP ash of flue ash, flue dust), and flue gas flow.

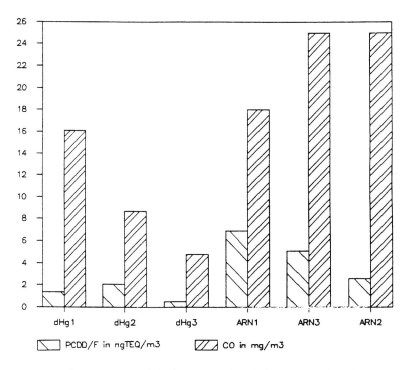

FIGURE 8   CO and dioxin concentrations in flue-gas (11 vol% O_2).

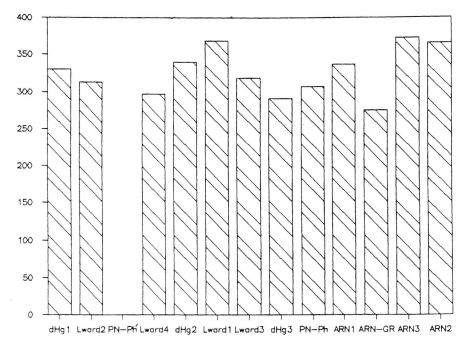

FIGURE 9 NO_x concentrations in flue-gas (mg/m$_0^3$, at 11 vol% O_2). Lward - MWI plant Leeuwarden, PN-Ph - Philips plant, Eindhoven.

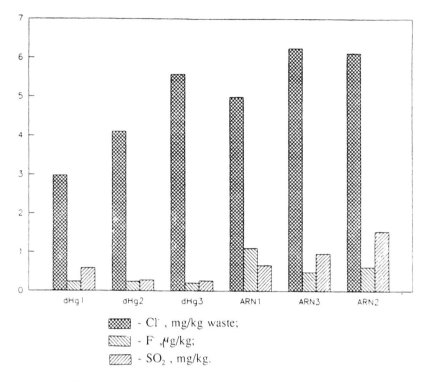

FIGURE 10 Halogens and sulphur dioxide in flue-gas (11 vol% O_2).

The emissions of CO, PCDD/PCDF (Fig. 8) were relatively low in both series of tests (MWI The Hague and ARN) in comparison with an average emission level at Dutch incineration plants in the period around 1990. In both series the combustion of combined processed waste gives the lowest CO emission. The fluctuations in CO emissions decreased with the increase in processing degree. The CO emission is by far a measure of regular combustion, therefore it is clear that waste processed to RDF burns more regularly than integral waste. In any case, the absolute difference between the CO values at different installations is higher than at the same incinerator burning different types of waste. CO emission seems be more dependent on the incinerator than on waste composition. The same can be noticed about PCDD/F emissions.

Since 1993 a new, strict legislation for emission control in Holland became valid. It prescribes emission limits of 50 mg/m_0^3 for CO and 0.1 ng/m_0^3 TEQ for PCDD/F (at 11 vol% O_2). All the new-built MWI plants and the most of the old ones adhere to the limits presently.

Waste processing before incineration does not seem to influence NO_x emissions. In 12 test runs the average NO_x emissions were between 280 and 380 mg/m_0^3 (Fig. 9). According to the current emission limit for NO_x of 70 mg/m_0^3, additional de-NO_x measures should be taken at all MWI plants of stoker grate type.

Figure 10 shows emission factors (emission per kg waste) for halogens and sulphur dioxide. There is no clear trend for the SO_2 and fluoride emissions in connection with the processing grade of waste. Chloride emission rises after processing, sometimes by 40%, due to the increase of plastics in the mix (mostly PVC).

Heavy Metals in Flue-Gas and Residues

There is an opposite trend in mercury emissions to the flue-gas after a pretreatment of MSW as curbside separation and mechanical processing only. After curbside separation the mercury emission is higher than in a reference case (integral waste), and after mechanical processing lower. After a separate collection of kitchen/yard waste a "thickening" effect in the rest fraction takes place, while during mechanical processing, small heavy particles containing mercury pass through the sieve together with "undersize" organic fraction. The concentrations of cadmium, lead and zink in the flue-gas have the same trend as the concentration of particulates (dust). Figure 11 gives the total emission factor of each heavy metal (in the residues and flue-gas together). Waste processing leads to an increase in heavy metals in the bottom ash (slag) but their leachability decreases. The same thickening of heavy metals in the flue ash takes place after waste processing. Actually, according to the new legislation, the concentration of heavy metals in the flue ash (with or without waste pretreatment) is high enough to be regarded as hazardous waste.

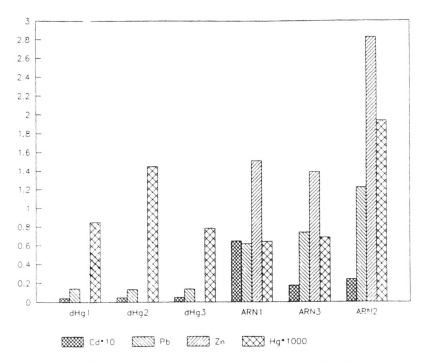

FIGURE 11   Total Cd, Pb, Zn, Hg in combustion residues (bottom ash, flue ash) and flue-gas (g/kg waste).

Energy Aspects

The energy balance for every test run was made from the following energy streams:

1) Energy supply: - waste
                           - combustion air
                           - feed water or cooling water
2) Energy discharge: - flue-gas
                           - flue ash on ESP
                           - bottom ash
                           - steam or cooling water discharge
                           - boiler radiation

As previously mentioned, the calorific values of different waste fractions were determined from the energy balance. Due to uncertainty in gas flow measurements there are uncertainties in calorific values, nevertheless they are preferred to the values calculated from waste analyses.

From the incineration tests in The Hague can be concluded that the boiler efficiency increases with the increase of calorific value, thus at a high degree of waste processing. This can occur due to lower heat losses through the bottom ash (see Kiers and Wardenaar, (1992)).

Process Management

The most significant and unequivocal advantage of RDF combustion for an incinerator is a noticeably better controllable incineration process. This means that the process is more stable and average carbon monoxide emissions are lower. A stable process is the best way to reduce CO emissions for which no flue-gas cleaning is possible. Probably as a consequence of a higher calorific value the properties of the slag become more favourable from the point of its leaching emissions. Nevertheless, a conclusion can be drawn that RDF combustion does not seem to reduce the size and the costs of the necessary flue-gas cleaning equipment and therefore remains expensive. Also, the more stringent legislation for reuse of incineration residues ("The General Administrative Order on Building Materials") would possibly make an extra processing of the residuals necessary in the very near future. Such procedures as vitrification, immobilization and washing-out of slags could offer a perspective and are at the moment the aim of research.

Consequences of Waste Processing for Waste Management

The results show that both types of waste treatment through source separation or mechanical processing, both aiming at RDF, have more or less the same score from the point of view of weight reduction, emissions or processing costs. Through a separate collection or a mechanical processing of waste one can drastically reduce the throughput of waste to be incinerated (Tab. VI). The relative incinerating capacity of the oven in tons, however, decreases because of the higher heating value of the ameliorated waste and to a lesser extent the increasing boiler efficiency. The thermal capacity of the boiler (steam production per hour) is not infinite and has a definite limit value. Working

TABLE VI

Different processing schemes compared: consequences for different aspects of an MSWI plant 1 – integral MSW combustion (without pretreatment); 2 – after curbside separation of kitchen/yard waste, 3 – (2), followed by mechanical processing to RDF.

Aspects of an MWI plant	1	2	3
Plant Supply, t	100%	65%	39%
Energy content, MJ	100%	75%	60%
Flue-gas cleaning capacity, m^3/h	100%	75%	60%
idem, m^3/t MSW	100%	115%	154%
Capacity grate/boiler, t/h	100%	81%	69%
Thermal capacity boiler, MJ/t	100%	125%	177%
Population served, %	100%	123%	145%

above the thermal capacity, though not so disastrous technically, is not efficient from an energy point of view. When planning incineration of other types of waste in an existing incinerator one has to take into consideration its calorific value, the thermal capacity of the boiler and the maximum boiler efficiency.

Taking the balance, separate collection or mechanical processing of municipal solid waste can lead to an enlargment of the area (or the number of households) served. One should, however, take into consideration that the landfill- or composting capacity must be increased too. Through the increase in calorific value of RDF one can expect an increase, rather than a decrease in costs of incineration (per ton waste). Still at lower throughput, the maintenance costs of a plant remain the same, making the gate fee higher.

CONCLUSIONS

A universal scenario for incineration tests on international scale can be made on the basis of the current evaluation study. Some advice to improve the efficiency of possible experiments in future can be given, for instance making pre-calculations from a sorting analysis on expected emissions, residues and energy factors. The most complete measurements (for mass and energy balance) should be carried out. To make the results representative, the test should be several days in length. More information should be obtained on the composition of other categories of waste than MSW (for example, industrial waste).

The evaluation study of large-scale incineration tests has shown that combustion of RDF prepared in different ways (through curbside separation of kitchen and yard waste or/and mechanical processing) improves the quality of burning and is therefore attractive. RDF remains attractive, not so much als a fuel but as a means to improve thermal waste processing. Considering that curbside separation is becoming still more common (in 1994 93% of Dutch households collected their kitchen and yard waste to make compost), we must pose the question on whether mechanical processing being remunerative. At the moment the ARN is the only MWI plant in the Netherlands which combusts RDF (in 1995 the share or RDF is estimated to be 7.5% of total MSW to be incinerated).

Different new MWI plants have been recently built and there are several planned; however, they are all designed for MSW with a calorific value about 10 MJ/kg i.e. without the compostable fraction. RDF plants with a processing facility seem to be too expensive. Also, recycling of plastics and paper is becoming more popular.

During the last two years (1993–94), the quality of emission abatement at Dutch MWI plants has been enormously improved after introducing new, strict legislation. At the moment almost all the existing plants are equipped with advanced scrubbers (acid and alkaline), activated carbon filters for dioxin control and SCR or SNCR de-NO_x installations. Moreover, the so called "primary" measures (opposite to "tail-end" measures) have been taken, such as: improving of an oven/boiler geometry and streaming pattern, an optimum primary/secondary air ratio and flue-gas velocity in combination with a good mixing on the grate, in order to reduce formation of flue dust, carbon monoxide and dioxins.

The current evaluation of emissions during the long-term incineration tests in 1988–90 has thus lost some of its current value in comparison with the situation before 1993. The source separation and mechanical processing of MSW proved to be more important for waste management in general in terms of creating more incineration capacity.

REFERENCES

Afvalverwerking in Nederland, gegevens 1993 (Waste Processing in The Netherlands, 1993); ISBN: 90-5484-093-5.
Aktieprogramma GFT, statusrapportage GFT-verwerking, oktober 1994. GFT-Infopunt, den Haag (1994) (Action Program Kitchen and Yard Waste, Status Report of Processing, October 1994, The Hague, 1994).
Cardon, A. L. and Pfeiffer, A. E. (1992) Developments Relating to Combined Heat and Power Production in Waste Incineration. (Paper presented at a Conference "Energy Economy – 92", Maastricht, 16 Sept. 1992).
Hesseling, W. M. F., Kalf, M. C. and Pfeiffer, A. E. (1993) Evaluatie van de Resultaten van Grootschalige Afvalverbrandingsproeven in Nederland. (Evaluation of the Results of Large-Scale Incineration Tests in The Netherlands, in Dutch). KEMA-report in preparation, 93-2027.
Kiers, A. and Wardenaar, J. (1992) Hoger Thermisch Rendement en Lager CO-uitstoot Mogelijk. (A Higher Thermal Efficiency and a Lower CO Emission Possible, in Dutch). *Misset's Milieu Magazine*, Nov. 1992, 50.
Mot, E. and Wormgoor, J. W. (1992) Municipal Solid Waste Conversion to Energy. A Summary of Current Research and Development Activities in The Netherlands. TNO Report 92-262.
Rijpkema, L. P. M. (1992) Survey of Municipal Solid Waste Combustion in Europe. TNO Report 92-304, pp. 95–96.

Low Pollution Technology in Incineration and Slagging of Municipal Solid Waste

T. SUZUKI, R. KITAMURA and T. ITO

Kobe Steel, Ltd. 3-18, Wakinohamacho 1-chome, Chuo-ku, Kobe Japan

Abstract—Municipal solid waste processing technologies have been developed for controlling dioxin emissions and for slagging the ash remaining from fluidized-bed incineration. The reduction in concentration of CO in the exhaust stream can be considered an index of improvement in combustion in reducing the level of dioxins. The secondary combustion characteristics of incinerators were investigated using a cylindrical combustion test furnace and the computer simulation code, FLUENT. The effects of secondary combustion air injection parameters such as using radial or tangential injection, and varying injection velocity and primary gas temperature have been investigated through these experiments. The experimental results were then compared with those of the simulation. In Japan, the space available for disposal of incineration ash is severely limited. In addition, the incineration ash from municipal solid waste includes dioxins and heavy metals. The slagging of ash has been conducted in a commercial size slagging combustor in order to reduce the volume of waste and prevent the leaching of heavy metals into water as well as decompose the dioxins in the ash.

Key Words: Dioxins; incineration; carbon monoxide; secondary combustion; slagging; heavy metals

INTRODUCTION

The amount of municipal solid waste (MSW) is increasing every year. The area available for landfill or disposal in Japan is very limited. Therefore, incineration of the MSW is common and 73% of MSW is incinerated in Japan. Recently, it was realized that incineration poses some pollution problems such as the presence of dioxins in the exhaust gas and treatment of incineration ash to render it non-toxic. The methodology for reducing the concentration of dioxins are reducing the level of CO, holding combustion temperature higher than 800 °C, long residence times and high excess air combustion in the incinerator. CO emission control is especially important in the suppression of dioxins and NO_x. Because two-staged combustion is adopted, high concentrations of CO are emitted at times in most types of fluidized-bed incinerators. The Japanese government has issued regulations recommending that CO emission be reduced to under 50 ppm for new continuous operation incinerators, 100 ppm for existing continuous operation incinerators and new semi-continuous operation incinerators and 200 ppm for existing semi-continuous operation incinerators. Secondary combustion is very important in controlling the CO emissions. The effects of secondary combustion air injection methods such as radial and tangential injection, injection velocity and primary gas temperature in reducing the level of CO have been investigated through experiments and computer simulations. The incineration of MSW decreases weight by 17%. However, the shortage of the landfill area is still a big problem in Japan. The specific gravity of the ash is less than 1.0, and therefore requires a large landfill area. The ash includes dioxins and heavy metals, causing severe problems with secondary pollution such as water leaching of the heavy metals and dioxins. The slagging of ash has been investigated in order to increase its specific gravity and prevent secondary pollution. The slag produced has been shown to be suitable for use as a construction material in place of natural stone.

EXPERIMENTAL

Secondary Combustion

Figure 1 shows the schematic of the apparatus, consisting of a hot stove and a secondary combustion chamber. City gas is burned in the hot stove at an air ratio of about 0.9 (slightly fuel rich), producing a hot gas, referred to below as primary gas. The composition of the city gas is: CH_4 88%, C_2H_6 5%, C_3H_8 5%, C_4H_{10} 2% by volume. The primary gas contains 1 to 2% CO. The gas temperature is approximately $850°C$ and the mean velocity at the inlet of the secondary combustion chamber is approximately 0.7 m/s. These conditions are used in order to simulate the real primary gas

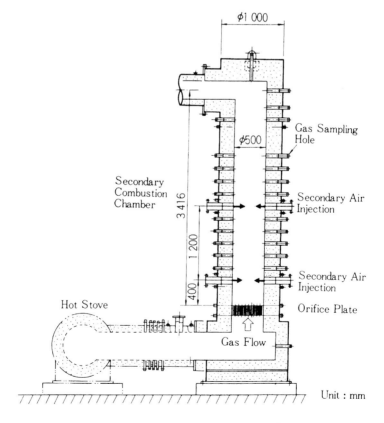

FIGURE 1   Secondary combustion experimental apparatus.

produced by the combustion and thermal decomposition of waste material in fluidized-bed municipal waste incinerators.

The primary gas was directed to the secondary combustion chamber and secondary air was injected into the chamber through four nozzles arranged either radially or tangentially (Fig. 2). The secondary air was not preheated. Its flow rate was 92 Nm^3/h, equivalent to an air ratio of about 0.8 to the city gas and thus producing a total air ratio ((primary air + secondary air)/stoichiometric air of city gas) of 1.7 in order to simulate typical conditions in an actual incinerator.

Gas composition and temperature were measured at various points in the secondary combustion chamber. The flow rate and velocity of the secondary air as well as the temperature and CO content of the primary gas were varied in order to determine their effect on CO concentration.

Slagging of MSW Incineration Ash

Figure 3 shows the slagging combustor test facility for slagging ash remaining from the incineration of municipal solid waste (MSW). The slagging combustor has been developed as a coal gasifier and a slagging combustor for sewage sludge ash by Suzuki *et al.* (1993). Combustion air at 450°C, produced by a hot stove, is fed to the slagging combustor. Primary air is supplied from the top of the combustor through a burner, and secondary air is tangentially injected into the combustor in the upper section of the combustor. The flow pattern in the combustor is controlled by manipulating the ratio of primary and secondary air in order to obtain high combustion performance and high slag capture rates. Butane gas fuel is injected into the strong turbulent swirling flow through the burner and quickly mixed with combustion air and burnt. MSW incineration ash up to 1200 kg/hr is supplied with primary air to the burner of the slagging combustor. The MSW incineration ash collides with the wall of the combustor due to high centrifugal forces. Because the wall temperature of the combustor is higher than the ash fusion temperature, the ash turns to molten slag and continuously flows down the combustor walls to a water cooled slag conveyer. The molten slag is cooled and

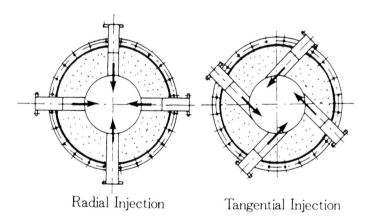

Radial Injection   Tangential Injection

FIGURE 2   Angles of secondary air injection.

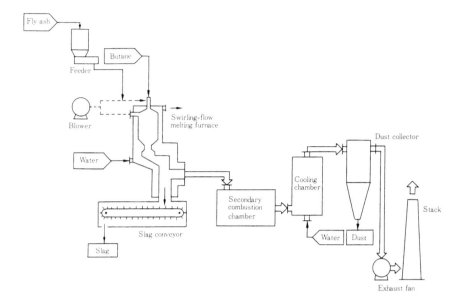

FIGURE 3   Schematic of slagging combustor test facility.

solidified on the conveyer, then dropped into a slag container. The exhaust gas is cooled in a cooling chamber and then cleaned by a bag filter.

Table I shows the results of a chemical analysis of the MSW incineration fly ash. The ash was collected in a fabric bag filter at a fluidized-bed incineration plant.

RESULTS AND DISCUSSION

Secondary Combustion

1. *The effect of secondary air flow rate*   Results were compared for secondary air flow rates of 46 and 92 Nm^3/h (total air ratio of 1.4 and 1.7 respectively). Figure 4 shows that a lower excess of secondary air reduces the CO level for both radial and tangential injection. The CO reduction is likely due to the higher gas temperatures that result from less cold air being introduced. This indicates that the mixing of primary gas and injected secondary air is of essential importance and that a decrease in secondary air feed rate can reduce the CO level in an actual incinerator.

2. *The effect of secondary air injection velocity*   The original secondary air velocity of 19.4 m/s was altered to 3.2 m/s while holding the flow rate constant by using larger diameter air injection nozzles. Figure 5 shows the CO concentrations in the secondary combustion chamber for both cases. For both radial and tangential injection, the CO level is reduced with the higher velocity secondary air injection because of quicker mixing of the two streams. Especially when using radial injection, higher secondary air

TABLE I

Properties of ash

Item		A Sewage Sludge Ash	B Minicipal Solid Wastes Ash
SiO$_2$	%DB	37.38	19.46
CaO	% DB	25.68	23.85
Al$_2$O$_3$	% DB	7.63	22.97
Fe$_2$O$_3$	% DB	8.74	9.5
K$_2$O	% DB	0.49	1.45
Na$_2$O	% DB	0.19	2.79
P$_2$O$_5$	% DB	7.41	2.26
MgO	% DB	1.06	2.34
CaO/SiO$_2$	—	0.69	1.23
Melting Point	°C	1240	1255
Melt-flowing point	°C	1265	1275
Particle size	μm	50	21

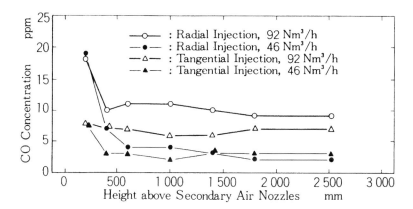

FIGURE 4   Effect of secondary air flow rate on CO concentration.

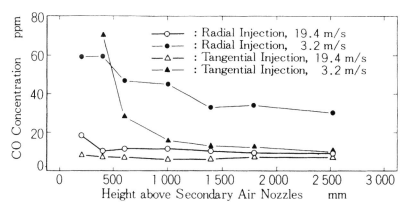

FIGURE 5   Effect of secondary air velocity on CO concentration.

velocity is believed to cause the four secondary air jets to collide more intensely at the center line of the chamber. This collision generates stronger turbulence, forcing the air flow downwards as well as upwards and forming a stronger recirculation zone, thus improving the mixing of the air and the primary gas.

3. *The effect of primary gas temperature*   The primary gas temperature was varied between 750°C and 850°C which is similar to the temperature range of the real primary gas in actual incinerators. As seen in Figure 6, CO emissions increase as the primary gas temperature decreases. The effect of the primary gas temperature on CO emissions is more apparent when using radial injection, and a primary gas temperature lower than 750°C would cause extremely high CO emissions in this case.

4. *The effect of CO content in primary gas*   The CO content of the primary gas was approximately 1% and 2% corresponding to that of the real primary gas in actual incinerators. In the case of radial injection, the CO emissions concentration was 9 ppm when the CO content of the primary gas was 1% while it was 6 ppm when the CO content of the primary gas was 2%. It is believed that higher CO content of the primary gas caused a greater release of reaction energy generated by the secondary combustion, thereby raising the average gas temperature in the secondary combustion chamber, and thus increasing the reaction rate and lowering CO emissions. As shown in Figure 7, the gas temperature was higher throughout the secondary combustion chamber in the case of CO = 2% than in the case of CO = 1%. With tangential injection, the results were similar; CO emissions of 8 ppm were measured in the case of CO content of 1% and 5 ppm in the case of 2% CO.

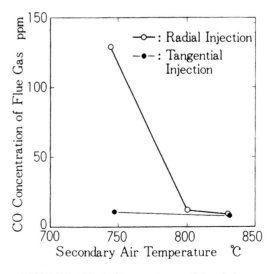

FIGURE 6   Effect of temperature on CO emission.

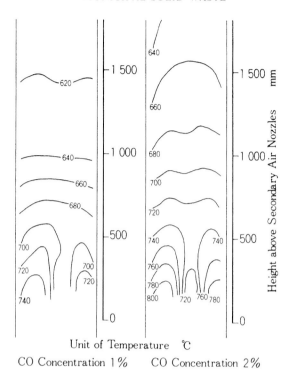

FIGURE 7   Contours of gas temperature in the secondary combustion chamber.

5. *Numerical simulation*   A numerical simulation was conducted using FLUENT for the secondary combustion in the lower part (0 to 1200 mm from the orifice plate, − 400 to 800 mm above the secondary air nozzles) of the secondary combustion chamber of the test furnace. Average values of temperature, flow rate, velocity and gas composition of primary gas, and temperature, flow rate, velocity and injection angle of secondary air were input into the simulation.

The predicted contours of O_2 concentration generally showed good agreement with the experimental results, and especially good agreement was obtained in the case of tangential injection using a fine grid (approx. 38,000 node points) and the Algebraic Stress Model to account for turbulence (Fig. 8). CO concentrations were, however, not very accurately predicted. This is likely due to the fact that the global Arrhenius reaction rate for CO oxidation are not very accurate. Many reported global equations for the reaction rate were simulated, including equations by Hottel *et al.* (1965), Howard *et al.* (1973) and Dryer and Glassman (1973). The equation reported by Gazhal (1971) was found to give the best results for this particular simulation. A parameter study was conducted on the pre-exponential factor and activation energy in order to improve the prediction accuracy of Gazhal's equation. Figure 9 shows the CO concentration predicted by the modified equation. The contours show a reasonably good agreement with the experimental data in the average CO level at the exit.

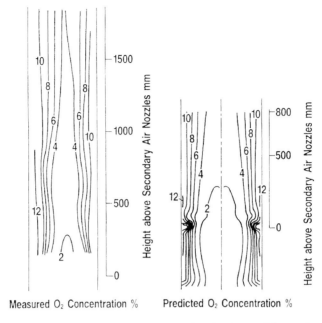

Measured O₂ Concentration %   Predicted O₂ Concentration %

FIGURE 8 O_2 contours in the secondary combustion chamber (tangential secondary air injection).

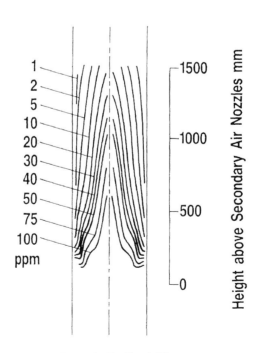

FIGURE 9   Predicted CO contours.

In order to further improve the accuracy of the simulation of the CO oxidation reaction, a number of promising, but slightly more complex calculation strategies are currently being evaluated. These calculations will be used in the future in order to further optimize the incinerator design to reduce the level of CO and other pollutants in the exhaust stream.

Slagging of MSW Incineration Ash

1. *Numerical simulation* A computer simulation was conducted in order to scale up the slagging combustor from an ash feed rate of 250 kg/hr to 1200 kg/hr. The computer model was developed using FLUENT. Figure 10 shows prediction of the streamlines and temperature distribution in the combustor. A ratio of primary to secondary air of 4:1 and a ratio of butane gas feed to ash feed to 2850 kcal/kg-ash were assumed. This figure shows that a strong recirculation zone is formed in the center of the both size combustors. There are no differences in the streamlines between the small combustor and the large combustor. However, the high temperature zone of the large combustor is wider than that of the small combustor. These predictions indicate that the slag capture rate of the small combustor and the large combustor are virtually identical and that the fuel consumption per ash feed rate is improved in the large combustor.

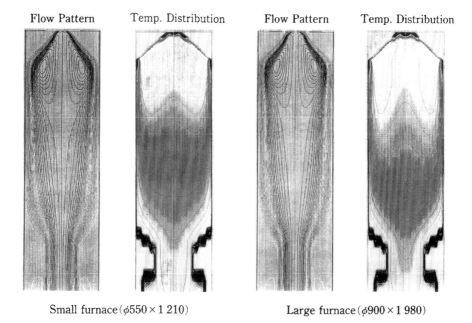

Flow Pattern   Temp. Distribution   Flow Pattern   Temp. Distribution

Small furnace ($\phi 550 \times 1\,210$)   Large furnace ($\phi 900 \times 1\,980$)

FIGURE 10   Prediction of streamlines and temperature distribution in the combustor. (See Color Plate I).

2. *Combustor performance*   Table II shows the experimental data from a large slagging combustor and a small slagging combustor. The slag capture rate of MSW incineration ash are lower than that of sewage sludge incineration ash because the MSW incineration ash is composed of more low temperature volatiles such as K_2O and Na_2O and finer particles than that of sewage sludge incineration ash. Recycling of the ash which escapes from the combustor and is captured in the bag filter back into the slagging combustor is required for the MSW incineration ash slagging. As the slagging combustor is scaled up holding the ratio of ash feeding rate to combustor volume, the heat loss in the slagging combustor decreases. As a result, the fuel consumption per ash feed rate decreases by 25%. NO_x emissions are under 150 ppm in both the large and the small slagging combustors.

3. *Dioxins decomposition*   Table III shows chemical analysis data of the level of dioxins during MSW incineration ash slagging. The operation data is shown in Table IV. The concentration of dioxins contained in the slag is 0.0012 ng TEQ/g, nearly the limit of detection. The slag produced in the slagging combustor is confirmed to be non-toxic. The balance of dioxins in Table III is shown in Figure 11. More than 98% of the dioxins in the ash are decomposed by slagging.

4. *Leach of heavy metals*   Table V shows leaching test results for the slag. Even leaching values obtained by a low pH method are below the detection limit, and therefore the slag is harmless in water.

TABLE II

Experimental data

Combustion chamber size mm	Large scale furnace $\phi900 \times 1980$		Small scale furnace $\phi550 \times 1210$
Ash Type	A Sewage Sludge Ash	B Municipal Solid Waste Ash	A Sewage Sludge Ash
Feed Rate kg/h	1200		250
Butane	90		25
Flow Rate Nm^3/h			
Air Temp. °C	450		450
Air Ratio for Combuster	1.2		1.2
Slag Capture %	93.4	82.6	95.0
Heat Loss %	14.2		18.0
Fuel Consumption kcal/kg	2125		2850
Combustion Load kcal/m^3h	2.02×10^6		2.46×10^6
Exhaust Gas			
NO_x ppm 12% —O_2	88.1	146	80
Temp. °C	1410	1435	1450

TABLE III

Analysis results of dioxins

Item Sample	Fly ash ng/g	Slag ng/g	Dust ng/g	Exhaust gas ng/Nm^{3}O$_2$12%
T$_4$CDD$_s$	9.7	0.0070	5.0	1.6
P$_5$CDD$_s$	11	0.0040	7.2	2.1
H$_6$CDD$_s$	24	0.011	6.5	4.0
H$_7$CDD$_s$	41	0.046	14	22
O$_8$CDD$_s$	17	0.020	4.6	12
Total PCDD$_s$	100	0.087	37	42
T$_4$CDF$_s$	44	0.0050	1.3	6.5
P$_5$CDF$_s$	34	0.0050	1.5	11
H$_6$CDF$_s$	28	0.0080	2.7	20
H$_7$CDF$_s$	39	0.036	11	65
O$_8$CDF$_s$	11	0.0077	6.7	43
Total PCDF$_s$	160	0.062	23	150
Total PCDD$_s$/PCDF$_s$	260	0.15	60	190
TEQ (International)	2.8	0.0012	0.45	1.8

TABLE IV

Operation data

Item		Run
Ash feed rate	(kg/h)	100
Butane gas volume	(Nm^3/h)	27
Air volume	(Nm^3/h)	900
Air ratio	(−)	1.2
Furnace wall temperature	(°C)	1,296
Furnace outlet temperature	(°C)	1,472
Furnace inner pressure	(mmH$_2$O)	− 3.9
Slagging ratio	(%)	80.4
CO	(ppm−O$_2$12%)	2.9
NO$_x$	(ppm−O$_2$12%)	117
SO$_x$	(ppm−O$_2$12%)	< 2
HCl	(mg/Nm^{3}O$_2$12%)	785

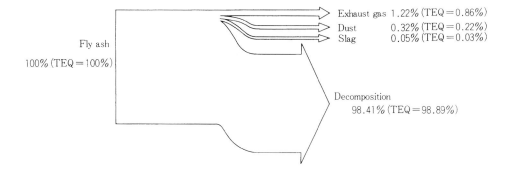

FIGURE 11   Balance of dioxins.

TABLE V

Leaching test results of slag

Analysis method Items	Standard method (Japan)	Low pH method (pH$_4$)	Leaching standard for landfill
T-Hg	< 0.0005	< 0.0005	< 0.005
Pb	< 0.01	< 0.01	< 3
Cd	< 0.01	< 0.01	< 0.3
Cr^{6+}	< 0.02	< 0.02	< 1.5
As	< 0.01	< 0.01	< 1.5
Org-P	< 0.1	< 0.1	< 1
PCB	< 0.0005	< 0.0005	< 0.003
CN	< 0.01	< 0.01	< 1

CONCLUSION

Municipal solid waste processing technologies have been developed for controlling CO (dioxin) emissions and for slagging the ash remaining from fluidized-bed incineration. Obtained results are summarized as follows:

Secondary Combustion

(1) Mixing of primary gas and injected secondary air is of essential importance for decreasing CO. CO level is reduced with the higher injection velocity and tangential injection of secondary air because of quicker mixing of the two streams.
(2) CO emissions increase as the primary gas temperature decreases. A primary gas temperature lower than 750°C would cause extremely high CO emissions when the radial injection is used.
(3) The predicted contours of O$_2$ concentration generally showed good agreement with the experimental results. CO concentrations were, however, not accurately predicted. This is likely due to the fact that the global Arrhenius reaction rate for CO oxidation is not accurate.

Slagging of MSW Incineration Ash

(1) Scale up method of a slagging combustor have been established. The slag capture rate of the large combustor and the small combustor are virtually identical and the fuel consumption per ash feed rate is improved by 25% in the large combustor.
(2) More than 98% of the dioxins in the MSW fly ash was decomposed by slagging.
(3) The leaching of heavy metals into water of the slag was prevented by slagging. The slag produced in the slagging combustor is confirmed to be non-toxic.

REFERENCES

Dryer, F. L. and Glassman, I. High Temperature Oxidation of CO and CH$_4$. *Fourteenth Symposium (International) on Combustion*, The Combustion Institute, 987–1003 (1973).
Gazhal, F. P. H. (1971) Kinetic of Carbon Monoxide Oxidation in Combustion Products, *M.S. Thesis in Chemical Engineering*, Massachusetts Institute of Technology.

Howard, J. B. *et al.* Kinetics of Carbon Monoxide Oxidation in Post Flame Gases. *Fourteenth Symposium (International) on Combustion*, The Combustion Institute, 975–986 (1973).

Hottel, H. C. *et al.* (1965) Kinetic Studies in Stirred Reactors: Combustion of Carbon Monoxide and Propane. *Tenth Symposium (International) on Combustion*. The Combustion Institute, 111–121.

Suzuki, T. *et al.* (1993) Development of a Low-NO_x Emission Slagging Combustor. *Combust. Sci. and Tech.*, **93**, 111–128.

Treatment of Scrubber Effluent from a Multiple Hearth Incinerator

J. P. JONES[a] and A. KANTARDJIEFF[b]   [a]*Department of Chemical Engineering University of Sherbrooke Sherbrooke, Quebec, Canada J1K 2R1;* [b]*Ekokan Inc. Rock Forest, Quebec, Canada J1N 1X7*

Abstract—This study investigated the feasibility and costs of applying a complete physical treatment to effluent from the scrubbers of a sludge incinerator. The study demonstrated the technical feasibility of obtaining storm water objectives for the effluent using a combination of chemical treatment, sedimentation and filtration, the costs of such treatment were evaluated. These costs are considered to be so high as to not justify the implementation of full physical-chemical treatment. It is recommended to recycle effluent to the head of the treatment station.

Key Words: Sludge; incineration; scrubber; suspended solids; coagulants

INTRODUCTION

The use of multiple hearth incinerators for disposing of sewage sludge is widely practiced. Such incinerators operate at relatively low temperatures and are liable to produce considerable fly ash which is removed by venturi scrubbers and multiple plate scrubbers. However the water from the scrubbers is itself polluted and must be treated. This problem may be particularly significant for a municipality since it will generally require dischargers to its sewage system to meet certain effluent standards.

Figure 1 shows the flow diagram for the principal elements for the multiple hearth incinerator under study. Solid, water, combustion gas flows are indicated.

Recently we undertook the study of the scrubber effluent treatment system for a multiple hearth sludge incinerator, for a municipal treatment station with a capacity of 1 400 000 m^3/d. The following parameters were not within the requirements for discharge to the storm sewage system: suspended material, phosphates, pH, metals (mercury, lead, zinc), fluorides, total cyanides, phenols.

The effluents standards are quite different for storm water and for sanitary effluent. The scrubber effluent was currently being discharged directly to the receiving body and thus should have met storm sewer standards although no decision had yet been made as to whether this scrubber effluent should be treated to storm sewer or sanitary sewer standards. Table I gives the concentration for wash water and process water. The problem could reasonably be divided into three:

- suspended solids and phosphates;
- heavy metals;
- cyanides and phenols.

711

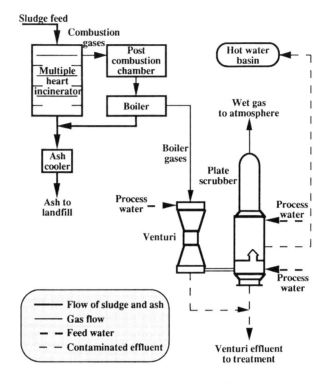

FIGURE 1   Flow diagram for multiple hearth incinerator.

TABLE I

Mean concentrations for wash water from the incinerators

Parameters	Units	Venturi	Process	Scrubber	Storm sewer effluent objectives
pH		5.9(13)	7.2 (12)	6.6 (13)	6.0 à 9.5
S.S.	mg/L	635 (15)	19 (14)	38.8 (15)	30
Total phosphates	mg/L	30.3 (15)	0.47 (14)	0.81 (15)	1
Fluorides	mg/L	1.5 (05)	0.28 (05)	0.29 (05)	2
Total cyanides	mg/L	0.47 (15)	0.078 (14)	0.55 (15)	0.1
Phenols	mg/L	12.8 (15)	10.6 (14)	8.7 (15)	20
Total Barium	mg/L	0.71 (15)	0.11 (14)	0.12 (15)	1
Total Cadmium	mg/L	59 (15)	1.74 (14)	3.16 (15)	100
Total Chromium	mg/L	250 (15)	8.14 (14)	12.1 (15)	1000
Total Copper	mg/L	811 (15)	42.9 (14)	105.9 (15)	1000
Total Tin	mg/L	322 (10)	41.8 (09)	42.1 (10)	1000
Total Mercury	mg/L	7.9 (15)	0.49 (14)	0.53 (15)	1
Total Nickel	mg/L	129 (15)	22.7 (14)	18.7 (15)	1000
Total Lead	mg/L	479 (15)	22.1 (14)	39.9 (15)	100
Total Zinc	mg/L	2.8 (15)	0.19 (14)	0.22 (14)	1

( ) indicates number of determinations.

There were a number of stages to the investigation:
- evaluation of the scope and magnitude of the effluent problems;
- literature evaluation of treatment alternatives;
- determination of standard practices in other multiple hearth incinerators of similar capacity;
- laboratory evaluation of coagulation, precipitation and oxidation techniques.

Table II shows the frequency and the extent that effluent standards were not met during a winter season. Figure 2 shows the frequency of excursions from the objectives for cyanide. Similar curves in Figure 3 and Figure 4 for mercury and suspended solids show continuous violation of the treatment objectives.

The EPA (1990) reported that particle emission and therefore heavy metals in the scrubber effluent are a function of number of operating parameters for the incinerator especially the operating temperature, and the sludge characteristics which themselves depend on the season. The EPA report also states that most sludge incineration facilities recycle the venturi and plate scrubber effluent to the head of the works. In Germany, many works apply treatment including:

- neutralization and sedimentation
- evaporation
- biological treatment
- finally recycle to the head of the works.

PROBLEM DEFINITION

The aim of this piece of work was to evaluate the practicability of three options. The options themselves were the following:

- return some or all of the scrubber effluent to the headworks without treatment;
- provide neutralization and simple sedimentation and return the effluent to the headworks;
- treat the scrubber effluent completely in order to meet storm water objectives and thereby be able to discharge directly to the receiving water.

The first option of discharging to the headworks without treatment was felt by the project team to be the most interesting but there were considerable political objections. The scrubber effluent did not meet sanitary effluent objectives and therefore would be unacceptable if the incineration facility was thought of as a discharger to the sewer system. Considerable efforts were made to demonstrate that this was a widely practiced and acceptable technique in other constituencies.

The second option was experimented at a laboratory scale and at small pilot scale on site, and it was hoped to demonstrate that sanitary effluent objectives could be met so as to be able to put forward this solution.

The final option was experimented in order to find ways of treating to objectives required for storm sewer discharge. This was thought to be technically feasible but the cost needed to be evaluated in order to determine its practicability.

TABLE II

Effluent winter 1990

	11/02/90	13/02/90	Effluent (before filtraion) 19/02/90	21/02/90	23/02/90	Objectives Storm sewers	Objectibves sanitary sewers
pH	7.3	7.4	8.8		6.3	6.0–9.5	6.0–10.5
Suspended solids mg/L	685	1620	1590	930	540	30	–
Total Phosphorus Mg/L	22	57	70	68	30	1	–
Fluorides mg/L	1.33	2.05	1.43	1.2	1.57	2	–
*Total Cyanides mg/L	0.09	0.66	0.17	0.4	0.06	0.1	10
Phenols μg/L	11	12	50	10	11	20	1000
Total barium mg/L	0.7	2.1	1.5	1.3	0.7	1	–
Total cadmium μg/L	54	100	67	50	47	100	2000
Total chromium μg/L	390	840	400	340	180	1000	5000
*Total copper μg/L	840	780	1300	1430	660	1000	5000
Tin μg/L	200	200	1100	740	280	1000	5000
*Total mercury μg/L	3.7	8.6	3.9	6.1	4.8	1	50
Total nickel μg/L	100	990	97	125	66	1000	5000
*Total lead μg/L	730	95	1380	1000	660	100	2000
*Total zinc mg/L	3.3	4.6	3.3	2.2	1.3	1	10

*Does not meet storm water objectives.

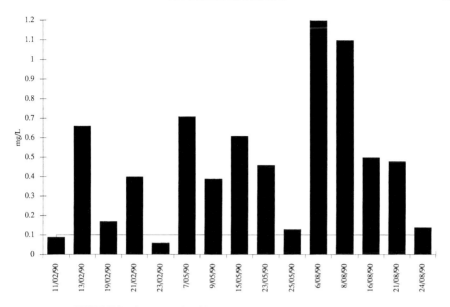

FIGURE 2   Concentration history of cyanide in effluent of scrubbers.

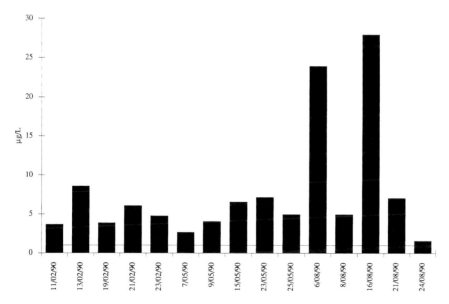

FIGURE 3   Concentration history of mercury in effluent of scrubbers.

CURRENT PRACTICE

Information was obtained for all sludge incinerators using multiple hearth incinerators for works treating more than 50 USMGD (200 000 m^3/d). Some incinerators (notably in Quincy, Mass.) are no longer in operation because of air pollution restrictions.

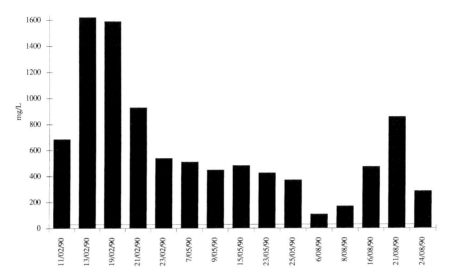

FIGURE 4   Concentration history of suspended solids in effluent of scrubbers.

Table III describes the current practice at 12 installations. Generally accepted good practice is to recycle scrubber effluent to the head of the works.

EXPERIMENTAL WORK

The laboratory treatment of scrubber effluent was evaluated by optimizing the most fundamental process first and then adding treatment to solve any remaining problems. Occasionally it was necessary to backtrack because the optimum at one stage had to be modified to include an additional treatment.

Simple Sedimentation

Simple sedimentation was applied in order to evaluate the removal of suspended solids and iron. Generally these two parameters were correlated since iron from the ferric chloride used in physical-chemical treatment process constituted an important fraction of the suspended solids.

Table IV shows the quality of the supernatant obtained. Suspended solids and iron have reached the required level while phosphorus and zinc have not. The pH is obviously unaffected and out of limits.

pH Modification

Literature (Grohmann *et al.*, 1988; Patterson, 1971) indicates clearly that most metals are precipitated as hydroxides and in order to have such precipitation, the pH must be raised probably to values above 9 and therefore possibly to values above the

TABLE III

Current practice at sludge incinerators

Localisation	Type of incinerator	Supplier	Flow (USMGD)	Type of treatment	Current practice
BUFFALO (New York)	Multiple hearth	—	178	None	Recycle to secondary treatment
CINCINNATI (Ohio)	Multiple hearth		107	None	Recycle to headworks
CLEVELAND (Ohio)	Multiple hearth	Nicholls	126	None	Lagoon
COLUMBUS (Ohio)	Multiple hearth	Nichols Krauss	92	None	Lagoon
DETROIT (Michigan)	Multiple hearth	Nicholls	698	None	Recycle to headworks
INDIANAPOLIS (Indiana)	Multiple hearth	Nicholls	126	None	Recycle to headworks
MINNEAPOLIS SAINT-PAUL (Minnesota)	Multiple hearth	—	250	None	Recycle to headworks
NEWROCHELLE (New York)	Multiple hearth	Zimpro	16	None	Recycle to headworks
PITTSBURG (Pennsylvania)	Multiple hearth	Zimpro	200	None	Recycle to headworks
QUINCY (Massachussets)	Multiple hearth	—	128	—	Out of service
SAINT-LOUIS (Missouri)	Multiple hearth	Nicholls first, rebuilt Hanken	140	None	Effluent used for hydraulic transport of ash. The overflow is sent into the receiving stream. There is considering recycle to the headworks.
TWO BRIDGES (New Jersey)	Fluidized bed	Dorr Oliver	70	None	Ash lagoon

TABLE IV

Results of tests for simple sedimentation

Parameters	Untreated Effluent	Supernatant (2)	% removed	Storm water objectives
Temperature (°C)	20	20	—	
pH	4.1	4.1	—	6–9.5
SS (mg/L)	731	24	97%	30
Iron (mg/L)	24	1.2	95%	17
P(1) (mg/L)	24.5	12	51%	1
Zn (mg/L)	4.5	3.5	22%	1

(1) expressed as $O-PO_4$.
(2) 55 minute sedimentation.

objectives for storm water. Lime (either slaked or unslaked) is the most common way to raise pH.

Table V demonstrates the usefulness of this and shows good removal of zinc at pH's above 6.5. Lime is also useful for precipitating phosphorus. However a pH as high as

TABLE V

Effect of pH modification with lime on several control parameters

pH Parameters	4.1 (*)	6.5	7.0	8.0	8.5	9.0	9.5	10.0	10.5
SS (mg/L)	24	4	4	4	4	2	2	2	2
Iron (mg/L)	1.2	0.39	0.40	0.35	0.42	0.28	—	—	—
P total (mg/L)	12.0	7.1	6.1	6.2	5.8	3.9	3.3	2.1	1.2
Zn (mg/L)	3.8	0.44	0.54	0.76	0.22	0.46	—	—	—

(*) Supernatant.

10.5 did not allow us to attain the objective for phosphorus. The removal of phosphorus probably as hydroxylapatite was evaluated for large lime doses as the lowest value obtained 1.2 mg/L is still slightly above the objective for storm water.

Addition of a Coagulant

The use of a coagulant was then investigated in combination with lime. Although it was possible to reduce phosphates below the objectives with sodium hydroxide, lime was the preferred neutralisation agent. Table VI shows the reduction of phosphorous with 75 mg/L of lime. The phosphorus concentrations are well within the objectives. A very important advantage over the use of lime alone is the reduced cost but also but perhaps more important is that 75 mg/L of lime maintains the pH between 9.3 and 9.5 and acid is not required to reduce the pH after precipitation. The proprietary product Charge pack 55 was attractive from an economic point of view.

The concentrations of other metals were evaluated in a confirmatory test which will be discussed later. Mercury represented a special problem since the concentration was high compared to the objectives. Mercury in a sewer system can come from a variety of sources including industrial use, dental use and certain pesticides.

There is a large body of knowledge about removal of mercury from effluents including the use of sodium sulphide, polyaluminium sulphate silicate, sodium boron hydride and several others. Several of these techniques were successfully applied. The mercury is recovered with the precipitated hydroxides.

Cyanide and Phenol Removal

Cyanide and phenol can both be removed with oxidation. In industrial processes, cyanide is generally removed by alkaline chlorination. Phenol however is not generally

TABLE VI

Performance of the best coagulants for a lime dose of 75 mg/L

Coagulant (dose) Parameters	Alum 48% (50 ppm)	Charge pack 55 (25 ppm)	Charge pack 55 (30 ppm)	Aluminex 2 (50 ppm)
P total (mg/L)	1.3	0.80	0.40	0.77

treated with chlorine because there is a danger of forming chlorophenols which can produce bad tastes in drinking water.

Hypochlorite was very effective in removing cyanides while the hydrogen peroxide effectively removed phenols and cyanides. These results are not presented here but will be available in a forthcoming publication.

CONFIRMATION OF FINDINGS

Three full tests were run to determine whether the results obtained in a somewhat piecemeal fashion could be confirmed in a full test. These tests included a complete battery of analytical results performed using standardized procedures. The removal of cyanides and phenols was not investigated during these tests. The conditions in the full tests are described below.

Test FT-A:

- increase the pH to 9.5 with lime
- add 30 mg/L polyaluminium chloride (PAC)
- add 1 mg/L high molecular weight anionic flocculant (Drew floc 2270)
- add a proprietary coagulant for heavy metals (MP7 from Drew Chemicals)
- filter the supernatant

Test FT-B:

- increase the pH to 9.5 with lime
- add 30 mg/L polyaluminium chloride (PAC)
- add 1 mg/L high molecular weight anionic flocculant (Drew floc 2270)
- add reduction agent $NaBH_4$ for mercury
- filter the supernatant

Test FT-C:

- increase the pH to 9.5 with lime
- add 30 mg/L polyaluminium chloride (PAC)
- add 1 mg/L high molecular weight anionic flocculant (Drew floc 2270)

Table VII shows the results obtained for the three full tests. The results show small discrepancies from the objectives for mercury and lead. These problems can be eliminated in the full scale pilot studies to follow.

COST COMPARISON

Based on the experimental findings, an evaluation was established for the costs for options 2 and 3 - simple sedimentation and complete physical-chemical treatment. Figure 5 shows the flowsheet for complete treatment. These costs for both capital and operating are shown in Tables VIII and IX. The costs for option 1 were not evaluated in detail since they were much lower and depended on the size of the existing pumping

TABLE VII

Final characterization of tests

Parameters (units)	Influent	Treated Test FT-A	Treated Test FT-B	Treated Test FT-C	Objectives
SS (mg/L)	1 044	5	6	8	30
pH	5.63	9.23	9.38	9.66*	6.0–9.5
Fe total (mg/)	10.2	0.65	1.15	1.14	17
P total (mg/L)	22.5	0.50	0.50	0.70	1
Zn (mg/L)	3.28	0.01	0.04	0.04	1
F (mg/L)	2.4	5.1*	4.9*	4.8*	2
Ba (mg/L)	5.6	< 0.2	< 0.2	< 0.2	1
CN (mg/L)	0.02	< 0.01	< 0.01	< 0.01	0.1
Cu (μg/L)	1 340	180	120	420	1000
Cd (μg/L)	230	10	10	10	100
Cr (μg/L)	370	40	20	< 10	1000
Sn (μg/L)	< 500	< 500	< 500	< 500	1000
Hg (μg/L)					
● before filtration	6.9	3.2*	1.0	0.4	1
● after filtration	—	0.2	0.2	0.3	
Ni (μg/L)	40	20	40	60	1000
Pb (μg/L)					
●before filtration	970	50	110*	50	100
●after filtration	—	< 10	< 10	45	
Phenols (μg/L)	47	< 5	< 5	16	20

*Parameters which do not meet objectives.

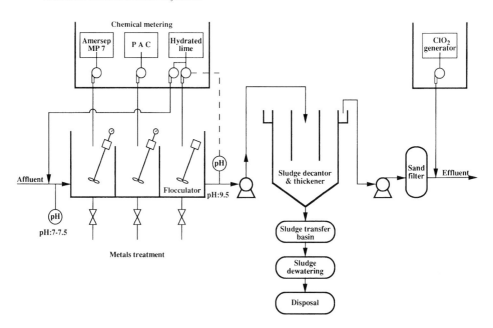

FIGURE 5   Physical-chemical treatment to meet storm sewer objectives.

TABLE VIII

Preliminary estimate of construction costs ($Can 1992)

Item	OPTION 2 Simple sedimentation	OPTION 3 Physical-chemical sedimentation
1. Site preparation	100 000.00$	200 000.00$
2. Civil work	1 000 000.00$	1 500 000.00$
3. Mechanical process work	2 900 000.00$	4 500 000.00$
4. Electricity & ventilation	550 000.00$	1 100 000.00$
Subtotal	**4 500 000.00$**	**7 300 000.00$**
Unforeseen	450 000.00$	700 000.00$
Subtotal	**5 000 000.00$**	**8 000 000.00$**
Incidental costs	2 387 500.00$	2 000 000.00$
TOTAL	**7 387 500.00$**	**10 000 000.00$**

TABLE IX

Preliminary estimate of annual operating costs ($Can 1992)

Item	OPTION 2 Simple sedimentation	OPTION 3 Physical-chemical treatment
1. Personnel	150 000.00$	200 000.00$
2. Chemicals	190 000.00$	1 530 000.00$
3. Sludge treatment	1 400 000.00$	1 630 000.00
TOTAL	**1 740 000.00$**	**3 360 000.00$**

station. In fact, this option had zero cost if the existing pumping station is big enough. The operating costs for complete physical-chemical treatment are twice those for the simple sedimentation.

CONCLUSIONS

This study has demonstrated the possibility of treating scrubber effluent to storm sewer effluent objectives for all parameters except fluorides. Occasional difficulties with mercury and lead can be solved in future experimentation.

This study also demonstrated the possibility of treating scrubber effluent to sanitary sewer requirements at considerably less cost.

RECOMMENDATIONS

Based on experiences in other sludge incinerators, it is recommended that option 1 be retained. The costs for complete physical-chemical treatment are not justified by the environmental advantages which may in fact be negative. This study has only evaluated the operating costs and not compared all the environmental costs. But a well informed judgement would indicate the wisdom of recycle to the head of the works.

ACKNOWLEDGEMENTS

The authors acknowledge the essential contributions of Jacques Pezet and Urgel Béchard.

REFERENCES

EPA (1990) Municipal sludge combustion technology. Center for environmental research information. *Technomic Publishing Co.*, Cincinnati, Ohio.
Grohmann, A. N., *et al.* (1988) Chemical treatment of flue-gas ashing liquids. In *Pre-treatment in chemical and wastewater treatment*, Spurguru lag, Berlin.
Patterson, J. W. (1971). Wastewater treatment technology. *Ann Arbor Science Publishers, Ann Arbor*, Michigan.

Gas Pollutants from Detonation and Combustion of Industrial Explosives

J. CAMPOS[a], A. PIRES[a], J. C. GOIS[a] and A. PORTUGAL[b]   *Laboratory of Energetics and Detonics* [a]*Mec. Eng. Dep.,* [b]*Chem. Eng. Dep. - Fac. of Sciences and Technology University of Coimbra - 3000 Coimbra - Portugal*

Abstract—The structure and fundamental thermodynamic equations of THOR computer code, to calculate the combustion and detonation products of industrial explosives, is presented. The most common industrial explosives in Portugal are ammonium nitrate - fuel oil compositions (anfo), dynamite and ammonium nitrate based emulsion explosives. The combustion and detonation products (CO_2, CO, H_2O, N_2, O_2, H_2, OH, NO, H, N, O, HCN, NH_3, NO_2, N_2O, CH_4 gases and two kinds of solid carbon - graphite and diamond) are calculated for the minimum value of Gibbs free energy, using three well known equations of state - BKW, H9 and H12. Detonation experiments are described and gas analysis discussed. Measured pollutants concentrations (CO, CO_2, NO and NO_2), as a function of volume of explosion chamber, prove the dependence of expansion mechanisms on CO and NO formation and recombination and validate theoretical predictions. Incineration of explosives in a fluidised bed is described. Products composition from isobare adiabatic combustion of selected explosives has been measured and correlated with previous calculations.

Key Words: Pollutants; detonation; combustion; industrial explosives; code THOR; chemical equilibria; chemical composition; minimum Gibbs energy; equation of state

INTRODUCTION

The potential hazards of fumes, from blasting operations in underground mines, have long been recognised. When an industrial explosive detonate small (1 to 5 wt-pct) but significant amounts of CO_2, CO, NO and NO_2 are liberated as detonation products.

These detonation products can be recombined with the environmental air, but under limited ventilation conditions can present a severe toxic fume hazard to mine workers. It is always convenient to remember that NO_2 has a lethal toxicity of about 200 ppm and a threshold limit value of about 5 ppm, similar to that of hydrogen cyanide.

A detonation wave, initiated by detonator in a cylindrical charge of explosive, shows its velocity increasing until to reach a constant and stable value. The stability of detonation is very important in industrial explosives to design applications and reduce pollutants, increased at non stationary regimes, when there is a transition from deflagration to detonation. The pollutants concentration are also dependent of initial density of explosive and of its position inside explosion chamber, because their formation are dependent of the expansion of detonation products (vd. Carbonel *et al.*,1980). Beyond this normal use of explosives, there are also large amounts of energetic substances which cannot be used because their life time is outdated or they are not within the minimal quality requirements (vd. ICT Conference, 1992). This is the case of explosive from old ammunitions, generally TNT compositions, and some explosive compounds in industrial production. There are many possibilities of eliminating them, but the most convenient is the combustion at isobaric conditions in a fluidised bed (vd. Pires et al.,1993).

There is a lack of information concerning tests, procedures and theoretical predictions of pollutants concentrations in fumes from detonation and combustion operations with industrial explosives. All these problems are also related to the kinetics of CO and NO formation and recombination during the expansion of detonation fumes. Many research institutions (Bureau of Mines - vd. Chaiken et al.,1974; Cerchar - vd. Carbonel et al.,1979, 1980) have soon developed tests, in different volume explosion chambers, in order to understand the relationship between mass of explosive and environmental air. Contrary to initial expectations, the NO_x concentrations (obtained by the sum of NO and NO_2 concentrations) in large experimental galleries, seems to be in good agreement with theoretical calculations with computer codes with adapted equations of state (vd. Chaiken et al.,1974). The classical approach to calculate detonation and combustion characteristics of reacting systems and predict pollutants, is to consider two ideal combustion regimes: isobare adiabatic combustion and Chapman-Jouguet detonation.

This paper presents the structure and fundamental thermodynamic equations of THOR computer code (vd. IEPG Reports, 1989, J. Campos, 1991) to calculate the thermodynamical properties and final composition of combustion products (gas and solid components). Experimental results of CO, CO_2, NO and NO_2 concentrations in detonation products of industrial explosives are correlated with those predict by our computer code, using three well known thermal equation of state (EoS): BKW, H9 and H12 EoS (vd. Campos, 1991).

INDUSTRIAL EXPLOSIVES - INITIAL COMPOSITION AND DENSITY

The most common industrial explosives in Portugal are ammonium nitrate - fuel oil compositions (anfo), and dynamite explosives, representing an annual production of 8000 tones for open air and underground mines applications. Recently ammonium nitrate based emulsion explosives are more and more used in those industrial applications.

In this study three types of explosives were selected:

– one composition of anfo, an ammonium nitrate - fuel oil composition, with 6 wt-pct of fuel oil, with initial density 870 kg/m^3,
– one composition of dynamite explosive, formed by 30 wt-pct of nitroglycerine, 6 wt-pct of DNT, 60 wt-pct of ammonium nitrate and 4 wt-pct of amidon, with density 1400 kg/m^3,
– one composition of emulsion explosive, formed by an aqueous solution of 10 wt-pct of water, of ammonium and sodium nitrates, respectively 72 and 10 wt-pct, emulsified with oils, wax and emulsifiers, 5.5 wt-pct, with hollow glass spheres as sensitizer, 2.5 wt-pct, with density 1170 kg/m^3.

THEORETICAL CALCULATIONS

The theoretical study starts with the prediction of detonation characteristics using THOR code, based on theoretical work of Heuze et al., 1985, 1989, and later modified (vd. IEPG Reports, 1989, 1990, Campos, 1991) in order to calculate the composition and thermodynamic properties of explosive compositions, for Chapman-Jouguet detonation conditions. Several equations of state are used, namely BKW, Boltzmann, H9, H12 and JCZ3 (vd. Campos, 1991). The results have then been compared within themselves and with results of different codes in open literature (Quatuor Code from Heuze *et al.*,1985, 1989, using the BKW EoS the TIGER Code (vd. Chaiken *et al.*,1975) and Mader Code, 1979, and using the KHT EoS the Tanaka Code, 1983).

Chemical Equilibria Conditions

The classical combustion system is generally a CHNO system. In our computer code it is possible to consider up to m atomic species and to form n chemical components with these atomic species. Among these n chemical components, m are considered "basic" chemical components and $n–m$ "non basic". The selection "ab initio" of the "basic" chemical components depends on the equivalence ratio r of the mixture, related to the stoichiometry ($r = 1$), and they are those which are expected to have significant concentrations in final products composition.

For a CHNO system it has been selected (vd. Manson, 1976; Heuze, 1989) the "basic" components:

- CO_2, H_2O, O_2 and N_2 for poor mixtures ($r < 1$),
- CO_2, H_2, H_2O and N_2 for rich mixtures ($r > 1$) of initial low density and C(s), CO_2, H_2O and N_2 for rich mixtures of initial high density (initial condensed or solid components).

A matrix (n, m) can then be created with each line obtained by the expression of its (n) component as a polynomial expression of chosen m "basic" components.

The mass balance yields a linear system involving m equations. In order to solve the problem it is necessary to add more $(n–m)$ equations. These $n–m$ equilibrium equations are determined by the method of Lagrange multipliers or the equilibrium constants (vd Brinkley, 1947, White *et al.*,1958). Consequently the system of equations is formed by m linear mass balance equations and $(n–m)$ non linear equilibrium equations.

In order to determine the chemical concentration of the n components, for imposed P and T conditions, two methods can been used:

- the chemical affinity method, proposed by Heuzé *et al.*, 1985,
- solving at first the system composed by the m 'basic' components, and secondly adding one by one more components, optimising the relative concentration inside the group related to the same atomic specie, for the minimum value of global Gibbs free energy $G = \Sigma x_i \mu_i$, being the Gibbs free energy of each component $\mu_i = G_{oi}(T) + R T \ln P + R T \ln (x_i)$. The values of $G_{oi}(T)$ is the Gibbs free energy as a function of temperature. They can be obtained from JANAF Thermochemical Tables, 1971, and from polynomial expressions of Gordon and McBride, 1971. This second method is slower than the first, but avoid numeric problems (vd. Campos, 1991).

The solution of the composition problem involves simultaneously:

- the thermodynamic equilibrium, obtained with the mass and species balance, and the equilibrium condition $G = G_{min}(P, T, x_i)$, previously described, generally applying to the condensed phase the model proposed by Tanaka, 1983,

– the thermal equation of state (EoS),
– the energetic equation of state, related to the internal energy $E = \Sigma\, x_i\, e_i(T) + \Delta e, e_i(T)$ calculated from JANAF Thermochemical Tables, 1971, and from polynomial expressions of Gordon and McBride, 1971,
– the combustion regime, being $P_b = P_o$ constant for the isobare adiabatic combustion (equal final and initial total enthalpy $H_b = H_o$), and the Chapman-Jouguet condition for the detonation regime (mass, momentum and energy balances and $dp/dV]S = (p - p_o)/(V - V_o)$).

The selection of components are dependent of atomic initial composition. For a classical CHNO system it is considered the equilibrium compositions of CO_2, CO, H_2O, N_2, O_2, H_2, OH, NO, H, N, O, HCN, NH_3, NO_2, N_2O, CH_4 gases and two kinds of solid carbon (graphite and diamond). It is possible to include more species in final products composition, like S, SO, SO_2, using data from JANAF Thermochemical Tables, 1971, and polynomial expressions of Gordon and McBride, 1971.

Theoretical Results

Theoretical predictions of detonation velocity D and pressure P_{CJ} were made using code THOR with three chosen equations of state BKW, H9 and H12 EoS, assuming a Chapman-Jouguet model of detonation and thermodynamical equilibrium of solid and gas detonation products (vd. Tab. I). Obtained results are in a good agreement with other calculations (vd. Johnson *et al.*, 1983) proving the validity of chosen EoS (vd. Figs. 1 and 2).

Products composition of detonation of selected explosives are shown in Figures 3 to 5. NO_x concentrations are more important for dynamite than for ANFO and emulsion explosives. CO concentrations have almost the same value for dynamite and emulsion explosive, but lower than for ANFO explosive. The theoretical results show NO concentrations decreasing and became negligible, and the CO concentrations increasing, when the global equivalence ratio of explosive is higher than 1.

Products composition characteristics of combustion of each explosive show almost the same results independently of EoS (vd. Campos, 1991). The calculated temperatures

TABLE I

Theoretical and experimental results of D and P of industrial explosives

Explosive	EoS	Theoretical		Experimental	
		$D_{CJ}[\text{ms}^{-1}]$	P_{CJ} [GPa]	$D_x[\text{ms}^{-1}]$	P [GPa]
Dynamite	BKW	6935	17.7	5900	13.2
	H9	7218	17.9		
	H12	7163	17.2		
Emulsion	BKW	6485	12.4	5340	12.0
	H9	6281	11.2		
	H12	6444	11.5		
Anfo	BKW	5274	6.8	3300	–
	H9	5230	6.3		
	H12	5180	6.1		

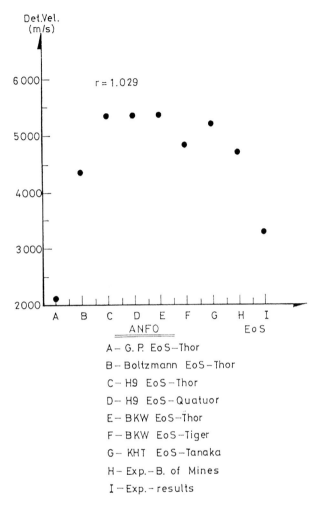

FIGURE 1   Detonation velocity of ANFO explosive.

are 1994, 2241 and 2459 K respectively for emulsion, ANFO and dynamite explosives. The products composition are shown in Figures 6 to 8. The CO and NO emission levels are higher than those calculated for detonation regime, in good agreement with theoretical predictions (vd. Campos, 1991).

EXPERIMENTAL PROCEDURE AND GAS ANALYSIS

Detonation Experiments

There are several problems inherent to laboratory conditions for detonation studies:

– the small quantities of explosive difficult to achieve stable detonations,

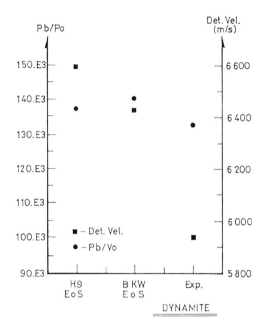

FIGURE 2   Detonation velocity and pressure of dynamite explosive.

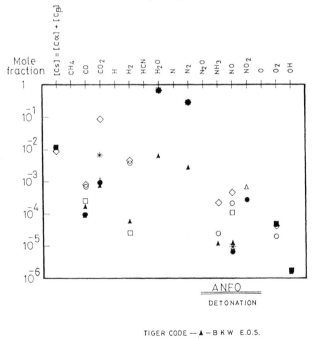

FIGURE 3   Composition of detonation products of ANFO.

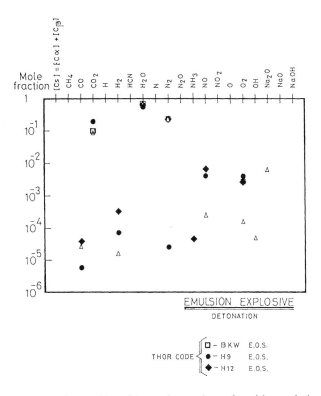

FIGURE 4   Composition of detonation products of emulsion explosive.

– the degree of influence of the explosive confinement, affecting the critical diameter and composition of initial expanded products,
– the expansion of detonation products and recombination with the existing air.

In order to reduce these inconvenients the used explosive is pre-packaged inside a 30 mm diameter steel tube, in order to maintain uniform and reproducible charge density and size (vd. Fig. 9). To evaluate detonation stability, the explosive charge has four contact probes formed by a very thin insulated electrical wire, short-circuited with the metal confinement by shock wave. These short-circuits actuate electronic time counters of 1 ns resolution and allow, changing distances between probes, the measurement of local mean velocity of detonation (vd. Sellam, 1986).

In these experiments two explosions chambers are used - a cylindrical closed chamber of $1.5 \, m^3$ (vd. Fig. 10) and a large explosion chamber of $19 \, m^3$ (vd. Fig. 11).

Typical obtained results are presented in Fig. 12, for dynamite, emulsion and anfo explosives, with strong steel and weak plastic confinement (L being the distance measured from initiation terminal where the detonator is fixed). Anfo requires dynamite booster for initiation. For the same charge diameter, final stable value of D is generally lower with weak confinement than with a strong one. The maximum stable value of detonation velocity, D_x (vd. Tab. I), can be obtained performing experiments

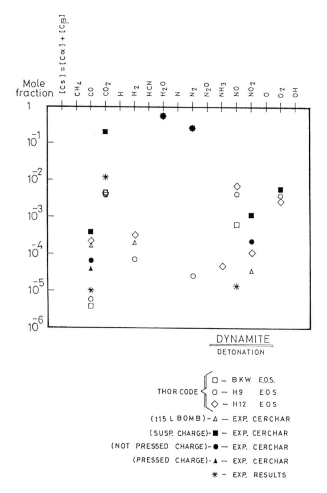

FIGURE 5   Composition of detonation products of dynamite.

with different diameters and extrapolating the obtained results to an infinite diameter. Detonation pressure of explosives was measured with shock induced polarisation (SIP) on 1 mm thickness PMMA plate (vd. Hauver, 1965) fixed at the end of the explosive charge. Acoustical approach, applied to the difference of shock impedance of PMMA and explosive detonation products (vd. Fisson, 1976), allows to obtain experimental detonation pressure (vd. Tab. I). It was impossible to measure detonation pressure of anfo explosive with this method, due to its low value and large scale of explosive heterogenities. The results of D_∞ and P are lower than the predicted theoretical values (vd. Tab. I). This proves the non ideal detonation behaviour of detonation observed in these kind of explosives, which agrees with the results shown generally in bibliography. The exhaust gas is sampled in cylindrical bottles or conducted directly from explosion

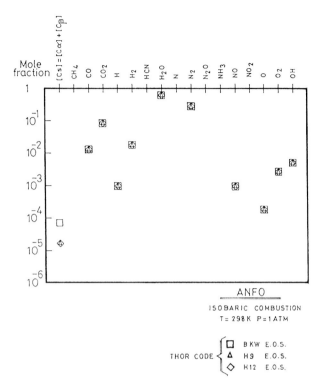

FIGURE 6   Composition of combustion products of ANFO.

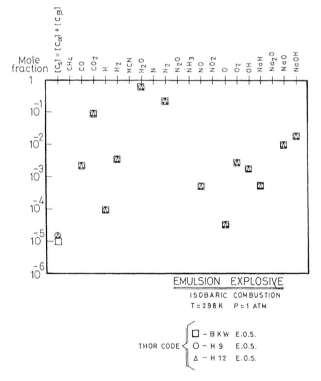

FIGURE 7   Composition of combustion products of emulsion explosive.

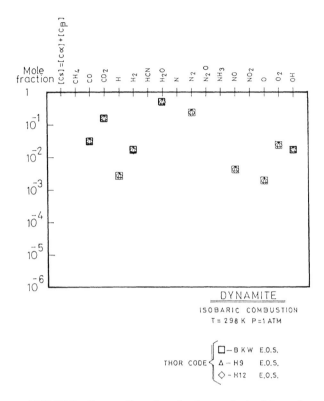

FIGURE 8   Composition of combustion products of dynamite.

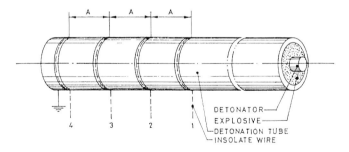

FIGURE 9   Cylindrical explosive charge.

chambers to CO, CO_2, NO, NO_2 measurement equipment. The composition of the exhaust gases, from recombination of air with the expanded detonation products, was measured. Gas chromatography (Hewlett Packard) with two columns and a switching system is also used due to the broad composition of the gases analysed (vd. Campos *et al.*,1991). The presented experimental results (vd. Figs. 3 to 5) are obtained by the extrapolation to zero from measured pollutants concentrations, as a function of initial

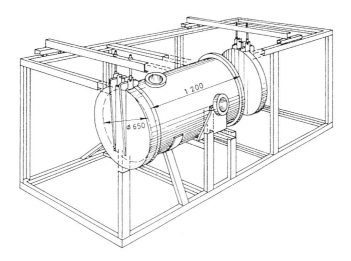

FIGURE 10   Cylindrical closed chamber of 1.5 m³.

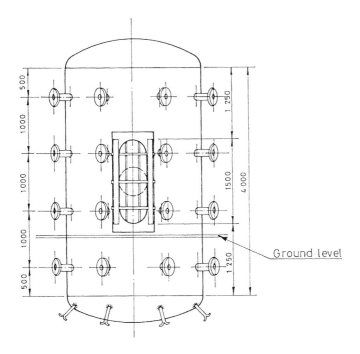

FIGURE 11   Large explosion chamber.

mass fraction of air/explosive inside explosion chamber (vd. Campos *et al.*, 1991).
Obtained results are in good agreement with experimental values found in literature,
principally presented by American Bureau of Mines (Chaiken *et al.*,1974) and by
Cerchar (Carbonel and Bigourd, 1980, Carbonel *et al.*,1980) and with previous calcula-

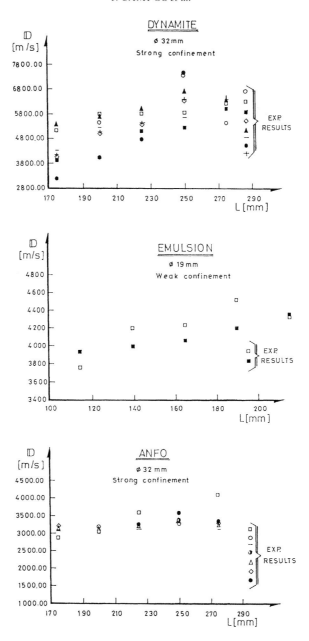

FIGURE 12   Detonation stability of selected explosives.

tions. Measured pollutants concentrations, as a function of volume of explosion chamber, prove also the dependence of expansion mechanisms on pollutants formation, related to kinetics of NO and CO formation and recombination.

Presented theoretical calculations of NO and CO concentrations does not necessary imply that NO and CO remain frozen in a real situation of detonation. It is possible that the reaction,

$$2\,NO + O_2\,(air) \rightarrow 2\,NO_2$$

can occur during and after expansion of detonation products of condensed explosives. This fact generally explains the difference between calculated and measured NO/NO_2 ratios cited in open bibliography (vd. Chaiken *et al.*, 1974). The occurrence of a similar reaction between calculated CO and air during expansion,

$$2\,CO + O_2\,(air) \rightarrow 2\,CO_2$$

can also explain the difference between calculated and measured CO concentrations. Finally, it is of considerable importance, from a hazards point of view, the formation of solid carbon, as CO or CO_2, in the expanded detonation products, due to the temperature decrease.

Combustion Experiments

Equipment for incineration of explosives (vd. Pires *et al.*, 1993) is based on the fluidised bed combustion of a mixture of explosive and sand, feed by a twin screw extruder (vd. Fig. 13) and warmed by an initial combustion of propane/air gaseous mixture.

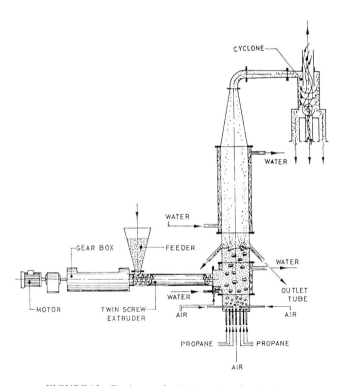

FIGURE 13   Equipment for incineration of explosives.

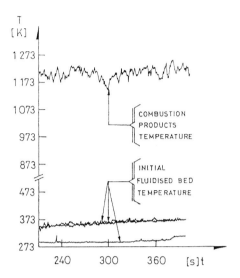

FIGURE 14   Fluidised bed combustion temperature as a function of time.

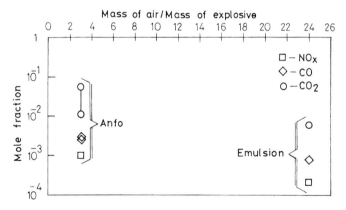

FIGURE 15   Measured concentration of NO_x, CO and CO_2

Consequently it is possible to have a fine control of mass flow and combustion temperature (vd. Fig. 14). Experimental procedures for gas analysis from combustion experiments are almost the same of detonation experiments, conducting directly the fumes from cyclone (vd. Fig. 13) to CO, CO_2, NO, NO_2 measurement equipment. The composition of the exhaust gases (vd. Fig. 15) show experimental results, for combustion regime of equivalence ratio to stoichiometry $r \approx 2$, in a quite good agreement with calculations. These results prove not only the dependence of explosive/sand mass concentration but also of gas or air dilution.

CONCLUSIONS

Theoretical calculations, using THOR code with BKW, H9 and H12 EoS and assuming the minimum value of Gibbs free energy, allow to predict performances and products composition (CO_2, CO, H_2O, N_2, O_2, H_2, OH, NO, H, N, O, HCN, NH_3, NO_2, N_2O, CH_4 gases and two kinds of solid carbon - graphite and diamond), for the selected three types of industrial explosives. The obtained results show a good agreement with other calculations presented in literature. Detonation experiments, evaluating the detonation stability and measuring CO, CO_2, NO, NO_2 products concentrations, show a good agreement with previous calculations and proves the validity of chosen BKW, H9 and H12 EoS. NO_x concentrations are more important for dynamite than for ANFO and emulsion explosives. CO concentrations have almost the same value for dynamite and emulsion explosive, but lower than for ANFO explosive. Measured pollutants concentrations, as a function of volume of explosion chamber, prove the dependence of expansion mechanisms on CO and NO formation and recombination.

Products composition from isobare adiabatic combustion of selected explosives has been calculated. Results show NO_x and CO higher concentrations for combustion than for detonation regime. Experimental incineration of mixtures of explosive/sand, in a fluidised bed warmed by an initial combustion of propane/air gaseous mixture, shows a stable initial combustion temperature of 1200 K. Its final value is dependent of mass fraction explosive/sand mixture. The composition of exhaust gases seems to be in good agreement with calculations.

REFERENCES

Brinkley, Jr. S. R. (1947) Calculation of the Equilibrium Composition of Systems of many Constituents. *J. Chem. Phys.*, n° 15, pp. 107.

Campos, J. Thermodynamic Calculation of Solid and Gas Combustion Pollutants Using Different Equations of State. *First International Conference on Combustion Technologies for a Clean Environment*, 1991, Vilamoura, Algarve.

Campos, J., Portugal, A. and Gois, J. C. Toxic Fumes From Industrial Explosives. *First International Conference on Combustion Technologies for a Clean Environment*, 1991, Vilamoura, Algarve.

Carbonel, P., Bigourd, J. and Dangreaux, J. Fumées de Tir. *Proceedings of 18th International Conference "Research on security of mines"*, 1979, Dubrovnik, 7–14 October.

Carbonel, P. and Bigourd, J. (1980) Méthodes d' Essais Pour Evaluer la Toxité des Fumées de Tir. *Propellants and Explosives*, **5**, 83–86.

Chaiken, R. F., Cook, E. B. and Ruhe, T. C. (1975) *Toxic Fumes From Explosives: Ammonium Nitrate - Fuel Oil Mixtures*. Report of Investigation n° 7867 – Pittsburgh Mining and Safety Research Center, Pittsburgh, Pa., U. S. A.

Fisson, F. (1976) "Étude de Grandeurs Caracteristiques de la Détonation d'Explosifs Liquides", *PhD Thesis*, E.N.S.M.A, Poitiers, France.

Gordon, S. and Mc Bride, B. J. (1971) *Computer Program For Calculation of complex Chemical Equilibrium Compositions, Rocket Performance Incident and Reflected Shocks and Chapman-Jouguet Detonations*. Report NASA SP 273, NASA Lewis Research Center.

Hauver, G. E. (1965) "Shock-Induced Polarisation in Plastics. Experimental Study of Plexiglas and Polystyrene", *Journal of Applied Physics*, **36**(7), 2113–2118.

Heuze, O., Bauer, P, Presles, H. N. and Brochet, C. The Equations of State of Detonation Products and Their Incorporation Into the Quatuor Code. *Proceedings of the eighth Symposium (International) on Detonation*, 1985, pp. 762–769, Albuquerque Conventional Center, New Mexico.

Heuzé, O., Presles, H. N. and Bauer, P. (1985) Computation of Chemical Equilibria. *J. Chem. Phys.*, n° 83 (9), pp. 4734–4735.

Heuzé and Olivier (1989) *Calculo Numérico das Propriedades das Misturas Gasosas em Equilibrio Termodinâmico*. Universidade de Coimbra, Portugal.

ICT 23rd International Annual Conference, 1992. *Waste Management of Energetic Materials and Polymers*. Fraunhofer Institut fur Chemische Technologie, Karslruhe, Federal Republic of Germany.

IEPG - Portuguese Research Group - Campos, J., Luz, Paulo and Marques, C. (1989, 1990) *BKW- KHT- H9 Equation of State Calculations* Report of Progress of Technological Area 25, Collaborative Technical Programme 1, Portugal.

JANAF, 1971. *Thermochemical Tables* - 2nd Edition. National Bureau of Standards, Washington DC.

Kihara, T. and Hikita, T. Equation of State for Hot Dense Gases and Molecular Theory of Detonation. *4th Int. Symposium on Combustion.*, 1953, **59**, 458. Baltimore, The Williams and Wilkins Co.

Johnson, J. N., Mader, C. L. and Goldstein, S. (1983) "Performance Properties of Commercial Explosives". *Propellants, Explosives, Pyrotechnics*, n° 8, pp. 8–18, U.S.A.

Mader, C. L., (1979) *Numerical Modelling of Detonations.* University of California Press, Berkeley, U. S. A.

Manson, N. (1976) *Cours de Hautes Temperatures.* Ecole Nationale Superieure de Mecanique et d' Aerotechnique, Poitiers, France.

Pires, A. and Campos, J. Clean Combustion of Energetic Materials in a Fluidised Bed. *Second International Conference on Combustion Technologies for a Clean Environment*, 1993, Lisboa.

Tanaka, K. (1983) *Detonation Properties of Condensed Explosives Computed Using the Kihara-Hikita-Tanaka Equation of State.* Report from National Chemical Laboratory for Industry, Ibaraki, Japan.

White, W. B., Johnson, S. M., Dantzig, G. B. (1958) Chemical Equilibrium in Complex Mixtures. *J. Chem. Phys.* **28**, 751.

Natural Gas Vehicles for a Clean Environment
An Overview and the Trends for the Future

ALDO BASSI   *SINTESI AB Srl Via G.B. Bodoni, 2 20155 Milan - Italy*

Abstract—The Natural Gas represents one of the most attractive alternative fuel for transportation, because of its large availability in the world, its partial renewability (biogas), good engines performances, and cleaner exhaust gas properties.

Some evident drawbacks stunt the diffusion of conversion of the retrofit for the existing vehicles: cylinders installation, safety certification, exhaust gas aftertreatment.

Strategic, economic and ecologic reasons could push the Governments and then the Manufacturers to increase the diffusion of Natural Gas Vehicles.

The following items are then shortly illustrated:
1) Natural Gas composition and properties as fuel for internal combustion engines in road transportation
2) Classification of Natural Gas Engines
3) Gasoline Engines converted to Natural Gas (performances and emissions)
4) Diesel Engines converted into Natural Gas (performances and emissions)
5) Natural Gas Vehicles (some considerations and trends)

INTRODUCTION

The natural gas has an effective potential as engine fuel for the future. Although the natural gas has been in use for many years in some countries and tried in many applications, there are new reasons to reconsider this fuel.

The natural gas is an abundant material in the world and can be obtained also by agricultural wastes as renewable energy resource, so there is a strong interest in its clean air properties, its lower energy costs, its renewability and reduced dependence on petrol over the world availability.

Moreover, the technologycal progress is overcoming problems met by present potentials users in the present environment from an ecologic and economic point of view.

An analysis on the advantages and disadvantages of natural gas as a motor vehicle fuel and on the opportunities and problems in the natural gas vehicles leads to the following key points:

- determine the effective economic advantages of natural gas as a motor vehicle fuel to potential customers (according to each particular national situation)
- increase the technical efforts by manufacturers to improve environmental benefits of natural gas as vehicular fuel
- diffuse and develop the fuelling stations and increase the customers demand
- establish standards and certification of natural gas composition and properties as engine fuel
- establish standards and certification programs for NGV equipments (safety)

It is generally accepted that the natural gas is a potential clean fuel, but it is evident that the huge research and development efforts made by manufacturers on gasoline and diesel engines have dramatically improved the performances and emissions of gasoline and diesel oil themselves; the emission advantages of NG over traditional fuels could disappear, if appropriate technical efforts were not applied to gas engines development, to accomplish the environmental standards legislations more and more stringent, in competition with gasoline and diesel oil fuels.

1. NATURAL GAS COMPOSITION AND PROPERTIES AS FUEL FOR INTERNAL COMBUSTION ENGINES IN ROAD TRANSPORTATION

The natural gas is a mixture of several gases.

The primary component is the methane, which tipically makes up $75\% \div 90\%$ of total volume. The other components include hydrocarbons, inert gases, such as nitrogen, helium and carbon dioxide. The natural gas, obtained from agricoltural wastes, besides the mentioned components, contains also hydrogen sulfide.

The composition of the natural gases available in the world differs considerably on the basis of the source and on the refinery process of the gas.

The table of Figure 1.1 represents the average of the gas composition in U.S.A. and in Europe.

The fact that the natural gas composition is quite variable around the world creates some difficulties for the engineer in defining and in tuning up the fuel feeding systems of the I.C. engines (i.c. carburettors or fuel injection systems), to certificate their perform-ances and exhaust pollution characteristics, typically affected by gas composition.

Certainly there are important problems in defining a standard composition of gas and in assuming its availability over the world. The Figure 1.2 represents the NG composition recommended by the Natural Gas Vehicles Coalition (USA) for the emission test certificate. But for the vehicles on the road, presumably it is very difficult or unfeasible to obtain or diffuse availability of a standard composition. Perhaps the solution can be found more easily in the sophisticated electronic engine management, which can take into account the different composition of the gas injected into the engine and consequently adjust the feeding regulation parameters, because natural gas must be considered a Flexible Fuel. Moreover the direct combustion detection sensors could be a suitable method to approach this problem, but important R and D efforts can be made in this field by engine manufacturers, before releasing production.

In the table of Figure 1.3 are represented the main characteristics of methane as the principal constituent of natural gas.

From this table we can observe that the natural gas has a high ignition temperature, higher than the standard gasoline, and this feature can be considered very good as antiknock characteristic for the total efficiency of the engine (RON = 130), but it can be considered also poor, because it is very difficult to ignite the air-methane mixture. In fact the air methane mixture requires a very high external amount of energy, typically released by a spark-plug, to overcome the complex chemical phenomena occurring in

COMPONENT	U.S.A	ITALY	HOLLAND	RUSSIA
HYDROCARBONS				
Mole percent				
Methane	92.21%	99.63%	89.44%	93.27%
Ethane	3.78%	0.07%	3.28%	3.32%
Ethylene	0.00%			
Propane	0.91%	0.04%	0.69%	0.83%
Propylene	0.00%			
Butanes	0.47%		0.29%	0.37%
Pentanes	0.10%		0.09%	
Hexane and higher	0.04%			
Total NMHC	5.30%			
Total C_{3+}	1.52%			
Total C_{4+}	0.61%			
Total C_{5+}	0.14%			
Ethane % of NMHC	77.56%			
Percent of total organic carbon				
Methane	88.37%			
Ethane	6.95%			
Propane	2.45%			
C_4	2.22%			
Total non-methane	11.63%			
Ethane % of NMHC	67.66%			
INERT GASES (Mole percent)				
CO_2	0.59%	0.01%	0.70%	1.00%
O_2	0.05%			
N_2	1.84%	0.25%	5.51%	0.91%
Total inerts	2.48%			
OTHER COMPONENTS (Mole percent)				
CO_2	0.00%			
H_2	0.01%			
WOBBE NUMBER	1348	1430	1330	1350

FIGURE 1.1   Natural Gas Composition (U.S.A. and EUROPE).

combustion chamber of the engine before that complete combustion itself occurs with the desired heat release. For the same reason the aftertreatment as in catalytic converter of the unburned methane hydrocarbons in the exhaust gases is very difficult.

Just to clearify this aspect, in Figure 1.4 is represented the sequence of the chemical reactions taking place in the gap of spark-plug of the combustion chamber during the ignition phase. Notice that the first two reactions are endothermic, requiring a very

Methane...88%   + 0.5% (*)
Ethane..8%   + 0.3%
C_3 and more complex HC_S...............4%   + 0.2%
C_5 and more complex HC_S...............0.5%   max
Total unsaturated HC.......................0.5%   max

Hydrogen..0.1%   max
Carbon monoxide............................0.1%   max

Wobbe number................................1350   + 20 (Btu/scf)

Absence of liquid over the whole range of temperatures and pressures encountered in the engine and in the fuel supply system

(*) expressed as % of total present organic carbon

FIGURE 1.2   Natural Gas Composition recommended by the NGV Coalition (U.S.A) for the Emission Test Certificate.

high activation energy to produce radicals CH_3, responsible for the following esothermic reactions.

Because of its high activation energy, the flame speed in laminar flow in the natural gas-air mixture is very low, lower than that of other hydrocarbons. So, the longer duration of combustion reduces the efficiency of the engine, partially compensated by an increased spark timing advance. The greater amount of heat rejected through the combustion chamber walls in the longer combustion time is the responsible of the reduced engine efficiency. To overcome this problem, turbolent combustion chambers have been designed by some manufacturers and are strongly recommended to increase the flame speed into the combustion chamber (see also Fig. 4.3).

On the contrary, the limits of ignition of methane-air mixture (Fig. 3.2) in a spark ignition engine determine a very wide region, in which combustion can occur. So very lean mixtures can be adopted in the engine with significant results in the clean exhaust properties (see also chapter 3).

On the other hand, the high autoignition temperature of natural gas in the air is a very good property, when we consider the safety requirements of the vehicles; in fact it

- Formula	CH_4	First of the alkans family
- Molecolar weight	16.043	gr/mole
- Critical pressure	46.04	bar
- Critical temperature	185.63	°K
- Boiling point	-161.52	°C
- Latent heat	121.86	Kcal/Kg
- Heat power (inf) (0°C)	35,861	KJ/m^3
	11,984	Kcal/Kg
- Heat power (sup)	9,530	$Kcal/m^3$
- Specific weight (0°C 1 bar)	0.7174	Kg/m^3
- Density respect to air	0.554	
- Specific weight - vapour	1.819	Kg/m^3
- liquid	422.62	Kg/m^3
- Point of flammability	530	°C
- Stoichiometric ratio air/methane	9.53	m^3/m^3
- Octane number (Research method)	130	RON
- Wobbe number - sup	12,800	$Kcal/m^3$
- inf	11,520	$Kcal/m^3$

	in air	in oxygen
- Limits of ignition	< 5%	< 5%
(20°C - 1 bar) % vol	> 15%	> 60%
- Autoignition temperature	580 °C	555 °C
(1 bar)		
- Combustion flame speed	0.4 m/sec	0.36 m/sec
(laminar flow)		

FIGURE 1.3   Main Characteristics of Methane.

reduces significatively the fire hazard. Moreover, the lower density of the methane, respect to the air, is another important characteristic reducing burst and fire hazard: any accidental leakage is rapidly dispersed in the air. This aspect allows the N.G.

#	Reaction	Description	Classification
1	$CH_4 + M \longrightarrow CH_3 + H + M$	- Endothermic reaction that needs a fixed amount of activation energy - M catalytic element (.....)	Chain preparation
2	$CH_4 + X \longrightarrow CH_3 + XM$	- Endothermic reaction that needs activation energy - X reaction element (H, O, HO....)	Chain preparation
3	$CH_3 + O_2 \longrightarrow CH_3O + O$	- Highly endothermic reaction	Chain Start
4	$CH_3O + M \longrightarrow H_2CO + H + M$	- Weakly esothermic reaction - low speed - M catalytic element	H_2CO = Formaldheide CH_2O = Aldheide radical
5	$H_2CO + X \longrightarrow HCO + XH$	- Weakly esothermic reaction - high speed if X = OH $\longrightarrow H_2O$, if X = H $\longrightarrow H_2$	H_2CO Chain carrier HCO Chain carrier
6	$HCO + M \longrightarrow H + CO + M$	- Highly esothermic reaction - M catalytic element	Chain propagation
7	$CH_3 + CH_3 \longrightarrow C_2H_6$	- Ethane formation endothermic reaction	Extinction phase preparation
8	$CO + OH \longrightarrow CO_2 + H$	- Esothermic reaction	Extinction phase preparation
9	Contrary to the 2	- Esothermic reaction	Flame extinction phase
10	Contrary to the 1	- Esothermic reaction	Flame extinction phase

FIGURE 1.4   Methane Combustion (Chemical Reactions Sequence) (*).

Vehicles users to park their cars in any garage with only an appropriate ventilation system.

One of the most important reasons for which NG is called clean fuel lays in the lower carbon dioxide emissions.

In Figure 1.5 are represented the amounts of CO_2 emitted during the combustion of several fuels to generate 10.000 Kcal. The hydrogen of course is the best fuel, because no carbon is involved in its combustion, but we can find the methane as the best fuel before the other traditional fuels.

But for a complete analysis about CO_2 emissions as greenhouse effect we can take into account the total carbon dioxide emissions involved during the complex process from the primary source to the road covered, including CH_4 emissions, reported as CO_2 equivalent. As a summary for this analysis, in Figure 1.6 are expressed some numbers that incorporate all the CO_2 emissions as total fuel cycle and their influence on the greenhouse effect. The natural gas used into the spark-ignited engines gives

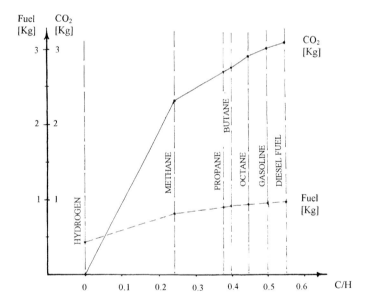

FUELS	C/H ratio	Heat power Kcal/Kg	Kg fuel x 10.000 Kcal	Kg CO_2 x 10.000 Kcal
H_2 HYDROGEN	0	28900	0.346	0
CH_4 METHANE	0.25	11984	0.834	2.27
$C_3 H_8$ PROPANE	0.375	11103	0.900	2.70
$C_4 H_{10}$ BUTANE	0.4	10955	0.912	2.763
$C_8 H_{18}$ OCTANE	0.444	10640	0.939	2.901
$C_{10} H_{22}$ DECANE	0.454	10595	0.944	2.926

FIGURE 1.5   Emission of CO_2 to create 10.000 Kcal as a Function of Fuel C/H Ratio (*).

Fuels cells (solar-hydrogen)	-90 to -85
Ethanol from cellulose	-75 to -40
Hydrogen (nuclear electricity)	-70 to -10
LPG	-30 to -10
Electric vehicle (natural gas)	-25 to -10
Natural gas vehicle	-20 to 0
Electric vehicle (current USA energy mix)	-20 to 0
Methanol from natural gas	-10 to +8
Ethanol from corn	-10 to +35
Electric vehicle (coal)	+25 to +50
Methanol from coal	+30 to +70
(Source: De Luchi 1991)	

FIGURE 1.6   Greenhouse Gas Emissions relative to Gasoline, total Fuel Cycle, advanced Technology Vehicles, CO_2 equivalent.

a $- 20\% \div 0$ advantage respect to the gasoline, but if it is used with electric propulsion system the advantages rises to $- 25\% \div - 10\%$. There is a concern however that leakages from the infrastructure of the natural gas systems contribute substantially to the greenhouse effect because of the high light absorption of methane (Fig. 1.6).

Among the characteristics of natural gas as fuel for transportation uses we have to consider also its transportability. For this aspect the main critical property of the methane as principal constituent of NG is its very low energy density and its very low boiling point ($- 161°C$) compared with other fuels. With these two characteristics (see Fig. 1.3) natural gas may be stored on-board a vehicle either as a compressed gas in high-pressure cylinders or as a cryogenic liquid.

The current maximum working pressure for compressed natural gas cylinders is 200 bar, obtaining the maximum volumetric energy content about one-fifth respect to the diesel fuel and one-fourth to the gasoline (see also Fig. 5.2).

In the liquified form in the cryogenic tank, the volumetric energy is about a-half that of diesel fuel. So the on-board storage characteristics represent the main drawbacks of the natural gas in road transportation applications, even if some significant improvements have been made and are in progress by specialized manufacturers.

The low density of natural gas has also a negative effect on the power output of the engine. The stoichiometric mixture of natural gas and air occupies about 10% more volume than a same energy content gasoline-air mixture, resulting in a 10°C penalty in engine power output. This in natural gas engines can be overcome mainly by compression ratio increase, allowed by high octane number of the methane.

2. CLASSIFICATION OF NATURAL GAS ENGINES

The use of compressed natural gas as a fuel for transport engines is still in the development stage, however experience has shown promising results in certain applications. In the table of Figure 2.1 are represented the development lines of actual CNG engines.

Three basic options are available for the use of natural gas in I.C. engines.

The first one is represented by gasoline engines converted into natural gas. This conversion is quite easy, because the combustion cycle (otto cycle) remains the same. Natural gas like gasoline requires an electric spark-plug energy to be ignited. In this mode no change on the engine structure is required and the engine can be returned to gasoline operation by a switch on the dashboard on the vehicle.

The engine has to be fitted by a gas carburettor or a gas injection system to prepare the gas-air mixture in its inlet manifold. A pressure reducer leads the gas from high pressure cylinder (200 bar) into circuit at atmospheric pressure suitable for feeding the carburettor. The spark-plug advance can be adjusted for gas operation by a small electronic device, added to ignition system.

This conversion is practically operated for the cars or light duty trucks, which normally operate with gasoline fuel. This kind of operation called "bi-fuel" has been the most popular retrofit in the world for fourty years.

The second basic option is represented by diesel engines converted into natural gas. This conversion can be made in other two ways: dual-fuel operation and full-gas operation.

In the dual-fuel operation the combustion cycle remains as diesel compression ignited, in the full-gas operation the combustion cycle is completely changed into otto-cycle with spark-plug ignition system, added to the engine together with the gas mixer, removing diesel injectors.

In the dual-fuel mode the quantity of diesel fuel supplied in each cycle is reduced as "pilot flame", and the natural gas mixed in the carburettor with the intake air makes up the total fuel input to the engine.

The benefits of dual-fuel conversion include the possibility to go back to 100% diesel operation, potential increase in power over diesel and economics.

In the spark ignition mode all diesel injection equipment is removed from the engine, the combustion chamber modified and a spark ignition system and gas carburettor are added to the engine.

The benefits of full gas conversion include the reduction of polluted gases, reduction of noise, potential for improved economics. Due to major changes in the structure of the base engine, there is no practical possibility to go back to diesel-fuel operation.

Both dual-fuel and full-gas conversions are normally applicated to heavy trucks and busses, where we usually find diesel engines. These kind of conversion are not yet well developed, they are limited so far to test fleets.

The third line represents the engines expressly designed for natural gas; they are today at the research stage in the specialized research centers in the world. So far vehicles manufacturers and specialized factors do not find a valid business reason to devote financial support to this kind of engine, because there is not enough market demand. They prefer to follow the first two development lines of Figure 2.1, because of the reduced financial resources required, and quite good market potential.

The fourth line in Figure 2.1 indicates the on-board storage systems for natural gas, that represents the critical point for power-train system in the vehicle.

748

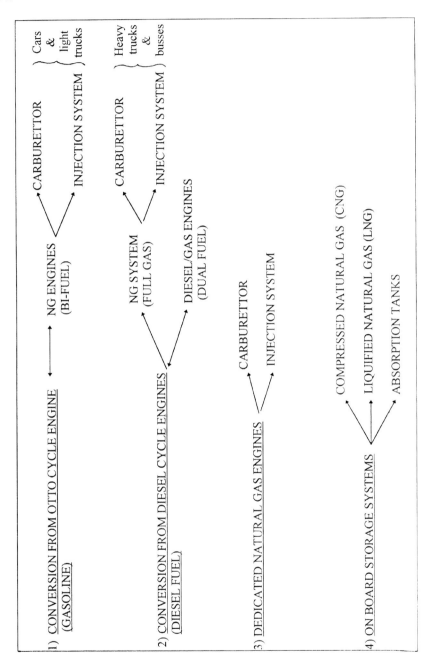

FIGURE 2.1   Natural Gas Power-Trains Classification (*).

Three are the systems feasible today; the first one (compressed natural gas) is by far the most popular; the other two – cryogenic and absorption method – are in the development stage in the early phase. All these on board storage systems have no influence on the layout of the engines, but represent one of the most important element for the natural gas engine diffusion.

3. GASOLINE ENGINES CONVERTED INTO NATURAL GAS

In spark-ignition engines, fuel and air enter the cylinders simultaneously during the intake stroke, so in naturally aspirated engines the maximum amount of energy, which can be taken in per cycle, depends on the cylinder volume and on the energy content of the air/fuel mixture.

Since gasoline enters the cylinder in the atomized liquid state, it occupies only a small fraction of the cylinder volume. When the engine is converted into natural gas, the gasous fuel occupies a much larger fraction. As a consequence, the power capacity of a given spark-ignition engine is more than 10% lower when operating on natural gas, when only the energy content of the intake mixture is taken into account. Moreover, lean mixtures need more space for the same amount of energy (see Figs. 3.1, 3.2).

The higher compression ratio can be adopted to partially overcome this problem; the higher octane number (NO 130) of the natural gas allows compression ratios above

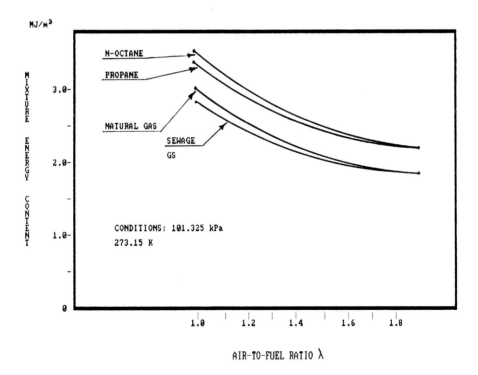

FIGURE 3.1   Energy Content of some Mixtures versus Air-to-Fuel Ratio (*).

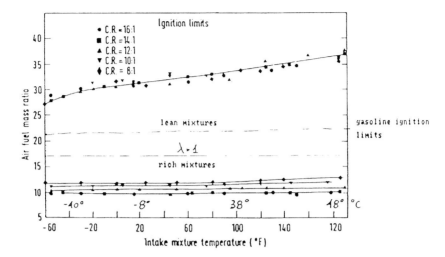

FIGURE 3.2   The Ignition Limits of Methane-Air Mixtures in a Spark Ignition Engine for a Range of Intake Mixture Temperatures and Compression Ratios (C.R.) (*).

15:1, increasing the total efficiency of the engine then its power output. But other design elements, like frictional losses, combustion chamber shape, spark-plug position etc., determine an optimum compression ratio for practical engines in 13:1.

Of course, we can not forget that turbocharger can be adopted to obtain the engine power required, taking into account the complexity and the cost of the engine and its lower overall efficiency.

In spark-ignition engines, the combustion process is initiated by a spark from the spark-plug. The ignition process is rather complex (see Fig. 1.4), being an interaction between spark energy, initial reactions of the small mixture kernel involved, heat exchange with the electrodes of the spark-plug as well as with the kernel boundaries and local turbolence levels. Natural gas mixtures have a wide ignition range (see Fig. 3.2) which is beneficial because the ignition becomes less sensitive to the air-fuel ratio.

After the ignition, the flame path within the cylinders is related to many factors, such as shape of combustion chamber, running speed, swirl and squish properties and mixtures composition.

Many tests have shown that in spark-ignition engines, natural gas engines burn somewhat slower than gasoline. In order to obtain the proper combustion phase angle, the ignition has to be further advanced for natural gas than for gasoline.

Earlier ignition results, in more recompression of the partially burned charged, a loss of efficiency. For this reason, design features to increase flame speed and reduce combustion are very important, specially for lean burn engines.

Several combustion systems have been developed to achieve these goals: compact chamber toward sphere shape to reduce size, "Nebula" chamber with the swirl (see Fig. 4.3), to increase turbolence, etc. These systems can be adopted only in dedicated natural gas engines.

If we consider only the conversion of gasoline engines into natural gas as retrofit, no modification of combustion chamber is adopted, mainly because the engine has to maintain the running availability with gasoline, and higher conversion costs.

In Figure 3.3 is represented a modern retrofit CNG for gasoline engine (BI-FUEL) equipped by electronic injection system and catalytic converter. A switch on the dashboard allows the driver to choose gasoline or CNG fuelling.

In Figure 3.4 is shown a typical arrangement of retrofit CNG for gasoline engines (courtesy by OMVL). The compressed natural gas coming from the gas cylinder (1) is reduced at the atmosphere pressure by a pressure-regulator (4) and meterd by an electrovalve (20) or stepped motor, then mixed with the air in the mixer (6). An electronic microcomputer, sensing oxygen content in the exhaust gases and other engine parameters, operates the electrovalve or stepped motor in optimised air-fuel ratio.

In the table of Figure 3.5 are summarized some of the bi-fuel retrofit emission data. These figures refer to gasoline converted engines without any modification, so they are not representative of what can be obtained by an engine dedicated to natural gas with

(catalytic converter version)

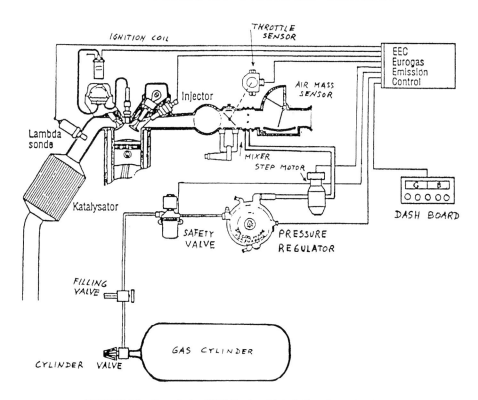

FIGURE 3.3   Retrofit for CNG Engine (Bi-fuel) Gasoline – CNG (*).

752

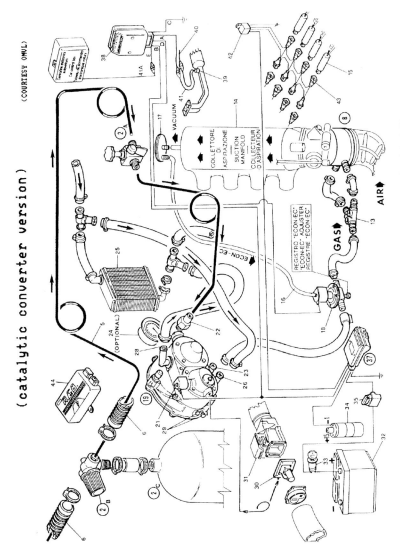

(catalytic converter version)

(COURTESY OMVL)

FIGURE 3.4   Layout arrangement of retrofit gasoline – CNG (Bi-fuel) (*).

	CO	HC + NO$_x$	PM	EVAP
	gr/km	gr/km	gr/km	gr/prova
1991/1994	2.72	0.97		2
1996/1997	2.20	0.50	-	2
1999/2000	1.50	0.20	-	-
VW Passat 2.0 CNG	1.37	3.74 (2.58+1.16)	-	-
FIAT UNO Arcobaleno	0.18	3.08 (2.18+0.90)	-	-
FIAT TIPO 3 WAY CAT	0.13	0.67 (0.61+0.06)	-	-

FIGURE 3.5   Emissions Limits ECE Cycle and some Emission Data of Vehicles CNG converted (*).

much higher compression ratio, optimised tuning, the catalytic converted optimized for natural gas emissions, etc.

Some considerations anyway can be made on the pollutant emissions of natural gas engines: poor fuel-air mixing can lead to increased emissions, many carburettors have poor mixtures preparation. This problem can be overcome by injecting measured quantities of fuel into the individual cylinders.

Natural gas, as gaseous fuel, performs better than liquid fuels during cold engine starts and warm-up. Exhaust gases may be cleaned by the use of a catalytic converter which converts substances such as HC, CO, NO$_x$ and a number of other unregulated and undesired emissions. In the "three way" catalytic converter, which is most used type, the simultaneous oxidation and reduction process requires full control of the air-fuel ratio. This is achieved by measuring the oxygen content of the gas entering the converter with a "lambda sensor". A signal from this oxygen sensor is then used to control air/fuel ratio.

Because, as shown in the first chapter, the methane-air mixtures are difficult to ignite, the reduction of MHC (Methan Hydro Carbons) is very difficult to achieve in the catalytic converter, then dedicated catalytic converters have to be identified and tested.

Development of the gasoline fuelled lean-burned engine has been a method of minimizing certain types of exhaust emission (see Fig. 3.6). Typically an air surplus of around 10 % is necessary for the complete combustion, but with natural gas 20 ÷ 30 % or more can be adopted.

The drawbacks of lean-burn engines is poor driveability and high HC emissions, caused by slow flame spread during the combustion.

Another important device able to reduce pollutants as NO$_x$ is External Gas Ricirculation (EGR). High EGR rates would be used at low and moderate loads to reduce NO$_x$, while in full power operations the engines would operate without EGR, thus achieve the same power output of the base gasoline engine.

An appropriate amount of EGR improves quite well the methane combustion.

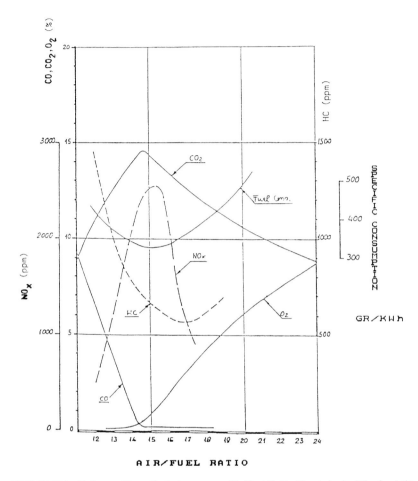

FIGURE 3.6   Pollutant Gases Emissions versus Air/Fuel Ratio (Spark-ignited Engines) (*).

All these considerations and many others lead to statement that quite a lot of resources have to be spent in the research and development programs in order to improve the natural gas engine performances and emissions versus its competitors gasoline engine, achieving the best results. But it is not clear if somebody will devote the necessary resources to the gas engine development.

4. DIESEL ENGINES CONVERTED TO NATURAL GAS (PERFORMANCE AND EMISSIONS)

In compression ignition engines (diesel cycle) air is compressed in the cylinder and the diesel fuel is injected into the hot air at the top of compression stroke.

Compression ignition engines have always to operate with a high excess-air ratio in order to ensure the complete combustion. In this type of engine, the addition of a gaseous fuel to the intake air can even increase the power output, or if diesel fuel is reduced at the idle delivery, the power until maximum rate can be achieved by natural gas addition.

This type of conversion, maintaining diesel flame as ignition pilot, is called dual-fuel.

In the dual-fuel compression-ignition engines, where diesel oil serves to ignite the main charge of natural gas and air, the combustion process is more complicated and its course depends on the relative amount of natural gas present. This process evolves through several intermediate stages until finally a progressive combustion results, resembling the process in a spark-ignition engine. Generally, the combustion process in dual-fuel operation occupies a larger crank angle internal than the combustion process in straight-diesel operation, but shorten than normal for a straight natural gas engine.

In Figure 4.1 is represented a typical diesel-gas system retrofit: it can be fitted on the diesel engine without modifying engine structure and layout. A mixer has to be added to air manifold and an actuator has to be applied into diesel pump, reducing diesel-fuel supply at idle rate, when dual-fuel operation is required. The mechanical linkage connects the pedal with both fuel pump and mixer.

The engine retains the ability to operate a 100% diesel fuel if necessary, meanwhile the percentage of $20 \div 40\%$ of diesel fuel to CNG ($80 \div 60\%$) is achieved in dual-fuel operation.

Igniting a lean natural gas charge, by injecting diesel-fuel into it, is similar to prechamber ignition.

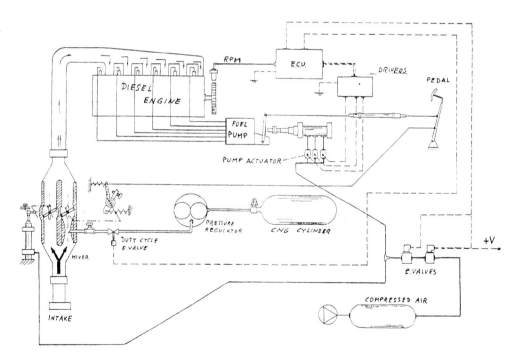

FIGURE 4.1   Diesel-Gas Retrofit Systems (ETRA System) (*).

Because of the high energy content and widespread distribution of the diesel pilot fuel, the gas ignites readily, and good combustion is assured at air-fuel ratios up to the bulk flammability of the gas.

The pilot fuel does contribute some hydrocarbons particulate matter, and probably a significant amount of NO_x.

Even with these aspects, however, NO_x emissions for a diesel/gas engine can be $20 \div 60\%$ lower than for a diesel, and full load thermal efficiency is often improved. Particulate emissions are usually low, but CO and HC, especially methane, are higher than those of a diesel.

Another method to operate the diesel engine conversion into natural gas is to convert it as spark-ignited engines, for heavy duty vehicles and busses.

In this case the conversion requires heavy changements of engine structure and layout, and of course the engine cannot retain the ability to operate with diesel fuel, because the engine is no more diesel cycle but spark-plug ignited.

The conversion will involve replacing the diesel injectors with a spark-plug and the diesel pump, and governor system will be eliminated. A specific ignition system with a distributor will be fitted. The gaseous fuel will be supplied through the inlet manifold with a mixer and the engine will be controlled by an electronic management system. The high compression ratio of the original diesel engine will be reduced to 13:1 almost bacause of the knock properties of the gaseous fuel, which, even better than gasoline, cannot withstand the higher compression ratios like those of diesel engine.

This conversion system is the most investigated by industrial vehicle manufacturers around the world, and according to many industrial and scientific operators presents the most favourable trends for the future (see Fig. 4.2), mainly for pollutant gases reduction.

Typically, these engines are for heavy duty vehicles as busses or trucks, which normally have a large cylinder unit displacement with important problems in combustion lead-time, because of the linear dimensions of combustion chamber.

Then the combustion chamber has to be designed and dedicated to "fast burn" combustion. Compact chamber or Nebula chamber (see Fig. 4.3) could be adopted to improve the combustion time, and overall efficiency of the engine.

The observations made on light duty engines are valid also for heavy-duty engines, because the combustion cycle is the same.

In Figure 4.4 are represented the emission limits of EURO1, EURO2 and EURO3 for heavy-duty engines, meanwhile in Figs. 4.5, 4.6, 4.7, 4.8 are indicated as histograms the typical emissions achievable in two certificate tests (13 model and transient) by different fuels: diesel, LPG, full natural gas, methanol, with different types of combustion: LB = lean burn and Stoichiometric mixtures. The main problem with natural gas emissions is in the HC emissions, composed by far by unburned methane.

The other pollutant gases from natural gas combustion are competitive against those of other fuels. The heavy-duty engines represent the main field for natural gas conversions, starting with busses or trucks fleets in urban areas, and in this direction are devoted the main efforts of manufacturers.

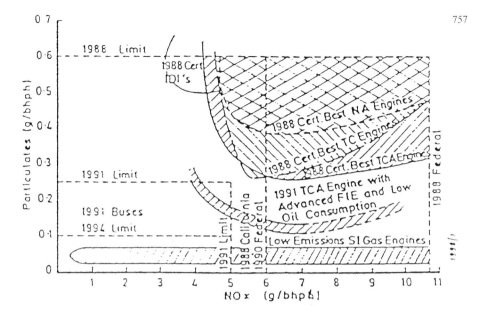

SOURCE: RICARDO U.K

FIGURE 4.2   Diesel NO$_x$ - Particulate Trade-off (Source RICARDO U.K.) (*).

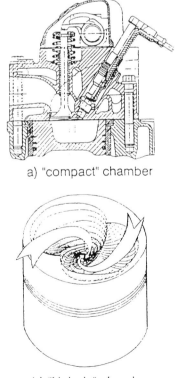

a) "compact" chamber

b) "Nebula" chamber

FIGURE 4.3   Two Types of "Fat Burn" Combustion Chamber" (Source RICARDO 1989) (*).

	CO gr/KW	HC gr/KW	NO gr/KW	PM gr/KW
EURO 1	4.5	1.1	8.0	0.37
EURO 2 ('95)	4.0	1.1	7.0	0.15
EURO 3 ('99)	0.6	0.6	4.0	0.15

FIGURE 4.4   Emission Limits ECE Euro (*).

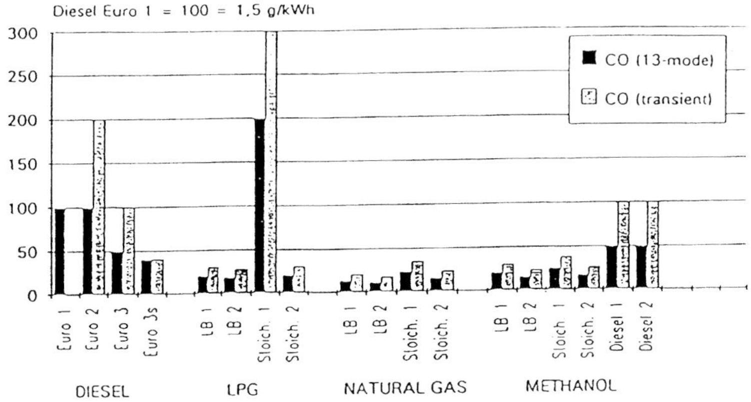

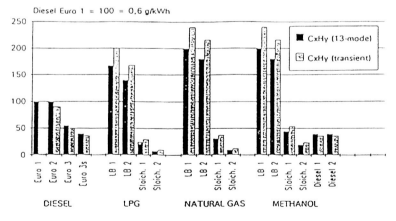

CO-emission in European 13-mode test. As baseline has been chosen 1.5 gr/kWh.
Transient correction from SAE paper 880715.

FIGURE 4.5   CO-Emission (*).

CxHy-emission in European 13-mode test. As basline has been chosen 0.6 gr/kWh.
Transient corrections according SAE paper 880715.

FIGURE 4.6 C_xHy-Emission (*).

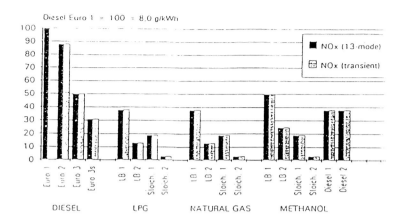

NOx-emission in European 13-mode test (8.0 gr/kWh). For transient operation, correction values are used from the SAE paper 880715.

FIGURE 4.7   NO$_x$-Emission (*).

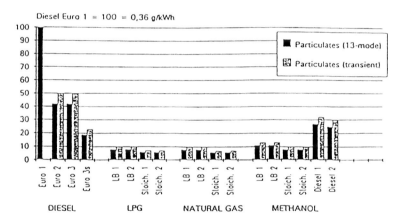

Particulate-emission in European 13-mode test. As basline has been chosen 0.36 gr/kWh. Transient corrections according SAE paper 880715.

FIGURE 4.8   Particulates-Emission (*).

5. NATURAL GAS VEHICLES (SOME CONSIDERATIONS AND TRENDS)

For automotive applications, the heating value of a fuel, i.e. the energy content per unit of volume, controls directly the required on-board storage space and driving range. The specific engine power depends rather on the volumetric energy content of the mixture of fuel and air, which enters the cylinders, than on the heating value of the fuel itself. The mixture properties control the combustion process in the engine, which in turn affects specific fuel consumption and driveability.

In Figure 5.1 are represented the heating values of automotive fuels in volume based and mass based. The value for gasoline, methanol, diesel oil and LPG are given for the fuel in the liquid state. The value for natural gas is given for a gas temperature of 15° and a storage pressure of 200 bar.

As a result of these considerations, and the technologies so far available for on-board natural gas storage, in Figure 5.2 the comparative weight of vehicle fuel and storage systems are represented.

In Figure 5.3 is shown the example of installation of CNG retrofit for a gasoline car (source OMVL). The typical performances for a car, that can be achieved with natural gas running, are: 250 ÷ 300 Km of driving range, 100 ÷ 120 Kg extraweight, 10% in output power. The main reasons for the customer, who perform its car into natural gas, are economic, depending upon local differential price between gasoline and CNG.

In the industrial vehicles, such as trucks and busses, the problem of cylinders installation differs quite for each solution, according to the base layout of the chassis and the driving range requested. It is necessary to take into account the maximum load allowable on the rear and front axes, because the commercial vehicles are typically designed with low margins in gross weight.

The potential benefits of large scale use of natural gas vehicles for urban air quality and economics are evident. Compared with gasoline and diesel vehicles, natural gas vehicles will emit smaller quantities of pollution gases, but they require noticeable efforts in R and D field.

Further research in a range of areas will be necessary if gas company and manufacturers of vehicles, engines and ancillaries are to take advantage of the potential for growth in NGV which exists in many countries.

Some of the topics which seem to require more study for the future are:

1) combustion investigations on fast burn modelling of combustion chamber design and fuelling mixing devices de sign (gas injectors)
2) effect of gas composition on emissions and performances
3) Electronic engine management system include air fuel ratio control for flexible fuel and emissions variable geometry of intake and turbocharger

FUEL TYPE	HEATING VALUE	
	Volume based GJ/m^3	Mass based MJ/Kg
Gasoline	28.5 ÷ 32.5	41 ÷ 44
Methanol	15.5 ÷ 16.8	19 ÷ 21
Diesel oil	14.9 ÷ 37.2	41 ÷ 42
LPG	24.5 ÷ 26.6	50 ÷ 51
Natural Gas	7.5 ÷ 10.5	38 ÷ 47

FIGURE 5.1 Lower heating values of automotive fuels (Approximate values, depending upon composition)

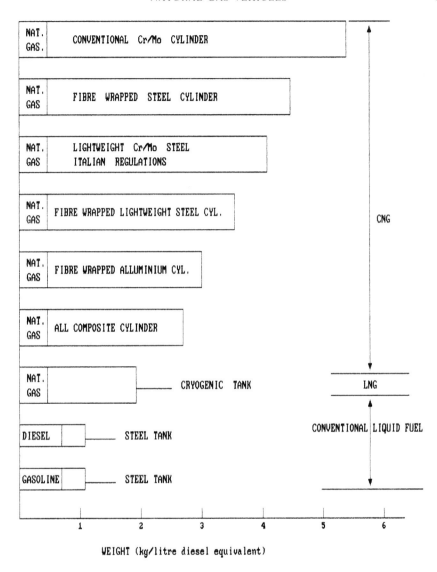

FIGURE 5.2   Comparative Weight of Vehicle Fuel and Storage Systems (*).

4) Catalytic materials dedicated to natural gas emissions
5) Spark ignition conversion of diesel engine (mainly for commercial vehicles operated as fleets)
6) Conversion dual-fuel of diesel engine for existing vehicles (emission reduction and economics improvement)
7) Dedicated natural gas engines and vehicles development

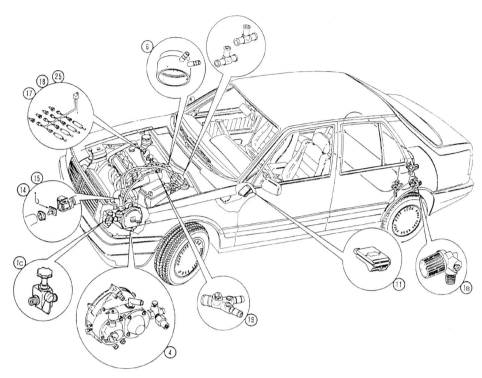

FIGURE 5.3   CNG Kit installation example on standard gasoline car (Source OMVL) (*).

REFERENCES

1. Mc Geer, P. (1982). "Methane - Fuel for the future" DURBIN - Plenum Press
2. Weaver, C. S. - "Natural Gas Vehicles - A Review of the State of the Art" 1989 SIERRA RESERCH, INC SACRAMENTO, CA USA.
3. "Natural Gas as Fuel in Public Transport Vehicles" European Seminar CEE THERMIE, MAY 1992.
4. "Gaseous Fuels - Technology, Performances and Emissions" *International Fuels and Lubricants Meeting and Exposition*, Baltimore, Maryland Sept. 25–28, 1989 SAE INT.
5. Cannon, J. S. (1989). "Drive for Clean Air" Natural Gas and Methanol Vehicles INFORM.

Effects of a Ring Crevice on Hydrocarbon Emission from Spark Ignition Engines

T. SAIKA[a], K. KOREMATSU[a] and M. KONO[b]   [a]*Department of Mechanical Engineering, Kogakuin University, Hachioji-shi, Tokyo, 192 Japan;* [b] *Department of Aeronautics, The University of Tokyo, Bunkyo-ku, Tokyo, 113 Japan*

Abstract—Control of the unreacted fuel would reduce total hydrocarbons from engines, therefore it is necessary to understand the mechanism of exhaust hydrocarbons from spark ignition engines. The authors of this paper would like to pay close attention to the effects of a ring crevice. The fuel-air mixture supplied to a spark ignition engine is compressed into the ring crevice in the compression stroke. The unreacted fuel within the crevice is exhausted in the expansion stroke, if the flame cannot enter the inside of the crevice. The purpose of this work is to separate the effects of the ring crevice on hydrocarbon emission from the other effects, so that the authors have proposed an element model for estimating the in-cylinder oxidation of the hydrocarbon from the crevice. To verify propriety of the theoretical model, the authos have also carried out the experiment with the linear crevices, different from a normal ring crevice, mounted on the piston top and have obtained the agreement of values between the experimental and the theoretical results.

Key Words: Hydrocarbon emission, HC oxidation model, ring crevice, spark ignition engine

INTRODUCTION

As the causes of hydrocarbon (HC) emitted from a spark ignition engine, the following five sources have been taken up (see e.g., Ferguson (1985) and Heywood (1988)).

- Fuel-air mixture in quench layers produced on a combustion chamber wall.
- Fuel-air mixture stored in crevice volumes when flame cannot enter the crevices.
- Absorbed fuel vapor in the lubricating oil during intake and compression strokes.
- Incomplete combustion of fuel-air mixture in a lean or a rich flammable limit.
- Fuel vapor compressed within deposits on a combustion chamber wall.

As Daniel (1957, 1962, 1967) has shown, a quench layer produced on a combustion chamber wall was believed to be a principal cause of HC emission. Daniel (1957) has revealed that quenching is generated on the wall surface when flame comes closer to the wall surface of the combustion chamber. Judging from the photograph, he has also revealed that the quench layer thickness at that time was approximately 50% of the quench distance. By calculating the mass of the HC from the quench layer thickness, surface area of the inside of the combustion chamber, and gas density at flame quenching, he came on to an idea that the quench layer is the most important factor to allow the mass to be placed on the same order as the one of the HC emitted from the engine. However, according to the analysis using a theoretical model (Adamczyk and Lavoie (1978); Hocks *et al.* (1981); Westbrook *et al.* (1981)), the generated quench layer is considered to be oxidized in a process of diffusion after being expanded. LoRusso *et al.* (1981, 1983) have endorsed the matter from the experimental results based on the spark ignition engine. Accordingly the quench layer has not been, at the present situation, regarded as an important factor any more as it was before.

When flame does not enter the crevice in the cylinder of the spark ignition engine, it is pointed out that the fuel-air mixture flowing from there becomes the HC emission. Many theoretical researches (Fendell *et al.* (1983), Lavoie *et al.* (1986); Schramm and Sorenson (1990)), experiments using spark ignition engines (Wentworth (1968, 1971a, 1971b); Tabaczynski *et al.* (1972); Haskell and Legate (1972); Furuhama and Tateishi (1972); LoRusso *et al.* (1980); Weis and Keck (1981); Namazian and Heywood (1982); Kaiser *et al.* (1984); Saika and Korematsu (1986)), and experiments employing constant volume combustion bombs (Adamczyk *et al.* (1981, 1983a, 1983b, 1989); Sellnau *et al.* (1981); Driscoll *et al.* (1984)) have been being conducted.

The biggest crevice existing in the spark ignition engine is the one among the ring-shaped crevices surrounded by a piston, a cylinder wall surface, and a piston ring. This is called a ring crevice. Also the ring crevice is sometimes called a piston top-land clearance. It is not unusual that the nomenclature is often shortened to a top-land crevice or is just called a ring-pack crevice. The screw thread of the spark plug and the space inside the electrodes of the spark plug are also regarded as crevices. Furthermore the crevice between the intake or exhaust port and the valve and the one on the gasket between the cylinder and the cylinder head are also available as crevices. As Tabaczynski *et al.* (1972) have shown the data of a six-cylinder engine, the ring crevice volume accounts for 2.9% of the clearance volume. In the meantime, the crevice volume of the spark plug accounts for 0.28% and the one of the gasket 0.34% according to the same data. Thus it is concluded that the factor that exercises greatest influence is the ring crevice.

Wentworth (1968, 1971a, 1971b) has verified that the total hydrocarbon emission can be restricted by reducing the HC emission from the crevice as a result of improvement of the ring crevice and piston. Haskell and Legate (1972) have pointed out that the HC is drastically decreased by widening the ring crevice width. Inferring from their theoretical study, Weiss and Keck (1981) have insisted that the volume of the ring crevice most contributes to the HC emission. With the use of a constant-volume combustion bomb from which all the crevices are eliminated, Adamczyk *et al.* (1981, 1983a, 1983b) have explained that the ring crevice exercised the greatest influence on the HC emission. The researchers have furthermore explained how the crevice volume and the position of the crevice will control the HC emission. Furuhama and Tateishi (1972) have made analysis of the gas in the ring crevice, and have made ascertained how the gas flows into and out of the ring crevice. Furthermore based on their theoretical observation, Fendell *et al.* (1983) have shown that control of the HC emission is possible by allowing the flame to propagate into the ring crevice.

In this paper, attention will be exclusively paid to the effect of the crevices. The authors (1986) have long been dealing with the flame propagation into the ring crevice and its entering condition using a constant-volume combustion bomb and a spark ignition engine, and they have successfully explained that the matter whether the flame enters the ring crevice will depend on the quench distance at the ring crevice. The purpose of this work is to clrearly separate the effects of the ring crevice from the other effects. For that purpose, the authors have proposed an HC model to clarify the in-cylinder oxidation process of the HC from the crevice. If an engine is warmed-up, about one-third of the HC will be burned up in the exhaust port and the exhaust manifold because of the high gas temperature (Caton *et al.* (1984)). It is important to estimate the in-cylinder oxidation amount even for the analyses and measurements of the oxidation in the exhaust port and the manifold with a fully warmed up engine. Furthermore to verify this model, the concentration of the HC emission from a spark ignition engine was measured and calculated by varying the crevice volume with the attachment of linear crevices on the piston top in addition to a normal ring crevice.

CALCULATION MODEL

The fuel-air mixture supplied to the spark ignition engine is compressed into the crevice in the compression stroke. The unreached fuel within the crevice is exhausted in an expansion stroke, even if the flame reaches the crevice entrance, so long as the flame cannot enter the crevice. Especially in a lean region, there exists a sufficient amount of oxygen. The unburned mixture can be oxidized because of high temperature within the combustion chamber. Here, the calculation model shown below is proposed to investigate the process of oxidation of the unburned mixture flowing the ring crevice.

When the flame does not enter the inside of the crevice in a lean region, what behavior does the unburned mixture compressed into the crevice show after flame quenching? Regarding the mixture flow into and out of the ring crevice, Namazian and Heywood (1982) have carried out visualization experiments with a special transparent square-cross-section single-cylinder spart-ignition engines and have estimated the HC flow rate. They have pointed out that in the flow out of the top-land crevice entrance during the expansion stroke, there is not only the flow expanding out of the ring crevice around the circumference of the piston but also a jet-type flow through the top piston-gap. The authors deals only with the flow from the ring crevice because of a larger area of the ring crevice. On the other hand Driscoll *et al.* (1984) have carried out the observation using high-speed schlieren photographs for a constant-volume combustion bomb, the unburned mixture compressed into the crevice flows out as a jet stream as soon as the pressure in the bomb begins to decrease. If the same phenomenon is created in the combustion chamber of the spark ignition engine, the high-temperature gas existing surrounding the jet stream is entrained into the jet stream. This might result in the oxidation of the unburned mixture. Such beings the case, the process of the unreacted mixture that flows out and is oxidized has been calculated by designing an element model using an empirical global rate expression. The authors also use the following suppositions for the element model.

- When combustion is completed in the combustion chamber, the unburned mixture is compressed into the crevice and is fulled. The mixture in the crevice is homogeneous. It temperature is the temperature of the cylinder wall surface, and its pressure is the cylinder pressure.
- The jet stream from the crevice is dealt with as the two-dimensional jet.
- In the expansion stroke after the combustion, the unburned mixture flows out from the crevice, as the pressure in the combustion chamber is decreased. The elements of

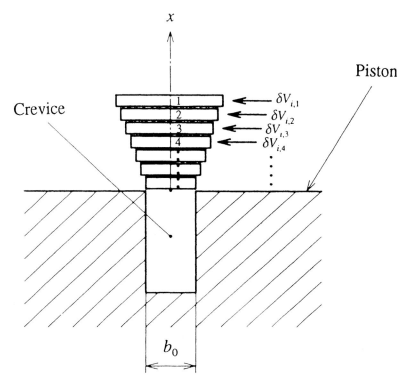

FIGURE 1   Sketch of the element model.

the mixture flowing from the crevice entrain the high-temperature burned gas for
a very short time, and the mixture is unstantaneously oxidized.
- The pressure history after the combustion is determined by considering the burned
  gas in the cylinder to bring about a polytropic change.

INITIAL CONDITION

The discharge flowing out from the crevice for an infinitesimally short time of δt is
presented for a polytropic change with the Equation (1).

$$\delta V_{i,j,0} = \left\{ \left(\frac{P_{i-1}}{P_i} \right)^{\frac{1}{Ku}} - 1 \right\} V_{cr} \tag{1}$$

Therefore, the Equation (2) give s the jet velocity, and the Equation (3) gives the element
mass. Supposing that the distance from the crevice entrance is x, the Equation (4) gives

the distance of the element from the crevice entrance.

$$u_{i,j,0} = \frac{\delta V_{i,j,0}}{A_{cr}\,\delta t} \tag{2}$$

$$\delta m_{i,j,0} = \frac{P_i\,\delta V_{i,j,0}}{R_u\,T_{wall}} \tag{3}$$

$$x_{i,j,0} = u_{i,j,0}\,\delta t \tag{4}$$

Let it be assume that the composition of the element in the crevice is that of the unreacted mixture, and the temperature of the gas within the crevice is the temperature of the cylinder wall. Therefore, we can obtain the initial concentrations with the following expressions from (5) to (9).

$$[CH_4]_{i-1,j} = \frac{\phi}{\phi + 9.52} \cdot \frac{P_{i-1}}{R\,T_{wall}} \cdot 10^{-6} \tag{5}$$

$$[O_2]_{i-1,j} = \frac{2.00}{\phi + 9.52} \cdot \frac{P_{i-1}}{R\,T_{wall}} \cdot 10^{-6} \tag{6}$$

$$[N_2]_{i-1,j} = \frac{7.52}{\phi + 9.52} \cdot \frac{P_{i-1}}{R\,T_{wall}} \cdot 10^{-6} \tag{7}$$

$$[CO_2]_{i-1,j} = 0 \tag{8}$$

$$[H_2O]_{i-1,j} = 0 \tag{9}$$

CALCULATION PROCEDURE

Figure 2 illustrates a calculation procedure. The Equation (10) gives the distance from the crevice entrance, and the discharge of the individual elements is expressed as a function of x. The Equation (11) and (12) give the jet velocity and discharge of an element using Goertler's solution, therefore the Equation (13) gives the entrainment discharge. The Equation (14) and (15) give the mass discharge of the entrainment and the element.

$$x_{i,j} = x_{i-1,j} + u_{i-1,j}\,\delta t \tag{10}$$

$$u_{i,j} = 3.39\,u_{i,j,0}\sqrt{x_{i,j}/b_0} \tag{11}$$

$$\delta V_{i,j} = 0.44\,\delta V_{i,j,0}\sqrt{x_{i,j}/b_0} \tag{12}$$

$$\delta V_{e,i,j} = \delta V_{i,j} - \delta V_{i-1,j} \tag{13}$$

$$\delta m_{e,i,j} = \frac{P_i\,\delta V_{e,i,j}}{R_b\,T_i} \tag{14}$$

$$\delta m_{i,j} = \delta m_{e,i,j} + \delta m_{i-1,j} \tag{15}$$

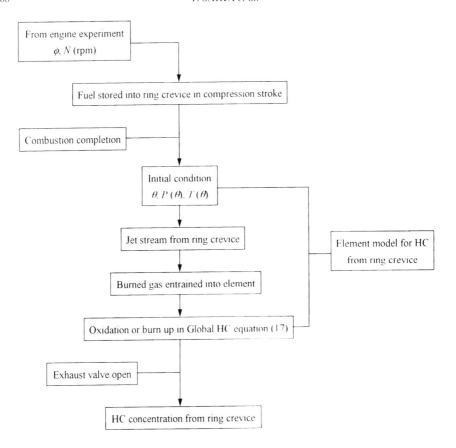

FIGURE 2   Computational sequence outlined.

Assuming the entrained gas is mixed with the element gas instantaneously, the average temperature of the element can be expressed with the Equation (16).

$$T_{a,i,j} = \frac{\delta m_{e,i,j} T_i + \delta m_{i-1,j} T_{a,i-1,j}}{\delta m_{i,j}} \tag{16}$$

Using the empirical global rate expression (17),

$$\delta[CH_4]_{i,j} = -C_R A \delta[CH_4]_{i-1,j}[O_2]_{i-1,j} \exp\left(-\frac{E}{R_u T_a}\right)\delta t \tag{17}$$

the composition changes due to oxidation for a time of δt are calculated with the Equations from (18)–(21),

$$[CH_4]_{i,j} = [CH_4]_{i-1,j} + \delta[CH_4]_{i,j} \tag{18}$$

$$[O_2]_{i,j} = [O_2]_{i-1,j} + 2\delta[CH_4]_{i,j} \tag{19}$$

$$[CO_2]_{i,j} = [CO_2]_{i-1,j} - \delta[CH_4]_{i,j} \tag{20}$$

$$[H_2O]_{i,j} = [C_2O]_{i-1,j} - 2\delta[CH_4]_{i,j} \tag{21}$$

where let it be assumed that the pre-exponential factor A and activation energy E are 6.7×10^{15}(mole/cc/s) and 95 (kJ/mole), respectively. Also C_R is an experimental constant, and let it be assumed that the constant is 0.1 as with Lavoie $et\ al.$'s case (1980).

VERIFICATION OF THE CALCULATION MODEL

Propriety of the calculation method described above is required to be verified by an experiment. Therefore, the experiment is conducted by mounting linear crevices different from the ring crevice on the piston top and by calculating the HC concentration which is supposedly increased owing to the settlement of these crevices. The three types of the crevices on the piston top are used. Figure 3 illustrates the location of them, and Table I shows these specifications. Also Table II shows the specifications of the spark ignition engine used for the experiment and the calculation. Completion of the combustion for individual equivalence ratios was judged from the diagrams of heat release ratio calculated from cylinder pressure diagrams. Table III indicates the crank angles, the cylinder pressures, and the cylinder wall temperatures just at completion of the combustion as initial conditions.

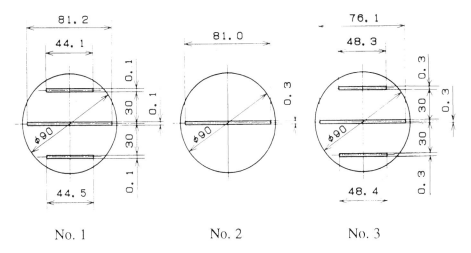

FIGURE 3   Details of crevices on the piston top.

TABLE I

Crevice specifications

No.	No. of Crevices	Width (mm)	Length (mm)	Depth (mm)	Volume (mm)
1	3	0.1	166.8	5.15	0.086
2	1	0.3	81.0	5.15	0.125
3	3	0.3	172.8	5.05	0.262

TABLE II

Test engine specifications

No. of cylinders	1
Operating cycle	4 stroke
Displacement	433 cc
Bore	90 mm
Stroke	68 mm
Compression ratio	6.0
Topland ring crevice volume	0.265 cc
Power at rated speed (Gasoline)	5.88 kW/3600 rpm
Fuel type	Methane

TABLE III

Summary of initial conditions at combustion completion

Equivalence ratio ϕ	Crank angle (°ATDC)	Cylinder pressure (MPa)	Wall temperature T_{wall}(°C)
0.65	11.3	1.30	153
0.70	11.0	1.40	163
0.75	10.4	1.50	173
0.80	9.5	1.60	183
0.85	7.9	1.70	193
0.90	5.3	1.76	203
0.95	2.3	1.86	212
1.00	1.8	1.88	222
1.05	3.2	1.84	232
1.10	5.9	1.76	242

Schramm *et al.* (1990) have proposed a model to estimate the influence of the ring crevice and lubricating oil film, and they have obtained good coincidence with the experimental results. However in their experiment, no sharp line is established to separate the influence of the ring crevice from the other influences. The author's peculiar crevices can define such a sharp line and allows the detailed verification of the calculation model to be made.

EXPERIMENTAL APPARATUS AND METHOD

Exhaust HC concentration has been measured by incorporating in the engine three types of the pistons whose linear crevice volume is different from each other and a usual piston having no linear crevice, i.e., four types in all to be operated. Since just 0.17% lowering of compression ratio is seen at the maximum with the addition of the crevice, almost the same operation condition is believed to be realized.

The experiment in this study is conducted in a lean flammable region, and no rich flammable region is dealt with. In the rich flammable region, there exist not only the HC from the ring crevice and the oil film but also the unburned fuel caused by the insufficiency of oxygen or the HC brought about from thermal dissociation even in the burned gas. With a purpose of avoiding such kinds of influences, the experiment was conducted mainly in a lean region.

Figure 4 illustrates a schematic diagram of the experimental apparatus. Using methane as fuel, the engine is operated under the condition of the constant rotational speed of 2000 rpm. The air flow rate is held almost constant. The absorption dynamic power of the DC electric dynamometer is 10 kW and, its driving dynamic power is 7 kW. Rotational speed can be controlled within the range from 0 to 4500 rpm.

As an exhaust gas analyzer an FID is used. The fuel supplied to the FID is mixed gas with a ratio of 40% hydrogen and 60% nitrogen, and 99.99% dry air is used as oxidizer. Methane-mixed nitrogen (44.6 ppm methane) is used for rectification. The length of the conduit for sampling is 5 m, and the sampling position of the exhaust gas as approximately 50 mm downstream from the exhaust valve. Accordingly the sampled specimen is always free from oxidation in the exhaust pipe and it is restricted to the oxidation in the combustion chamber. Before the specimen is sent to the FID, the specimen is dehydrated and soot is removed from it.

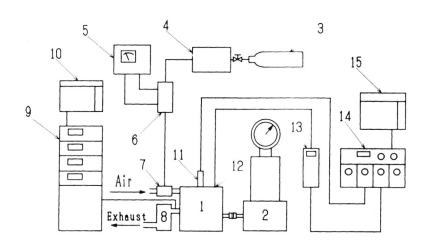

1. Test engine	9. Gas analyzer (FID)
2. Electric dynamometer	10. Pen recorder
3. Methane bomb	11. Pressure transducer
4. Surge tank	12. TDC pulse detector
5. Differential pressure gage	13. Signal conditioner
6. Laminar flow meter	14. Digital memory
7. Gas carburetor	15. X-Y recorder
8. Muffler	

FIGURE 4   Schematic diagram of the experimental apparatus.

COMPARISON OF EXPERIMENTAL AND THEORETICAL RESULTS

Figure 5 shows the HC concentration from the engine. From the results, it is noticed that the HC concentration becomes higher with the larger volume of the crevices in relation to the individual equivalence ratios. The broken line indicates the results for a piston without linear crevices. The HC emission exceeding the broken line is the HC from the attached linear crevices.

From Figure 6 to Figure 8 illustrate comparison between the experimental results. The experimental and the theoretical results are shown at the left sides of the bar graphs of the individual equivalence ratios, whereas the theoretical results are at the right sides. The concentration given with the broken line in the experimental results indicates the case where there is no crevice on the piston top. The same affair is shown in the theoretical results as well. Therefore the hatched part in the experimental results is obtained from creation of the crevice. The hatched part in the theoretical results denotes the reduced portion with the oxidation.

The experimental and the theoretical results, from which the reduced portions due to oxidation are removed, agree well with each other. From this it is estimated that part of the unreacted mixture flowing out from the crevice and oxidized is discharged from the combustion chamber to the exhaust pipe. Especially when there is so much of the unburned mixture from the crevice as with the case of the equivalence ratios in Figure 8 being 0.74, 0.82, it is understood that the amount of the oxidized mixture is also increasing. Therefore it is concluded that the oxidized amount of the unreacted mixture can be estimated using the element model.

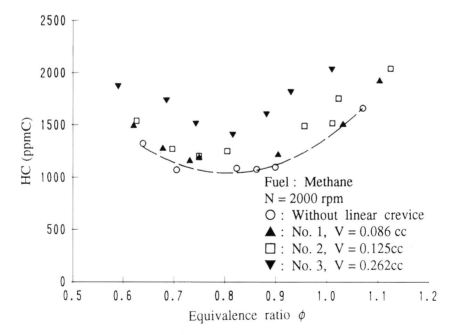

FIGURE 5   HC emission concentration from the test engine.

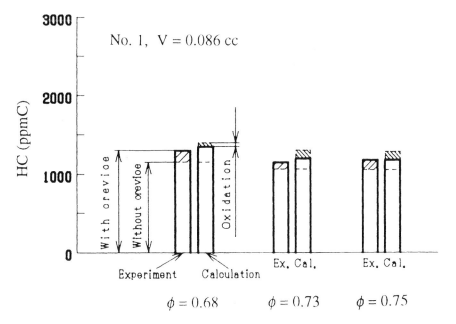

FIGURE 6   Results of calculations for the piston No. 1.

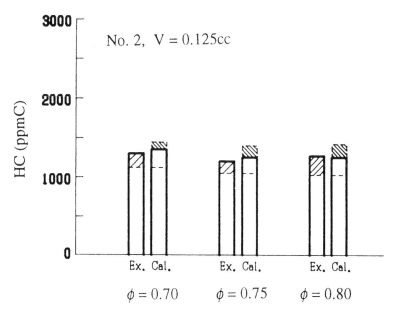

FIGURE 7   Results of calculations for the piston No. 2.

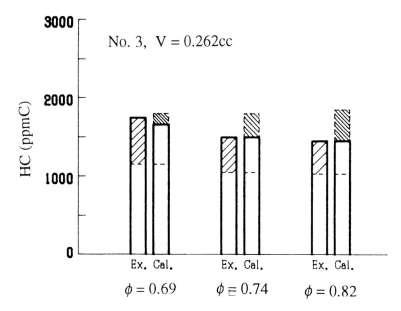

FIGURE 8   Results of calculation for the piston No. 3.

SEPARATION OF THE EFFECT OF THE RING CREVICE FROM THE OTHER EFFECTS

Using the element model, let the effects of the ring crevice on the HC emission be separated from the other effects. An equivalence ratio ranges from 0.65 to 1.10. As shown in Table IV, the crank angle, the cylinder pressure, and the wall temperature at completion of the combustion are obtained from the experimental results for the engine in the same case of the linear crevices. Figure 9 illustrates the shape of the ring crevice whose volume is 0.265 cc.

Table V and Figure 10 show the theoretical result. The solid line in Figure 10 indicates the experimental data for the equivalence ratios, which includes the HC from

TABLE IV

HC from straight crevices

Equivalence ratio ϕ	HC from crevice No. 1 (ppm)	HC from crevice No. 2 (ppm)	HC from crevice No. 3 (ppm)
0.65	183	388	818
0.70	137	332	703
0.75	104	262	557
0.80	81	208	441
0.85	67	173	368
0.90	64	171	363
0.95	55	156	332
1.00	54	156	332
1.05	76	186	395
1.10	101	221	468

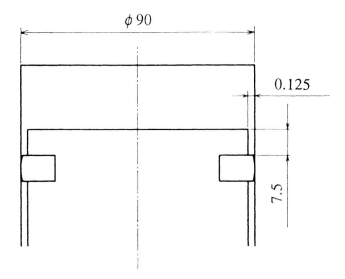

FIGURE 9   Details of the top-land ring crevice of the engine.

TABLE V

Ratio of HC from ring crevice to HC from cylinder

Equivalence ratio ϕ	HC from crevice (ppm)	HC from crevice (ppm)	HC ratio of ring crevice to cylinder (%)
0.65	541	1260	43.0
0.70	404	1142	35.3
0.75	305	1069	28.5
0.80	238	1041	22.9
0.85	196	1056	18.5
0.90	186	1117	16.7
0.95	157	1222	12.9
1.00	154	1371	11.3
1.05	223	1565	14.2
1.10	300	1803	16.6

the crevices and also the other sources. The broken line indicates the theoretical result of the HC from the ring crevice. The difference between the solid line and the broken line is therefore the HC from the other sources. The total HC from the combustion chamber might result mainly from the ring crevice and the oil film, because both poor combustion and carbon deposits are carefully removed from the experiment.

As a result, the maximum 43.0% of the HC is from the ring crevice at $\phi = 0.65$. Under the lean condition, the ring crevice has a serious effect on the HC emission in comparison with the other effects. When the equivalence ratio is more than 0.8, the fuel concentration is higher than at the lesser equivalence ratio. Despite the higher fuel concentration, the HC from the crevie is lesser. For example, at $\phi = 1.00$, the result indicates the minimum 11.3% of the HC from the ring crevice. In this region, the HC from the ring crevice might show lesser value, if the flame should happen to enter the

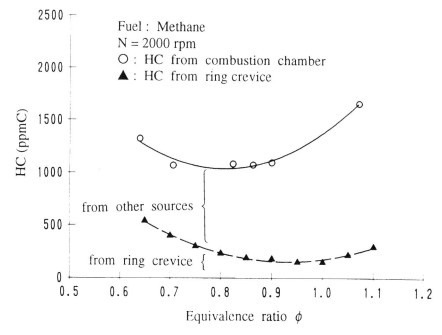

FIGURE 10   Results of calculation for the ring crevice.

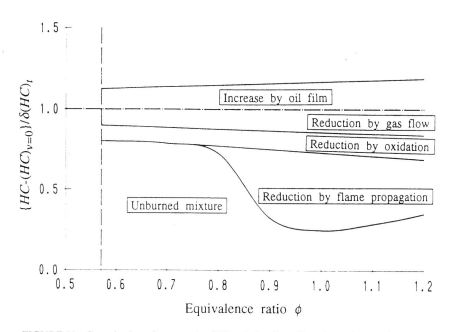

FIGURE 11   Organization of sources that HC emission from SI engine reduces or increases.

ring crevice. Therefore, in the neighborhood of the theoretical fuel-air concentration, the lubricating oil seriously influences the HC from the combustion chamber.

CONCLUSIONS

The authors have proposed the HC model for elucidating the in-cylinder oxidation process of the HC that is compressed into the ring crevice within a spark ignition engine in a compression stroke and flows out from there after flame arrival and quenching. To verify the theoretical method, they have carried out the experiment with the linear crevices, different from a normal ring crevice, mounted on the piston top, and have obtained the agreement of values between the theoretical and experimental results. Therefore, it is concluded that the oxidized amount of the unreacted mixture can be estimated using the model.

The authors also have applied the model to the ring crevice of the spark ignition engine to separate the effect of the ring crevice on the HC emission from the other effects, especially the effect of the lubricating oil. The result is the maximum 43.0% of the HC from the ring crevice at $\phi = 0.65$ and the minimum 11.3% at $\phi = 1.00$. Therefore, it should be noted that the effect of the lubricating oil film on the HC from the combustion chamber is greater than that of the ring crevice particularly in the neighborhood of the theoretical fuel-air concentration.

NOMENCLATURE

A	Pre-Exponential factor in global rate expression (mole/cc/s)
A_{cr}	Crevice entrance area (cm^2)
b_0	Crevice width (m)
C_R	Non-dimensional oxidation parameter
E	Activation energy in global rate expression (J/mole)
K_b	Average burned-gas specific heat ratio
K_u	Average unburned-gas specific heat ratio
m	Element mass (kg)
m_e	Burned gas mass entrained into element (kg)
P	Cylinder pressure (Pa)
R	Universal gas constant (J/mole/K)
R_u	Unburned gas constant (J/kg/K)
R_b	Burned gas constant (J/kg/K)
T	Cylinder temperature (K)
T_a	Average temperature in element (K)
T_{wall}	Cylinder wall temperature (= gas temperature in crevice) (K)
t	Time (s)
u	Jet velocity (m/s)
u_e	Jet velocity just after issuing from crevice (m/s)
V_{cr}	Crevice volume (cc)

V Element volume (cc)
V_0 Element volume just after issuing from crevice (cc)
V_e Burned gas volume entrained into element (cc)
x Distance from crevice entrance (m)
ϕ Fuel-air equivalence ratio
$[CO_2]$ Carbon dioxide concentration (mole/cc)
$[CH_4]$ Methane concentration (mole/cc)
$[H_2O]$ Water concentration (mole/cc)
$[N_2]$ Nitrogen concentration (mole/cc)
$[O_2]$ Oxygen concentration (mole/cc)

Subscript

0   Initial condition of individual elements
i Number of time
j Number of element

REFERENCES

Adamczyk, A. A. and Lavoie, G. A. (1978) Laminar Head-on Flame Quenching- A Theoretical Study, *SAE Paper* 780969.

Adamczyk, A. A., Kaiser, E. W., Cavolowsky, J. A. and Lavoie, G. A. An Experimental Study of Hydrocarbon Emissions from Closed Vessel Explosions. *Eighteenth Symposium (International) on Combustion*, (1981) p. 1695.

Adamczyk, A. A., Kaiser, E. W. and Lavoie, G. A. (1983a) A Combustion Bomb Study of the Hydrocarbon Emissions from Engine Crevices. *Combustion Science and Technology*, **33**, 261.

Adamczyk, A. A., Kaiser, E. W., Lavoie, G. A. and Isack, A. J. (1983b) Single-Pulse Sampling Valve Measurements of Wall Layer Hydrocarbons in a Combustion Bomb. *Combustion and Flame*, **52**, 1.

Adamczyk, A. A. (1989) Hydrocarbon Emissions from an Annular Crevice: Effects of Spark/Insert Position, Equivalence Ratio and Pressure. *Combustion Science and Technology*, **64**, 263.

Caton, J. A., Heywood, J. B. and Mendillo, J. V. (1984) Hydrocarbon Oxidation in a Spark Ignition Engine Exhaust Port, *Combustion Science and Technology*, **37**, 153.

Daniel, W. A. Flame Quenching at the Walls of an Internal Combustion Engine, *Sixth Symposium (International) on Combustion*, (1957) p. 886.

Daniel, W. A. and Wentworth, J. T. (1962) Exhaust Gas Hydrocarbons–Genesis and Exodus. *SAE Paper* 486 B.

Daniel, W. A. (1967) Engine Variable Effects on Exhaust Hydrocarbon Composition (A Single-Cylinder Engine Study with Propane as the fuel). *SAE Paper* 60124.

Driscoll, J. F., Mirsky, W., Palaniswamy, S. and Liu, W. Experimental Study of the Flow Field Produced by a Wall Crevice in a Combustion Chamber. *Twentieth Symposium (International) on Combustion*, (1984) p. 209.

Fendell, F., Fink, S. and Feldman, P. (1983) A Note on Unburned Hydrocarbon Emissions from Automotive Engines. *Combustion Science and Technology*, **30**, 47.

Ferguson, C. R. (1985) *Internal Combustion Engines, Applied Thermosciences*, John Wiley and Sons, pp. 403–410.

Furuhama, S. and Tateishi, Y. (1972) Concentration in Gas of Piston Top-land. *Transactions of SAE-Japan*, No. 4, 3(in Japanese).

Haskell, W. W. and Legate, C. E. (1972) Exhaust Hydrocarbon Emissions from Gasoline Engines–Surface Phenomena, *SAE Paper* 720255.

Heywood, J. B. (1988) *Internal Combustion Engine Fundamentals*, McGraw-Hill Book, pp. 601–619.

Hocks, W., Peters, N. and Adomeit, G. (1981) Flame Quenching in Front of a Cold Wall Under Two-Step Kinetics. *Combustion and Flame*, **41**, 157.

Kaiser, E. W., Rothschild, W. G. and Lavoie, G. A. (1984) Storage and Partial Oxidation of Unburned Hydrocarbons on Spark-Ignited Engines–Effect of Compression Ratio and Spark Timing. *Combustion Science and Technology*, **36**, 171.

Lavoie, G. A., Lorusso, J. A. and Adamczyk, A. A. (1980) Hydrocarbon Emissions Modeling for Spark Ignition Engines. *Combustion Modeling in Reciprocating Engines*, Prenum Press. p. 422.

Lavoie, G. A., Adamczyk, A. A., Kaiser, E. W., Cooper, J. W. and Rothschild, W. G. (1986) Short Communication–Engine HC Emissions Modeling: Partial Burn Effects. *Combustion Science and Technology*, **49**, 99.

LoRusso, J. A., Lavoie, G. A. and Kaiser, E. W. (1980) An Electro-Hydraulic Gas Sampling Valve with Application to Hydrocarbon Emissions Studies. *SAE Transactions*, **89**, 304, *SAE Paper* 800045.

LoRusso, J. A., Kaiser, E. W. and Lavoie, G. A. (1981) Quench Layer Contribution to Exhaust Hydrocarbons from a Spark-Ignited Engine. *Combustion Science and Technology*, **25**, 121.

LoRusso, J. A., Kaiser, E. W. and Lavoie, G. A. (1983) In-cylinder Measurements of Wall Layer Hydrocarbons in a Spark Ignited Engine. *Combustion Science and Technology*, **33**, 75.

Namazian, M. and Heywood, J. B. (1982) Flow in the Piston-Cylinder-Ring Crevices of a Spark-Ignition Engine: Effect on Hydrocarbon Emissions, Efficiency and Power, *SAE Paper* 820088.

Saika, T. and Korematsu, K. (1986) Flame Propagation into the Ring Crevice of a Spark Ignition Engine *SAE Transactions*, **95**, 497, *SAE Paper* 861528.

Schramm, J. and Sorenson, S. C. (1990) A Model for Hydrocarbon Emissions from SI Engines. *SAE Paper* 902169.

Sellnau, M. C., Springer, G. S. and Keck, J. C. (1981) Measurements of Hydrocarbon Concentrations in the Exhaust Products from a Spherical Combustion Bomb. *SAE Paper* 810148.

Tabaczynski, R. J., Heywood, J. B. and Keck, J. C. (1972) Time-Resolved Measurements of Hydrocarbon Mass Flowrate in the Exhaust of a Spark-Ignition Engine. *SAE Transactions*, **81**, *SAE Paper* 720112.

Weis, P. and Keck, J. C. (1981) Fast Sampling Valve Measurements of Hydrocarbons in the Cylinder of a CFR Engine. *SAE Paper* 810149.

Wentworth, J. T. (1968) Piston and Ring Variables affect Exhaust Hydrocarbon Emissions. *SAE Paper* 680109.

Wentworkth, J. T. (1971a) The Piston Crevice Volume Effect on Exhaust Hydrocarbon Emission. *Combustion Science and Technology*, **4**, 97.

Wentworkth, J. T. (1971b) Effect of Combustion Chamber Surface Temperature on Exhaust Hydrocarbon Concentration. *SAE Paper* 710587.

Westbrook, C. K., Adamczyk, A. A. and Lavoie, G. A. (1981). A Numerical Study of Laminar Flame Wall Quenching. *Combustion and Flame*, **40**, 81.

Partial Disconnection of Some Cylinders of Work as a Way of Diminishing Diesel Engine Exhaust Gases Toxicity

S. POSTRZEDNIK and A. CIESIOLKIEWICZ   *Silesian Technical University*
Institute of Thermal Technology - Division of Internal Combustion
Engines ul. Konarskiego 18, 44-100 Gliwice, Poland

Abstract—Serially produced in Poland diesel engines (type: SW400, 6C107) have got, besides number advantages which cause their wide industrial application, unsatisfactorily exhaust gases toxicity characteristics.

In the idle running and low engine load considerable reduction of toxic components in diesel engine emission, can be obtained through partially disconnection of some cylinders of work. Results of investigations presented in paper show the positive effects of applied method and achieved solution. Total emission of toxic components (carbon monoxide CO, nitrogen oxides NO_x) can be significantly reduced, more than 20% of emission value for serially produced engine. Diesel engines with in such a way corrected working system can be used in the underground mining transport and in the bus service.

Key Words: Diesel engines, toxicity diminishing, clear environment.

INTRODUCTION

Reduction of the toxic emission substances of diesel engines can be achieved through Tandara (1991), Bayhan and Postrzednik (1993):

a. better preparation of the feed medium (fuel-air mixture) relating to their thermodynamic parameters (Postrzednik and Żmudka, 1991), homogeneity and quality (e.g. atomization of the fuel),
b. assurance of optimal working conditions of individual cylinders (Ciesiolkiewicz and Postrzednik, 1992) by introducing design changes in the fuel dosage system and automatic matching of the start of fuel injection angle to the instantaneous running parameters,
c. improvement in combustion process in the cylinders (Nagai and Kawakami, 1990), owing to the introduction of proper combustion chamber and fuel injection system,
d. use of special (Lausch *et al.*, 1993) emission control systems (catalyst, afterburner).

The idea – 'partially disconnection of some cylinders at part load' – has been developed for spark ignition engines, with the aim to offset their poor efficiency at part load.

Diesel engines are very efficient at part load, but this technique can be used for the purpose of pollutants emission control. In the idle running and low engine load considerable reduction of toxic components in diesel engine emission, as shown by the investigations carried out by the authors (Postrzednik *et al.*, 1992) can be obtained through partially disconnection of some cylinders of work, while the working cylinders are more loaded as normal. The number of operating cylinders is adapted to an instantaneous engine load (the minimal number is in the idle running).

The following effect is to be achieved: the fuel injection cut-off by some fuel pump sections, what (in idle running and low engine load) leads automatic to the disconnection of some cylinders of work.

Disconnection of the fuel injection was realized by the correction of control side edge of plunger. It causes – depend on engine load – diversification of the beginning of injection and fuel charge mass. Automatic change of the injection advance angle, in dependence on the instantaneous engine load and fuel charge, was realized by the proper shaping of the top edges of the injection pump plungers.

These modifications are based on lengthwise channel widening in a few injection pump-piston, thereby enable the disengagement of some sectors of the engine cylinders (for a very small portion of fuel) and bevel of the upper edges of the pump-piston, causing continuous changes of the start of fuel injection angle depending on the fuel portion.

The proposed treatment is called correction of pump-pistons of fuel injection pump. The number of cylinders working and the angle of injection advance depend on the instantaneous running parameters of the diesel engine (fuel charge). Investigations were carried out on a slow suction 6-cylinder diesel engine SW400, with direct fuel injection.

EFFECT OF THE QUANTITY OF FUEL CHARGE MASS ON COMBUSTION PROCESS INDICES

The basic parameter of the injection system of internal combustion engine is the quantity of fuel charge m_j, kg/cycle, supplied into one cylinder in one working cycle.

The actual fuel charge m_j, kg, can be determined from the flux $\dot{m}_f$, kg/s, of fuel consumption, and the engine speed $\dot{n}_s$, 1/s, as

$$m_j = ((s\,\dot{m}_f)/\dot{n}_s z) \tag{1}$$

where: z - number of cylinders, $s = 2$ for 4-stroke engine.

The quantity of fuel charge m_j determines directly the achieved engine load, expressed by the engine torque M_0 – for the investigated diesel engine SW400 it has been shown on Figure 1.

The amount of fuel charge m_j results directly from the geometric properties of the pump-piston section of injection pump (diameter and working stroke of pump-piston) and the volumetric efficiency of the pump, taking into consideration fuel leakage in the pump system.

The increase of the fuel charge m_j causes diminishing of the air excess λ- Figure 1. The air excess λ influences the combustion conditions and the exhaust gas composition.

The decrease of the air excess (at $\lambda \geqslant 1$) leads to the combustion temperature rise and vice versa. The high combustion temperature and presence of the free excess oxygen in the combustion chamber are conducive to nitrogen oxides NO_x formation (Schoubye, 1992). By primary methods, the amount of nitrogen oxides NO_x formed in the combustion process decreases by (Bayhan and Postrzednik, 1993):

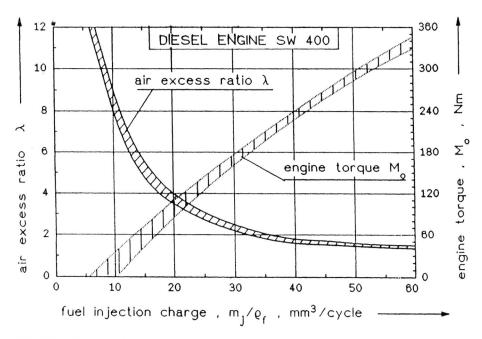

FIGURE 1   Range of air excess ratio and torque of diesel engine SW400 as a function of fuel injection charge.

— increasing the amount of combustion air,
— reducing the combustion maximum pressure by delayed fuel injection,
— part recirculation of exhaust gas,
— injection water into the combustion chamber or
— using water emulsified fuel.

Formation of toxic components (carbon monoxide CO, nitrogen oxides NO_x, hydrocarbons C_mH_n, soot) during combustion process results from imperfect combustion (formation of CO, C_mH_n) and from the level of the fuel combustion temperature (formation of nitrogen oxides NO_x).

It is known, that increase of air excess λ provides to reduce the nitrogen oxides NO_x emission and instantaneous increase of the carbon monoxide CO emission. It has been confirmed (Ciesiolkiewicz and Postrzednik, 1992) by the investigations of the diesel engine SW400.

The influence of the fuel charge m_j on the content of chosen components in exhaust gases of the diesel engine SW400 are shown on Figure 2. The function $CO = f(m_j)$ has a minimum at a fuel charge value $m_{j,0}$, where the values of air excess λ for the range of CO_{min} are $\lambda = 2 \div 3$.

During engine operation at idling (at $m_j < m_{j,0}$) the content of carbon monoxide CO is relatively high, even though excess air is considerable (λ reaching up to 10). This is probably caused by the low combustion temperature in the cylinder (low rate of chemical reactions) and lower rate of fuel injection (low rotational speed of injection

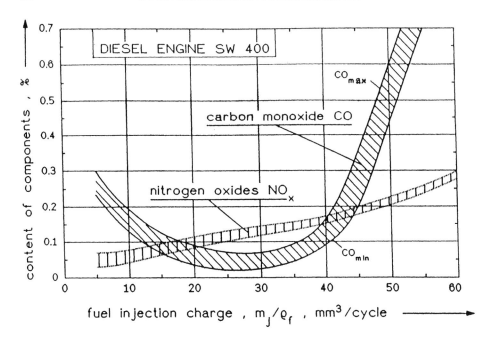

FIGURE 2   Content of exhaust gas components (CO, NO$_x$) of the investigated diesel engine.

pump) and worse atomization of the fuel, at lower intensity of charge swirl and so on.

There are another reasons for the carbon monoxide CO rise at the higher fuel charges (for $m_j > m_{j,0}$); the air excess ratio λ lowers, the oxygen concentration O_2 decreases and local there is not enough of free oxygen to react with fuel vapours totally (formation of carbon monoxide CO), and because the combustion temperature is high proceeds the thermal decomposition of the fuel portions (hydrocarbons C_mH_n and soot).

Increasing the excess air ratio λ results in increase of discharge losses too. Conversely, reducing discharge losses require reduction of λ, which in turn lead to increase in imperfect combustion losses and results in increase of CO in discharge gases. To avoid losses the most appropriate combustion conditions require minimization of the sum of both losses. This exists in a case, where negligible traces of carbon monoxide CO are noted and then the working conditions of the cylinders are optimal. This corresponds to part-loading of engine (average fuel portion) at average engine speed.

THE OPERATION PROCEDURE FOR MINIMIZATION OF CARBON MONOXIDE EMISSION

On the basis of obtained results, one notice the existence of the engine working range, whereby emission of carbon monoxide CO remains constant, considerably low (below 0.05%). The essence of the method is controlling of the multi cylinder engine at low load and idling so as to put only some of its cylinders in operation, in a range favorable for clean exhaust gases. Due to this, regulation of the fuel charge to some of the cylinders is proposed, which can lead to partially disconnection of some cylinders of work. The fuel injection cut-off by some fuel pump sections leads automatic to the disconnection of some cylinders of work.

In this case the operating cylinders must be somewhat more loaded as normally (minimal), so they will work at the optimum of fuel charge $m_{j,0}$ (at minimal content of carbon monoxide CO) and owing that considerable reduction of toxic components in diesel engine gases can be obtained.

The technical realization of this concept of switching off injection is based on design diversification of the engine injection system. The beginning of fuel charging and the minimum fuel charge for individual cylinders are diversified and so adapted, that the number of working cylinders increase automatically with increased load.

Practically, it leads to lengthwise channel widening of injection pump-piston in a few sectors of the injection pump, with shortening of the control edges – Figure 3a. The amount of lengthwise widening (so called correction ratio of pump-piston) corresponds to the disengaged cylinder, and can be determined experimentally.

The general rate of reducing carbon monoxide concentration CO-on Figure 4 has been shown the results for the investigated diesel engine SW400 – depend on the number of disengaged cylinders, which can increase as the engine load is reduced down to idling-just enough to put three (two) cylinders in operation. In this way, the working

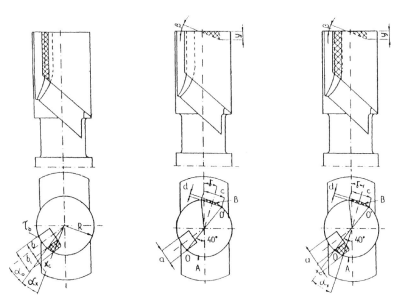

a. correction of control side edges   b. adopting of injection advance angle   c. complete plunger correction

FIGURE 3   Correction procedure of injection pump enabling disconnection of some engine cylinders of work: a. correction of control side edges, b. adopting of injection advance angle, c. complete plunger correction.

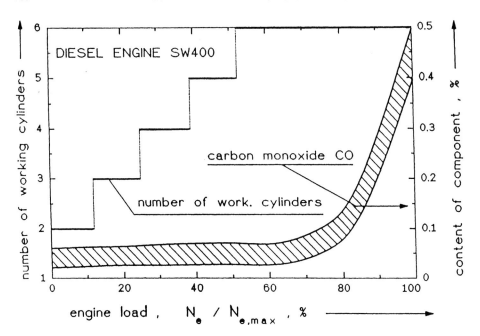

FIGURE 4 Number of working (disconnected) cylinders and effects of application of injection pump correction.

cylinders receive greater fuel charge but in the range of smallest content of carbon monoxide CO in exhaust gases.

The disengaged cylinders compress and blow out clean air to the exhaust manifold forming additional load for the working cylinders and causing addition dilution of gases leaving the exhaust system. To avoid it the oblique exhaust gas recirculation is proposed.

AUTOMATIC ADOPTING OF THE INJECTION ADVANCE ANGLE

The injection advance angle should be in keeping with the fuel charge mass m_j. Increase of the fuel charge mass m_j should accompany growth of the injection advance angle φ_a, what is connected and results from the fuel combustion law (Postrzednik *et al.*, 1992) in the engine combustion chamber.

The influence of the injection advance angle φ_a on the content of the toxic substances in exhaust gases is presented on the Figure 5 and Figure 6. The measurements were provided for the diesel engine SW400 on the operation (full load at the engine speed $\dot{n}_s$) characteristics. Automatic change of the injection advance angle, in dependence on the instantaneous engine load and fuel charge, can be realized – Figure 3b, by the proper shaping of the top edges of the injection pump plungers.

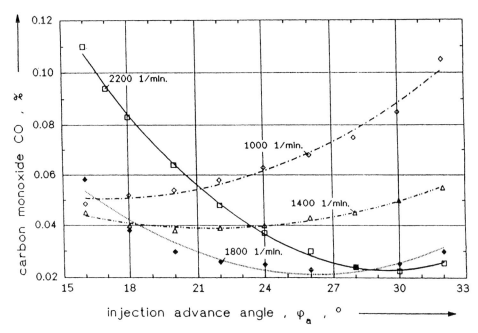

FIGURE 5   Effect of the injection advance angle and operation parameters on the content of carbon monoxide.

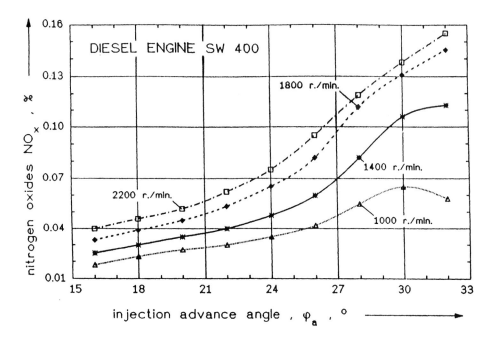

FIGURE 6   Effect of the injection advance angle and operation parameters on the content of nitrogen oxides.

EFFECTS OF THE GLOBAL INJECTION PUMP CORRECTION

The essence scheme of the global injection pump correction, which encloses and leads to:
– partially disconnection of some cylinders of work,
– adopting of the injection advance angle,
is shown on the Figure 3c.

The expected effects which should be achieved through this modification can be determined from the results presented on the Figures 4, 5 and 6.

Results of preliminary investigations show the positive effects of applied method and achieved solutions. Total emission of toxic components (carbon monoxide CO, nitrogen oxides NO_x) can be significantly reduced, more than 20% of emission value for serially produced engines SW400.

CONCLUSIONS

Investigation carried out revealed, that in idle running and low diesel engine load considerable reduction of the exhaust gases emission can be achieved through partially disconnection of some cylinders of work. In such a way the number of operating cylinders is adapted to an instantaneous engine load. Through partially disconnection of some cylinders of work and structural modifications of the injection pump the best effects can be achieved in the idle running and at a low engine load.

Significant reduction of the pollutants in exhaust gases (carbon monoxide CO, nitrogen oxides NO_x) is expected in the whole operating range of the investigating engines.

There are some new problems connected with the technique partially disconnection of diesel engine cylinders, for instance:

– thermal load and efficiency of the diesel engine,
– quality of the combustion during the transient period,
– power, torque characteristics and the behaviour of the engine under transient conditions,

which will be the main objective of investigations in the next stages.

Serially produced diesel engines (SW400, 6C107), besides their numerous operation benefits for a wide range of industrial application, will be on this way characterized as environment friendly at all and low exhaust gas toxicity engines.

REFERENCES

Bayhan, M. and Postrzednik, S. (1993) Methods for Reduction of NO_x Emissions in Internal Combustion Engines and Connection to Sound Control. *ENSEC '93 - Conference "Energy Systems and Ecology"*, Cracow.
Ciesiolkiewicz, A., Postrzednik, S. and Siurek, J. (1992) Method of Diesel Engine Exhaust Gases Toxicity Diminishing in a Way of Correction of the Injection Pump. *International Science Conference on Internal Combustion Engines KONES '92*, Wroclaw-Szklarska Poreba.

Lausch, W., Fleischer, F. and Maier, L. (1993) Möglichkeiten und Grenzen von NO_x - Min-
    derungsmaßnahmen bei Viertakt - Großdieselmotoren. Motortechnische Zeitschrift, No. **2**, 54.
Nagai, T. and Kawakami, M. (1990) Study into Reduction of NO Emission in Medium - Speed Diesel
    Engines. Technical Papers ISME, Kobe 1990.
Postrzednik, S. and Żmudka, Z. (1991) Evaluating and Analysis of the Filling Ratio of the Internal
    Combustion Engine. 3rd International AUTEC Congress for Automotive Engineering, Stuttgart.
Postrzednik, S., Żmudka, Z. and Tiamiyu, K. M. (1992) Using the Measurement of the Exhaust Gas
    Composition of Diesel Engines for Determination of the Filling Ratio and Emission Factors of Toxic
    Substances. *International Symposium CIMAC '92*, Warsaw.
Schoubye, P. (1992) Reduction of NO Emissions from Diesel Engines. *International Symposium CIMAC '92*,
    Warsaw.
Tandara, V. (1991) Einfluss der okologischen Aspekte auf die Entwicklung der Verbrennungsmotoren. 3-rd
    International AUTEC Congress for Automotive Engineering, Stuttgart.

Efficiency of Spark-Ignition Engine Fuelled with Methanol-Gasoline Blends

ANDRZEJ KOWALEWICZ   *Politechnika Radomska im.gen. Kazimierza Pulaskiego Al. Boleslawa Chrobrego 45 26–600 Radom, Poland*

Abstract—Though many papers have been published on application of methanol as a fuel for S.I. engines, a review of which has been done by the present author (1993), the problem of engine overall efficiency has not been fully recognized yet.

In this paper some results of investigations focused on efficiency of S.I. engine fuelled with methanol-gasoline blends are presented. The experiments were carried out in Radom Technical University on engine of Polonez car produced in Poland in a large range of load and speed with the application of methanol-gasoline blends having from 0 to 60% of methanol fraction by volume.

Key Words: Alternative fuel, methanol engine

INTRODUCTION

Methanol was recognized a long time ago as a good fuel for spark-ignition engines mainly due to its high resistance to knocking combustion. Recent experiments have shown that engines fuelled with neat methanol or methanol-gasoline blends have less emissions of NO_x, CO, HC and particulates, although aldehydes emission is higher in comparison with fuelling with neat gasoline.

Also according to some publications in this case overall engine efficiency may by higher, although these results are incomplete because experiments were carried out not in the whole range of engine speed and load. That inclined the present author to perform the work, which is intended to explain fully the problem of overall efficiency of the engine fuelled with methanol-gasoline blends. The first stage of this work (reported herein) covers experimental investigation, the second one (now in progress) - theoretical approach for explanation of experimental results.

Recent review of methanol application as a fuel for C.I. engines was carried out by the present author for C.I. engines (1992). In these publications he also discussed the problem of necessary engine modifications for fuelling with methanol fuels. As far as S.I. engines are concerned, application of methanol-gasoline blends up to 5% by vol. of methanol needs not any modifications and up to 15% only small ones. Higher fractions of methanol in the blends demand some engine modifications (presented in Tab. I) due to two main reasons of another nature:

— methanol calorific value is lower than that of gasoline,
— methanol causes corrosion of engine materials and degradation of oil lubrication properties.

TABLE I

Fuel systems and engine modifications for methanol-gasoline application

Fuel systems	
Fuelling with Methanol-Gasoline Blends (with addition of Solubility Improver)	Carburettor system or Injection System
Dual Fuel System	Two Separate Carburettors or Injection Systems (Stratified Charge)
Engine Modificatins Turning of Carburettor or Injection System Adaptation of Cold Start-up and Warm-up Devices Modification of Carburettor or Injection Fuel System to Avoid Vapour Locks in Fuel Lines Changing Materials in Fuel System	

Also there are some shortcomings with phase separation of gasoline and methanol in presence of water, due to which some third agent should be added to the blend. Problems related to application of methanol-gasoline blends to S.I. engines and possible solutions are given in Table II.

Methanol prices in comparison with other fuel costs of introduction of methanol fuels into the market were presented by Sperling (1990). At present methanol fuel prices in Europe and USA are higher than these of conventional fuels, but in the case of mass production of methanol the prices will equalize.

At Radom Technical University the work on application of methanol as a fuel for internal combustion S.I. engines (as well as C.I. engines) is being performed for several years. Some results of this work focussed on efficiency of S.I. engines are reported herein.

TABLE II

Problems related to application of methanol-gasoline blends and possible solutions

Problem	Solution
Stability of Methanol-Gasoline Blend	• Additives Preventing Phase Separation (Solubility Improvers)
Prevent Phase Separation	• "Dry" Handling
Insufficient Resistance of Structural Materials to Methanol Corrosion	• Anticorrosive Coatings of Fuel System Surface
Higher Volatility at the Beginning of Evaporation	• Modification of Gasoline Composition
Higher Fuel Mass-Flow Rates	• Enlargement of Cross-Sectional Areas or Nozzle Orifice of Fuel System

OBJECTIVES OF THE WORK

The objectives of the investigations reported in this paper are:
- to measure the overall efficiency of the S.I. engine fuelled with methanol-gasoline blends,
- to find the dependance of the overall engine efficiency on the methanol fraction in the blend.

Additional objective of the paper is to discuss the increase of thermal efficiency of the engine fuelled with methanol-gasoline blends in comparison with that of engine fuelled with neat gasoline and to propose further work (theoretical mainly) on this problem.

EXPERIMENTS

Test Stand

The experiments were performed on Polonez engine. The objective of the experiments was to measure engine overall efficiency in the case of fuelling with methanol-gasoline blends. The schematic of the test-stand in Figure 1. Production version of Polonez engine ($V_{ss} = 1,492 \, \text{dm}^3$, $N = 55,2 \, \text{kW}$ at $n = 5200$ rpm) was modified to methanol

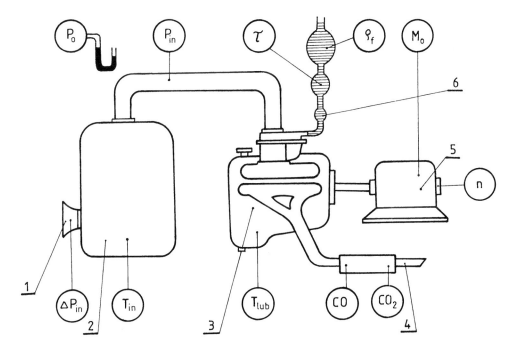

FIGURE 1   Schematic of test-stand:1-lemniscate flow meter, 2-inlet surge tank, 3-engine tested, 4-exhaust duct, 5-torque meter, 6-fuel consumption meassuring vessel.

fuelling: instead of standard fuel nozzle, with constant cross-section area, a special one with continuously controlled cross-section of flow-area was installed. The engine operated only in the range of partial load and speed, so for the sake of simplicity one throttle was fixed. Except these modifications no others were made. Engine was fuelled with methanol-gasoline blends prepared just before experiments.

Measurements and Computations

The following quantities were measured:

– engine speed and load by the use electronically controlled hydraulic brake HP 75 (Polish production),
– fuel consumption by the use of volumetric method,
– air flow with the use of lemniscate flow-meter,
– CO and CO_2 concentrations in the engine exit manifold by the use of AVL Model 465.

The following methanol-gasoline blends were applied: M10, M20, M30, M40, M60, where the number denotes methanol volumetric fraction X in the blend and for comparison, reference fuel-neat gasoline (ON 94).
    For each fuel blend load characteristics were made for each constant speed of: 2000, 2500, 3000 and 4000 rpm (altogether 20 characteristics).
In order to exclude the influence of other parameters except these of the fuel:

– ignition advance was kept unchanged for all the fuel blends,
– load characteristics for one constant speed were made in the same day,
– the content of CO and CO_2 in exhaust gases was measured, in order to control air excess ratio of the mixture.
– all experiments were done for the same thermal state of the engine (temperature of the lubricating oil was fixed constant).

For computations the following formulae were used. Air excess ratio:

$$\lambda = \frac{m_{air}}{m_f L_{th}} \tag{1}$$

Where L_{thb} is stoichiometric air for the blend, expressed as follows:

$$L_{thb} = mL_{thm} + gL_{thg} \tag{2}$$

Air flow rate was normalized to standard athmosferic temperature and pressure as follows:

$$m_{air} = m_{air\,meas} \frac{1013}{p} \sqrt{\frac{T}{288}} \tag{3}$$

Methanol mass fraction m and gasoline mass fraction g in the blend were computed from the following formulae:

$$m = \frac{X\rho_m}{X\rho_m + (1-X)\rho_g} \tag{4}$$

$$g = 1 - m \qquad (5)$$

where (at 20 °C.):

$\rho_m = 790 \text{ kg/m}^3$,

$\rho_g = 740 \text{ kg/m}^3$,

X - methanol volumetric fraction in the blend.

Calorific value of the blend:

$$W_u = W_{ug} g + W_{um} m \qquad (6)$$

For computation $W_{ug} = 42 \times 10^3$ kJ/kg and $W_{um} = 19.8 \times 10^3$ kJ/kg were taken.

Engine efficiency:

$$\eta = \frac{1}{\dfrac{m_f}{N} W_u} 100\% \qquad (7)$$

RESULTS

As the result of experiments the load characteristics of the engine were obtained, from which it is evident that the higher the content of methanol in the blend, the higher the engine efficiency – Figure 2. Maximum of engine efficiency fuelled with M60 fuel reaches 31,0% whereas for neat gasoline M0 only 28,5% (i.e. 8,8% gain for M60 fuel). This positive difference in efficiency of the engine operating at lower load than for $\eta = $ max is also evident. The same results may by drown from speed characteristic, Figure 3. In the range of speed from 2000 rpm to 3500 rpm for $M_o \approx M_{o\eta = max}$ the higher methanol content in the blend, the higher the engine efficiency.

During the tests, the engine was fuelled with all the fuels operated at similar air excess ratio $\lambda \approx$ const. The lowest difference between λ for M0 and this for any blend was observed in the case of the lowest methanol content M10 (Fig. 4), the highest — for the highest methanol content M60 (Fig. 5). Higher values of λ are for M60 than for M0 (neat gasoline). Emission of CO_2 for methanol fuel is lower in comparison with gasoline: the higher the methanol content, the lower the CO_2 emission. Emission of CO is similar.

DISCUSSION

The increase in engine efficiency with increase of methanol fraction in the blend results from the following reasons.
1. First of all, according to Hirano et al. (1981), methanol-air mixtures burn faster than gasoline-air ones and due to that for fuelling with methanol more heat is evolved near TDC resulting in higher pressure than for gasoline. This fact results in closer

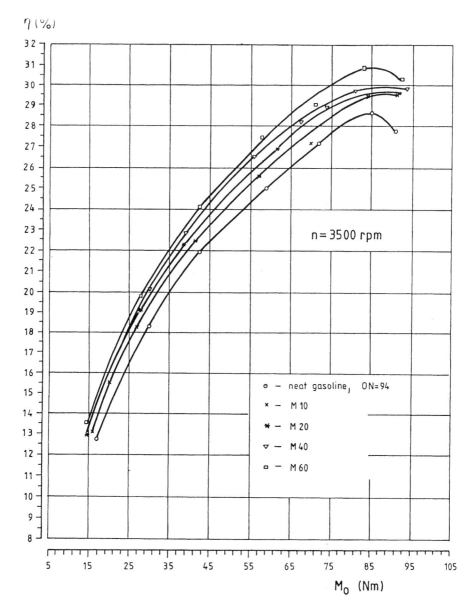

FIGURE 2   Overall efficiency of the Polonez engine fuelled with neat gasoline and methanol fuels: M10, M20, M40 and M60 vs load. Engine speed $n = 3500$ rpm.

proximity of the real thermodynamic cycle to theoretical one (i.e. Otto cycle). Better results may by obtained when ignition is a little retarded (and optimized) than for gasoline fuelling.

2. Lower compression pressure measured e.g. for S.I. engine by Różycki (1992) and for C.I. engine by Kowalewicz and Luft (1993) for fuelling with methanol fuels, resulting in lower work of compression.

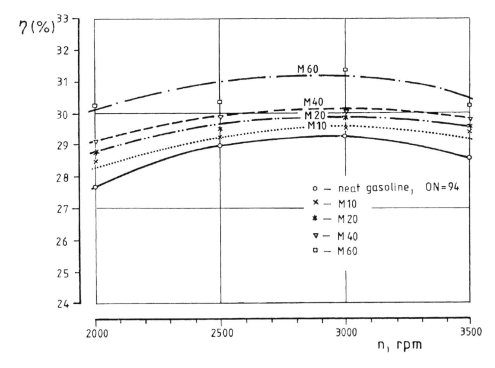

FIGURE 3   Maximum overall efficiency of Polonez engine vs speed for fuelling with neat gasoline and
methanol-gasoline blends respectively. On each curve torque is constant and very close to maximum
($M \approx 85\cdots90$ Nm).

3. During combustion of methanol more moles of gases evolve than for hydrocarbon
   fuels due to higher molar conversion factor, what results in higher pressure during
   expansion stroke and – in consequence – greater work of expanding gases.
4. Lower bulk temperature level due to lower compression pressure and lower
   combustion temperature results in:

   – lower heat loss to the walls, i.e. lower cooling losses,
   – lower exhaust gas temperature, i.e. lower exhaust losses.

5. Due to faster combustion of methanol-air mixtures less fuel is unburnt, i.e. there are
   lower losses of incomplete combustion, what was investigated by Trzaskowski and
   the present author (1990).

CONCLUSIONS

It has been experimentally examined in the large range of the engine load and speed
that methanol engine is more efficient than gasoline one.
The higher methanol content in the blend, the higher engine efficiency.

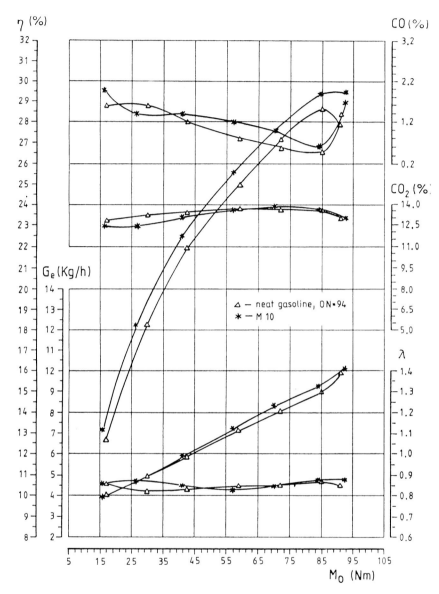

FIGURE 4   Load characteristics of the Polonez engine with gasoline and M10 fuel. Engine speed $n = 3500$ rpm.
- gasoline, ON = 94.
- M10 fuel.

The higher methanol content in the blend, the lower CO_2 emission, what means that methanol fuel contributes less to greenhouse effect.

It is expected that a further increase in the engine efficiency may be obtained when engine spark timing is optimized for each content of methanol in the blend.

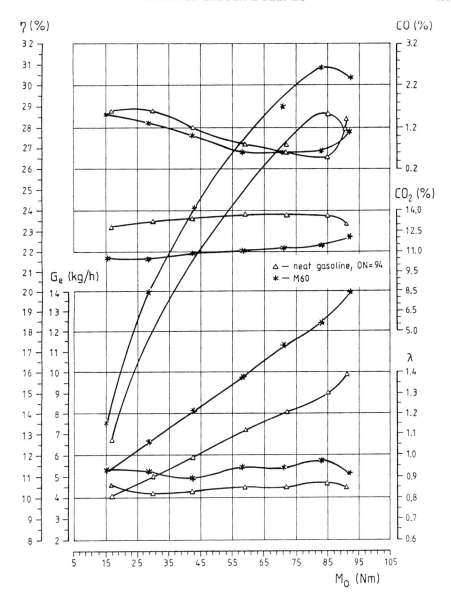

FIGURE 5   Load characteristics of the Polonez engine with gasoline and M60 fuel. Engine speed $n = 2500$ rpm -gasoline, ON $= 94$ -M60 fuel.

More work is needed to clarify the causes of higher efficiency of the engine fuelled with methanol in comparison with fuelled with gasoline.

ACKNOWLEDGEMENT

The paper is based on the diploma work of Mr. A. Gielniewski performed with the kind help of Mr. M. Gola under the author's direction in Radom Technical University.

NOMENCLATURE

G_e	–	fuel consumption, kg/h
g	–	gasoline mass fraction in the blend
L_{th}	–	stoichiometric air/fuel ratio, air kg/fuel kg
M_o	–	engine torque, Nm
m	–	methanol mass fraction in the blend
m_{air}	–	mass flow rate of air
m_f	–	mass flow rate of fuel
N	–	engine power, kW
n	–	engine speed, rpm
p	–	atmospheric pressure, bar
T	–	ambient temperature °C
W_u	–	caloritic value of fuel, kJ/kg
η	–	engine overall efficiency
ρ	–	density

Subscripts

b	–	blend
g	–	gasoline
m	–	methanol

REFERENCES

Hirano, M., Oda, K., Hirano, T. and Akita, K. (1981) Burning velocities of methanol-air-water gaseous mixtures. *Combustion and Flame*, 1981, **40**, 341–343.

Kowalewicz, A. (1990) Methanol as a fuel for C.I. engines. *International Symposium on Small Diesel Engines*. Warsaw'92.

Kowalewicz, A. (1993) Methanol as fuel for spark ignition engines: a review and analysis. *Proc. Instn. Mech. Eng.*, **207**, D00291, 43.

Kowalewicz, A. and Luft, S. (1993) Methanol fuelled with heat recovery. *Twenty-sixth ISATA, Aachen*, Sept. 1993.

Rózycki, A. (1991) Measurement of pressure vs C.A. for different fractions of methanol in methanol-gasoline blends. Work of I.C. Engines *Dept. Radom Technical University* (Unpublished)

Trzaskowski, A. and Sperling, D. (1990) New transportation fuels. University of California Press, Berkeley. *M.A. Thesis* carried out under direction of the present author in Dept. of I.C. Engines of Radom Technical University.

Measurement of the Exhaust Emissions from Otto-Engines Equipped with a Three-Way Catalyst Regarding the Secondary Emission of Nitrous Oxide

CHRISTOPH WEHINGER and ROLAND MEYER-PITTROFF *Department for Energy- and Environmental Technique for the Food Industry Technical University Munich D-85350 Freising-Weihenstephan-Germany*

Abstract— This paper presents measurements of vehicular nitrous oxide emission taken on a chassis dynamometer. The tested vehicles were equipped with a closed-loop three-way-catalyst. Tests were carried out in neutral gear and at constant velocities with defined load of the dynamometer. The N_2O-concentration was measured with a gas-chromatograph online in the hot, moist exhaust gas. Measurements were taken in front of as well as behind the catalytic converter. The N_2O-concentration maxima in front of the catalytic converter reached approx. 5 ppm N_2O, those behind the catalytic converter approx. 35 ppm N_2O. The emitted amount of N_2O was dependent on the air/fuel-ratio, on the catalyst inlet temperature and on mileage.

Key Words: Nitrous oxide, three-way-catalyst, NSCR-method, secondary emission

INTRODUCTION

The present technical standard for reducing vehicular emissions is the closed-loop three-way-catalyst (TWC). With the non-selective catalytic reduction-procedure (NSCR-procedure) the pollutants carbon monoxide (CO) and hydrocarbons (HC) are minimized through oxidation reactions. At the same time, nitrogen oxides (NO_x) are reduced. Therefore, strict adherence to the stochiometric combustion, which is ensured by the closed-loop TWC, is essential. Besides the desirable reduction of the pollutants CO, NO_x and HC, new pollutants, the secondary emissions, as hydrogen sulfide (H_2S) and nitrous oxide (N_2O) are formed at the catalyst surface. The formation of N_2O was proved by investigations of Prigent and de Soete (1989), Zajonts *et al.* (1991), Engler *et al.* (1991), Cho *et al.* (1989) and Mc Cabe and Wong (1990) with catalytic converters in laboratory reactors, by measurements on engine dynamometers and with vehicles on chassis dynamometers.

There is uncertainty concerning the amount of N_2O emitted by traffic. Different numbers have been published. Dasch (1992) determined a worldwide annual vehicular emission of 200.000 t N_2O/a. On the base of a total anthropogenuous N_2O-emission of $6,1 \cdot 10^6$ t N_2O/a, the proportion of traffic amounts to 3,3 %. De Soete (1990) specifies the worldwide vehicular N_2O-emission as 900.000 t N_2O (as N)/a. This equals $1,4 \cdot 10^6$ t N_2O/a. Relative to the total anthropogenuous N_2O-emission of $6,1 \cdot 10^6$ t N_2O/a mentioned by Dasch, the proportion of traffic amounts to 23,0 %.

Measurements concerning the vehicular N_2O-emission do exist, but they were taken during special driving cycles, where no constant operating conditions were reached.

With these cycles the determination of the parameters leading to the formation of N_2O is not possible. To obtain constant operating conditions, tests in neutral gear and at constant velocities of 50, 80, 100 and 120 km/h with defined dynamometer-load were carried out.

IMPORTANCE OF NITROUS OXIDE

N_2O is a trace gas that contributes to the greenhouse effect and thus to the global warming of the earth. The N_2O-contribution to the greenhouse effect is approx. 5 %. Furthermore, it is involved in catalytically destroying the ozone layer by forming NO-radicals.

The sources of N_2O can be divided into biogenuous and anthropogenuous sources. The most important biogenuous source is the natural activity of bacteria in soils and oceans (nitrification and denitrification). Important sources of the anthropogenuous N_2O-formation are:

– Decomposition of nitrogen fertilizers in soils
– Combustion of fossil fuels
– Reduction procedures for nitrogen oxides (NSCR- and SNCR- procedure)

The present concentration of N_2O in the atmosphere is estimated at 311 ppb by the German Department for Environment (Umweltbundesamt) (1992). The annual growth-rate is about $0,2 - 0,3$ % and is due to anthropogenuous sources.

FORMATION OF N_2O

N_2O is formed as an intermediate product on the surface of the catalytic converter during the reduction of NO by CO.
NO reacts with the N-atom, which results from the NO-decomposition:

$$NO + N \rightarrow N_2O \tag{1}$$

The reaction of N_2O-decomposition is:

$$N_2O \rightarrow N_2 + O \tag{2}$$

The oxygen atom reacts with CO to form CO_2. Equations (1) and (2) are only two special reactions out of a complex adsorption-, desorption- and reaction-mechanism. Prigent and de Soete (1989) and Cho et al. (1989) describe this mechanism with the following equations in summarized form:

$$NO + CO \rightarrow 1/2\ N_2 + CO_2 \tag{3}$$

$$2\ NO + CO \rightarrow N_2O + CO_2 \tag{4}$$

$$N_2O + CO \rightarrow N_2 + CO2 \tag{5}$$

Equation (3) describes the reduction of NO by means of CO. The formation of N_2O according to equation (4) occurs for temperatures below the light-off-temperature. The light-off-temperature characterizes the temperature, where a reduction-rate of 50% is reached. Cho *et al.* (1989) determine the NO-light-off-temperature at a Rhodium-catalyst to be approx. 210°C.

The concentration of N_2O is increasing on account of the formation corresponding to equation (4). With increasing temperature, the decomposition of N_2O starts according to equation (5). At higher temperatures NO and CO form N_2 and CO_2 the desired reduction, equation (3), is active. Laboratory investigations of Shelef and Otto (1968), Cho and Shanks (1989) and Dasch (1992) showed N_2O-maxima on a Platinum catalyst at 300°C, on a Rhodium catalyst at 280°C and on a Pt/Rh-catalyst at 230°C.

MEASUREMENT TECHNIQUE

For the analysis of N_2O a gas chromatograph (GC) Sichromat 1, produced by Siemens, was applied. The N_2O-concentration was measured online in the hot, moist exhaust gas. The detection limit for N_2O is 1,5 ppm. Before starting the tests, the calibration of the GC was checked with calibration gas (26 ppm N_2O in N_2, accuracy ± 2 %). The operating conditions of the GC are listed in Table I.

For the measurement behind the catalytic converter, the exhaust gas was taken directly after leaving the exhaust pipe, for the measurement in front of the catalytic converter, it was taken from the exhaust manifold. The gas was brought to the analyzers by heated transfer lines. Figure 1 shows the measuring system.

EXPERIMENTS

The influences of the following parameters concerning the N_2O-emission were investigated on a chassis dynamometer:

– Age of the catalytic converter (mileage)
– Velocity
– Air/fuel-ratio
– Catalyst inlet temperature

TABLE I

Operating conditions FOR the GC

carrier gas	hydrogen
column temperature	40°C
columns	GSQ, 30 m
	molecular sieve 5 Å, 20 m
detector	2 thermal conductivity
	detectors
detector temperature	120°C

TABLE II

Used analyzers and technical data

measuring instrument	manufacturer	measured component	measuring range mg/m^3
Flame-ionisation-detector Compur FID	Bayer Diagnostic GmbH, München	$C_{organic}$	0–100·000 6 measuring ranges
NDIR-multicomponent-process-photometer MCS100	Bodenseewerk Perkin Elmer GmbH, Überlingen	NO*	O–1500
		CO*	0–750
		NH_3	0–700
		SO_2	0–3000
		N_2O	0–500
			% by vol.
		CO_2^*	0–20
		H_2O	0–20
Oxygen analyzer Oxynos 100	Rosemount GmbH & Co, Hanau	$O_2^†$	0–25

*For the components CO, NO and CO_2 a non-dispersive-infrared analyzer (NDIR) was applied.
†For O_2, a special oxygen analyzer was used. For the detection of HC a flame ionization detector.

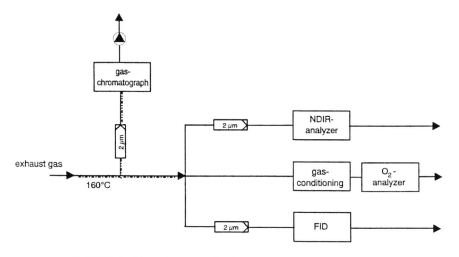

FIGURE 1   Measuring system for determining exhaust gas emissions.

The influence of the age of the catalytic converter was analyzed by testing cars of identical type but different mileages. The vehicles and their technical data are listed in Table III.

The cars were examined without prior control of functioning of the closed-loop TWC, so that real traffic conditions are reflected. The proposed inspections were made.

To determine the influence of velocity the experiments were carried out in neutral gear and at constant velocities at defined load of the dynamometer for 50, 80, 100 and 120 km/h. These velocities are supposed to simulate real conditions for city, country and highway traffic.

TABLE III

Tested vehicles, equipped with a closed-loop TWC and technical data

manufacturer	model	type	power	engine displacement	mileage
			kW	cm³	km
BMW	1992	520i, 24 valves	110	1991	16.000
	1991	520i, 24 valves	110	1991	77.000
	1991	520i, 24 valves, automatic	110	1991	95.000
VW	1992	Golf	55	1781	13.000
	1991		40	1272	100.000
Mercedes-Benz	1991	300SE, 24 valves	170	3199	16.000

The adjustment of the dynamometer-load was chosen according to the regulations of the present legal test cycle 91/441/EC in the European Community for the license of new automobiles. The correlation of dynamometer load and velocity is shown in Figure 2 for the tested cars.

The test was divided into two phases, a warm-up phase and a measuring phase. During the warm-up phase stable conditions were reached. A motor-oil temperature of 90°C was chosen as indicator for stable conditions. The following measuring phase lasted 300s. During this phase the revolutions per minute, the oil temperature, exhaust gas temperature and the measured values of the NDIR- and the FID-analyzer were registered.

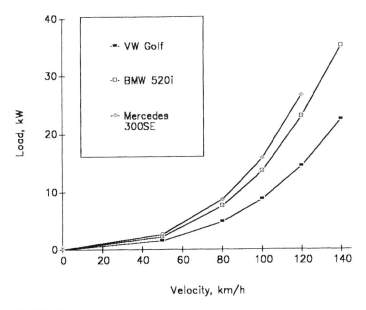

FIGURE 2   Correlation of dynamometer load and velocity for the tested cars.

RESULTS

All concentrations mentioned below are on the basis of dry gas.

Influence of Mileage and Velocity

The highest N_2O-concentration in front of the catalytic converter for all three BMW vehicles was 4 ppm at a velocity of 80 km/h for the vehicle with a mileage of 16.000 km. The maximum emission of 6 ppm N_2O for the VW vehicles was obtained at a velocity of 80 km/h for a mileage of 100.000 km.

The correlations of concentrations measured behind the catalyst and velocity are shown in Figure 3 for the BMW vehicles with different mileages. The presented concentrations are mean values of 2–5 single measurements.

In neutral gear the N_2O-emission is small at approx. 3 ppm, the emission is mounting with increasing velocity. Maxima are reached at a velocity of 50 km/h, independent of mileage. In the order of increasing mileage the highest N_2O-concentration is 12 ppm, 34 ppm and 21 ppm. For higher velocities, the concentration is decreasing. From 100 km/h on the maximum is approx. 5 ppm N_2O.

For the vehicle with a mileage of 77.000 km, the N_2O-concentration is higher than for the vehicle with a mileage of 95.000 km. The same effect has been observed for the CO-emission. The NO-emission, however, is lower. A reason for this effect could be a stronger thermal ageing of the Platinum surface caused by the operating conditions. A chemical or physical poisoning of the catalyst caused by fuel additives is also possible.

This effect did not occur with the other tested cars. For the models of VW and Mercedes Benz (MB), the correlation of N_2O-concentration and velocity has the same

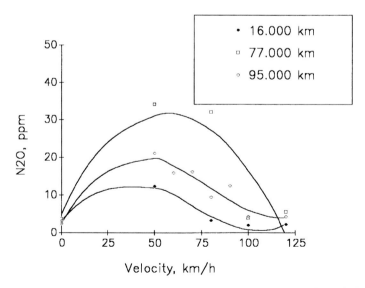

FIGURE 3   Correlation of N_2O-concentration and velocity for different mileages (BMW 520i).

graph as shown in Figure 3. The N_2O-maximum of 19 ppm for the Golf with a mileage of 13.000 km was reached at a velocity of 100 km/h, the one for the Golf with a mileage of 100.000 km at 42 ppm at a velocity of 50 km/h. The maximum of 10 ppm for the Mercedes-Benz was measured at a velocity of 80 km/h, but in neutral gear a relatively high concentration of 9 ppm N_2O was obtained.

Influence of the Air/fuel-Ratio

The most important influence parameter for engine-combustion is the air/fuel-ratio, expressed in terms of the Lambda value λ. The Lambda value is defined as quotient of supplied air quantity and stochiometric air quantity. With a computer program, written for the analysis of the tests, the calculation of the Lambda value λ was examined. Therefore, the equation of Brettschneider (1979) was used.

Figure 4 shows the correlation of N_2O-concentration and Lambda value λ for the BMW vehicles with different mileages. For all three automobiles, marked N_2O-maxima, dependent on λ, appeared. For the vehicle with a mileage of 16.000 km, the maximum is 8 ppm N_2O at $\lambda = 1,000$. For the car with a mileage of 77.000 km, the highest N_2O-concentration of 40 ppm is reached at $\lambda = 1,003$. For the car with a mileage of 95.000 km a concentration of 30 ppm N_2O is reached at $\lambda = 0,998$. All the vehicles show N_2O-concentration maxima close to the stochiometric gas mixture of $\lambda = 1,00$.

The correlation of N_2O-emission and Lambda value is confirmed by the results obtained by measuring the cars of VW and the one of Mercedes-Benz. The maxima appeared at approx. 12 ppm for $\lambda = 1,001$ (Golf, 13.000 km), at 40 ppm for $\lambda = 1,004$ (Golf 100.000 km) and at approx. 10 ppm for $\lambda = 1,001$ (MB 300SE).

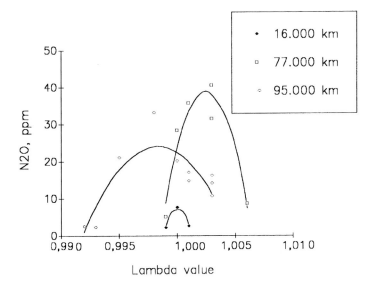

FIGURE 4   Correlation of N_2O-emission and Lambda value λ for different mileages (BMW 520i).

Influence of the Catalyst Inlet Temperature

The exhaust gas temperature at the inlet of the catalytic converter is about 300°C in neutral gear (680 revolutions/minute) and 850°C with full load (5500 revolutions/minute) for the BMW vehicles. It has been assumed, that the correlation between catalyst inlet temperature and engine load, expressed in revolutions/minute, is a linear one. This assumption could be proved indirectly by measuring the exhaust gas temperature after leaving the exhaust pipe. To achieve this, a thermocouple was used. A linear correlation of the measured temperature and engine load was established.

By means of the measured revolutions/minute the velocity was related to catalyst inlet temperature. The resulting correlation of N_2O emitted at different velocities and catalyst inlet temperature is shown in Figure 5 for the BMW vehicles with different mileages. There is a marked maximum for the N_2O-emission at a gas temperature of about 480°C, independent on mileage. The N_2O-formation starts at about 300°C, reaches the maximum and is decreasing for higher temperatures, according to the decomposition of N_2O. At a catalyst inlet temperature of about 650–700°C a complete decomposition of N_2O is reached.

In comparison to the above mentioned laboratory investigations the N_2O-maximum occurs at much higher temperatures. The difference is about 250 K. Engine dynamometer-tests performed by Prigent and de Soete (1989) confirm this difference. The N_2O-maximum appeared at a catalyst inlet temperature of 360°C with a Pt/Rh-catalyst. The difference could be due to the altered atmosphere of real exhaust gas and synthetic gas mixtures used for the laboratory experiments.

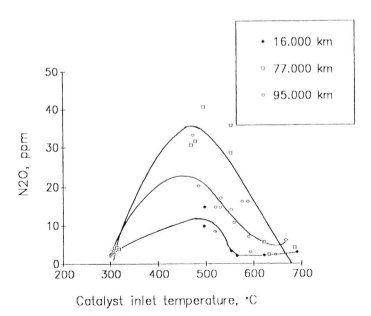

FIGURE 5   Correlation of N_2O-emission and catalyst inlet temperature for different mileages (BMW 520i).

SUMMARY

The influence of the parameters as velocity, age of the catalytic converter, air/fuel-ratio and catalyst inlet temperature on the N_2O-emission was investigated on a chassis-dynamometer. Therefore, tests in neutral gear and at constant velocities with defined dynamometer-load were performed with cars of identical type but different mileages.

The N_2O-concentration was determined with a gas chromatograph online in the hot, moist exhaust gas in front of and behind the catalytic converter.

The highest N_2O-concentration in front of the catalytic converter was 4 ppm N_2O for the BMW vehicles and 6 ppm N_2O for the VW vehicles.

The N_2O-concentration behind the catalytic converter shows a marked maximum dependent on the velocity. For the BMW with a mileage of 77.000 km the maximum yield is 34 ppm for a velocity of 50 km/h, for the Golf with a mileage of 100.000 km 42 ppm N_2O are reached for a velocity of 50 km/h. All vehicles but one showed that the N_2O-emission mounts with the increasing age of the catalytic converter, assuming that higher mileage corresponds to higher usage.

The influence of the air/fuel-ratio concerning the N_2O-emission was determined. Maxima near the stochiometric gas mixture ($\lambda = 1,0$) were analyzed. Analysing the influence of the catalyst inlet temperature on the N_2O-concentration a clear maximum at a temperature of about 480°C is evident.

The investigations confirm the formation of the secondary emission N_2O at the catalyst surface. N_2O represents an intermediate product of the NO-reduction in the process of forming N_2. The N_2O-amount emitted is dependent on operating conditions like engine load, air/fuel-ratio and mileage.

REFERENCES

Brettschneider, J. (1979) Berechnung des Luftverhältnisses von Luft-Kraftstoff-Gemischen und Einfluß von Meßfehlern auf λ. In *Bosch, Technische Berichte*, **6**, **4**, pp. 177–186.

Cho, B. K., Shanks, B. H., and Bailey, E. (1989) Kinetics of NO Reduction over Supported Rhodium Catalysts: Isotopic Cycling Experiments. *Journal of Catalysis*, **115**, 486–499.

Dasch, J. M. (1992) Nitrous Oxide Emissions from Vehicles. *J. Air Waste Manage. Assoc.*, **42**, 63–67.

De Soete, G. G. (1990) What is burning about N_2O? *Ijmuiden Newsletters*, **1**, S. Doc. No. k 60/y/19.

Engler, B., Koberstein, D., Lindner, D. and Lox, E. (1991) The Influence of Three-Way Catalyst Parameters on Secondary Emission. In Crucq, A. (ed.), *Catalysis in Automotive Pollution Control II*, Elsevier Science Publishers B. V., Amsterdam, pp. 641–655.

Mc Cabe, R. W. and Wong, C. (1990) Steady-State Kinetics of the CO-N_2O Reaction Over an Alumina-Supported Rhodium Catalyst. *Journal of Catalysis*, **121**, 422–431.

Prigent, M. and de Soete, G. G. (1989) Nitrous Oxide N_2O in Engine Exhaust Gases- A First Appraisal of Catalyst Impact, *Society of Automotive Engineers*, SAE-Paper No. 0148–7191.

Shelef, M. and Otto, K. (1968) Appearance of N_2O in the Catalytic Reduction of NO by CO, *Journal of Catalysis*, **10**, 408–412.

Umweltbundesamt (1992) In *Daten zur Umwelt*, Erich Schmidt Verlag GmbH & Co., Berlin. Zajonts, J., Gutknecht, C. and Frey, V. (1991) Emission nichtlimitierter Abgasbestandteile aus Ottomotoren mit Katalysatoren. *Umweltbundesamt*, Zwischenbericht Nr. 10405508.

Experimental Study of NO_x Reduction by Impinging-Jet-Flame in a Closed Vessel with Opposed Dual Prechambers

IN-SEUCK JEUNG* and KYUNG-KOOK CHO   *Department of Aerospace Engineering, Seoul National University, Seoul 151-742, Korea*

Abstract—Series of experiments were carried out to elucidate NO_x reduction by employing impinging-jet-flame spouting into the main combustion chamber through the connecting orifice between the main chamber and the dual opposed prechambers in a closed vessel simulating automotive engine combustion. Different combustion chamber geometries and fuel types as well as various combustion processes were employed. High speed schlieren photographic visualization of flame propagation, peak chamber pressure, NO_x concentration were measured with varying volumetric ratio of the prechamber to the main chamber and diameter of connecting orifice between the chambers.

Results of experiments show that the case which has wider space at the central zone of main chamber, which may have better situation of well mixing, reduces much more NO_x emission than that of the case of narrower space, where the mixing process may be confined in somewhere locally, simultaneously achieving the high-load condition. But, type of fuel does little change combustion behavior or reduction of NO_x generation.

Key Words: NO_x, Impinging-Jet-Flame, constant volume bomb combustion

INTRODUCTION

Nitrogen oxides in the atmosphere contribute to photochemical smog, to the formation of acid rain precursors, to the destruction of ozone in the stratospere and to the global warming. The principal nitrogen oxides present in the atmosphere are nitric oxide (NO) and nitrogen dioxide (NO_2), collectively referred to as NO_x, and nitrous oxide (N_2O). A significant amount of the increased emissions is attributed to human activities, in particular to increased combustion of biomass and fossil fuels (Bowman (1992)).

Automotive engine combustion is one of major sources of NO_x emissions not only, but also CO_2 productions which also cause global warming. Over the past two decades, increasingly stringent regulations reducing allowable NO_x emissions from the automotives have been implemented, and CO_2 regulations are also being discussed, while improvement of fuel economy has been recently required. These requirements may be talked in other words as low NO_x, CO_2 formation simultaneously high-load combustion/unit fuel consumption or best fuel economy/unit power output or then minimum NO_x, CO_2 formation/unit power output.

But, it has been a general trend that high-load combustion is incompatible with low NO_x formation. To some extend, this trends is true for both premixed flames and diffusion flames, whether homogeneous mixtures or stratified mixtures, both

---

*Author for correspondence.

combustion in closed vessels and flowing systems. Recently, new combustion systems with various types of prechamber, swirl augumentation or swirl/tumble augumentation, and multi-spark systems, etc. have been developed to improve fuel economy with minimizing reduction of power output. In practice, examples of lean burn combustion or EGR reduce a fair amount of NO_x formation, but, on the other hand, it may sacrifice decrease of engine power output also.

Near stoichiometric burn operation should be favorable, as it can provide high-load operation, better fuel economy and besides low emission of CO and HC, if one can find a way to reduce NO_x formation simultaneously. Series of demonstration on high-load operation and low NO_x formation can be found in terms of the impinging jet/ spray/ flame (Miura, Tsukamoto, Kawagoe, Nakaoji, and Kaneko (1978), Fujimoto, Kaneko, and Tsuruno (1984), Jeong (1989), and Okajima and Kumagai (1990)). According to Fujimoto *et al.* (1984) and Okajima and Kumagai (1990), an original idea on the impinging-jet-flame proposed by Professor S. Kumagai was that 'some burnt gas is continuously supplied to the unburnt gas just adjacent to the flame front.' This idea may be referred to a certain kind of mixing process by intense turbulence, resulting improved better mixing and homogeneous reduced fluctuating rms value.

Here, we summarized previous experimental results of Fujimoto *et al.* (1984), who burnt methane-air premixed gas in a flat discal combustion vessel, those of Jeong (1989), who burnt propane-air premixed gas in a rectangular combustion vessel, and our present experiment burning propane-air premixed gas in a flat discal combustion vessel. All those equipped with dual opposed prechambers, which can produce a pair of impinging-jet-flame from the prechamber immersed in the main chamber that come into a head-on collision with each other impinging-jet at the central zone of the main chamber. And we attempted to draw a possible common explanation on the impinging-jet-flame characteristics.

EXPERIMENT

In order to investigate characteristic behavior of the impinging-jet-flame, a flat discal combustion vessel was constructed, and pressure measurement, NO_x measurement, and high speed schlieren photographic flame visualization were accomplished.

Figure 1 shows a view of the flat discal combustion vessel which consists of a main chamber and dual prechambers. A case when none of prechamber is installed, the case is said to be the case of *Laminar-Flame and Ordinary Turbulent-Flame Combustion*. And a case when only one prechamber is installed, the case is said to be the case of *Single-Jet-Flame Combustion*. And finally a case when dual opposed prechambers are installed, the case is said to be the case of *Impinging-Jet-Flame Combustion*. This vessel of aluminum, 120 mm in diameter, 40 mm in depth and about 425 cc in volume, has two optical glass windows on both sides for high speed schlieren photography. The main chamber is equipped with a piezoelectric pressure transducer, valves for gas intake/exhaust/sampling, a thermocouple, spark plugs on the cylindrical wall. Ignition was made coincidently at both prechambers with the home-made CDI sets. Present experiment was carried out changing the volumetric ratio of the prechamber to the main chamber (such as 8.4 %, 11.1 %, 13.2 %, 14.8 %, 17.5 %), and also changing the

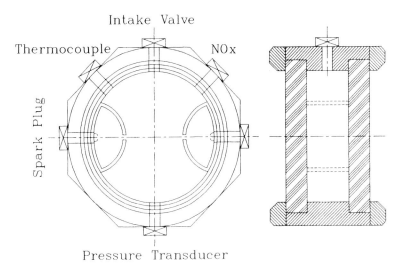

FIGURE 1   Flat discal combustion vessel.

diameter of the connecting orifice between the main chamber to the prechambers (such as 5.0 mm, 5.9 mm, 7.1 mm, 7.9 mm, 8.7 mm, 9.5 mm, 10.7 mm, 13.5 mm), but the distance from the orifice to the other opposed orifice was kept constant.

Notice that our combustion vessel used is very similar with that of Fujimoto *et al.* (1984), 116 mm in diameter, 38 mm in depth, and about 400 cc in volume. They changed the volumetric ratio as many cases as 4.13%, 6.0%, 8.25%, 12.0%, 16.7%, 24.0% and also changed the orifice diameter such as 3.5 mm, 5 mm, 7 mm, 10 mm, 14 mm.

Figure 2 shows a view of the rectangular combustion vessel whose size is 50 mm × 50 mm × 150 mm for the main chamber with varying volume of dual prechambers. Maximum size of each prechamber is 50 mm × 50 mm × 50 mm. Maxi-

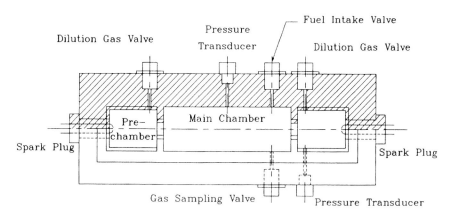

FIGURE 2   Rectangular combustion vessel.

mum total volume of this combustion vessel configuration is 625 cc. Jeong (1989), who used this rectangular combustion vessel, equipped variety of the prechamber volume relative to the main chamber volume, such as 15%, 20%, 25%, 30%, and changed the orifice diameter like 3.8 mm, 5.0 mm, 7.0 mm, 10.0 mm, 20.0 mm.

Premixed gas used was propane-air premixed gas of 3.8% volumetric mixture ratio or equivalence ratio of 0.95. (Notice that Fujimoto *et al.* (1984) used methane-air premixed gas of equivalence ratio 0.95, while Jeong (1989) used propane-air premixed gas of same equivalence ratio.) The equivalence ratio was calculated from their partial pressures of fuel and air. This premixed gas was charged into the combustion vessel initially at atmospheric pressure and room temperature.

An evacuated sampling bottle and a high pressure buffer bottle pressurized by argon gas were used to dilute sample exhaust gas, and diluted sampled gas was introduced to a chemiluminescence analyzer (Yanagimoto ECL-77A, Japan) through the sampling bottle.

More detailed informations on the experiments of other authors can be found from the references (Miura *et al.* (1978), Fujimoto *et al.* (1984), and Jeong (1989)). Summary on the details of used combustion chamber dimensions and used fuels are listed in Table I.

RESULTS AND DISCUSSIONS

Results on the relation between NO_x concentration and final maximum combustion pressure of different combustion features and on the related flame propagation photography are presented. Typical three different combustion features are categorized

TABLE I

Summary of the related investigations

Investigator	Fujimoto et al	Jeong	Okajima and Kumagai	Authors
Objective Engine	SI Engine	SI Engine	CI Engine	SI Engine
Configuration	Impinging-Jet-Flame from Prechamber	Impinging-Jet-Flame from Prechamber	Bosch-Type Direct Fuel Injection	Impinging-Jet-Flame from Prechamber
Fuel	Methane (CH_4)	Propane (C_3H_8)	Hexadecane $(C_{16}H_{34})$	Propane (C_3H_8)
Chamber	Flat Discal	Rectangular	Thick Discal	Flat Discal
Chamber Size (Main Chamber, Vm)	$116\phi \times 38$ mm (400 cc)	$50 \times 50 \times 50$ mm (375 cc)	$108\phi \times 60$ mm (700 cc)	$120\phi \times 40$ mm (425 cc)
(Prechamber, Vp)	4.13–24.0% of Vm	15–30% of Vm	No Prechamber	8.4–17.5% of Vm
Orifice Diameter	3.5–14 mm	3.8–20 mm	No Orifice	5.0–13.5 mm

(according to Fujimoto *et al.* (1984)), i.e. Laminar-Flame and Ordinary Turbulent-Flame Combustion System, Single-Jet-Flame Combustion System, and Impinging-Jet-Flame Combustion System.

Laminar-Flame and Ordinary Turbulent-Flame Combustion System

Figure 3 shows relation between NO_x concentration and final maximum combustion pressure for the case of laminar-flame and ordinary turbulent-flame combustion system of Fujimoto *et al.* (1984), Jeong (1989), and our present experiment. These results show that usual laminar flame propagation or ordinary turbulent flame propagation gives NO_x concentration increasing almost linearly wrt combustion pressure, if system combustion pressure is increasing. This is because the final combustion pressure is related to the average gas temperature within the vessel at the last burning period, and hot burnt gas temperature basically governs the NO_x concentration. This combustion system continues for a rather relatively long combustion period (usually 60 ms ca.), or burnt gas is staying long period under high temperature circumstance. When we recall that thermal NO_x formation reaction kinetics is very rate limiting and temperature sensitive, results of Figure 3 is reasonable as far as product gas is exposed to the high temperature circumstance for a long period. Notice that even different combustion vessel geometry or fuel type do not change basic feature of this system.

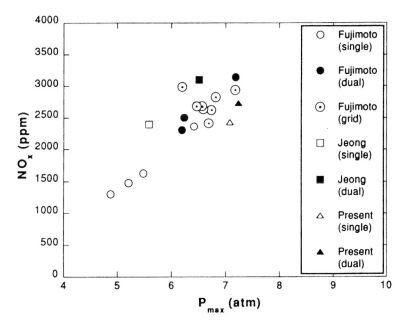

FIGURE 3 NO_x concentration wrt final maximum combustion pressure Pmax for laminar-flame and ordinary turbulent-flame combustion system.

   Single means laminar flame propagation in the main chamber at a single point spark, dual means laminar flame propagation in the main chamber at dual point sparks, and grid means turbulent flame propagation in the main chamber at a central spark with grid.

Single-Jet-Flame Combustion System

Figure 4 shows relation between NO_x concentration and final maximum combustion pressure for the case of single-jet-flame combustion system. Even though slope of NO_x concentration wrt final maximum combustion pressure for this case is slightly lower than that of the case in Figure 3, we still can see that NO_x concentration is increasing according to the increase of the final maximum combustion pressure. Slight reduction of NO_x concentration may be understood with the fact that single-jet-flame combustion system reduces total burning time (ca. 35 ms, by half to the case of laminar-flame and ordinary turbulent-flame combustion system), or burning velocity increases with turbulence intensity, and then mixing process of the hot burnt gas into the cold unburnt gas becomes a better improved mixing process resulting reduced peak fluctuating temperature.

This imagination can be noticable for the result of present experiment that NO_x concentration is reversely decreasing at the higher final maximum pressure region. In this region, it resembles the behavior of impinging-jet-flame combustion system explaining next.

Impinging-Jet-Flame Combustion System

Results of impinging-jet-flame combustion system are summarized in Figure 5. Notice that result of present experiment and Fujimoto *et al.* (1984) are very alike, both burning in a flat discal combustion vessel with different premixed gases. Both results show that NO_x concentration is drastically decreasing with increase of final combustion pressure.

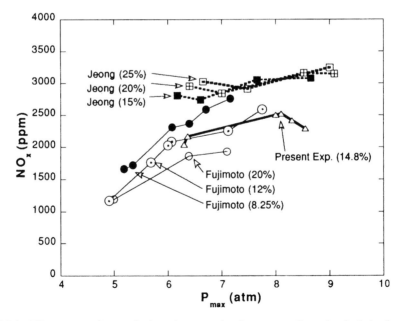

FIGURE 4 NO_x concentration wrt final maximum combustion pressure Pmax for single-jet-flame combustion system. Numbers in parenthesis denote volumetric ratio of the prechamber to the main chamber.

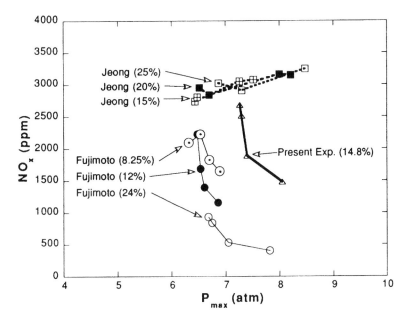

FIGURE 5 NO_x concentration wrt final maximum combustion pressure Pmax for impinging-jet-flame combustion system.
   Numbers in parenthesis denote volumetric ratio of the prechamber to the main chamber.

But, result of Jeong (1989) which burnt propane-air premixed gas in a rectangular combustion vessel does not show the reduction of NO_x concentration with increase of final combustion pressure, rather show the very similar trend as in the case of single-jet-flame combustion. The fastest burning can be attained by this combustion system. (Total burning time is ca. 30 ms.)

   Four different flame propagation processes can be seen from Figure 6 to Figure 8. Figure 6(a) shows a very typical laminar flame propagation process. Figure 6(b) shows one typical example of turbulent flame propagation of single-jet-flame combustion. This schlieren picture shows that spouting jet immersed into the main chamber would not become a violently active flame until very late period, but this jet would disturb the unburnt gas in the main chamber, which can promote the unburnt gas turbulence intensity in the main chamber. Combustion in the main chamber is starting from the opposite wall where spouting jet collides on, and speed of spouting jet becomes slower. When combustion starts to be a vigorous flame, shape of this flame is very similar to a band of wrinkled distorted, but connected flames. This flame proceeds to the main chamber side, but does not show the strong exchange or mixing of burnt gas with unburnt gas, as the established flame is a connected flame band that only the limited front surfaces exposed to the unburnt gas can exchange the hot burnt gas with the cold unburnt gas and keeps hot burnt gas behind the reacting flame front. Two typically different combustion progresses of impinging-jet-flame can be seen from the different flame propagations shown, first in Figure 7 for the burning in a rectangular combustion vessel which has such narrower space at the central zone of the main chamber where

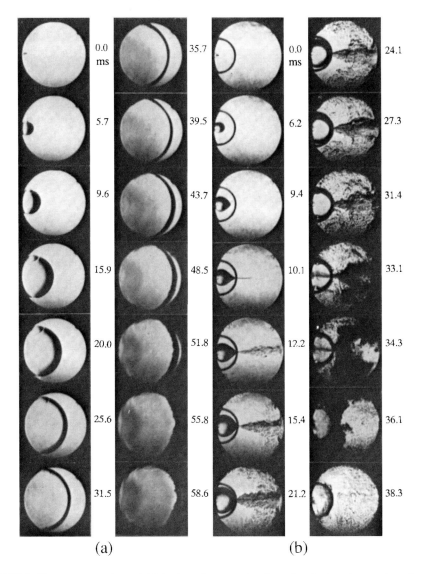

FIGURE 6   Flame propagations of (a) laminar-flame combustion of a single point spark and of (b) single-jet-flame combustion.

Volumetric ratio of the prechamber to the main chamber is 14.8% and orifice diameter is 5 mm for the case of (b).

a head-on collision is occurred by a pair of impinging jet flame spouting into the main chamber, and second in Figure 8 for the burning in a flat discal combustion vessel which has rather wider space at the central zone of the main chamber not much confining the turbulent mixing process in contrast to a relatively confining mixing process of impinging-jet-flame with the unburnt gas due to the geometry of a rectangular combustion vessel. Flames in the rectangular combustion chamber (Fig. 7) are also

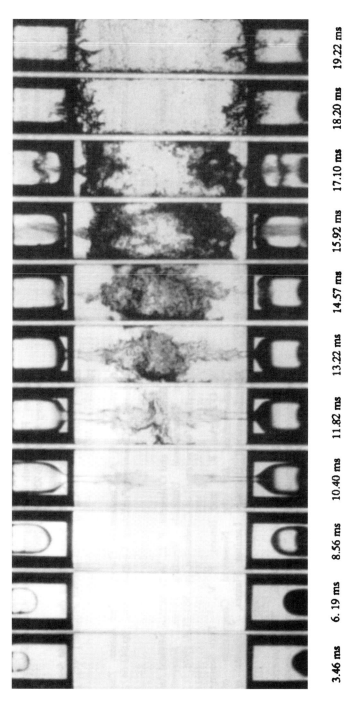

FIGURE 7 Flame propagation in a rectangular combustion vessel.
Volumetric ratio of the prechamber to the main chamber is 20%, and orifice diameter is 10 mm.

3.46 ms   6. 19 ms   8.56 ms   10.40 ms   11.82 ms   13.22 ms   14.57 ms   15.92 ms   17.10 ms   18.20 ms   19.22 ms

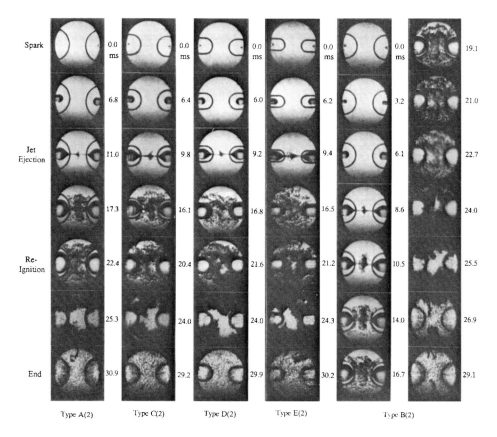

FIGURE 8   Flame propagation in a flat discal combustion vessel.
    Volumetric ratio of prechamber to main chamber is 17.5% for case A(2), B(2); 14.8%, C(2); 13.2%, D(2); 11.1%, E(2); 8.4%, and orifice diameter is 5 mm for all the cases.

formed two connected flame shapes similar to the single-jet-flame in the flat discal combustion chamber, and consequently show same trend as in the case of the single-jet flame combustion.

Different from the proceeding 3 cases, last case showing the impinging-jet-flame in the flat discal combustion chamber (Fig. 8) gives a different situation, that we can see the very distributed small flame pockets surrounded by the unburnt cold gas pockets, around the time of 20 ms marked as the moment of reignition. This situation suggests us the original idea on the impinging-jet-flame that some burnt gas is instantaneously and continuously supplied to the unburnt gas just adjacent to the flame front. Clearly, we can see this progress from the time of 14.0 ms of case B(2) in Figure 8 when combustion initiated in the main chamber until the time of 24.0 ms when vigorous combustion is almost finishing. This strong mixing processes are always seen from the all the cases of A(2)–E(2), which correspond to the maximum prechamber volume to the minimum prechamber volume.

These results suggest better improved mixing probably due to the pair of impinging-jet-flame introducing much flow of unburnt gas into the burnt gas and homogeneous reduced fluctuating rms value possibly due to the head-on collision of a pair of impinging-jet-flame may cause fast burning, reduced fluctuating temperature peak but rather homogeneous temperature distribution, and then much reduced NO_x concentration.

CONCLUDING REMARKS

Impinging-jet-flame combustion system in a closed combustion vessel simulating combustion at the top-dead-center-period of actual automotive engine combustion can give a fair amount of NO_x concentration reduction still maintaining the final combustion pressure at highest level near stoichiometric mixture burning. Present results are coincident with the previous results of Fujimoto et al. (1984).

It seems that fuel type used should not be even a minor influential factor on the characteristics of impinging-jet-flame, but geometry of combustion vessel may be a possible influential factor.

Here, influence of the combustion vessel geometry suggest, in a sense indirectly, that better improved mixing with homogeneous reduced fluctuating rms value may introduce fast burning resulting a highest pressure level, reduced fluctuating temperature peak, and then reduced NO_x concentration.

ACKNOWLEDGEMENTS

First author (ISJ) expresses his appreciation to the Korean Government for his financial support (MOE-1994 Professor Training Program Grant) to extend his sabbatical leave at the University of California-Irvine, where most of this manuscript work has been finalized.

Authors wish to acknowledge that early stage of this study was partly supported by a Grant-in-Aid of fiscal year 1989 (Subject No. 890203) from the Korea Science and Engineering Foundation and also by the Kia Motors Corp. Experiments of Dr. K. S. Jeong, Mr. H. B. Park, and Mr. N. K. Cho were contributed in many ways.

REFERENCES

Bowman, C. T. (1992) Control of combustion-generated nitrogen oxide emissions; Technology driven by regulation. *Twenty-fourth Symposium (International) on Combustion*, The Combustion Institute, 859–878.
Fujimoto, S., Kaneko, Y. and Tsuruno, S. (1984) Possibility of low-NO_x and high-load combustion in pre-mixed gases. *Twentieth Symposium (International) on Combustion*, The Combustion Institute, 61–66.
Jeong, K. S. (1989) A study on the combustion characteristics of premixed gases in a closed combustion chamber with/without pre-chambers. *Ph.D. thesis*, Seoul National University (in Korean).
Miura, S., Tsukamoto, T., Kawagoe, M., Nakaoji, S. and Kaneko, Y. (1978) NO_x formation with various combustion processes of premixed gas. *Mitsubishi Heavy Industries Technical Review*, 15–5, 1–11.
Okajima, S. and Kumagai, S. (1990) Experimental investigation of soot and NO_x reduction by impinging spray combustion in a closed vessel. *Twenty-third Symposium (International) on Combustion*, The Combustion Institute, 275–279.

CO_2 Mitigation through Mixed Steam and CO_2 Gas Turbine Cycles

J. DE RUYCK[a], G. ALLARD[b], D. BRÜGGEMANN[c] and PH. MATHIEU[d]

[a]*Vrije Universiteit Brussel, Belgium;* [b]*ALCE consultants, Belgium;*
[c]*RWTH Aachen, Germany;* [d] *Université de Liège, Belgium*

Abstract–A possible route for reducing the CO_2 emissions from power production on the long term consists in capturing and disposing CO_2 from power plant exhaust gases. Viable capture and disposal both call for new technological developments however, and the present project addresses the problem of viable CO_2 capture by examining mixed CO_2/steam based power cycles. The project is part of a Joule II 'Combined Cycle Project' which covers all Joule II activities on CO_2 recovery from coal and gas combined cycles.

Key Words: CO_2, gas turbine, oxyfuel combustion, evaporation cycle

INTRODUCTION

The use of semi-closed cycles running on CO_2 rather than on steam or gas is not new, but it never reached any realization phase. This is mainly due to i) the cost of separating oxygen, ii) the need for specially developed oxyfuel burners and iii) the need for specially designed turbomachines. The actual CO_2 mitigation problem is however a strong incentive to re-examine the idea of the CO_2 cycle, since a CO_2-rich excess gas can easily be captured. To achieve this, it is essential to apply the most recent techniques for improving power cycle efficiencies such as the use of combined cycles, cogeneration, fuel cells, integration of coal or biomass gasification, etc.

The present project proposes the application of Steam Injected (STIG, Kolp 1988) and Evaporative gas turbine cycles (HAT, Day 1993) in which Nitrogen is replaced by CO_2 in a semi-closed cycle, and to compare this with the other potential routes such as the CO_2 capture from stacks. The concepts of STIG and HAT are shown to yield efficient power production, which is essential for a viable CO_2 capture. Optimal cycle configurations are searched and a step towards realization is done by building a 100 kWe closed test cycle running on a mixture of steam and CO_2. An oxyfuel burner is designed and will be tested through CARS spectroscopy. On the medium term, the project can result into viable zero emission CO_2 demonstration cycles for specific applications such as power generation from CO_2-rich wells, enhanced oil recovery, industrial CO_2 production and others. On the long term the project can result into viable zero emission cycles for utility power production.

STATE OF THE ART

Basically, three approaches can be followed for zero emission power systems:

i) CO_2 can be recovered from the exhaust gases of existing systems (with mono-ethanolamine, alcanolamine, NaOH)

823

ii) Cycles can be designed to run on CO_2 as working fluid, requiring oxygen separation from air and an oxyfuel burner. Pressurized CO_2 is available from such a cycle and is easy to recover.

iii) The fuel can be reformed (natural-gas) or partially combusted to CO and shifted (solid fuels) towards H2 and CO_2 with subsequent separation. The resulting hydrogen can be burned either in a combined cycle, or in future fuel cell plants.

In gasification plants the third solution is probably the shortest cut to capture CO_2 (e.g. Seifritz, 1989), but recent studies show that the three routes may have comparable net efficiencies (Blok 1993). In other applications (combustion of natural-gas, enhanced oil recovery, industrial furnaces) the production of energy through a CO_2 cycle appears as a possible alternative, but it never reached any realization phase. The primary factor affecting the practicability of this approach appears to be the important energy requirement for oxygen separation, which is comparable to the energy requirement of the exhaust CO_2 removal approach (Hohman and Kwai, 1980, Albanese and Steinberg, 1980, Horn and Steinberg, 1981, Steinberg *et al.*, 1985–1987). Horn and Steinberg (1981) considered a CO_2 moderated, oxygen fired coal plant for CO_2 production in enhanced oil recovery and claimed this to be a viable option. In 1982 (not published), General Electric also presented a scheme for using a 60 MW combined CO_2 - steam cycle for enhanced oil recovery. The recovered CO_2 would be used for an increased output in oil wells, disposing the CO_2 at the same time. In this scheme a gas turbine would run on a CO_2 rich gas, which is cooled by a spray cooler in which the water is removed from the combustion products. GE considered the concept as viable in the case of enhanced oil recovery and did not consider the gas turbine modification as an insoluble problem. The project remained in a conceptual phase. No thorough cycle analysis was presented.

Experiments with flue-gas recirculation have already been conducted in the past. Recirculated gases can be used in modified boilers without significant effects on the (coal) combustion up to 50% CO_2 concentrations (e.g. Kumer *et al.*, 1987). Oxyfuel burners have been developed for particular applications such as NO_x reduction, glass melting furnaces, steel heating etc. ... Experiments with closed cycle gas cleaning have also been reported (e.g. Endo *et al.*, 1989). Turbines for special applications obviously do exist and apart from the development costs should not be a critical issue for CO_2 cycles. CO_2 turbomachinery is in use in CO_2 cooled nuclear plants, and very special designs are reported in the framework of CO_2 removal (e.g. Helgeson, 1985).

Hence, no critical issues appear in the realization of a CO_2 -based cycle. The steps to be performed are the design of an efficient cycle followed by a lab scale proof of concept.

PROJECT ORGANIZATION

The project has been organized in three phases:

– phase 1: feasibility and design studies (duration 1 year)
– phase 2: test rig construction (duration 6 months)
– phase 3: test rig diagnostics (duration 1 year)

Phase 1 started in December 1992 and the present paper reports on the first activities. Three institutions and one industrial partner join their efforts in the project and co-author the present paper. The VUB co-ordinates the project and will perform all thermodynamic studies and the thermodynamic design of the cycle during phase 1. In agreement with the other partners, the test rig will be located at this institute and the VUB will take care of the cycle diagnostics in phases 2 and 3. ALCE is a consultancy company which takes care of the engineering and the start up of the test rig and of the oxyfuel burner. RWTH Aachen has the required experience for the testing of the oxyfuel burner in the CO$_2$/steam atmospheres. The ULG has been assessing the know how of gas turbines and combined cycles for years and has close contacts with the gas turbine industry. The institute will take care of the turbomachinery required for the realization of a CO$_2$/steam cycle. Since no gas turbine testing is considered, the participation of the ULG will consist in the exploration of the gas turbine market for possible application of existing turbomachines in the present cycle concept. The ULG will explore past and present activities on CO$_2$ gas turbines, and define a first design for a small prototype CO$_2$ gas turbine.

COMBINING CO$_2$ AND STEAM

Combining steam and gas is recognized as an efficient route for power production. Combined cycles consist of two separate gas and steam cycles whereas the direct mixing of air and steam in a binary cycle is introduced as an alternative in the smaller power ranges (50 MW and below). In Steam Injected Gas Turbines (STIG) the steam is mixed with the compressed air, superheated in the combustor and expanded in the gas turbine. More innovative is the use of the *evaporation cycle* where water is evaporated at non-constant temperature (El Masri 1988, Day 1993, Berta *et al.*, 1990, De Ruyck *et al.*, 1991, Rosen 1993). In theory, the cycle efficiency can be improved by several points when evaporating at variable temperature through a reduction in exergy destruction when recovering heat.

A major drawback of mixed air and steam cycles is the cost of the feed water and its treatment, the water being lost in the atmosphere. This drawback may be overcome by condensing the important amount of water vaapor from the stack and by recycling this water into the cycle, as originally proposed by Cheng (1981) or recently by Nguyen *et al.*, (1992). Consequent lower stack temperatures result in an even better efficiency and in reduced water treatment costs.

PROPOSED CYCLE

The cycle proposed originally is shown on Figure 1. It shows the most simple evaporative cycle configuration, where low temperature heat recovery occurs through liquid water heating, and high temperature recovery through superheating of the CO$_2$/steam mixture. Water is condensed out of the exhaust CO$_2$/steam mixture. The CO$_2$ rich gas and liquid water are reconducted into the cycle after cold clean up. The recirculated water is evaporated into the CO$_2$ next to compression. Excess CO$_2$ rich

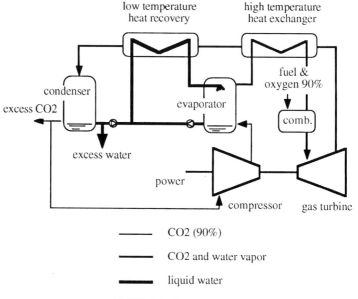

FIGURE 1   Proposed cycle.

gas can be retrieved near the bottom of the condenser or under pressure next to the compressor.

Little fundamental work is done about mixed gas and steam cycles and CO_2/steam mixtures behave in a way which is different from air/steam mixtures. A binary cycle concept is therefore being studied which is shown on Figure 2. This cycle consists of

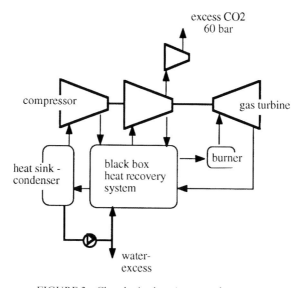

FIGURE 2   Closed mixed gas/steam cycle concept.

a single phase expansion, (separate) two phase compression, a heat recovery system including all recovery items and considered as a black box, a combustor operating on natural-gas or on carbon monoxide and a heat sink which is essentially a condenser (at varying temperature). The way the black box heat recovery system is to be realized is of no concern, provided no second law violation occurs. An excess CO$_2$ compressor is added to deliver the CO$_2$ at a pressure of 60 bars for disposal purposes. The oxygen production is not considered for the time being.

Simulations are performed through Aspen Plus, where a correct behavior of CO$_2$/water mixtures has first been investigated. Temperatures of 1200°C at turbine inlet and 40°C at condenser outlet are imposed for all simulations. Polytropic efficiencies of 88% are taken for compression and expansion. Mechanical efficiencies are 99%. Head losses are 2% are in all heat exchangers and 3% in the heat recovery system and 4% in the combustor. A temperature difference of 100°C (pinch) is imposed at the hot side of the heat recovery system. At cold side the exhaust temperature is kept at 10°C above saturation before entering the condenser, resulting in pinch temperature differences not lower than 40°C between the exhaust to the condenser and the recirculating water. This relative low temperature difference at cold side might lead to a second law violation inside the black box recovery system, which is watched by limiting its exergy destruction to about 3%. A blade cooling model has been taken into consideration in the turbine.

Results from this cycle are shown on Figures 3 to 5. Three ways of compression have been considered: single-stage compression, two-stage compression with intercooling to the heat sink, and two-stage compression with intercooling where the heat is recovered in the black box recovery system. The condenser and combustion pressures were varied to find an optimum and results are shown in the different figures. Full cost estimations and optimizations will be performed in a later phase. Figure 3 shows efficiencies versus specific power for the three compression options. Intercooling with heat recovery in the black box clearly shows the best performances. This concept

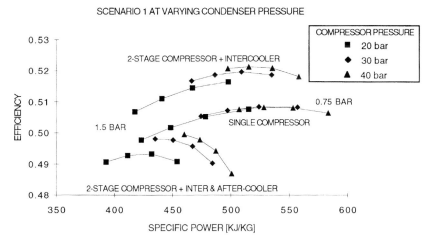

FIGURE 3   Characteristics of cycle concept.

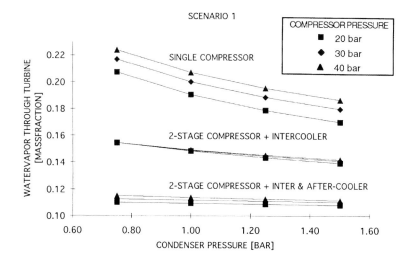

FIGURE 4   Required water injection in the cycle concept.

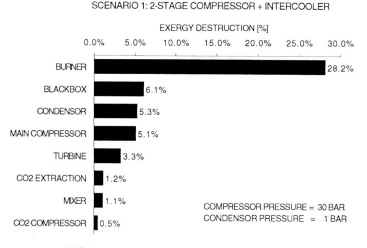

FIGURE 5   Exergy destructions in the cycle concept.

exactly corresponds to the HAT concept (Day 1993). A similar concept study has been made with a mixed open air /steam cycle for comparison (not shown). From this comparative study, the corresponding air cycle peaks at 52%, against 55% in the case of CO_2 *including* the delivery of excess CO_2 at 60 bar (*excluding* the oxygen production). As expected the optimization calls for high pressure levels. 40 bar is considered as a maximum, leading to optima of 55%. Mass specific powers are comparable to the those found in STIG and HAT cycles. The use of heavy gases however yields higher volume specific powers, but more stages are to be expected in the turbomachines. It is not clarified yet how the corresponding cost structure of the complete CO_2 cycle will be. Figure 4 shows the corresponding amounts of water to be recirculated in order to match the specified pinches. The correlation between efficiency and amount of water to be injected appears from this figure and amounts between 25 and 30% steam are to be expected in the final turbine designs.

The exergy destruction in the different components and exit flows is shown on Figure 5. Most of the destruction occurs in the combustor whereas all other components show destructions which are hard to reduce further, the corresponding exergy destructions being hard to reduce further. The extracted gas contains 99% CO_2 and 1% water vapour by mass (assuming pure oxygen). The exergy cost of extracting the CO_2 is not excessive, but obviously the cost of producing the oxygen is still to be taken into account. Additional gases such as argon or nitrogen in the oxygen supply would reduce the CO_2 content accordingly. The required oxygen production would cost about 9 points in efficiency when using natural-gas, resulting in a net cycle efficiency of 46%.

TEST RIG

The project partners agreed on the realization of a 100 kWth test rig. This test rig will essentially be used for the testing of oxyfuel burners in CO_2/steam atmospheres. The easiest way to produce such atmospheres is to realize the complete cycle concept, with exclusion of the gas turbine. Since the inclusion of the gas turbine is considered as too expensive and not relevant for the test rig, it will not be included but simulated by an expansion with aftercooling. The small scale allows the use of liquid oxygen supply and there is no need for an oxygen separation plant. The test rig will provide know how on the complete cycle and allow to prove the cycle concept.

A draft of the combustion chamber and cooler is shown in Figure 6. The pressure in the combustor is 10 bar. The burner is a dual flow burner. The flame temperature is controlled by addition of some working fluid (CO_2/steam) to the oxygen in the first flow and by the second circumferential flow. Swirl is obtained through tangential injection of the working fluid and the flame is of the recirculation type. The fuel injection occurs through circular injection holes. If necessary, a ceramic shield will be fitted to the tip of the burner. No extra active cooling is required. Two 30 mm thick borosilicate windows on both sides are mounted for flame diagnostics. The windows are cooled to 300°C (flame emissivity is about 0.9, peak flame temperature is 2000°C). The combustor is followed by a heat exchanger and expansion valve which replace the expansion engine. Materials used are 316L and titanium to withstand CO_2 corrosion.

CARS THERMOMETRY

The performance of burners is strongly depending on their flame field structures. These are determined by a number of physical quantities such as flow velocity, mixture composition and temperature. Furthermore, a complete description of turbulent combustion requires the knowledge of time fluctuations of these parameter fields.

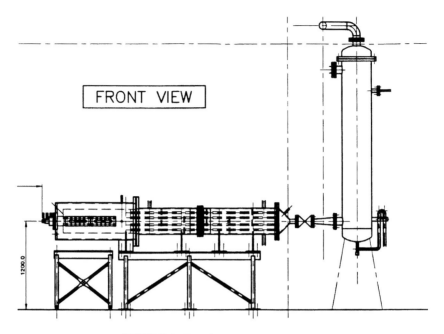

FIGURE 6   Test rig combustion chamber.

Since the combustion is pressurized, a non intrusive measurement method is to be used. Amongst these, Coherent Anti-Stokes Raman Spectroscopy (CARS) has been selected because of the strong CARS signal and promising signal to noise ratio's. CARS has already been applied in an industrial environment and is expected to be operational for the present test The fundamentals of CARS and numerous applications have been reviewed (Attal-Tretout *et al.*, 1990, Greenhalgh 1988, Eckbreth 1988) and shall not be repeated here. The conventional approach to CARS thermometry is based on the excitation, detection and analysis of the vibrational bands of nitrogen which is abundant in air-fed combustion. The study of oxyfuel burners with carbon dioxide as carrier gas requires the development of alternative variants.

A narrowband excitation of two CO_2 transition systems which show a very dense line structure and form an extremely narrow band is suggested. The signal ratio reflects the thermal distribution on molecular energy levels and is strongly depending on temperature. Fluctuations in the exciting dye laser pulses are corrected for by a comparison of line intensities to simultaneously generated signals from a reference cell containing CO_2 at known conditions. Accuracy and precision of this novel method is investigated by test cell experiments (Brüggemann *et al.*, 1993). Typical single pulse spectra are shown in Figure 7 and demonstrate the excellent signal-to-noise ratio. As shown in Figure 8 even small temperature variations are detected. The standard deviation of single pulse data is only 3 to 5% in the accessible temperature range. Although further studies are necessary the results seem to promise the applicability of this technique to real burners using a special mobile CARS system.

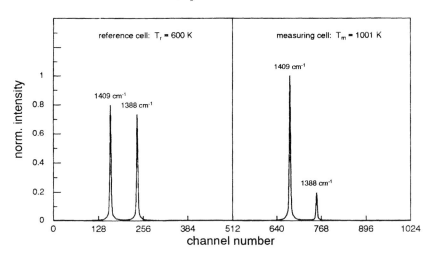

FIGURE 7   Typical single pulse CO_2 spectra simultaneously generated in test cells at 1000 K and 600 K, resp.

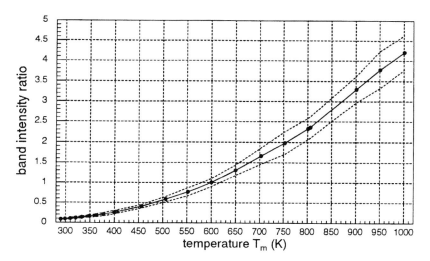

FIGURE 8   Temperature sensitivity and single pulse accuracy of the two-lines technique.

GAS TURBINE

In the JOULE project, the task of the Department of Nuclear Engineering and Power Plants (University of Liège, Belgium) is to address the issues on performances and design of a gas turbine raised by the use of either pure CO_2 or in a mixture with steam as working fluids instead of air. In a first step, the work is to provide an answer to the following questions:

– what happens if air is replaced by CO_2, CO_2/H2O or CO_2/H2O/Ar in an *existing*
industrial gas turbine ? Has such an engine already been designed, for instance in
nuclear reactors using CO_2 as coolant ?
– what are the required adaptations to fit an existing air gas turbine to the use of CO_2,
if that makes sense ?

To achieve this a technical questionnaire has been addressed to the constructors, which
is reproduced in annex A. In a second step, the target of the project would be the design
of a new gas turbine, using CO_2 as working fluid, firing oxygen and operating in
a semi-closed cycle. (see Fig. 9)

In this paper, the influences of the use of CO_2 instead of the nitrogen in air as
a working fluid are investigated. In Table I the physical properties of CO_2 are
compared with those of air.

At first glance, CO_2 is heavier (molecular weight is 44 instead of 28.9 kg/kgmol) and
less compressible than air. At ambient temperature, the specific heat ratio γ is lower for
CO_2 (1.3) than for air (1.4) and decreases more rapidly with temperature for CO_2 than
for air, especially between 15° and 300°C, the operational range of the compressor. The
specific heat capacity c_p is lower for CO_2 than for air and that on about the whole range
of temperatures. The density ρ of CO_2 is always the greatest, at any pressure and
temperature. The speed of sound a is lower for CO_2 and amounts to around 80% of
that of air at 15°C. These two latter properties will strongly influence the gas turbine
flow characteristics like the Mach numbers, the load of a stage, the choking condition,
the rotational speed, etc.

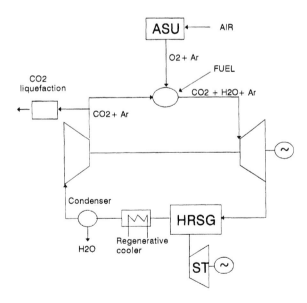

FIGURE 9   Semi-closed CO_2 gas turbine cycle inside a combined cycle. Dashed lines – gasification unit of
an IGCC cycle.

TABLE I

Comparison of physical properties of CO$_2$ vs. air at 15°C and 300°C respectively

	γ	c_p (kJ/kg/K)	ρ (kg/m^3)	a (m/s)	v (10^{-5}m^2/s)
CO$_2$ 15°C	1.30	0.84	1.85	266	0.8
300°C	1.22	1.06	0.92	364	2.9
AIR 15°C	1.40	1.0	1.21	340	1.5
300°C	1.38	1.05	0.61	476	4.8

To get a first idea of the influence of the use of CO$_2$, the performances of a class 150 MW gas turbine have been calculated using the Aspen Plus simulation code. The modelling of the gas turbine cycle and the numerical procedures as well as the detailed results will be published in future papers currently in preparation or submitted for publication.

The main trends when using CO$_2$ as working fluid in an industrial gas turbine are as follows. As CO$_2$ is heavier than air, the corresponding speeds of sound are lower so that the sonic condition is reached more quickly. The load per stage will be lower so that the number of stages has to be higher to provide an assigned pressure ratio. The size of the engine will increase. On the other hand, a thermodynamical study shows that the maximal efficiency is obtained for a pressure ratio lying above 100 while the optimal specific work is to be found at a pressure ratio lying between 30 and 40. The mean pressure level is hence higher than with air. So it seems that an existing gas turbine cannot be easily used when working with CO$_2$ and that a new CO$_2$ gas turbine has to be designed in order to provide such high pressure ratios. When introducing a recuperator the maximum efficiency is obtained for a much lower pressure ratio (6 to 10), which is the range of existing gas turbines. Hence *existing* engines might be considered when including a recuperator.

The GT manufacturer's answer to a questionnaire concerning the use of CO$_2$ in an existing GT.

In order to assess the possible use of CO$_2$ in existing gas turbines, a questionnaire (see Annex A) was submitted to GT manufacturers. The following preliminary considerations on the use of existing GTs are derived.

1. There is no simple and straightforward explanation of the behavior of a GT when running on CO$_2$ instead of air. Indeed, because of different density, heat capacity and isentropic exponent of CO$_2$ compared to air, the aerodynamic flow changes seriously when keeping the same geometry and rotational speeds. The calculations show that the limit of save load is reached earlier in the compressor than with air.

    In order to satisfy as close as possible the similarity rules, the Mach number has to remain unchanged. As the speed of sound decreases significantly (by 22 % at ambient temperature), the rotation speed has to be decreased so that the absolute velocity of the gas is at maximum sonic. In addition, as the speed of sound changes with temperature, the Mach numbers in the compressor and expander can depart significantly from those based on air. It follows that the coupling of compressor and expander has to be reconsidered. Also the pressure ratio changes with the decrease of rotational speed and the associated impact on the compressor behavior is to be examined.

   As a consequence, even if a compressor can operate with both fluids, an existing GT designed for air cannot operate as such with CO_2.

2. In a closed cycle, the condenser pressure or the lower pressure of the cycle can be under the atmospheric pressure. The operation of the compressor with a lower inlet pressure is theoretically possible. Again the aerodynamic flow has to be recalculated and possibly angles of blades to be adapted. Due to the different density of CO_2, the height of the blades has to be changed.

   As a result, an existing GT cannot be used as such.

3. Keeping the pressure ratio at its design value for air, the calculations show that the mass flow required to cool the first vane at the expander inlet will increase by about 30 % when operating on CO_2. When using typical temperatures of the most advanced GTs, it is shown that the vane cooling by CO_2 flowing in the existing geometry is at first sight satisfactory. However the whole thermo-mechanical field has to be recalculated and shown to satisfy the resistance criteria.

4. In principle, one can imagine to use a recuperator in an existing GT for CO_2 operation that allows to operate optimally at low pressure levels ($r_p \approx 6\text{--}7$), but the number of compressor stages should probably be increased and a special design of the combustion chamber is necessary because of the higher temperature of the working fluid entering the combustion chamber.

5. The use of a CO_2/H2O/rare gas mixture "equivalent" to air to satisfy the similarity rules (same isentropic exponent, Mach and Re numbers), can theoretically allow the use of existing GTs provided that H2O is in gaseous state.

   These last two points are under investigation.

   In conclusion, a redesign of the turbomachinery is unavoidable when using CO_2 as working fluid. However it is also possible to design a new engine using isolated components, i.e. an independent compressor (radial, screw compressor) and a steam-like expander.

ANNEX A: Questionnaire to gas turbine manufacturers

– Are pressure ratio's of 30 to 40 technically feasible when using CO_2 mixtures and what are the problems raised by high pressure levels ?

– If a CO_2/H_2O mixture can be realized with molecular mass and specific heat ratio close to those of air, can an existing gas turbine work with this fluid and what are the possible issues ?

– What happens if air is simply replaced by CO_2 in an existing gas turbine (compressor design, change of design point) ?

– What are the required adaptations to fit an air gas turbine to CO_2 ?

– What is the compatibility of the used materials with CO_2 (decarbonization, corrosion,..)

– What are the problems to burn natural-gas with pure oxygen in a CO$_2$ atmosphere
– Is the presence of CO to be expected (by decarbonization) CO can be burned, even during expansion. Due to the possible excess oxygen, combustion and expansion processes might be influenced (there is a process using partial combustion of CO in the combustion chamber and during the expansion; it is designed to carry out an expansion as close as possible to an isothermal expansion).
– CO$_2$ being less compressible and heavier than air, what are the mechanical blade loads and the number of stages in the case of a modification of an existing air gas turbine or in a new design of a CO$_2$ gas turbine.
– What about blade cooling by CO$_2$ (the conventional heat transfer coefficient is different, depending on the blade metal mean temperature and the corresponding gas temperature) ?
– How would the geometry of a CO$_2$ gas turbine look like ?
– What would be (approximately) the polytropic efficiency of compressor and expander ?
– What would be the cost to develop a large CO$_2$ gas turbine ?
– What would be the commercial cost of a CO$_2$ gas turbine ?

CONCLUSIONS

In the present project a step is made towards the realization of a semi-closed cycle running on CO$_2$ or on CO$_2$/steam mixtures. The different problems of cycle design, oxyfuel combustion, gas turbine and recirculation are addressed. A test rig is under construction to identify problems related to such type of CO$_2$ capture. A route for CO$_2$ mitigation is thus in development, which is believed to contribute in the global warming issue.

ACKNOWLEDGEMENTS

The present project is funded by the CEC DG XII, Joule II program. It is co-funded by the Flemish Regional Government. With co-operation of Mr. S. Bram (VUB), Mr. Bollig (RWTH Aachen), Y. Radelet (ULG) and Mr. Distelmans (ULG).

REFERENCES

Albanese, A. and Steinberg, M. (1980) Environmental control technology for atmospheric carbon dioxide. *Brookhaven National Lab., Upton NY.*
Attal-Trétout, B., Bouchardy, P., Magre, P., Péalat, M. and Taran, J. P. (1990) *Appl. Phys.*, **B51**, 17.
Berta, G. L., Durelli, E. and Prato, A. P. (1990) A special arrangement of hybrid gas turbines. *AICHE 25th IECEC conference proceedings*, **5**, 495.
Blok, K. (1993) Final Report of the Research Programme on Carbon Dioxide Recovery and Storage. Report nr 92063 Dept of Science, Technology and Society. Part of the Dutch Integrated research programme on carbon dioxide recovery and storage (SOP-CO$_2$, national programme).
Brüggemann, D., Wies, B. and Bollig, M. (1993) Improved CARS Thermometry by Narrowband Excitation of CO$_2$ Bands'. *Proceed. XIIth European CARS Workshop*, Villingen, 1993.
Brüggemann, D. and Bollig, M. (1993) Two-line CARS Spectroscopy for Thermometry with Improved Precision and Accuracy. *Internal report* (to be published).
Day, W. H. and Rao, A. D. (1993) FT4000 HAT With Natural-Gas Fuel. *Turbomachinery International*, jan/feb 1993, p. 22.

De Ruyck, J., Maniatis, K., Baron, G. and Pottie, K. (1991) A biomass fueled cogeneration plant based on an evaporative gas turbine cycle at the University of Brussels. *ASME COGEN-TURBO*, **6**, p 443.

Eckbreth, A. C. (1988) Laser Diagnostics for Combustion Temperature and Species. *Abacus*, Turnbridge Wells, 1988.

El Masri, M. A. (1988) A modified, High-efficiency, recuperated gas turbine cycle. *ASME journal of engineering for gas turbines and power*, **110**, april 1988, p. 232.

Endo, N., Yoshikawa, K. and Shioda, S. (1989) Reduction of impurity contamination in a working gas for closed-cycle MHD power generation. *Energy conversion Management*, **29**(3), 207–215.

Greenhalgh, D. A. (1988) Quantitative CARS Spectroscopy. *Advances in Non-Linear Spectroscopy*. ed. Clark and Hester, Wiley, p193.

Helgeson, N. (1985) Investigation of the feasibility of a biphase turbine for industrial process energy recovery. DOE/ID/12357-T1

Hohmann, R. P. and Kwai, A. H. (1980) Survey of methods for isolating and containing gaseous carbon dioxide as nonvolatile products. *Massachusetts Inst. of Tech.*, Oak Ridge, TN.

Horn, F. L. and Steinberg, M. (1981) Carbon dioxide power plant for total emission control and enhanced oil recovery. *Brookhaven National Lab.*, Upton NY.

Kolp, D. A. and Moeller, D. J. (1989) World's first full STIG LM5000 Installed at Simpson Paper Company. *ASME J. of Gas Turb. and Power*, **111**, april 1989, p. 200.

Kumer, R., Fuller, T., Kocourek, R., Teats, G. and Young, J. (1987) Tests to produce and recover carbon dioxide by burning coal in oxygen and recycled flue-gas: Black Hills power and light company customer service boiler 2, Rapid city, South Dakota. *Argonne National Lab.*

Mathieu, Ph. and De Ruyck, J. (1993) The CO_2 Gas Turbine Option for Recovery of CO_2 from CC and IGCC Plants'. *ASME COGEN-TURBO, Bournemouth, Sept.'93*, **8**.

Nguyen, H. B. and den Otter, A. (1992) Development of Gas Turbine Steam-Injection Water Recovery (SIWR) System. *ASME paper 92-GT-87*.

Rosen M. (1993) Evaporative Gas Turbine Cycles - A Thermodynamic Evaluation of Their Potential. *Internal report ISRN LUTMDN/TMVK, University of Lund, Sweden*.

Seifritz, W. (1989) A new mixed fossil/nuclear energy system for the production of electricity with zero emission of carbon dioxide. *Nuclear Technology*, **88**, 201–206.

Steinberg, M. and Cheng, H. C. (1985) Systems study for the removal, recovery, and disposal of carbon dioxide from fossil fuel power plants in the US. *Report Brookhaven National Lab.*, Upton NY nr 36428.

Steinberg, M. and Cheng, H. C. (1987) Advanced technologies for reduced CO_2 emissions. *Conference on prospects for mitigating claimatic warming by carbon dioxide control*, Boston, 11/2/1988.

The Design of a Coal-Fired, High-Temperature Furnace for an Advanced Combined-Cycle System

ERIC G. EDDINGS, MICHAEL P. HEAP, DAVID W. PERSHING,
   ADEL F. SAROFIM and PHILIP J. SMITH   *Reaction Engineering*
International Salt Lake City, Utah (USA)

Abstract—DOE's Office of Fossil Energy has initiated a project to develop a High Performance Power System (HIPPS) to produce electricity from coal with an overall thermal efficiency of 47% or higher and minimal pollutant emissions. One technology option is a combined cycle system that uses a high-temperature, high-efficiency gas turbine driven by a working fluid separately heated in a coal-fired high temperature combustor. This paper describes the trade-offs associated with the design of a coal combustor that will satisfy the constraints of the high efficiency power system. The design study considered two basic combustor types: a rich, well-mixed, physically-staged system operating at the optimum stoichiometry to minimize fixed nitrogen species prior to the addition of staging air, and a controlled-mixing, axial flame. The paper uses a comprehensive coal combustion model, limiting-case detailed gas-phase kinetic models, and bench-scale experiments to evaluate the attractiveness of each combustion system for the HIPPS.

Key Words: Coal, power generation, NO_x, modeling

INTRODUCTION

U. S. electric utilities operating coal-fired power production plants are continually being required to decrease emissions to lower levels. The 1990 Clean Air Act Amendments have prescribed stringent emission limits which must be met over the next few years. In order to assure sound economic growth, the U.S. must develop highly-efficient, environmentally-safe methods of power generation. Even with the most optimistic growth projections for conservation, nuclear and renewable energy, the Department of Energy (DOE) projects that the use of coal will double by the year 2030. The clean and efficient use of coal in power generation is becoming an increasingly challenging design task for the manufacturers of traditional coal-fired steam cycles.

The U.S. Department of Energy (DOE) recognizes the need for new developments and has initiated a program to promote the utilization of recent technological developments in the design and operation of coal-fired power generation systems. Advances in the areas of materials, NO_x control, and computer interfaces have led to new technologies which can be applied to provide cleaner, more efficient power generation. Because of the costs involved in the initial development of such an advanced system, DOE's Office of Fossil Energy has provided the financial support for the initial stages of development. The program has the goal of providing a new generation of High Performance Power Systems (HIPPS) to provide power from coal for the next century.

The primary objective of this program is to demonstrate the technical and economic feasibility of our advanced power concept for a HIPPS by: a) producing conceptual designs for key components; b) comparing the economics of our concept with alternative approaches, and c) preparing a development plan which will allow a commercial demonstration by the end of the century. In addition, the design must satisfy the following requirements established by DOE for commercial HIPPS facilities.

- A minimum conversion efficiency, coal pile to busbar, of 47%.
- NO_x less than 0.15 lbs (as NO_2) per million BTU of fuel input.
- SO_x less than 0.15 lbs (as SO_2) per million BTU of fuel input.
- Particulate matter less than 0.0075 lbs per million BTU of fuel input.
- All solid waste streams will be benign.
- Initially the fuels will be coal and natural gas with coal providing at least 65% of the heat input with a growth potential to 95%.
- The commercial plant will have a 65% annual capacity factor and generate electricity at a 10% lower cost than current conventional plants.

A design team, comprised of United Technologies Research Center, PowerTech International (Reaction Engineering International and PSI Technology Company), the University of North Dakota Energy and Environmental Research Center, Bechtel Corporation, Oak Ridge National Laboratories and United Technologies Turbo Power and Marine Division, has developed a conceptual design for a combined cycle plant that includes a high temperature combustor/air heater (HITAF) which will meet these specifications.

HIGH-PERFORMANCE COMBINED CYCLE

Preliminary studies have identified a baseline power plant concept consisting of two HITAF units, one gas turbine and a single steam turbine producing an overall plant output of about 268 MW at a projected heat rate of 6867 BTU/kWh. Studies by United Technologies Turbo Power and Marine Division and Bechtel Corporation have indicated that this size plant is in the range being considered by the utility industry for coal-fired combined cycle plants (250 MW–400 MW). Also, highly reliable and efficient steam turbine and heat recovery equipment can be obtained commercially at reasonable $/kW in this size range, thereby reducing the risk and the cost of commercial implementation.

Much of the development will be centered around the new High Temperature Advanced Furnace (HITAF) which will be used to preheat clean, pressurized air (the working heat transfer fluid). Both a radiant heat exchanger and a convective heat exchanger will be integrated with the design of the HITAF to provide maximal heat conversion to the working fluid. Since gas turbine efficiency increases with increasing inlet temperature, up to 35% of the thermal input to the overall cycle will consist of natural gas utilized in a duct burner. The duct burner is used to increase the temperature of the air exiting the HITAF heat exchanger from 1255 K to a turbine inlet temperature of 1644 K. The resulting turbine exhaust will have combustion products present due to the use of the duct burner, and will have a vitiated oxygen concentration of approximately 17.6% at a turbine outlet temperature of 811 K. A portion of this vitiated air stream will be used as combustion air in the HITAF and the remaining portion will be passed through a steam cycle for residual heat removal. The flue-gas from the HITAF will also be passed through the steam cycle for heat removal.

The environmental advantages of the proposed cycle include reductions in NO_x, SO_x and particulate emissions, as dictated by the program goals, and also reductions in the emission of CO_2 which is considered to be a 'greenhouse gas' and may be a contributor to global warming. Due to the higher overall conversion efficiencies, much less coal is utilized to produce the same amount of electricity thereby reducing the amount of CO_2 produced per kilowatt·hour of electrical energy. Also, the use of 35% natural gas in the cycle provides an additional reduction in CO_2 emissions over the use of coal. Natural gas, which is predominantly CH_4, provides a higher hydrogen-to-carbon ratio than coal (4:1 as opposed to typically less than 1:1); therefore, the products from combustion of natural gas tend to contain more H_2O and less CO_2 than coal.

COMBUSTOR DESIGN

The design of the HITAF unit must address the following major issues:

  Ensure complete combustion.
  Minimize the formation of nitric oxides.
  Produce ash particles that do not affect the performance of the air heater.

The selection of an appropriate HITAF design for implementation in the new combined cycle will require the determination of an optimal compromise of these issues. Two different design variations are under consideration during the initial stages of development and these designs are described in the following paragraphs. Both computational and experimental techniques are used to evaluate the combustion performance of these potential HITAF designs.

Long Axial Flame

The formation of nitric oxide from the combustion of coal has been shown to be predominantly from the oxidation of fuel-bound nitrogen [Pershing and Wendt, 1977]. In addition, the extent of initial fuel/air mixing has been shown to have a large impact on NO formation as shown by emissions from premixed flame versus an axial diffusion flame [Pershing et al., 1990; Heap et al., 1973]. These investigators attributed this effect to the varying degree of oxygen in contact with volatile nitrogen fractions of the coal and that the formation of N_2 from volatile nitrogen species is therefore maximized when the fuel/air contacting rate is minimized (using the axial diffusion flame).

In order to take advantage of this effect, one of the designs considered for the HITAF utilizes a long axial diffusion flame for in-furnace NO_x control. The long flame is naturally staged, providing a very fuel rich core with gradual addition of air by the coaxial secondary air stream. The long flame allows the persistence of the fuel-rich core for long residence times promoting the continued devolatilization of volatile nitrogen species in an oxygen-deficient environment.

Rich, Well-Mixed Flame

It is well known that NO_x formation in atmospheric-pressure continuous combustors is not limited by thermodynamic equilibrium considerations. The equilibrium concentrations of fixed nitrogen species (nitrogen species other than N_2) are very low under fuel-rich conditions and low NO_x combustor designs using a fuel-rich chamber are based on this fact. The primary chamber is operated fuel-rich to provide the optimal conditions for low fixed-nitrogen species at equilibrium, and the chamber is operated hot to drive the kinetics towards equilibrium. Burnout air is added downstream, and cooling is used to lower the gas temperature prior to air addition to minimize oxidation of remaining fixed-nitrogen species to NO. The cooling can be achieved by heat extraction or segmented air addition.

DESIGN APPROACH

Combustion Simulations

In order to be able to evaluate the performance of the two different design options, computer simulations were utilized to predict the full-scale combustor performance as a function of combustion parameters. The computer model utilized for the combustion simulations (JASPER) is a steady-state, axisymmetrical, computational-fluid-dynamics code which fully couples the effects of reacting gases and particles with the impacts of turbulence and radiation. The turbulence is simulated using a two-equation model (k-ε) for closure. The gas-phase equations are solved using an Eulerian framework and the coal-particle trajectories, solved in a Lagrangian framework, are coupled with the gas-phase equations through particle source terms in both mass- and energy-continuity equations. The gas-phase kinetics are assumed to be mixing-limited; therefore, the gaseous combustion is modeled using a statistical probability density function based on the mixture fraction of the inlet streams. The statistics of the mixing of the coal off-gas is similarly computed. Devolatilization of products from the coal particles, char oxidation, and particle swelling and fragmentation are included in the comprehensive simulation based on time-mean properties of the surrounding gas phase. Particle and gas-phase radiation are modeled using the discrete ordinates method.

Experimental System

Additional design evaluations were made utilizing a 100,000 BTU/hr combustion research facility at the University of Utah. The facility, shown in Figure 1, has a "U" configuration and is a down-fired facility allowing simulation of various aspects of the HITAF system. There are both natural gas and electric preheaters which allow variation in the degree of vitiation and in the level of preheat of the combustion air. The burner is located at the top of the left side of the "U"-shaped furnace in the drawing shown in Figure 1 and has a long vertical path over which long axial flames can be tested. The test facility is equipped with a multitude of access ports over the entire length of the combustion chamber allowing extensive testing of various staging and reburning configurations. In addition, the large single section on the right, or outlet side

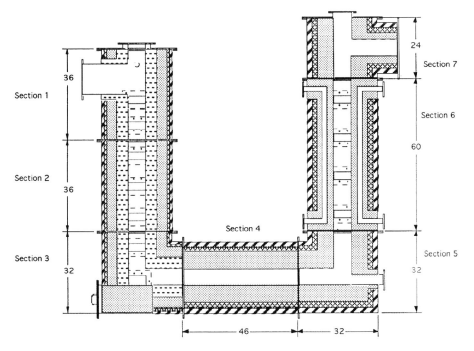

FIGURE 1   100,000 Btu/hr down-fired test facility (the burner fits on the top left section). All dimensions are in inches.

of the "U" facility is fitted with auxiliary burners in isolated combustion chambers allowing for the maintenance of an isothermal reaction zone for SNCR testing.

HEAT TRANSFER CONSIDERATIONS

A primary consideration for both the long flame and the rich, well-mixed design is their ability to effect the required heat transfer to the working fluid in the radiant heat exchanger. The issues of concern include the outlet temperature of the working fluid, the fireside wall temperature and the temperature of the flue gas at the outlet of the HITAF. The outlet temperature of the working fluid is important as the cycle is limited on the amount of natural gas that can be used to boost that temperature to the desired turbine inlet value of 1644 K. The fireside wall temperature is important as the design calls for a flowing slag layer and the walls must be hot enough to prevent the buildup of a very thick layer of ash prior to the onset of viscous flow. The deposit thickness on the radiant heat exchanger panels will have a direct impact on the heat transfer rate to the working fluid. The temperature of the flue-gas is important as it ultimately encounters the convective heat exchanger, and this heater is to be made of materials that cannot withstand temperatures in excess of 1255 K. In order to minimize the amount of cooling required prior to reaching the convective heater (accomplished either by flue gas recirculation or the use of water walls), the flue-gas exit temperature must be minimized. A parametric modeling study was therefore required to investigate the impact of variations in combustor type, ash deposition and combustor geometry on the ability to achieve these heat transfer constraints.

With respect to combustor type, it should be noted that the two designs have very different heat release patterns. The rich, well-mixed case has a fairly uniform heat release beginning very early in the axial length of the combustor. The long axial flame, however, has a very delayed heat release and may require additional surface area for heat transfer. A series of simulations were carried out investigating the effect of increasing combustor length for a fixed diameter (8 m) with both combustor types. The results, shown in Figure 2, indicate that an additional eight meters of length are required with the long axial flame to achieve the same heat exchanger outlet tempera-ture obtained with the well-mixed flame. This additional length corresponds to approximately 200 m^2 of additional surface area.

To quantify the impact of ash deposition on the heat-transfer effectiveness, various values of an overall wall heat transfer resistance were used with a well-mixed combus-tor simulation in order to compute the change in heat exchanger outlet temperature. The overall wall heat transfer resistance represents a combined resistance due to the liquid slag layer, the dry ash layer, a refractory layer, the tube wall in the radiant panel and a convective heat transfer coefficient within the tube. The results of these simulations are shown in Figure 3 where the air outlet temperature is plotted as a function of the wall resistance value used in each simulation. As is shown, the outlet temperature falls off rapidly with increasing resistance indicating the need to minimize deposit thickness. The value required to achieve the target outlet temperature, how-ever, does represent reasonable operating conditions for the HITAF indicating the feasibility of the proposed design.

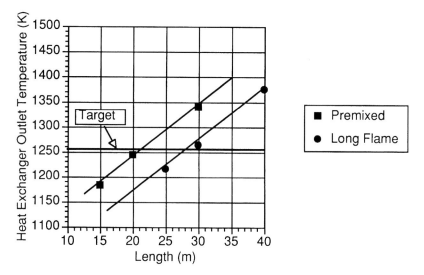

FIGURE 2   Effect of combustor length on radiant heat exchanger outlet temperature for the two flame types.

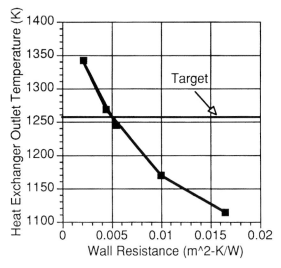

FIGURE 3   Radiant heat exchanger outlet temperature as a function of the overall wall heat-transfer resistance.

Changing the dimensions of the furnace can also affect the radiant heat transfer to the working fluid by altering the radiating flame volume, and the impact of such changes were investigated with a series of well-mixed combustor simulations. Variations in both combustor diameter and length for the well-mixed design were investigated and the results are shown in Figure 4. The computed heat exchanger outlet temperature is shown as a function of combustor length for three different diameters.

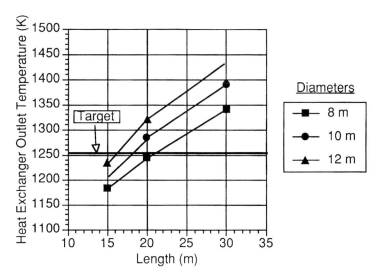

FIGURE 4   Effect of variations in combustor geometry on the radiant heat exchanger outlet temperature.

The required outlet temperature is indicated by the horizontal line and the usable geometries are those which yielded values above the target line. The wall resistance used for these calculations was the minimum acceptable value obtained from the previous set of calculations, 0.0055 m$^2 \bullet$ K/watt.

It should be noted that the geometrical configuration which is best suited to obtaining the maximum heat exchanger outlet temperature, is not necessarily the best configuration for the other heat transfer constraints. Recall that in addition to achieving the design heater outlet temperature, the fireside wall temperature must be high enough to maintain a flowing ash layer and that the temperature of the exiting flue-gas must be minimized. Figure 4 showed that increases in combustor geometry were favorable for maximum heater outlet temperature, primarily because of the corresponding increase in heat exchanger surface area. The same calculations, however, showed that increases in diameter and length had a tendency to decrease the fireside wall temperature. Similar decreases were observed for the flue-gas exit temperature. These results indicated that there should be an optimum geometry which would satisfy all three heat transfer constraints. The predicted flue-gas exit temperature as a function of length and diameter is plotted in Figure 5 along with the region of geometries achieving the heat exchanger outlet temperature and the region of geometries maintaining wall temperatures above the prescribed slagging value for the Utah bituminous coal. As shown in Figure 5, there is an optimum geometry of approximately 9 m diameter and 25 m length which minimizes the flue-gas exit temperature and still achieves the design heater outlet temperature and a flowing slag layer.

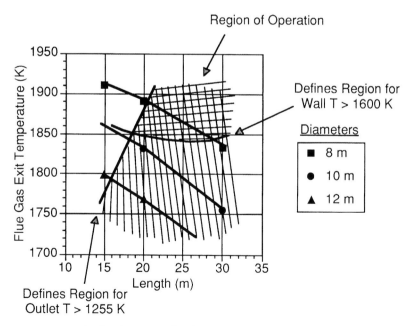

FIGURE 5   Superimposition of the three thermal design constraints as function of combustor geometry.

The results of these simulations indicate that both designs, the long axial flame and the rich, well-mixed flame, are capable of achieving the heat transfer constraints imposed upon the HITAF and also that there are trade-offs to be considered in terms of combustor size when selecting the design to be utilized for the HITAF application. There are additional trade-offs between the two designs related to NO_x formation which will be discussed in the next section.

NO_x CONSIDERATIONS

Kinetic and Equilibrium Calculations

The aerodynamic staging of long axial flames appears to be an effective means for controlling NO formation since it is the fluid mechanics which control the extent of mixing and the fluid mechanics are easily adjusted by burner design and operation. For the rich, well-mixed design, however, there are questions as to how fuel-rich does the combustor need to be to minimize fixed nitrogen species and how hot does the chamber need to be to effectively drive the kinetics toward the low equilibrium values. To answer these questions, a series of equilibrium and kinetic calculations were performed to determine if the stoichiometries and temperatures required represented feasible operating conditions.

Equilibrium calculations were performed with a Utah bituminous coal to determine the total fixed nitrogen production $(NO + NH_3 + HCN)$ as a function of stoichiometry. The results, shown in Figure 6, indicate a minimum in equilibrium TFN at a stoichiometric ratio of 0.6 (using vitiated combustion air); therefore, it would be desirable to operate the rich combustor near this stoichiometry to obtain the minimum possible TFN for this coal.

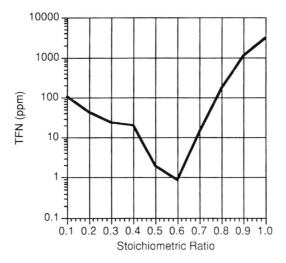

FIGURE 6   Total fixed nitrogen predicted at equilibrium using a bituminous coal burned in vitiated air.

There still remains the question, however, as to how hot the combustor needs to be before TFN decay can approach this equilibrium value. The maximum temperature possible for a given set of conditions is given by the adiabatic flame temperature; therefore, an understanding of the variation of this temperature with stoichiometry is required before performing detailed rate calculations of TFN decay. Figure 7 shows the adiabatic flame temperature for the bituminous coal as a function of stoichiometric ratio (again using the high-temperature, vitiated combustion air). It is apparent that peak flame temperatures occur near a stoichiometric ratio of 1.0 and that as stoichiometry is decreased, the flame temperature falls off rapidly. At a stoichiometry of 0.6 the adiabatic flame temperature has dropped to a value of approximately 1800 K from a maximum value of nearly 2350 K. Thus, a reduction in stoichiometry to provide a reduced equilibrium TFN production will also result in a considerably lower chamber temperature.

To investigate the residence times necessary for achieving significant TFN decay, the CHEMKIN kinetics program [Kee *et al.*, 1980] was used with the Utah bituminous coal. As the nitrogen chemistry is considered to be rate-limiting under these conditions, the equilibrium coal combustion products were used as program inputs with the addition of 600 ppm of HCN. The added HCN, which simulates the evolved product of volatile fuel-bound nitrogen, was allowed to decay using the Miller-Bowman [1989] mechanism set and the rate of decay as a function of temperature was determined. The results, shown in Figure 8, indicate that temperatures in excess of the adiabatic flame temperature (at SR = 0.60) are required to achieve substantial TFN decay in less than 1 second. If heat removal is utilized, which would be the case with the radiant heat exchanger, the rate of TFN decay drops off rapidly, with only a 30% reduction in TFN after 3 seconds for a temperature that is only 100 K less than adiabatic conditions.

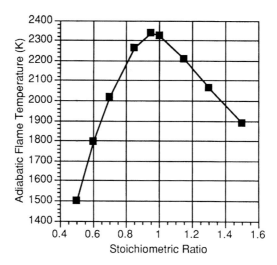

FIGURE 7   Adiabatic flame temperature as a function of stoichiometry for bituminous coal burned in vitiated air.

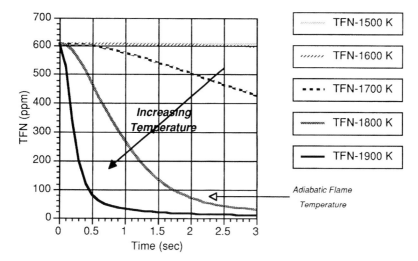

FIGURE 8   TFN decay as a function of temperature for bituminous coal in vitiated air with the addition of 600 ppm HCN.

These results indicate that the rich, well-mixed concept may require heat addition in order to provide an environment capable of minimizing NO_x, and as the purpose of the combustor is to provide heat to the working fluid, the rich, well-mixed concept appears to be a less feasible design from an NO_x standpoint.

Combustor Simulations

The combustion model (JASPER) was also used to evaluate the NO performance of the two different flame types. A comparison of the axial NO profiles predicted using the two different flames is shown in Figure 9. The figure illustrates the average NO values predicted at each axial location along the length of the combustor. As is shown, the NO emissions from the long flame case are predicted to be considerably lower than the well-mixed case. The delayed mixing of the naturally-staged long flame reduces fuel/air contacting in the early stages of the flame allowing for devolatilization of fuel-bound nitrogen in an oxygen-deficient environment. In addition, enough residence time is available within the fuel-rich core to allow the majority of the fuel nitrogen to react to form molecular nitrogen (N_2) prior to encountering significant amounts of oxygen. The oxidation of N_2 to NO does not occur by the same mechanism as the fuel-bound nitrogen and requires considerably higher temperatures for significant conversion.

Preliminary Experimental Results

The initial experimental studies focused on evaluating the ability of the long axial flame concept to produce low NO_x emissions under conditions representative of the proposed HITAF design. All of the experiments reported in this section were conducted with a Utah bituminous coal, firing at 90,000 BTU/hr. The secondary air was preheated to 616 K and was introduced axially; no swirl was used in these experiments.

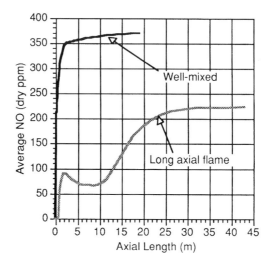

FIGURE 9   Predicted average NO profiles for the two flame types.

The primary air was held constant at approximately 15% of the stoichiometric requirement and was not preheated. The overall excess oxygen in the exhaust was held at 3%.

Research funded by the EPA at the International Research Foundation in the early seventies [Heap and Folsom, 1990] identified the importance of both near-burner fuel/air mixing and the point of flame stabilization on NO formation in pulverized-coal flames. Axial fuel injection tends to naturally retard near-field mixing and thereby reduce NO_x emissions, relative to fuel injection schemes which direct the fuel radially outward into the secondary air. However, if the point of ignition of the axial fuel jet is displaced, significant secondary air entrainment can occur prior to ignition, thus increasing the oxidation of fuel nitrogen within the core of the fuel jet. If ignition occurs at the injector tip, then entrainment is greatly decreased and NO_x emissions decrease.

In this study, varying amounts of natural gas were injected through five axial holes around the outside of the primary air/coal stream, to provide positive anchoring of the axial flame. As the data in Figure 10 indicate, this can have a major impact on NO_x emissions. With all axial secondary air and axial coal injection (0% natural gas), the coal flame is lifted approximately 30 cm from the fuel injector tip and the NO_x emissions are relatively high (870 ppm). Addition of small amounts of gas do not immediately pull the flame back onto the injector tip; however, once sufficient gas is added, the flame front moves back and the emissions drop dramatically. Further, once the flame has been located, only a small amount of gas is required to maintain ignition at this point. As the data indicate, 5% gas is quite sufficient to maintain this ignition and this value was used in all subsequent studies.

Similar tests were also conducted on the effects of primary and secondary air flow rate variations and, as expected, increasing the percent of total air introduced in the primary stream increased NO_x emissions significantly. Increasing secondary air (and hence overall excess air) also increased NO_x emissions but less dramatically.

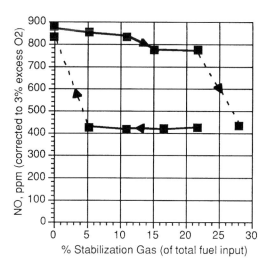

FIGURE 10 Measured NO emissions as a function of the percent of total fuel input introduced as pilot gas in the coal flame.

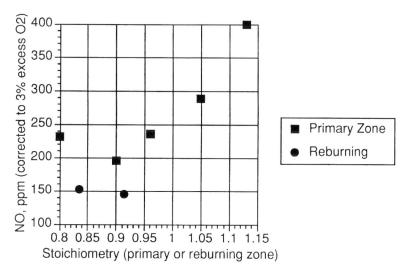

FIGURE 11 Measured NO emissions as a function of either primary zone stoichiometry or reburning zone stoichiometry.

Figure 11 shows the effects of staging and reburning on NO_x emissions with the gas-stabilized axial flames. Decreasing the primary zone stoichiometry from 1.15 (no staging air) to approximately 0.95 reduced the emissions from 400 ppm to just below 200 ppm. Similar reductions have been reported previously by other investigators for conventional flames [Chen *et al.*, 1986; Borio *et al.*, 1989] but the optimum stoichiometry was found at more fuel-rich conditions (0.6–0.7). This apparent shift in optimum stoichiometry is likely due to the fact that the gas-stabilized axial flames are naturally-staged due to the slow fuel/air mixing.

Figure 11 also shows the application of natural gas reburning to this system. For natural gas additions to the levels shown (reburning zone stoichiometries of 0.92 and 0.83), it was possible to reach emission levels below 150 ppm NO_x without the use of selective NO_x reduction agents. In this case, natural gas was used as the reburning fuel and was injected at the same point as the second stage air in the previous experiments (at port #1 in furnace section 3 in Fig. 1), about 2 meters downstream from the point of coal injection. In the reburning tests, the final burnout air (to a level of $\sim 3\%$ O_2 in the flue) was added at port #1 in section 4 (about 1 meter downstream from the point of reburning gas injection). At the minimum NO_x conditions, approximately 10% of the total fuel input was reburning gas.

The initial experimental results demonstrate the potential of the gas-stabilized axial flame to produce very low NO_x emissions under staged-combustion conditions. Future work will focus on optimizing the application of reburning and selective reduction agent addition to minimize the emissions from the overall process, and in order to reach NO_x emissions levels below 100 ppm .

CONCLUSIONS

Two different flame types were considered for use in a High-Temperature Advanced Furnace which is to be integrated into a new combined cycle for power generation. Computer simulations were used to quantify the impact of flame type (long axial flame vs. well-mixed rich flame), ash deposition and combustor geometry on both heat transfer constraints and NO formation. Limiting-case detailed kinetics calculations were also performed to evaluate the NO performance of the rich, well-mixed flame concept.

Based on these computational results, the long axial flame appears to have a greater potential for use as the HITAF design. Not only is in-furnace control of NO emissions more feasible with the long axial flame, but ash deposition on the radiant panels is also minimized, due to the predominantly downward flow direction of the combustion gases.

Experimental verification of the long flame concept also showed the viability of coupling the naturally-staged flame with additional NO_x control schemes, such as additional air staging and reburning, to provide for very low NO emissions from a HITAF based on the long axial flame concept.

ACKNOWLEDGEMENTS

Financial support for this work was provided from the Department of Energy under Contract #DE-AC22-92PC91155. The DOE Program Manager is Mr. Clifford Smith. The authors would like to acknowledge the contributions of their co-workers and the HITAF design team from United Technologies Research Center, PSI Technology Company, the University of North Dakota Energy and Environmental Research Center, Bechtel Corporation and Oak Ridge National Laboratories.

REFERENCES

Borio, R. W., *et al.* (1989) "Application of Reburning to a Cyclone Fired Boiler", Proceedings: 1989 *Joint Symposium on Stationary Combustion NO$_x$ Control*, San Francisco, CA, **1**, EPA-600/9-89–062a (NTIS B89–220529).

Chen, S. L., Heap, M. P., Seeker, W. R. and Pershing, D. W. Bench and Pilot-Scale Process Evaluation of Reburning for In-Furnace NO$_x$ Reduction, *Twenty-First Symposium (International) on Combustion*, The Combustion Institute, Pittsburgh, PA (1986).

Heap, M. P. and Folsom, B. A. (1990) "Optimization of Burner/Combustion Chamber Design to Minimize NO$_x$ Formation During Pulverized Coal Combustion", Ch. 9, Pulverized Coal Combustion: Pollutant Formation and Control, 1970–1980, EPA-600/8-90-049.

Heap, M. P., Lowes, T. M. and Walmsley, R. "Emission of Nitric Oxide from Large Turbulent-Diffusion Flames, *Fourteenth Symposium (International) on Combustion*", The Combustion Institute, Pittsburgh, PA. (1973)

Kee, R. J., Miller, J. A. and Jefferson, T. H. (1980) "CHEMKIN: A general-purpose, problem-independent, transportable, Fortran chemical kinetics code package", Sandia Rep. SAND80–8003.

Miller, J. A. and Bowman, C. T. (1989) "Mechanism and Modeling of Nitrogen Chemistry in Combustion", Prog. *Energy Combust. Sci.*, **15**, 287–338.

Pershing, D. W. and Wendt, J. O. L. "Pulverized Coal Combustion: The influence of Flame Temperature and Coal Composition on Thermal and Fuel NO$_x$", *Sixteenth Symposium (International) on Combustion*, The Combustion Institute, Pittsburgh, PA (1977).

Pershing, D. W., Heap, M. P. and Chen, S. L. (1990) "Bench-Scale Experiments on the Formation and Control of NO$_x$ Emissions from Pulverized-Coal Combustion", Ch. 8, Pulverized Coal Combustion: Pollutant Formation and Control, 1970-1980, EPA-600/8-90–049.

Coal Gasification Power Cycles Using Mixtures of CO_2, Steam and Argon

H. VERELST[a], T. VAN MIERLO[a], K. MANIATIS[a], G. P. SAKELLAROPOULOS[b], STÖSTÖM[c], and P. PILIDIS[d]

[a]Vrije Universiteit Brussel, Brussel, Belgium; [b]CPERI and University of Thessaloniki, Thessaloniki, Greece; [c] Kungl Teknisa Hogskolan (KTH) Stockholm, Sweden; [d]Cranfield Institute of Technology, Bedford, UK

Abstract—This paper investigates the use of a mixture of CO_2, steam and argon as gasification agent and as power cycle fluid. The process is based on the gasification of bituminous coal or lignite with a mixture of CO_2, steam and eventually argon.

Sulphur compounds and HCl are removed up to 97% by adding dolomite in the fluidized bed reactor. Eventually the rest H_2S is removed in an acid gas scrubber. Oxygen instead of air is used in the gasification process and the combustion of the fuel gas. The flue-gasses will contain almost pure carbon dioxide and water vapour and are ready for direct compression and safe disposal.

Dried bituminous coal or lignite is fed to the gasifier at a rate of 20 kg/s, corresponding with a thermal energy input around 560 MW. The net power produced in a semi open cycle, not taking into account the gas cleaning and using a simple gas turbine, is found to be around 149 MW, resulting in a global efficiency of 27%. Taking the oxygen production (95% O_2) into account, the global efficiency lowers to 20%. Adding a recuperative Rankine cycle and the CO_2 compression energy, the global efficiency of the cycle is 39.2%.

Key Words: Combined cycle, clean gas technology

INTRODUCTION

Emissions of carbon dioxide are considered as the major source of increase of the greenhouse effect. Greenhouse mitigation strategies are under development but it must be recognised that fossil fuels will continue to play a dominant role in electrical power generation for several decades. Until now very few initiatives have been undertaken to solve the problem seriously, namely a more rational and efficient capture of CO_2. Carbon dioxide is mainly considered as an inert gas resulting from combustion processes, this attitude must be reversed and CO_2 must be used whenever possible in power cycle processes. A mixture of CO_2, steam and argon can be used as gasifying medium and power cycle fluid. Until now such an approach has not been examined.

In this project a mixture of oxygen and argon is used in the gasification process. Furthermore, it is believed that the use of argon eventually can increase the global efficiency of the power cycle.

In the gasification step a mixture of CO, CO_2, H_2, steam, CH_4 and traces of H_2S is produced. SO_2 and HCl can be eliminated by adding dolomite in the gasifier fluidized bed unit. No NO_x will be formed since N_2 is already eliminated in the air separation and the operating temperature in the gasifier is below 1000°C. After combustion, and condensed water removal, the gas will contain almost pure carbon dioxide and is ready for safe disposal.

TECHNICAL DESCRIPTION

The process is based on the gasification of coal or lignite in a bubbling fluidized bed with a mixture of CO_2, steam and argon as fluidization agent. The process is showed schematically in Figure 1.

Lignite or coal is fed into a sand bubbling fluidized bed gasifier, adding dolomite. The fuel is pyrolyzed and the resulting char is gasified by the mixture of carbon dioxide and steam. The eventually produced sulphur compounds, mainly H_2S, and/or HCl will react with the dolomite.

The resulting ash is removed from the bed. Large particulates are removed in a cyclone while the remaining fine dust is eliminated in a high temperature ceramic filter. Other sulphuric components mainly hydrogen sulphide and COS are removed, after cooling down, in an absorber unit, upstream the oxyfuel burner.

A cryogenic column is used to carry out the air separation, producing a mixture of 95% oxygen and 5% argon, being less expensive than producing 99.5% oxygen (0.23 kWh/kg instead of 0.25 kWh/kg).

The oxygen-argon mixture is used for the combustion of the produced low heating value gas. At present, in the simulations, the power generation is considered as a black box, represented by a single gas turbine, since several alternatives are possible. The gas produced will contain almost pure carbon dioxide (with water vapour and some inerts in minor concentration).

Part of it can be compressed immediately and is ready for safe disposal under the ground or in the sea. The other part is recycled to the fluidized bed gasifier along with waste steam from the power generation.

SIMULATIONS

Process Description

Simulations are performed using the process simulator ASPEN+. The first calculations were carried out using a simplified flowscheme (Fig. 2).

The gasifier has to be presented by two theoretical units, one where a decomposition of the coal in its elements is carried out, followed by a unit, starting from the elementary coal composition, leading to all possible thermodynamical equilibrium products. Experiments to provide the necessary kinetics for lignite char gasification in CO_2 are carried out at KTH in Sweden.

The feed-CO_2 stream (CO_2GASIF), containing about 90% CO_2 and 10% steam, is compressed up to 15 bars, as well as the necessary oxygen, whose flowrate is regulated so that the gasifier works quasi adiabatically.

Particulates are removed from the productgas using a cyclone. The ceramic filter is not included for the moment in the simulations. Oxygen is also fed to the burner at a flowrate to have a complete conversion of CO in CO_2 without having an oxygen excess. Part of the produced CO_2 will be recycled, after cooling and removal of the condensed water, to keep the temperature of the produced gas, going to the gasturbine, at its maximal possible value of 1200°C.

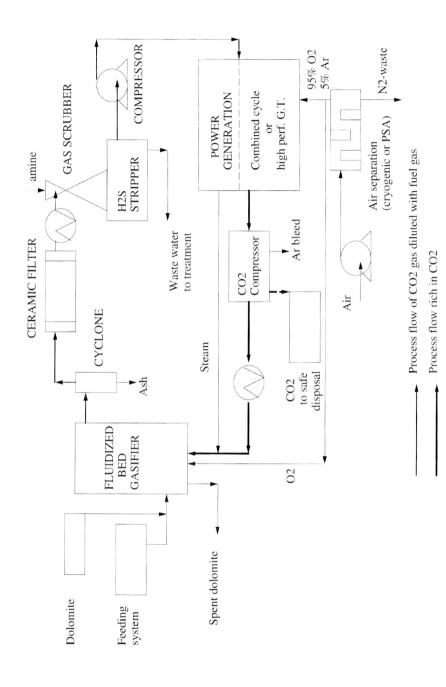

FIGURE 1   Scheme of the process.

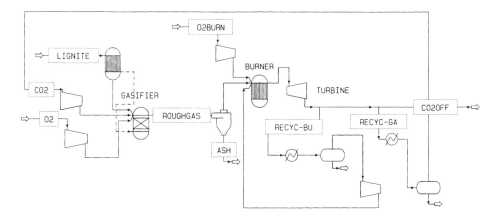

FIGURE 2   Aspen + flowscheme for the power cycle.

The product gas from the burner (HOTGAS) is fed to a gas turbine unit, followed by a waste heat boiler, producing steam for a Rankine cycle. The consumed and produced work for compressors and turbine are combined, resulting in a net power stream.

Gasifier

For reasons of simplicity, all calculations were made, as already mentioned, treating the fluid bed gasifier as a equilibrium reactor.

The standard case, to which all different sensitivity studies have to be referred, was operated at 15 bar, 950°C, with a CO_2 recycle flowrate of 20 kg/s.

The coal, with a heating value of 28.0 MJ/kg, is fed to the gasifier at a flowrate of 20 kg/s, corresponding to a thermal energy input around 560 MW. The composition of the dry coal used for the simulation is presented in Table I.

As fluidization and gasification medium a mixture of mainly CO_2, with 10% of steam as moderator was used. From Figure 3, where the produced CO, CO_2 and H_2 flowrates (from the gasification) as function of the recycled CO_2 flowrate are presented, it is clear that, for a complete conversion of the coal, a minimum flowrate of about 20 kg/s of

TABLE I

Composition of the bituminous coal

elementary components	weight(%)
ash	11.03
C	68.68
H	5.01
N	1.26
Cl	0.21
S	4.42
O	9.39

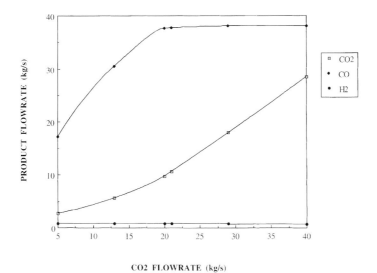

FIGURE 3   Influence of the CO_2 flowarate.

recycled CO_2 is needed. On the other hand, a higher flowrate of recycled CO_2 doesn't increase the produced CO-amount any further, leading to a lower heating value of the produced gas. The composition of the produced gas, with a heating value of 9.3 MJ/kg, is presented in Table II.

A study of the temperature influence (Fig. 4) in the gasifier, suggests a working temperature between 900 and 950°C, corresponding to values used for efficient steam gasification.

In all cases the necessary oxygen for the gasification process was kept at a minimum value, in a way that the gasifier operates quasi adiabatically.

Taking into account the minimum fluidization (u_{mf}) velocity for the fluid bed (sand particles, $\rho_s = 2.67 \ 10^3$ kg/m^3, $d = 250$ μm) of about 0.05 m/s (Leva) and the maximum loading capacity of a coal gasification fluid bed being about 750 kg/m^{2}h, the value of 20 kg/s (at 950°C and 15 bar) corresponds to a velocity equal to 3 times u_{mf}. This is in practice too low for an efficient fluid bed operation.

In a sensitivity study (Fig. 5), where the influence of the pressure on the equilibrium conversion is examined, it is clear that at pressures below 30 bar, no effect can be

TABLE II

Composition of the produced gas

		molar fraction
CO	37.1 kg/s	61%
H_2	0.84 kg/s	17%
CO_2	13 kg/s	13%
CH_4	0.12 kg/s	0.3%
Total sulphur		1.2%
Total chlorine		500 ppm

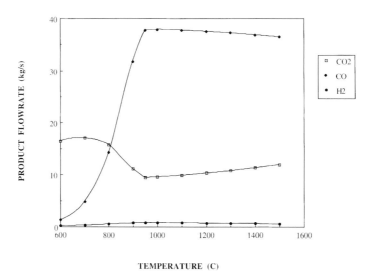

FIGURE 4   Influence of the gasifier temperature.

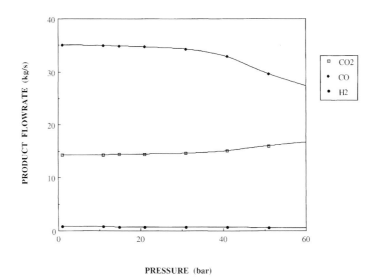

FIGURE 5   Influence of the gasifier pressure.

identified. This suggests to operate a CO_2 gasification process at rather low pressures, since at such conditions the gas velocity is substantially higher although the CO_2 mass flowrate can be kept low enough to obtain a gaseous fuel which still has a reasonable heating value of above 9 MJ/kg (increasing the CO_2 to C ratio to much above the stoechiometric value leads to a too low heating value).

In the real practical case dolomite will be added, resulting in a catalytic activity for the char decomposition, a removal of the chlorine and a sulphur removal up to 97%.

THE POWER CYCLE

Introduction

After the gasification, the produced gas is treated in an ash and dust removal system, presented in the model by a single cyclone, and, if necessary, cleaned further, after cooling with heat recuperation, in a desulphurisation unit for removal of H_2S (not included in this scheme). The gas is, eventually after further compression, sent to a burner, operated at 1200°C (maximum allowed temperature for the expansion turbine), where it is expanded till 1 bar. The expanded gas, between 700 and 800°C (depending on the inlet pressure) is cooled down in a countercurrent waste heat boiler, producing steam for a Rankine cycle.

Finally, the gas consisting mainly of CO_2, water vapour and eventually argon, is cooled further till ambient conditions (heat-sink). The condensed water is separated from the gas stream, which is split in two recycle streams necessary for operation of the fluid bed gasifier and as diluent to keep the burner outlet temperature at the maximum possible value of 1200°C, and an off gas stream of nearly pure CO_2, ready for compression (till 60 to 100 bar) and safe disposal.

Influence of the Oxygen Purity

If an oxygen purity of 99.5% is desired, a cost of 0.25 KWhr/kg has to be included. For an oxygen purity of 95% 0.23 KWhr/kg is needed. Since 13.7 kg/s O_2 is needed in the gasifier and 29.3 kg/s in the burner, a supplementary energy input of 35.6 MW is needed for 95% oxygen and 38.9 Mw is needed for 99.5% oxygen.

Not only the oxygen cost is related to the oxygen purity, but also the offgas compression cost due to the ballast of argon. When using 99.5% oxygen, 18.8 MW is needed to compress the offgas till 100 bar. For 95% oxygen 19.7 MW is needed.

The oxygen purity has also an influence on the power produced by the gasturbine and the power produced by the recuperative Rankine cycle. This will be discussed later.

Rankine Cycle

The hot outlet stream of the gasturbine is fed to a countercurrent heatexchanger where steam is produced, used in a recuperative simple Rankine cycle (Fig. 6). The pressure in the boiler amounts to 60 bar; the pressure in the condenser is 0.1 bar (around 40°C). The Rankine cycle contains a single expansion.

Efficiency of the Cycle

The thermal energy input of the cycle is 560 MW. When using 88% efficiency for the gas turbine and 78% efficiency for the compressors, the produced power is 149 MW, giving a global efficiency of 27% (without a waste heat recuperation Rankine cycle).

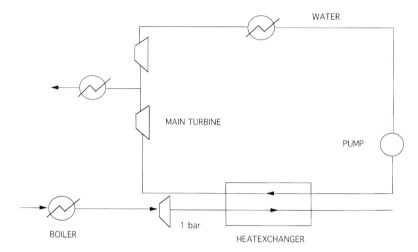

FIGURE 6   Rankine cycle.

The oxygen production will decrease of course the efficiency. When using 99.5% oxygen, 38.9 MW is needed, decreasing the global efficiency to 19.5% . The efficiency will be increased by using a mixture of 95% O_2 and 5% argon which has a power demand of 35.6 MW (20.2% efficiency). Further the CO_2 compression energy for safe disposal has to be taken into account, as well as the gas cleaning (ceramic filter and H_2S scrubber), operations which will decrease the global efficiency further.

When taking the Rankine cycle into account, the global efficiency will be increased up to 39.2%. Table III gives an overview of the energy costs and benefits of the combined cycle.

Another improvement of the cycle can eventually be the use of a mixture CO_2 and steam, enriched in argon, as working fluid (higher c_p/c_v ratio). However simulations showed that the netpower increased with the amount of argon in the cycle, but the global efficiency decreases due to the need of an extra gas separation unit. So no other working fluid as the produced fuel itself has to be used in the cycle.

TABLE III

Efficiency of the cycle

	95% oxygen	99.5% oxygen
O_2 production	35.6 MW	38.9 MW
CO_2 compression	19.7 MW	18.8 MW
turbine cycle	149.1 MW	147 MW
Rankine cycle	126.3 MW	127 MW
	220.1 MW	216.3 MW
efficiency	39.2%	38.5%

CONCLUSIONS

A mixture of carbon dioxide, steam and argon can be used as gasifying agent and as power cycle fluid. SO2 and HCl are removed using dolomite in the fluidized bed gasifier. H_2S is removed in a physical or chemical absorption unit. No NO_x will be formed since oxygen is used instead of air for all combustion processes. The optimal working conditions of the gasifier are around 950°C and a recycle CO_2 flowrate of 20 kg/s.

Cryogenic air separation produces a mixture of 95% oxygen and 5% argon. By this, the global cycle efficiency will be decreased, but after combustion the gas will contain only CO_2, steam and inerts and no additional cleaning is needed. Part of the produced gas is recycled to the gasifier and the burner. The other part is liquefied and disposed.

The improvement of the process lies in the use of CO_2 instead of steam as the main gasification agent, together with the use of nearly pure oxygen as oxidant, producing an off-gas, containing only CO_2 and steam, together with some inerts. The advantage of this integrated gasification power cycle is that to obtain a zero-emission, only a desulphurisation (no NO_x) is needed, with a direct CO_2 capture.

ACKNOWLEDGEMENT

This project is funded by the CEC DG XII, JOULE II program under contract PL 920567.

REFERENCES

Allam, R. (1985) Air separation and Integration in an IGCC Process. *Air Products and Chemicals.*
Elliot, M. (1981) Chemistry of Coal Utilization. John Wiley and Sons. Canada.
Espie, D. M., Mandler, J. A., Miller, D., O'Connor, D. and Allam, R. J. (1992) The Unique Challenges of Cryogenic Distillation Column Control for Integrated Coal Gasification Combined Cycle Applications. *Air Products and Chemicals.*
Probstein, R. F. and Hicks, R. E. (1985) Synthetic Fuels. Mc Graw Hill. Singapore.
Sens, P. F., Mc Mullan, J. T. and Williams, B. C. (1991) Prospects for Zero Emission Coal-Fired Power Plants. *IEA International Conference on Technology Responces to Global Environmental Challenges.*
ASPEN + Solids Manual (1988).

Concept of Boiler Efficiency Assessment Expert System

N. AFGAN and M. G. CARVALHO   *Instituto Superior Técnico Lisboa, Portugal*

Abstract—Boiler Expert System (Afgan *et al.*, 1991) includes the efficiency assessment module based on the second law efficiency defined by exergy balance of the boiler furnace.

The coal fired boiler exergy analysis has served as a tool to define the domain interest for the respective module of the boiler expert system. Main emphasis is given to the exergy degradation and exergy losses of the boiler furnace. Using this approach in domain definition a set of objectives is derived which are used in the description of the respective situation to be recognized. Knowledge base is structured in accordance with the problem definition with particular reference to the assessment of quality of the coal, excess air, coal humidity, flue gas temperature and unburned carbon.

Diagnostic system is based on the on line reading of operation data, including: exit steam pressure and temperature, steam flow rate, flue gas temperature, CO content in the flue gases. Parameters are converted into semantic symbols to be used in the definition of respective actual situation.

Particular attention was devoted to the heat transfer exergy degradation because the temperature and heat flux distribution in the boiler could be obtained with respective mathematical models of the boiler furnace. In this respect, it may be shown that future development of the numerical models of the boiler furnace describing combustion, fluid flow and heat transfer will be strongly related to the need of the expert system development.

INTRODUCTION

Efficiency assessment for any technical system has been the main requirement for its design, operation and maintenance. It is immanent to this requirement to include the thermodynamic approach based on the first and second law of thermodynamics. The energy efficiency is based on the first law of thermodynamics which is defined by the energy balance of the system. The estimate part of energy considered as the useful energy for the respective system is the end result of the energy efficiency assessment. This means that the part of energy considered as the useful energy divided by the total energy input is the efficiency of the system.

The second law of thermodynamic introduced by Carnot (1824) has described the quality of the energy conversion system by introducing the second law of efficiency. In this respect it was defined that the energy quality could be taken as the parameter for the justification of the energy system. In order to determine quality of energy conversion Rant (1960) has introduced the definition of exergy as a part of energy that can be fully converted into any other kind of energy. However, it must be kept in mind that exergy results from the possibility of an interaction between the matter under consideration and the common components of the environment (Bosnjakovic, 1965).

BOILER EXERGY ANALYSIS

In order to analyze the efficiency of the boiler system let us take it as a control volume system at the steady state for which the energy and exergy balance equations can be written, respectively as

(Energy in) = (Energy in steam produced) + (Energy losses)

(Exergy in) = (Exergy in steam produced) + (Exergy losses) + (Exergy degradation)

In the efficiency point of view, the input is converted to product ratio as
  Energy efficiency

$$\eta = \frac{\text{Energy in steam produced}}{\text{Energy input}} = 1 - \frac{\text{Energy losses}}{\text{Energy input}}$$

Exergy efficiency

$$\varepsilon = \frac{\text{Exergy of the steam produced}}{\text{Exergy input}} = 1 - \frac{\text{Exergy losses}}{\text{Exergy input}} - \frac{\text{Exergy degradation}}{\text{Exergy input}}$$

The main irreversible phenomena that occur in the steam generator are as follows (Szargut et al., 1988):

a. Irreversible combustion
b. Irreversible heat transfer between combustion gases and working fluid
c. Irreversible heat transfer through the wall to environment
d. Irreversible heat transfer with mineral products
e. Irreversibility due to unburned fuel
f. Rejection of hot combustion product gases into the environment.

In order to define the exergy losses due to the irreversible phenomena in steam generator, let us make exergy balance of the steam generator (Morran, 1982 and Kotas, 1980). Figure 1 shows the exergy balance for the steam generator.

$$B_{ch,F} + B_{Air} = B_{steam} + B_{Air} - B_{Air,in} + \delta B_{comb} + \delta B_{ht} + \delta B_w + \delta B_c + \delta B_t$$

where

$B_{ch,F}$	-	exergy of the fuel
B_{Air}	-	exergy of the air
B_{steam}	-	exergy of the steam
B_{comb}	-	exergy degradation in combustion
δB_{ht}	-	exergy degradation by heat transfer
δB_w	-	exergy loss by heat transfer through the wall
δB_c	-	chemical exergy loss by combustible losses in flue gases
δB_t	-	thermal exergy loss by flue gases

Dividing this equation with $B_{ch,F}$, we will obtain

$$1 = \frac{B_{steam}}{B_{ch,F}} + \frac{\delta B_{comb} + \delta B_{ht}}{B_{ch,F}} + \frac{\delta B_w + \delta B_c + \delta B_t}{B_{ch,F}}$$

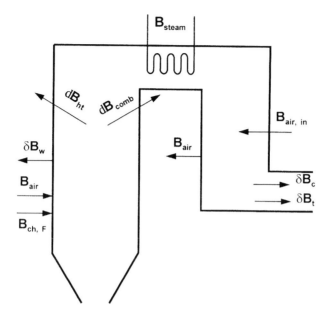

FIGURE 1   Exergy balance for steam generator.

or

$$1 = \varepsilon + \Delta\varepsilon_{dest} + \Delta\varepsilon_{loss}$$

so that

$$\varepsilon = \frac{B_{steam}}{B_{ch,F}}$$

$$\Delta\varepsilon_{dest} = \frac{\delta B_{comb} + \delta B_{ht}}{B_{ch,F}}$$

$$\Delta\varepsilon_{loss} = \frac{\delta B_w + \delta B_c + \delta B_t}{B_{ch,F}}$$

where

ε	-	second law efficiency
$\Delta\varepsilon_{dest}$	-	exergy destruction
$\Delta\varepsilon_{loss}$	-	exegy losses

Second Law Efficiency

The second law efficiency for the steam generator is given as the ratio of the exergy increase of water as it passed from the water to the steam to exergy input with the fuel

(Kotas, 1980 and Ahem, 1980)

$$\varepsilon = \frac{B_{\text{steam}}}{B_{\text{ch},F}} = \frac{D(i_s - i_w)}{B_{\text{ch},F}} = \frac{D[c_{\text{pw}}(T_{\text{sat}} - T_w) + r + c_{\text{ps}}(T_s - T_{\text{sat}})]}{B_{\text{ch},F}} \left(\frac{T_m - T_0}{T_m}\right)$$

where

D	-	steam flow rate
$c_{\text{pw}}, c_{\text{ps}}$	-	specific heat of water and steam, respectively
T_{sat}, T_w, T_s	-	temperature of saturated steam, water and superheated steam, respectively
r	-	latent heat
$B_{\text{ch},F}$	-	chemical exergy of the fuel
T_m	-	average temperature of the working fluid

In order to determine chemical exergy of the fuel, let us take the following consideration. The fuel combustion in the steam generator could be evaluated as a series of exothermic chemical reactions between reactants giving stoichiometrically defined products. It is customary to report the low heating value in kJ/kg along with the analysis of fuel. Various formulas have been proposed for evaluating LHV from the ultimate analysis of fuel.

For this analysis it is accepted that LHV is determined by Mendelev formula

$$\text{LHV} = 4,187\,[81\,x_c + 300\,x_H - 26(x_o - x_S) - 6(9\,x_H + x_w)]$$

where

x_c, x_H, x_o, x_S, x_w - mass fraction of carbon, hydrogen, oxygen, sulfur and water in the fuel

The equation for the chemical exergy of the fuel is adapted with assumption that contribution of sulfur, ash and water are neglected, then we can write that

$$B_{\text{ch},F} = F(\text{LHV} + h_{gs}X_w)\beta$$

where

$$\beta = 1.0437 + 0.1896\,\frac{x\text{H}_2}{x_C} + 0.0617\,\frac{x\text{O}_2}{x_C} + 0.0428\,\frac{x\text{N}_2}{x_C}$$

where

$x_C, x_{H_2}, x_{O_2}, x_{N_2}$ - mass fraction of carbon, hydrogen, oxygen and nitrogen in the fuel.

Exergy Destruction

The exergy degradation in the boiler furnace is composed of two parts. One relates to the exergy losses due to combustion process and the second part is the result of the heat transfer from combustion products to the working fluid in the furnace

$$\Delta\varepsilon_{\text{des}} = \frac{\delta B_{\text{comb}} + \delta B_{\text{ht}}}{B_{\text{ch},F}}$$

The exergy destruction in the combustion process could be determined if the combustion process is considered in two steps. Namely, the first step is the chemical reaction which defines exergy losses due to the chemical part of the combustion process. The second step consists on the heating part of combustion process and takes place at isobaric and adiabatic condition.

The exergy loss of the combustion process could be defined as the difference between exergy of fuel and exergy of the flue gases determined with the same reference state. For the energy system we introduce the following simplification, so that

$$\delta B_{comb} = F\, T_0 (S_R - S_F)$$

where

T_0 - environment temperature
S_R - entropy of combustion reactants
S_F - entropy of combustion products

so that

$$\delta B_{comb} = \sum_i (m_i S_i)_g - \sum_j (m_j S_j)$$

where

m_i, m_j - mass of the species in reactants and products of combustion gases, respectively.
S_i, S_j - specific entropy of reactants products species

A large part of the total heat generated in the steam generator by combustion is transferred to the furnace wall by radiation. It is reasonable to assume that exergy loss due to radiation heat transfer has to be taken into consideration (Bejan, 1980 and 1982). Also, a substantial part of the exergy loss is caused by the heat conduction through the fouling layer on the boiler surface wall. If it is assumed, that combustion products are at the gas temperature T_g and radiation heat transfer in the steam generator is defined by Newton law with respective gas and surface temperature difference taking into account fouling layer heat conduction resistence on the boiler surface, the exergy degradation due to heat transfer in the boiler is then:

$$\delta B_{ht} = A_s\, T_0 \left[\frac{T_g - T_{sat}}{\dfrac{1}{\alpha_1} + \dfrac{\delta_1}{k_1} + \dfrac{\delta_0}{k_0} + \dfrac{\delta_2}{k_2} + \dfrac{1}{\alpha_2}} \right] \left(\frac{1}{T_{sat}} - \frac{1}{T_g} \right)$$

where

α_1 - heat transfer coefficient by radiation
α_2 - heat transfer by evaporation
$\delta_1, \delta_2, \delta_0$ - are the thickness of outer and inner layer on the surface and thickness of tube respectively
$k_1, k_2\, k_0$ - are thermal conductivity of the outer and inner layer and tube, respectively

Exergy Losses

The exergy losses in the steam generator are defined as

$$\Delta \varepsilon_{loss} = \frac{\delta B_w + \delta B_c + \delta B_t}{B_{ch, F}}$$

δB_w is the exergy loss due to heat transfer through outside wall of the steam generator. It could be defined as

$$\delta B_w = Q_{out} \left(\frac{1}{T_0} - \frac{1}{T_{wall}} \right)$$

Q_{out} -   heat transferred to the environment
T_{wall}, T_0 -   temperature of the wall and environment, respectively

δB is the chemical exergy loss by flue gases and combustive gases in flue gases

$$\delta B_c = F \left\{ R T_0 \left[\ln \frac{Z_{CO_2}}{Z_{CO_2}^0} + 2 \ln \left(\frac{Z_v}{Z_v^0} \right) + 10.53 \ln \left(\frac{Z_{N_2}}{Z_{N_2}^0} \right) + 0.8 \ln \left(\frac{ZO_2}{Z_{O_2}^0} \right) \right] \right.$$

$$\left. (\Delta G + R T_0 \ln Z_{CO}) \right\}$$

where

$Z_{CO_2}, Z_v, Z_{N_2}, Z_{O_2}, Z_{CO}$ -   molar fraction of CO_2, vapor, N_2,
                                              O_2 and CO in flue gases
$Z_{CO_2}^0, Z_v^0, Z_{N_2}^0, Z_{O_2}^0$ -   molar fraction of CO_2, vapor, N_2
                                              and O_2 in environment

δB_t is the thermal exergy loss by flue gases and is defined as

$$\delta B_t = [h(T) - h(T_0) - T_0(s(T) - s(T_0))] CO_2$$
$$+ 2 [h(T) - h(T_0) - T_0(s(T) - s(T_0))]_v$$
$$+ 10.53 [h(T) - h(T_0) - T_0(s(T) - s(T_0))] N_2$$
$$+ 0.8 [h(T) - h(T_0) - T_0(s(T) - s(T_0))] O_2$$

where

$h(T), h(T_0)$ - enthalpy of the species in flue gases at the temperature T and T_0
$s(T), s(T)$ - entropy of the species in flue gases at the temperature T and T_0.

EXPERT SYSTEM CONCEPT

The efficiency assessment expert system concept for the steam generator is designed as the on-line system, with data obtained by the respective acquisition of operating parameters. It is a real-time-sensor based detection and diagnosis system, with the aim of handling temporal data on in the specific domain related to the key area of interest. It is a framework for real time fault detection and diagnosis using temporal data. Data reading is scheduled in regular time increments, validated and stored in the respective buffer. The boiler expert system (Afgan *et al.*, 1991) with the module for the efficiency assessment is object-attribute-value oriented system with objects specification based on the exergy balance equation. In this respect, determination of the respective diagnostic parameters are derived accordingly.

The essential part of this expert system concept is the selection of the diagnostic parameters and its reading as a set of the values describing actual situation of the system. Its temporal monitoring will be stored for the specific time dependent pattern recognition. The same diagnostic parameters will be used for the description of the respective malfunction and sets of the parameters related to the individual situation to be used in performing fault diagnosis.

Selection of Diagnostic Variables

Second law efficiency

As it was shown the second law efficiency is function of operating parameters defining energy efficiency of the boiler multiplied by relative temperature difference between the average temperature of working fluid in the boiler and temperature of environment
    So that

$$\varepsilon = \eta_e \frac{T_m - T_0}{T_m}$$

From the expression for the second law efficiency it follows that

$$\varepsilon = \frac{D[c_{pf}(T_{sat} - T_{w,in}) + h_{gh} + c_{ps}(T_{sup} - T_{sat})]}{FLCV} \frac{T_m - T_0}{T_m},$$

In order to determine this quantity the following parameters have to be known

$$D_{in}, T_{w,in}, p, F$$

assuming that c_{pf}, c_{ps}, LCV are constant within the range of parameters considered.
    These are diagnostic variable for the second law efficiency.

Exergy destruction

The exergy destruction comprises two parts, namely: exergy destruction due to combustion and exergy degradation due to the heat transfer in boiler. If we assume that combustion contribution to the exergy destruction is small in comparison to the heat transfer part, it follows that

$$\Delta\varepsilon_{deg} = \frac{A_s T_0}{LCV} \left[\frac{T_g - T_{sat}}{\frac{1}{\alpha_1} + \frac{\delta_1}{k_1} + \frac{\delta_0}{k_0} + \frac{\delta_2}{k_2} + \frac{1}{\alpha_2}} \right] \left(\frac{1}{T_{sat}} - \frac{1}{T_g} \right)$$

If it is assumed that $\alpha_2 \gg \alpha_1 = \text{const}$ and $\delta_0, \delta_2, k_1, k_0, k_2 = \text{const.}$ the expression for the exergy degradation will depend only on δ_1 and T_g. This leads us to the conclusion, that with assumption that the main diagnostic parameters for the exergy destruction due to heat transfer are

$$T_g, \delta_1, T_{\text{sat}}$$

Exergy losses

As it was shown the exergy losses are composed of three parts. One, related to the exergy loss due to heat transfer from the boiler outside surface to the environment. It is very low in magnitude and it will be not taken into consideration in this analysis. The second part is combustion product chemical exergy losses in the environment.

$$\Delta \varepsilon_{\text{EL}_{\text{comb}}} = \frac{N_{\text{cp}}}{\text{LCV}} \left\{ \left\{ R\, T_0 \left[\left(\ln \frac{Z_{\text{CO}_2}}{Z_{\text{CO}_2}^0} \right) + 2 \ln \left(\frac{Z_v}{Z_v^0} \right) + 10.53 \ln \left(\frac{Z N_2}{Z_{N_2}^0} \right) \right. \right. \right.$$

$$\left. \left. \left. + 0.8 \ln \left(\frac{Z_{O_2}}{Z_{O_2}^0} \right) \right] + (\Delta G + R\, T_0 \ln Z_{\text{CO}} \right\} \right\}$$

The third part of the exergy losses in the boiler corresponds to the flue gas thermal exergy at the outlet of the boiler. As it was defined, it is

$$\varepsilon_{\text{EL}_{\text{ht}}} = \left[\frac{N_g}{\text{LCV}} h\, T - h\, T_0 - T_0(s\, T) \right]$$

$$+ 2 [h(T) - h(T_0) - T_0(s(T) - s(T_0))]_v$$

$$+ 10.53 [h(T) - h(T_0) - T_0(s(T) - s(T_0))]\, N_2$$

$$+ 0.8 [h(T) - h(T_0) - T_0(s(T) - s(T_0))]\, O_2$$

$$+ (\Delta G + RT_0 \ln Z_{\text{CO}})$$

where

$$T = T_{\text{out}}$$

For this analysis we will assume that only Z_{CO_2}, Z_{O_2}, Z_{CO} and T_{out} are parameters effecting change in the exergy losses in the boiler. This will lead us to the diagnostic parameters for the exergy losses as follows

$$Z_{\text{CO}_2}, Z_{O_2}, Z_{\text{CO}}, T_{\text{out}}$$

KNOWLEDGE BASE

Facts Object-Attribute-Value Structure

The knowledge base is structured by the objects oriented methodology (Karni and GalTzur, 1990 and Armor, 1987), using objects second law efficiency, exergy destruction and exergy losses. The objects are defined as follows

Object	SLE					
Attribute	D	T_{in}	T_{sup}	p	F	ε
Values	0	0	0	0	0	0
	$+\Delta$	$+\theta_i$	$+\theta_s$	$+\pi$	$+\phi$	$+\Delta\varepsilon$
	$-\Delta$	$-\theta_i$	$-\theta_s$	$-\pi$	$-\phi$	$-\Delta\varepsilon$

Object	ED			
Attribute	T_W	T_{sat}	δ_1	$\Delta_{\varepsilon_{deg}}$
Values	0	0	0	0
	$+\theta$	$+\theta_s$	$+\Delta\delta_1^1$	$+\Delta_{\varepsilon_{deg}}$
	$-\theta$	$-\theta_s$	$-\Delta\delta_1^2$	$-\delta\Delta_{\varepsilon_{deg}}$

Object	EL				
Attribute	Z_{CO_2}	Z_{O_2}	Z_{CO}	T_{out}	$\Delta_{\varepsilon_{loss}}$
Values	0	0	0	0	0
	$+\xi CO_2$	$+\xi O_2$	$+\xi CO$	$+\theta_{out}$	$+\delta\Delta_{\varepsilon_{loss}}$
	$-\xi_{CO_2}$	$-\xi_{O_2}$	$-\xi_{CO}$	$-\theta_{out}$	$-\delta\Delta_{\varepsilon_{loss}}$

where

$\theta_i\,\theta_s\,\theta_w\,\theta_g\,\theta_{out}$ - are relative temperature difference of the receptive diagnostic temperature

$$\theta_j = \frac{T_j - T_j^0}{T_j^0}$$

T_j^0 - standard value for the respective temperature.

$\xi_{CO_2},\,\xi_{O_2},\,\xi_{CO}$ - relative concentration differences for the respective species

$$\xi_i = \frac{Z_i - Z_i^0}{Z_i^0}$$

Z_i^0 - standard concentration of the species i

π - relative pressure difference between measured pressure and standard value

$$\pi = \frac{p - p^0}{p^0}$$

p^0 - nominal pressure in the steam generator

ϕ - relative fuel consumption difference between measured and standard value.

$$\phi = \frac{F - F^0}{F^0}$$

F^0 - fuel consumption for the nominal power of the boiler

ε -   relative second law efficiency

$$\delta\Delta\varepsilon = \frac{\varepsilon - \varepsilon^0}{\varepsilon^0}$$

ε^0 -   second low efficiency for the nominal power

$$\delta\Delta\varepsilon_{\mathrm{deg}} = \frac{\Delta\varepsilon_{\mathrm{deg}} - \Delta\varepsilon_{\mathrm{deg}}^0}{\Delta\varepsilon_{\mathrm{deg}}^0}$$

$\delta\Delta\varepsilon_{\mathrm{deg}}^0$ -   exergy degradation at nominal power
$\delta\Delta\varepsilon_{\mathrm{deg}}$ -   relative exergy losses difference

$$\delta\Delta\varepsilon_{\mathrm{loss}} = \frac{\Delta\varepsilon_{\mathrm{loss}} - \Delta\varepsilon_{\mathrm{loss}}^0}{\Delta\varepsilon_{\mathrm{loss}}^0}$$

$\delta\Delta\varepsilon_{\mathrm{loss}}^0$ -   exergy losses at nominal power

Knowledge Base Structure

Knowledge base is organised as the Object-Attribute -Value structure. As the object in this structure is EFFICIENCY. The EFFICIENCY object is two attributes: second law efficiency SLE and Causes for the change of SLE, having attribute degradation of efficiency ED and losses of efficiency EL. SLE is defined with respective attribute: Flow rate D, Inlet temperature T_{in}, Superheat steam temperature T_{sup}, Pressure p, Fuel consumption F. EL have a following attributes: Wall temperature T_w, Saturation temperature T_{sat}, Fouling layer thickness δ_1. EL have the following attributes: Concentration of CO_2, Concentration of O_2, Concentration of CO and Temperature of the outlet gases T_{out}. Schematic structure of the knowledge base is given on Figure 2.

In the terminology common for the knowledge engineering the Efficiency object can be described in the LISP language as

$$EF(E(D(0_1 + \Delta, -\Delta)\, T_{\mathrm{in}}(0_1 + \theta_I, -\theta_I)\, T_{\mathrm{sup}}(0_1 + \theta_s, -\theta_s)\, p(0_1 + \pi, -\pi)$$

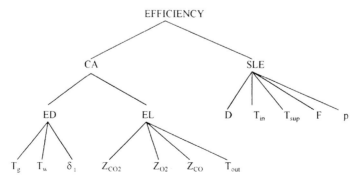

FIGURE 2   Schematic structure of the knowledge base.

$$F(0_1 + \phi, -\phi,)) CA(ED(T_w(0_1 + \theta_w, -\theta_w) T_{sat}(0_1 + \theta_s, -\theta_s)\delta_1(0_1 + \Delta\delta_1, -\Delta\delta_1))$$

$$EL(Z_{CO_2}(0_1 + \xi_{CO_2}, -\xi_{CO_2})Z_{O_2}(0_1 + \xi_{O_2}, -\xi_{O_2})Z_{CO}$$

$$(0_1 + \xi_{CO}, -\xi_{CO}) T_{out}(0_1 + \theta_{out}, -\theta_{out}))))$$

The rules are organized to recognize the expert assessment for the specific causes of a boiler malfunction. Specifically the following situations are taken into consideration: fuel quality, air humidity, excess air and fouling of the heat transfer surfaces.

The rules are designed in the IF/THEN form and organized generically in order to comply with the multiple situation recognition.

Recommendation Structure

The recommendation structure is made as an expert advice to every recognized specific situation. At the moment this part of the expert system is not very elaborated and serves only as an example to demonstrate its feasibility. In the later stage of the system development more attention will be devoted to its demonstration. It will include several levels of expertise, namely: warning for immediate action; planing actions for improvement; assessment of the cost for the respective situation; maintenance advices. Each of these groups of recommendation is related to the respective situation when it is recognized as the cause for the changes in the second law efficiency, exergy degradation and exergy losses.

CONCLUSIONS

The second law efficiency model of the boiler is an efficient tool to be used in the effectiveness assessment of the boiler system. It was shown that it could be used as a model for the conceptual design of the expert system. Particular quality of this approach relies on the selection of appropriate diagnostic variables which are relevant to the description of the specific exergy degradation or losses. As the exergy flux in the boiler is closely related to the specific events leading to the malfunction, its use in the description of different situations of interest gives the possibility of a quality approach to the assessment of the system.

Particular attention was devoted to the heat transfer exergy degradation because the temperature and heat flux distribution in the boiler could be obtained with respective mathematical models of the boiler furnace. In this respect, it may be shown that future development of the numerical models of the boiler furnace describing combustion, fluid flow and heat transfer will be strongly related to the need of the expert system development. Specific emphasize should be given to the development of the numerical codes with generic capability to be used in the design of the knowledge base of the boiler expert system. Through the development of this concept, we have learned that the major difficulty in the further development of boiler expert system is the deficiency of the reliable field data to be used in the knowledge base as the description of the respective situation to be retrieved by expert shell.

In spite of the fact that the present expert systems is in its conceptual form, this paper has shown that the use of the expert system in a power plant will mainly depend on the existence of an additional diagnostic system to the one normally used in the operation control.

Great potentiality of the expert system in the boiler surveillance for its efficient operation is opening a new challenge for its further development. In this respect we feel that this was a firs step in the right direction.

REFERENCES

Afgan, N. H., Bloch, A. G., Radovanovic, P., Zhurovlev, Yu. and Gorb, E. I. (1991) Boiler Expert System. ICHMT FORUM on *Expert System and Mathematical Modeling in Energy Engineering*, Erlangen.

Afgan, N. H., Radovanovic, P., Radanovic, Lj., Bloch, A. G. and Gorb, E. I. (1991) An Expert System for Boiler Surface Fouling Assessment. ICHMT FORUM on *Exergy System and Mathematical Modeling in Energy Engineering*, Erlangen.

Ahem, J. E. (1980) The Exergy Method of Energy System. John Willey and Sons, Inc, Toronto.

Armor, A. F. (1987) Expert System: An Opportunity for Improved Power Plant Performance. Proceedings 1987 Conference on Expert System Application in Power Plant, Boston.

Bejan, A. (1980) *Second Law Analysis in Heat Transfer. Energy*, **5**(8/9).

Bejan, A. (1982) *Second Law Analysis in Heat Transfer and Thermal Design. Advances in Heat Transfer*, **15**, 1–58.

Bosnjakovic, F. (1965) Technical Themodynamics. Holt, Rinhart and Winston, N.Y..

Carnot, S. (1824) Refléctions sur la Puissance Motrice du Feu et sur les Machines. Progrès à Développer cette Puissance. Paris, Badulier.

Karni, R. and Gal-Tzur, A. (1990) Paradigms for Knowledge-based System in Industrial Engineering. *Artificial Intelligence in Engineering*, **5**(3).

Keenan, J. H. and Shapiro, A. H. (1947) On the History and Exposition of the Laws of Thermodynamics. *Mechanical Engineering*, **69**, Nov., pp. 915–921.

Kotas, T. (1980) Exergy Analysis of Thermal Power Plant. *Int. J. Heat Fluid Flow*, **2**(4).

Morran, M. J. (1982) Availability Analysis. Prentice - Hall, Inc., Englewood Cliffs.

Rant, Z. (1960) Exergy Diagrams for Combustion Gases. Brennst. *Warme Kraft*, **12**(1) 1–8.

Szargut, J., Morris, D. R. and Steward, F. R. (1988) Exergy Analysis of Thermal, Chemical and Metallurgical Processes. *Hemisphere Pub. Corp.*, Washington.

Gas Reburning. Influence of Different Variables on the NO_x Reduction

R. BILBAO[a], M. U. ALZUETA[a], A. MILLERA[a], J. LEZAUN[b] and J. ADANEZ[c]

[a]*Departamento de Ingeniería Química y Tecnologías del Medio Ambiente. Facultad de Ciencias. Universidad de Zaragoza. 50009-Zaragoza. Spain.* [b]*ENAGAS. Centro de I + D. Autovía de Madrid. Apartado 354. 50080-Zaragoza. Spain.* [c]*Instituto de Carboquímica (CSIC). Plaza Paraíso, 4. 50004-Zaragoza. Spain.*

Abstract—Reburning with natural gas is a simple method to reduce NO_x emissions produced in pulverized coal boilers. In this work, an experimental study of the reburning zone has been carried out in a system which allows the simulation of several operating conditions, introducing different gas mixtures and injecting natural gas. The influence of different variables has been analysed. The variables studied have been the temperature, the gas residence time, the concentration of oxygen and NO proceeding from the primary combustion zone and the amount of natural gas introduced as reburning fuel. The effect of these variables has been considered in order to know the NO reduction and the gas composition exiting from the reburning zone.

Key Words: NO_x reduction, reburning, natural gas, operating conditions

INTRODUCTION

Combustion in pulverized coal boilers is a technology widely extended throughout the world. This technology usually generates environmental problems due to emissions of particles, SO_2 and NO_x. Owing to these problems, the environmental legislation in industrialised countries is more and more restrictive in regard to emission limits for pollutant compounds. This work concentrates on the reduction of the NO_x emissions.

Several methods can be used in order to reduce NO_x emissions. Combustion modifications, such as low NO_x burners, air staging and reburning, can be taken into account. Other methods such as downstream injection of selective reducing agents (particulary ammonia and urea) or the use catalysts, can produce significant additional reductions (Bergsma, 1985; Hjalmarsson, 1990). Reburning with coal (Burch *et al.*, 1991), natural gas (Seeker *et al.*, 1985; Folsom *et al.*, 1987; Kilpinen *et al.*, 1990) or other hydrocarbons (Chen *et al.*, 1985) has been demonstrated to be an effective technique, being a simple method that can be applied in pulverized coal power stations with low investment costs. Furthermore, this method allows the simultaneous use of other SO_2 reduction methods.

The reburning process divides the combustion boiler into three zones, Figure 1.

In the primary zone, most of the fuel is burned with a slight air excess. The typical combustion products are produced in this zone, including the NO_x. The reburning fuel is injected into the reburning zone, creating a fuel-rich zone, where the NO_x formed in the primary zone are reduced by the action of hydrocarbon radicals. Additional

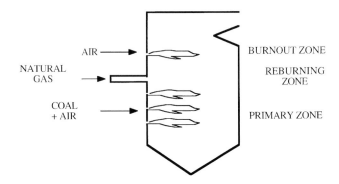

FIGURE 1   Reburning process in a pulverized coal boiler.

combustion air is added in the burnout zone to obtain complete combustion and to oxidize any remaining fuel fragments.

Reburning with natural gas allows us to make use of a fuel with a high heating value and with little or no nitrogen and sulphur content. This method has been studied by other authors (Chen *et al.*, 1985; Toqan *et al.*, 1987; Glarborg, 1990; Mereb and Wendt, 1990), but due to different coals and operating conditions used in pulverized coal boilers, it is necessary to perform additional studies which allow the resolution of some questions not yet resolved.

In this context, the knowledge of the influence of different variables on the process is very important. With this aim, an experimental installation that allows the study of the reburning process has been designed and built. The influence of different variables (natural gas concentration, temperature, residence time, etc.) on the NO reduction has been studied, introducing into the reactor different flue gas mixtures which simulate the gas composition exiting from the primary zone in a pulverized coal boiler.

EXPERIMENTAL METHOD

The experimental installation, Figure 2, consists basically of a gas feed system, a reaction system and a gas analysis system.

The reaction system includes a ceramic reactor heated by an electric furnace that allows us to reach temperatures up to $1500°C$. The reactor is an alumina tube of 20 mm inside diameter and 2500 mm in length.

A gas consisting of O_2, N_2 and CO_2 was prepared and bubbled into a water container until saturation in order to reach the desired moisture at a given temperature. The gas was mixed with NO and natural gas, and the mixture was fed into the reburning reactor.

The contents of NO_x, O_2, CO and CO_2 were measured by continuous analyzers. O_2, CO, CO_2 and other hydrocarbons were also determined by gas chromatography. The NH_3 concentration was determined by passing a known volume of gas through an aqueous acid solution, which was subsequently analyzed using the Nessler colorimetric

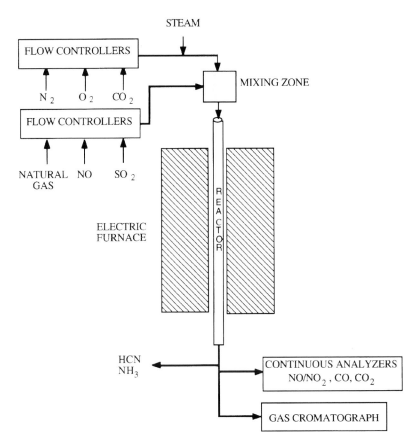

FIGURE 2   Experimental system.

method (Clesceri *et al.*, 1989). The HCN concentration was also determined by collecting a gas sample in a basic solution and analysing it by the barbituric-pyridine colorimetric method (Rodier, 1981; Clesceri *et al.*, 1989).

Previous Experiments

Some previous experiments have been performed in order to know the behaviour of the experimental system. The objective of these experiments was to determine longitudinal temperature profiles inside the reburning reactor and to determine the possible formation of thermal NO at different flow rates and compositions of N_2 and O_2.

Temperature profiles   For different system temperatures and flow rates of N_2, the temperature profiles inside the reburning reactor have been measured. A thermocouple was introduced at different depths of the reactor and the longitudinal temperature profiles were determined for different system temperatures and flow rates.

As an example, the temperature profiles obtained, for a flow rate of 600 Nl/h of N_2 and system temperatures of 1200 and 1400°C, are shown in Figure 3.

The profile shape is very similar for all temperatures, and there is a central zone of approximately 800 mm in length where the temperature can be considered to be constant. This length has been chosen as the reaction zone. Moreover, in this zone, the influence of flow rate on the temperature profiles is not appreciable.

Formation of thermal NO   A part of the NO_x emissions produced in the combustion process is generated at high temperatures, from the nitrogen and the oxygen contained in the air used in the combustion.

Thermal NO formation has been studied at temperatures between 1200–1500°C, for different total flow rates and O_2 concentrations. Figure 4 shows the thermal NO concentration obtained at different temperatures for a flow rate of 900 Nl/h and a oxygen percentage of 2%. It can be observed that the thermal NO formation begins to be appreciable at temperatures higher than 1350°C, and only at temperatures near to 1500°C, is this NO formation important. Therefore, at the temperatures used in this work, the thermal NO formation can be considered to be negligible. Similar trends are obtained for other flow rates and O_2 concentrations.

Reburning Experiments

The influence of temperature, gas residence time and inlet concentrations of NO, O_2 and natural gas has been analysed.

In all the experiments, fixed concentrations of CO_2 (20%) and steam have been used. The flow rate and all the compositions are in dry basis, and the steam is added to the total flow rate representing an extra 6%.

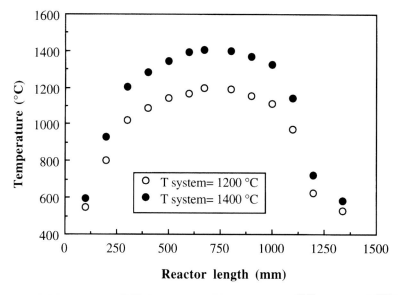

FIGURE 3   Longitudinal temperature profiles inside the reactor. Flow rate = 600 Nl/h.

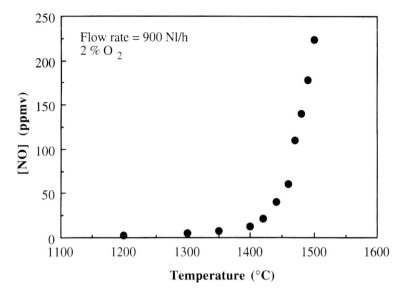

FIGURE 4   Thermal NO formation.

Depending on the type of experiments, the variables have been modified in the following ranges:

- Inlet natural gas concentration: 1–4.5%. The natural gas has an average composition of 90.5% CH_4, 8.5% C_2H_6, 0.5% C_3H_8, 0.4% N_2 and 0.1% C_nH_m ($n = 4$ to 7).
- Inlet O_2 concentration: 0–5%.
- Mean gas residence time: 98–280 ms.
- Inlet NO concentration: 500–1200 ppmv.
- Reburning temperature: 1200–1500°C.

RESULTS AND DISCUSSION

Some significant results obtained in these experiments are shown in this work. These results are presented as the influence of the different variables.

Influence of Natural Gas Concentration

An example of the influence of the percentage of natural gas on the NO reduction for different oxygen concentration values (1.5–2.5%) is shown in Figure 5, when the temperature is 1200°C.

A maximum NO reduction appears for a given natural gas concentration. Several authors (Myerson, 1974; Chen *et al.*, 1985; Fujima *et al.*, 1990) have obtained an optimum natural gas concentration, but this optimum depends strongly on the operating conditions (Glarborg, 1990). The maximum can be explained because, in these conditions, for low natural gas concentrations not enough hydrocarbon radicals

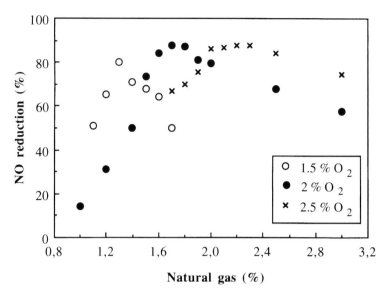

FIGURE 5   NO reduction versus natural gas concentration, for different O_2 concentrations. Temperature = 1200°C. Flow rate = 900 Nl/h. $[NO]_{inlet}$ = 900 ppmv.

are formed, and for high natural gas concentrations, the CO_2 formation could be favoured.

The value of natural gas concentration, for which the highest NO reduction is obtained, diminishes when the oxygen concentration diminishes. In all the cases, the maximum corresponds approximately to the same O_2/natural gas ratio of 1.2, Figure 6, which means the same stoichiometric value exists in the reburning zone. This O_2/natural gas ratio has been described in the literature, where different values of this ratio appear, which depend strongly on the type of coal and its heating value (Knill and Morgan, 1989).

The influence of the natural gas concentration on the output concentration of HCN must also be considered, because the HCN produced in the reburning zone can afterwards be oxidized to NO in the burnout zone. The results obtained of output concentrations of NO, HCN and the sum of both (TFN) are shown in Figure 7 for a temperature of 1200°C. The NH_3 obtained is negligible. The HCN increases as the natural gas concentration increases. The NO presents a minimum which corresponds to the maximum above mentioned NO reduction.

The total fixed nitrogen (TFN) also presents a minimum for a natural gas concentration of 1.7% when 2% of O_2 is used. Similar tendencies are observed at other temperatures. Figure 8 shows some results obtained at 1400°C. It can be observed that the total fixed nitrogen also presents a minimum. In the light of these results, and considering the conditions used in pulverized coal boilers in Spain, concentrations of 1.7% natural gas and 2% O_2 were selected in order to study the influence of other variables.

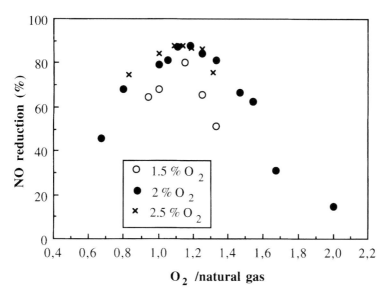

FIGURE 6 NO reduction versus O_2/natural gas ratio, for different O_2 concentrations. Temperature = 1200°C. Flow rate = 900 Nl/h. = 900 ppmv.

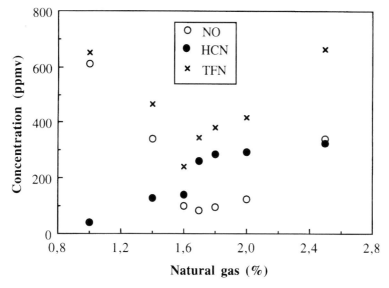

FIGURE 7 Output concentrations of NO, HCN and TFN versus natural gas concentration. Temperature = 1200°C. Flow rate = 900 Nl/h. $[O_2]$ = 2%. $[NO]_{inlet}$ = 900 ppmv.

Other results that can contribute to explain the NO_x reduction are the CO and CO_2 concentrations obtained in the exit gas. The values depend on the O_2/natural gas ratio used. An increase in this ratio causes the CO quantity to reach a maximum for the same ratio value, as happens in the NO reduction. This effect can be observed in Figure 9.

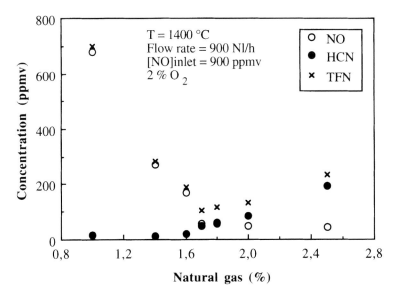

FIGURE 8   Output concentrations of NO, HCN and TFN, for different natural gas concentrations.

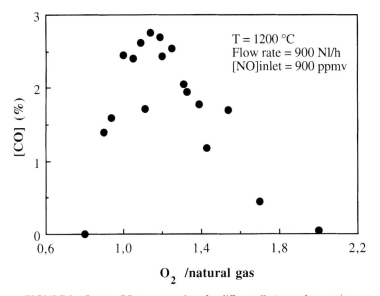

FIGURE 9   Output CO concentrations for different O_2/natural gas ratios.

This maximum could be explained because CO was a buffer of oxygen, which would vary with the different conditions, one of them being the reburning stoichiometry (Myerson, 1974).

The CO_2 output concentration increases as the O_2/natural gas ratio increases, Figure 10. When the results are analysed, it is necessary to take into account several

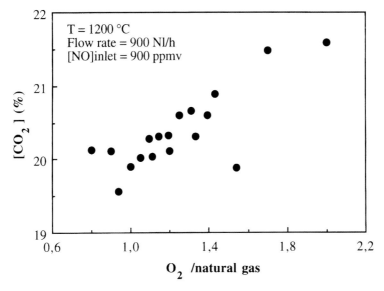

FIGURE 10   Output CO$_2$ concentrations for different O$_2$/natural gas ratios.

simultaneous effects. First, the favoured formation of CO$_2$ when the natural gas concentration increases for a fixed O$_2$ concentration. Another effect would be the decrease of the concentration of hydrocarbon radicals when the O$_2$ concentration increases, for a given natural gas concentration. Finally, the combustion reaction of natural gas would be displaced to the CO$_2$ formation, when O$_2$ and natural gas concentrations increase.

Influence of Oxygen Concentration

The oxygen quantity that enters the reburning zone is a very important value. It logically depends on the air excess used in the primary combustion zone.

   The NO reductions obtained, at 1200°C and 1.7% of natural gas, for different oxygen concentrations are represented in Figure 11. When low oxygen concentrations are utilized the NO reduction is very poor, but it increases sharply when the oxygen concentration approximates to the value of 2%, and afterwards decreases gradually. The same tendency was observed for other natural gas concentrations, but at higher temperatures this effect is less noticeable. The increase in the NO reduction can be explained because the quantity of oxygen begins to be enough to provide sufficient reducing free radicals (Myerson, 1974). However, when the oxygen concentration is high, NO formation rather than reduction is favoured (Glarborg, 1990), because the oxygen concentration is high enough to destroy the hydrocarbon radicals.

   When the influence of O$_2$ concentration is considered on the TFN generated in the reburning zone, a slight variation of the HCN concentration is observed, the NO originated being that which determines principally the amount of TFN obtained, Figure 12.

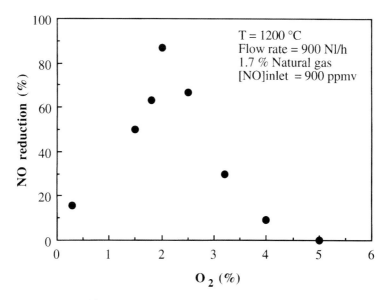

FIGURE 11   NO reduction versus O_2 concentration.

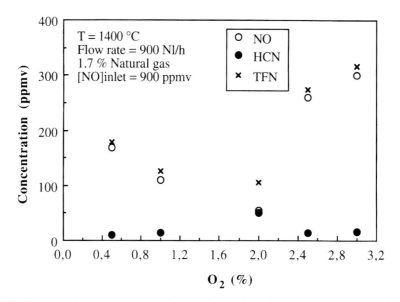

FIGURE 12   Output concentrations of NO, HCN and TFN, for different O_2 concentrations.

Influence of Residence Time in the Reburning Zone

An example of the influence of different gas flow rates on the output concentration of nitrogenous species is shown in Figure 13. The corresponding values of the gas

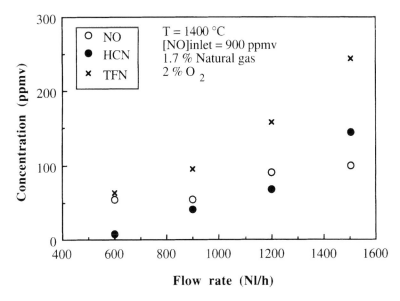

FIGURE 13   Output concentrations of NO, HCN and TFN, for different flow rates.

residence time studied in the reburning zone are 246, 164, 123 and 98 ms, for the experiments performed at 1400°C.

It can be observed that NO, HCN and logically TFN increase as the flow rate increases or the residence time diminishes. The NO concentration suffers only a small variation, but the HCN undergoes an important change for the range of residence times studied. These effects produce an appreciable increase on the total TFN obtained.

Influence of Inlet NO Concentration

It is logical to think that the NO concentration exiting from the reburning zone will depend on the NO concentration value from the output of the primary combustion zone.

The results obtained of output concentrations of NO, HCN and TFN for different inlet NO concentrations at 1400°C are shown in Figure 14. The effect of inlet NO concentration is not very important for the output concentrations of HCN, but this effect is more pronounced on the TFN. Therefore, it can be concluded that in order to obtain important NO_x reductions and lower TFN concentrations, it is advantageous to start with inlet concentrations of NO which were not too high (Kilpinen et al., 1990), even when the optimum operating conditions are chosen in relation to NO_x reduction.

Influence of the Temperature

The influence of the temperature has been studied through the results obtained at three different temperatures (1200, 1300 and 1400°C).

Figure 15 shows the NO reduction in the reburning zone and Figures 16 and 17, the HCN and TFN concentrations obtained in the output gas respectively for the temperatures studied.

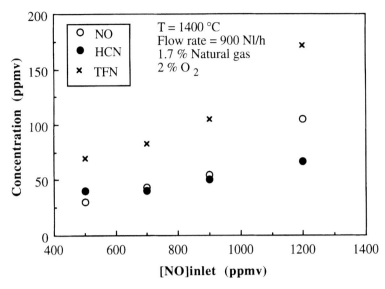

FIGURE 14   Output concentrations of NO, HCN and TFN, for different inlet NO concentrations.

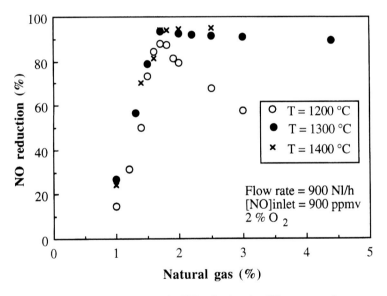

FIGURE 15   Influence of the temperature on the NO reduction, for different natural gas concentrations.

It can be observed that the NO reduction increases and the output HCN concentration decreases when the reburning temperature is higher. Logically, the TFN quantity in the exiting gas of the reburning zone decreases when the temperature increases.

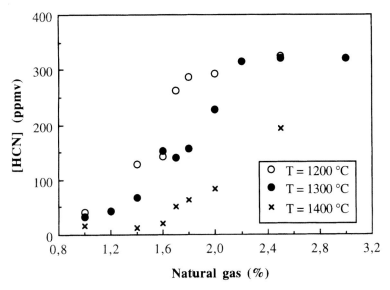

FIGURE 16 Output HCN concentration, for different temperatures and natural gas concentrations. Flow rate = 900 Nl/h. $[O_2] = 2\%$. $[NO]_{inlet} = 900$ ppmv.

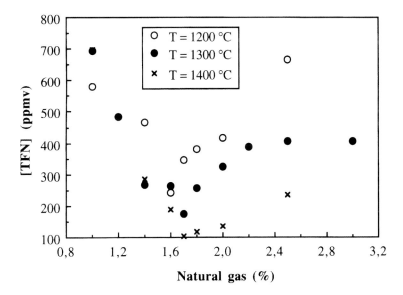

FIGURE 17 Output TFN concentration, for different temperatures and natural gas concentrations. Flow rate = 900 Nl/h. $[O_2] = 2\%$. $[NO]_{inlet} = 900$ ppmv.

CONCLUSIONS

• Significant NO reduction is obtained when natural gas is used as reburning fuel. This reduction presents a maximum when the natural gas and oxygen concentrations are varied.

- Appreciable quantities of HCN can be obtained at the exit of the reburning zone.
- As the gas residence time in the reburning zone increases, a decrease of NO and HCN concentrations in the exit gas is observed.
- An increase in the reburning temperature causes a decrease in NO and HCN concentrations in the exit gas.
- Given the results obtained in this work, the application of the reburning process to industrial boilers using adequate conditions could allow the achievement of good results in the reduction of NO_x emissions.

ACKNOWLEDGEMENTS

The authors express their gratitude to the ENAGAS, ENDESA and SEVILLANA companies, to OCIGAS, OCIDE and CICYT (Project AMB92-0888) for providing financial support for this work and also to MEC (Spain) for a research grant awarded to M.U. Alzueta.

REFERENCES

Bergsma, F. (1985) Abatement of NO_x from coal combustion. Chemical background and present state of technical development. *Ind. Eng. Chem. Process Des. Dev.* **24**, 1–7.
Burch T. E., Tillman, I. R., Chen W. Y., Lester, T. W., Conway, R. B. and Sterling, A. M. (1991) Partitioning of nitrogenous species in the fuel rich stage of reburning. *Energy and Fuels*, **5** (2), 231–237.
Chen, S. L., Clark, W. D., Greene, S. B., Heap, M. P., Moyeda, D. K., Overmoe, B. J., Pershing, D. W., Seeker, W. R. and Silcox, G. D. (1985) Controlling pollutant emissions through the supplemental use of natural gas. *Topical Report No. 5083-251-0905. Gas Research Institute*, Chicago.
Clesceri, L. S., Greenberg, A. E. and Trussell, R. R., Eds. (1989) *Standard methods for the examination of Water and Wastewater*, APHA-AWWA-WPCE, Washington, Chap. 4, 20–42 and 111–126.
Folsom, B., Bartok, W. and Heap, M. (1987) Field evaluation of gas reburning-sorbent injection technology. *Report No. GRI-87/0208. Gas Research Institute*, Chicago.
Fujima, Y., Takahashi, Y., Kunimoto, T. and Kaneko, S. (1990) Field application of MACT. *Proceedings of the Reburning Workshop, Örenäs Slott*, Sweden, 7–24.
Glarborg, P. (1990) Unresolved questions in natural gas reburning. *Proceedings of the Reburning Workshop, Örenäs Slott*, Sweden, 314–325.
Hjalmarsson, A. K. (1990) NO_x control technologies for coal combustion. *Report No. IEACR/24. IEA Coal Research*, London.
Kilpinen, P., Hupa, M., Glarborg, P. and Hadvig, S. (1990) Kinetic parametric study of NO reduction chemistry in reburning. *Proceedings of the Reburning Workshop, Örenäs Slott*, Sweden, 191–203.
Knill, K. J. and Morgan, M. E. (1989) The effects of process variables on NO_x and nitrogen species reduction in coal fuel staging. *Proceedings of the Joint EPRI/EPA Symp. on Stat. Combust. NO_x Control*, **2**, 1652–1669.
Mereb, J. B. and Wendt, J. O. L. (1990) Reburning mechanisms in a pulverized coal combustor. *Proceedings of the 23rd Symposium (International) on Combustion*. The Combustion Institute, Pittsburgh, 1273–1279.
Myerson, A. L. (1974) The reduction of nitric oxide in simulated combustion effluents by hydrocarbon-oxides mixtures. *Proceedings of the 15th Symposium (International) on Combustion*. The Combustion Institute, Pittsburgh, 1085–1092.
Rodier, J. (1981) *Análisis de las Aguas*. Ed. Omega S. A. Barcelona, 137–149 and 541–549.
Seeker, W. R., Chen, S. L., Clark, W. D. and Heap, M. P. (1985) Controlling pollutant emissions through the supplemental use of natural gas. *Report No. GRI-86/0258. Gas Research Institute*, Chicago.
Toqan, M. A., Teare, J. D., Beér, J. M., Radak, L. J. and Weir, A., Jr. (1987) *Reduction of NO_x by fuel staging. Proceedings of the Joint EPRI/EPA Symp. on Stat. Combust. NO_x Control*, **2**, 35/1–35/18.

Development of an Emissions Control System for a Coal-fired Power Plant

T. LUCAS[a], J. ABADÍA[a], J. J. CASARES, J. A. SOUTO and
V. PÉREZ-MUÑUZURI
Departamento de Ingeniería Química, Universidad de Santiago de Compostela, Spain
[a]*ENDESA, As Pontes de García Rodríguez, La Coruña Spain*

Abstract—An economically feasible alternative to flue-gas desulphurization for coal-fired power plants burning local coals with high sulphur content is mixing them when possible with clean imported coals. The aim of the emissions control system presented here is to optimize the use of clean coals, holding as a constraint the limit that guarantees air quality. It must be taken in consideration that power plants operate with great inertia: the changes in operating conditions and the mixture of coals have only a delayed effect over emissions as changes must be implemented in steps to avoid any irreversible damage on the units. SO_2 concentration, used as a legal measure of air quality, varies over a wide range due to meteorological changes.

As a result, no feedback control system over the emissions may be used satisfactorily for distances of less than 100 km from the emitting source. The alternative is to use models capable of predicting SO_2 ground level concentration (glc) for different emission levels, over a certain time to come, and to plan the operation of the power plant according to those emission levels estimated as acceptable. This prediction is based on the application of three models: a combustion model, a weather forecasting model and an atmospheric diffusion model.

Key Words: Emission control, Weather forecasting, Atmospheric diffusion.

INTRODUCTION

A sensible exploitation of natural resources, fossil fuels for generation of electricity, must comply with the established limits of air quality. Sulphur dioxide is an atmospheric contaminant that, under certain circumstances and specific meteorological conditions, has toxic effects on the environment. Its concentration is a commonly used measure for the definition of air quality.

Whenever coal with high sulphur content is to be burned in a power plant so that the limits of air quality are to be satisfied, two generic alternatives exist:

- Separation of the contaminant, from the fuel itself or from the flue-gases.
- Reduction of the emissions of contaminant by mixing or substituting the local high sulphur coal with other compatible coals with a very low sulphur content.

For an already operating power plant with a limited life expectancy, the costs associated with the design and installation of a flue gas desulphurization unit might be too high. Therefore, for a problem of this kind, the second alternative will be more profitable if clean coal can be bought at competitive prices. Nonetheless, the simple substitution of local coal by clean coal would not allow an optimum use of the resources.

Restricting ourselves to the case of SO_2, spanish and european legislation in air quality establish limits both for emissions and ground level concentration. For the majority of locations and under most meteorological situations, a limit on emissions guarantee the observance of the directive for ground level concentrations although, the existence of single sources of significant magnitude, and also specific meteorological conditions, may cause an impact of the plume on the ground that may go beyond the legal limits for SO_2 glc.

To avoid the presence of these rare episodes, it is necessary to predetermine the maximum emission allowed for any meteorological condition that, as a consequence, will define an optimum mixture of coals. But, the changes in operating conditions of a power plant, including the modification in the quantity of clean coal used, have only a delayed effect on the emissions, because all changes must be done stepwise in a adequately operated unit. Therefore, feedback control (Figs. 1a–b) cannot be suggested, and it is required a control system based on the predictions of a model that plans the activity of the power plant following its estimations (Fig. 1c).

SYSTEM MODELING

The modeling of the processes that go from the supply of coal to the power plant to the presence of contaminants in the air, can separated in two basic stages:

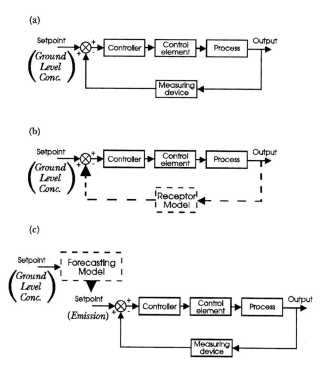

FIGURE 1 Emissions control strategies: (a) Feedback control; (b) Feedforward control; (c) Forecasting control.

- Coal combustion.
- Dispersion of flue-gases in the atmosphere.

Although the modeling of the combustion process in coal fired boilers is not fully solved yet, the emission control system does not require a detailed knowledge of the conditions inside the boiler. More specifically, it is necessary to know only some of the characteristics of the flue-gases, namely: composition, temperature and pressure; and coal consumption.

An overall mass balance for the boiler provides, with sufficient approximation, the composition and flow of the flue-gases. Emission temperature, in spite of being a decisive parameter in the estimation of the plume rise, does not suffer great changes with respect to the difference between this temperature and ambient temperature. A historical average value for this temperature has been adopted as a valid hypothesis.

To predict the dispersion of flue-gases in the atmosphere it is necessary to know first the future movement of air in the surroundings of the emitting source and subsequently, to simulate the behavior of the plume under those conditions. Hence, two new models external to the combustion process are required:

- A meteorological prediction that allows the calculation of the evolution in time of the three dimensional wind fields over the area under consideration.
- An atmospheric diffusion model, that will supply an adequate representation of the plume behavior, for various meteorological conditions.

MASS BALANCE OVER THE BOILER

A mass balance over the boiler of a coal-fired power plant requires a set of data that can be summarized by,

- Coal consumption.
- Coal composition: ultimate analysis, moisture and ash content.
- Average composition of air and its relative humidity.
- Excess air.
- Pressure and temperature of flue-gases in the stack.

The parameters to be used as variables for the control are coal consumption and composition, and excess air supplied. These parameters can be adjusted within certain limits. Average composition of air and its humidity, and the temperature and pressure of flue-gases are taken as constants and equal to the average historical values for the power plant under study. From this information, and for different mixtures of coals and excess air supplied, the mass balance gives an SO_2 concentration in the flue-gases, and flue-gas flow at the preset temperature and pressure, for the stationary period considered. These data are sufficient to characterize the emission and to be able to simulate the behavior of this gas mixture in the atmosphere.

Two fundamental considerations must be taken into account when solving the mass balance,

1. The global reactions that take place and their degree of completion: It has been assumed that all reactions are simple and complete so the products are formed following the stoichiometry of the reactions. Therefore the products of the coal combustion will be: CO_2, H_2O, and SO_2, plus N_2, CO_2, H_2O, Ar and the excess of O_2 all these supplied with the air used as comburent.
2. The state equations for the mixture of gases resulting from the combustion process: For the temperatures and pressures at the stack of a power plant the two-term virial equation was considered due to its simplicity and precision:

$$B = \sum_i \sum_j (y_{gi} y_{gi} B_{ij}) \tag{1}$$

$$Z = 1 + \frac{BP}{RT} \tag{2}$$

The assumption of constant temperature and pressure for the stack gases would at first glance give the impression of constant molar volume for the mixture, but the adoption of the virial equation that considers also the composition, allows for the calculation of thermodynamic properties of mixtures of different composition.

WEATHER FORECASTING MODEL

The preparation of a meteorological prediction twenty four hours in advance and for a limited area requires, on one hand to consider the overall evolution of the atmosphere outside this area (boundary conditions) and, on the other, to be able to describe the flow of air within the boundaries with special reference to the topography.

The first part of this problem can only be solved by organizations, national or international, integrated in a network sharing information of meteorological sensors and macroscale forecasting models. In Spain, the National Meteorological Institute, INM, is capable of supplying meteorological predictions of wind and temperature at various levels with high reliability up to three days, but over a horizontal grid with a resolution of 68 km. These predictions are available at 3 or 6 hours intervals. Unfortunately, this information is insufficient for a more precise description of the flow of air in the atmospheric boundary layer, in the surroundings of a power plant. So, a meteorological prediction model of limited area, of dynamic type, has been developed capable of generating 3-D wind fields with a horizontal resolution of 1 km; this model uses as boundary conditions the predictions of the INM, but complements these predictions with the description of microscale phenomena, such as the effects of the topography over the flow of air, within its equations.

The Limited Area Model

The dynamic model is based on the finite difference solution of the hydro-thermodynamic and turbulent kinetic energy equations. The hydrostatic part of the equations has been solved using the approximation of Yamada and Mellor (1975), as adopted by Enger (1990). The topography of the area is added to the model by means of a system of coordinates that follow the ground; the new coordinate η is defined as,

$$\eta = s\frac{z - z_g}{s - z_g} \tag{3}$$

where s is the height in the model and z_g is the topographic height. Following Yamada (1985), a constant s was assumed equal to the sum of the maximum height of the ground z_g^{max}, and the highest level solved. z_g is a function of the position and, therefore, of the rectangular coordinates x and y.

In this new system of coordinates, and using the hydrostatic approximation, the horizontal components of the wind can be written as,

$$\frac{\partial U}{\partial t} = -U\frac{\partial U}{\partial x} - V\frac{\partial U}{\partial y} W\frac{\partial U}{\partial z} - u\frac{\overline{\partial u}}{\partial x} - v\frac{\overline{\partial u}}{\partial y} - w\frac{\overline{\partial u}}{\partial \eta} - \theta\frac{\partial \Pi}{\partial x} \tag{4a}$$

$$+ g\frac{(\eta - s)}{s}\frac{\partial z_g}{dx} - fV_g + fV$$

$$\frac{\partial V}{\partial t} = -U\frac{\partial V}{\partial x} - V\frac{\partial V}{\partial y} W\frac{\partial V}{\partial z} - u\frac{\overline{\partial u}}{\partial x} - v\frac{\overline{\partial u}}{\partial y} - w\frac{\overline{\partial u}}{\partial \eta} - \theta\frac{\partial \Pi}{\partial y} \tag{4b}$$

$$+ g\frac{(\eta - s)}{s}\frac{\partial z_g}{\partial y} + fU_g - fU$$

where f is the Coriolis parameter expressed as,

$$f = 1.45 \cdot 10^{-4} \sin\phi \ s^{-1} \tag{5}$$

The horizontal changes of pressure Π are negligible for the area considered. The mean turbulent flows, vertical and horizontal, are very small compared with the advection terms and, therefore, have not been considered in the computations. Enger *et al.* (1993) include similar simplifications.

The vertical component of the velocity is obtained from the incompressible form of the continuity equation,

$$\frac{\partial U}{\partial x} + \frac{\partial V}{\partial y} + \frac{\partial W}{\partial \eta} = \frac{1}{s - z_g}\left(U\frac{\partial z_g}{\partial x} + V\frac{\partial z_g}{\partial y}\right) \tag{6}$$

The set of partial differential equations (4) and (6) is solved numerically by a forward in time and upstream in space finite differences method. Accordingly, the advection terms of equations (4) for the x direction are rewritten as,

$$\frac{\partial \varphi}{\partial t} = -U\frac{\partial \varphi}{\partial x} \Rightarrow \left\{ \frac{\varphi_i^{t+1} - \varphi_i^t}{\Delta t} = \begin{array}{l} -u_i^t\frac{\varphi_{i+1} - \varphi_i}{\Delta x} \text{ if } U_i^t \leqslant 0 \\ \\ -u_i^t\frac{\varphi_i - \varphi_{i-1}}{\Delta x} \text{ if } U_i^t > 0 \end{array}\right. \tag{7}$$

Δx and Δy are assumed constants and equal to 1000 m. For each level η the number of grid points is 61×61. A Runge-Kutta method of 4th order with variable step was used for integrating equations (4) with respect to time; minimum acceptable tolerance was 10^{-6}. The integration procedure for equations (4) is repeated for each height considered, and equation (6) is used at each point of the grid to calculate the flux of vertical wind *W*. The levels in height are not necessarily equally spaced so this flux $\partial W/\partial\eta$ has been calculated by finite differences following the expression,

$$\frac{\partial W}{\partial \eta} = \frac{1}{2}\left[\frac{W_{i+1} - W_i}{\eta_{i+1} - \eta_i} + \frac{W_i - W_{i-1}}{\eta_i - \eta_{i-1}}\right] \qquad (8)$$

Now, equation (4) can be rewritten in matrix form as,

$$\begin{bmatrix} b_1 & c_1 & & & & \\ a_2 & b_2 & c_2 & & & \\ & & \cdot & & & \\ & & & \cdot & & \\ & & & & \cdot & \\ & & & a_n & b_n \end{bmatrix} \begin{bmatrix} W_1 \\ W_2 \\ \cdot \\ \cdot \\ \cdot \\ W_n \end{bmatrix} = \begin{bmatrix} R_1 \\ R_2 \\ \cdot \\ \cdot \\ \cdot \\ R_n \end{bmatrix} \qquad (9)$$

where, from equation (4),

$$R_i = \frac{1}{s - z_g}\left(U\frac{\partial z_g}{\partial x} + V\frac{\partial z_g}{\partial y}\right) - \frac{\partial U}{\partial x} - \frac{\partial V}{\partial y} \qquad i = 1,...,n \qquad (10)$$

The gradients are solved using equation (8) whenever is possible. The other coefficients of equation (9) are given by the following expressions,

$$a_1 = 0 \quad \text{and} \quad a_i = \frac{-1}{(\eta_i - \eta_{i-1})} \qquad i = 2,...,n \qquad (11)$$

$$c_i = \frac{1}{(\eta_{i+1} - \eta_i)} \qquad i = 2,...n-1 \quad \text{and} \quad c_n = 0 \qquad (12)$$

ATMOSPHERIC DIFFUSION MODEL

The description of the transport of a plume in the atmosphere includes phenomena of different nature: plume rise, advection, diffusion within the plume and transformation and deposition of the species present in the plume, Casares and Souto (1991). The last two phenomena, however, are usually considered independently of the rest of the processes as in the solution of the atmospheric diffusion problem; the first challenge is to know where the contaminants are at any time and only afterwards to study the transformations they may suffer in the route they follow. Once these phenomena are identified, the modeling of the atmospheric flow is based on the equations of conservation of mass, energy and momentum. The way these equations are solved can be rather different, and two general approaches exist:

- The eulerian solution, where a fixed reference system is established (usually in the emitting source).
- The lagrangian solution that uses two systems of coordinates, one fixed (the emitting source) for the overall movement of the contaminants (advection) and a mobile one

that moves with the plume, for the description of the turbulent diffusion within the plume.

From the analysis of a great number of existing formulations, it was chosen a model that allows the simulation of the evolution of the plume from a power plant with a reasonable computation time. Such model is a Lagrangian model based on a Gaussian formulation that allows for the lack of homogeneity in the atmosphere and the topography of the zone under study.

Typically, the lagrangian models use puffs, as volumes or discrete structures of emitted material, to represent the plume. Normally, one of these puffs of contaminants (Fig. 2) consists of: the center of the puff, the amount of material in the puff, and the distribution of material with respect to the center of the puff. This last element will give its shape and size, and is generally associated to a Gaussian function whose parameters σ_y y σ_z increase with the distance covered.

The fundamental problem in the discretization of the plume as a sum of puffs lies in the shape and size approximation of the puffs to the shapes and sizes the actual plume may adopt. As the Gaussian distribution has proved inadequate, other alternatives have been tried that can be divided in two groups:

a) Implementation of a non-Gaussian distribution numerically generated: it is the procedure followed in the particle-in-cell models.
b) Definition of a puff with several centers, at each of which a Gaussian distribution, with differing σ_y and σ_z, is assigned.

The Gaussian distribution allows a number of mathematical simplifications that are not possible with any other type of distributions. The consideration of several centers

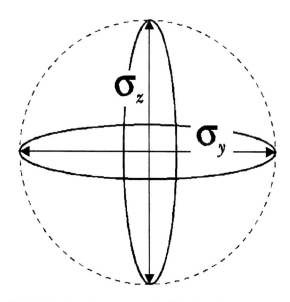

FIGURE 2   Gaussian structure (*puff*) for the plume representation.

with a Gaussian distribution defining an adaptive puff was proposed by Sheih (1978), although this author distributed six centers so that the meteorological horizontal and vertical variations affect the puff with a similar resolution. Field observations (Pooler and Niemeyer, (1970)), have demonstrated the vertical gradients, specially wind and temperature, are the cause for the distortion of the plume.

Adaptive Puff Model

The basic element of the lagrangian adaptive puff model (Ludwig *et al.* (1989)) is the discrete representation of the plume by puffs (Fig. 3) described by five points at different heights. The material emitted, represented by the puff, is distributed among the five points so that the total amount of material Δq in a layer of a depth Δz is given by a Gaussian function defined as:

$$\Delta q = \frac{\Delta z}{2\pi\sigma_z}\exp\left[-\frac{\tilde{Z}^2}{2}\right] \tag{13}$$

where $\tilde{Z}$ is:

$$\tilde{Z}(z) = \int_{z_0}^{z}\frac{d\xi}{\sigma_z(\xi)} \tag{14}$$

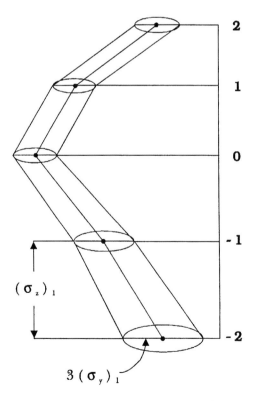

FIGURE 3   Adaptive Lagrangian puff.

The nondimensional height $\tilde{Z}$ allows fixing a vertical scale determined by the integral of $1/\sigma_z$ from the central point to the point considered, giving a measure of the equivalent number of standard deviations between the center of the adaptive volume and z. The values of $\tilde{Z}$ and σ_y may be linearly interpolated. The material located more than $2\sigma_z$ above or below each center, or at more than $3\sigma_y$ of the horizontal axis, is less than 5% of the total, so it is ignored for the calculations.

The equation for the distribution of the concentration, c, for the whole puff, from the total amount of contaminant, Q_c, assigned is given by,

$$c = \frac{Q_c}{(2\pi)^{2/3}\,\sigma_y^2\,\sigma_z}\,\exp\left[-\frac{1}{2}\left(\frac{y}{\sigma_y}\right)^2\right]\exp\left[-\frac{\tilde{Z}^2}{2}\right] \tag{15}$$

The use of the Lagrangian formulation that adapts to the different meteorological conditions of the atmospheric layers allows a great flexibility when handling the spatial variability of the meteorological conditions. The five points that describe an individual adaptive puff, can move with the wind maintaining the relative positions. The vertical separation of the points can be adjusted according to various turbulence parameters taking into consideration the atmospheric stability at the selected levels, whilst the horizontal dispersion is represented by the horizontal standard deviation σ_y. This increased flexibility requires a computational effort up to five times greater that the traditional Gaussian model.

The model is structured around four basic items:

Optimal generation of adaptive puffs   The objective is to obtain the minimum number of adaptive puffs that represent a discrete model of the plume. To get this number, it is necessary to consider the wind velocity and the turbulent atmospheric diffusion at the top of the stack. The model generates a new group of puffs every 10 min. The number N of puffs emitted is selected in such a way that the space between the first two puffs emitted over the period of 10 min. would be approximately equal to $\sigma_{y0}/2$ at the end of the interval, being σ_{y0} the standard deviation of the traverse Gaussian distribution in the puff at the top of the stack. The amount of material, Q_c, is the same for each puff of the group emitted over the period of 10 min., so that the emission of gases over that period is equally distributed over all the puffs generated in that time.

Advection   The movement of the points that define each adaptive puff is calculated numerically using interpolated conditions at the start of the time intervals $\Delta t = 2$ min. The horizontal movement is obtained by a simple discrete approximation:

$$x_{t+\Delta t} = x_t + U\Delta t \tag{16}$$

$$y_{t+\Delta t} = y_t + V\Delta t$$

where U and V are the wind components in the west (X direction) and south (Y direction) respectively, calculated by the weather forecasting model (prediction) or by measurements (simulation). The vertical movement for each of the points is obtained as the sum of the vertical movement of the air (W), the plume rise (w_p) and the vertical diffusion (w_d), that is to say,

$$z_{t+\Delta t} = z_t + (W + w_p + w_d)\Delta t \tag{17}$$

The vertical component of the atmospheric wind, W, is obtained by the weather forecasting model, or by measurements.

To calculate the contribution of the plume rise, w_p, Briggs correlations were used. The parameter that determines the plume rise is called buyancy, F, and defined as:

$$F = \frac{gv_e D^2 (T_e - T)}{4T_e} \tag{18}$$

The plume rise is calculated using four different expressions, depending on the stability and the wind velocity. To quantify the stability, a parameter, s_p, from the Monin-Obukhov expressions is used. This variable is directly related to the vertical temperature gradient (Tab. I).

The inlet velocity, v_I, is a rather important parameter to account for the different behavior of the plume in heavy or light wind situations. It represents the inlet velocity of the plume into the atmosphere as a result of the combined effect of the mechanical and thermal drives. For stable conditions, v_I is defined as,

$$v_I = 0.14 F^{1/4} s_p^{1/8} \tag{19}$$

For neutral or unstable conditions, v_I becomes,

$$v_I = \frac{1.6 F^{1/3}}{Z_{máx}} (10 H_e)^{2/3} \tag{20}$$

The calculation of the vertical component due to the plume rise is done by means of the four expressions presented in Table II.

The contributions of the turbulent diffusion for the transport of pollutants in the traverse and vertical dispersion of the plume are calculated by the exponential form $\sigma_{y,z} = a_{y,z} X^{b_{y,z}}$ of the Pasquill-Gifford equations, both for the vertical σ_z and horizontal σ_y dispersions, where X is the travel distance, and $a_{y,x}$ and $b_{y,z}$ are empirical parameters. Then, w_p is calculated as a function of σ_z.

Optimization of the number of puffs   As Ludwig *et al.* (1977) have shown, an overlapping of Lagrangian elements with Gaussian distribution of the concentration is sufficient to physically make them undifferentiable and to be grouped in one, only if

TABLE I

Quantitative definition of atmospheric stability

	Stable	Neutral or Unstable
$s_p = \dfrac{g}{\theta}\dfrac{d\theta}{dz}$	$> 4.2 \cdot 10^{-2} \dfrac{v_H^2}{H_e^2}$	$\leqslant 4.2 \cdot 10^{-2} \dfrac{v_H^2}{H_e^2}$

TABLE II

Briggs' equations adapted for plume rise vertical component, w_p, estimation

	Stable	Neutral or unstable
$v_H > v_I$	$w_p = \dfrac{2.6}{\Delta t}\left(\dfrac{F}{100 s_p H_e^2 v_H}\right)^{1/3}[(x_t + v_H \Delta t)^{2/3} - X_t^{2/3}]$	$w_p = \dfrac{1.6 F^{1/3}}{v_H \Delta t}[(X_t + v_H \Delta t)^{2/3} - X_t^{2/3}]$
$v_H \leqslant v_I$	$w_p = \dfrac{5.3}{\Delta t}\left(\dfrac{v_I}{10 H_e}\right)^{2/3}\left(\dfrac{F}{s_p^{3/2}}\right)^{1/4}[(t + \Delta t)^{2/3} - t^{2/3}]$	$w_p = \dfrac{1.6 F^{1/3}}{v_I \Delta t}[(X_t + v_H \Delta t)^{2/3} - X_t^{2/3}]$

they are separated by a distance equal or smaller than the value of σ_y for one of them. Under these conditions the material inside both adaptive puffs can be summed up in one, and to this puff are assigned the coordinates to its five points and the values of σ_y corresponding to the arithmetic means of the values of the original puffs. On the other hand, an adaptive puff is eliminated when its five points are beyond the model domain. When some of the points are beyond the region for which there have been defined values of the meteorological variables, the values used are those of the point of the meteorological grid closest, and the puff is kept.

Calculation of ground-level concentration   A serious problem of Lagrangian models is the calculation of the Eulerian concentrations of the contaminants. Typically, this problem is solved adding and averaging for all the period considered the contributions of all the volumes that represent the plume at each of the locations, whose coordinates are referred to one point (normally, the emitting source). This task, however, can be rather inefficient and would require high computation times even higher that the computation of the evolution of the puffs of contaminant. For the adaptive puff model, the list of specified receptors is reorganized so that only the puffs at a distance smaller than $3\sigma_y$ of each receptor must be consider with regard to the contribution to that receptor. Subsequently, for a certain time interval, the adaptive puffs that affect the receptors are obtained.

The contribution to the concentration in each receptor, and for each adaptive puff, is obtained from equation (15), using the values of z, σ_y and $\tilde{Z}$ that correspond to the height of the puff at which is the receptor. Also the contribution by reflection of the adaptive puff on the ground must be considered. Hence, in the calculation of the contribution of each adaptive volume to each receptor two possibilities should be considered:

- If the height of the receptor (z_r) is between z_i and z_{i+1}, the model will use equation (15) to calculate the contribution of sector i to the receptor, ignoring the other three sectors, unless the condition of the following paragraph is verified.
- If the point $i + 1$ of the volume is below the surface of the ground, that is, $z_{i+1} < \tilde{Z}$, and the vertical coordinate resulting from the reflected projection of the receptor over the ground, $-z_r$, is located between z_r and z_{i+1}, equation (15) is used to

calculate the contribution of the reflection, but using $z = -z_r$ as the value in the equation; this contribution is added to the direct contribution.

EXPERIMENTAL VALIDATION OF THE SYSTEM

Initially, experimental data on emission, meteorology and ground level concentration were recorded during three days, 22 to 24 October 1991, to analyze the capabilities of the diffusion model to simulate actual episodes of fumigation.

The experimental procedure used was based on a permanent grid of remote meteorological and environmental sensors, and the area considered was limited to the WSW sector up to 30 km from the Power Plant.

The most relevant automatic measurements for the control system of SO_2 emissions, considered nowadays at As Pontes Power Plant, are showed at Table III.

Also, free-flying helium balloons were used up to 3000 m high, because the emission is produced from a very high stack (356.5 m over ground level), to verify changes in wind direction in height and detection of temperature inversions higher that the top of the meteorological station tower.

Five radiosondes were made during the selected three days, with measurements of wind direction and velocity, temperature and moisture. These vertical profiles were extrapolated to the nearest hour, grouping the measurements with the observed variations at ground level. These measurements were interpolated and homogenized by a simple meteorological diagnostic model, using the classical divergence-null condition.

On the 22nd, the average height of the inversion layer was about 2300 m (msl), but wind velocity reached 20 m/s. The plume, emitted at 700 m (msl), generated fumigation around the three F's stations in WSW direction (Fig. 4); the high wind velocity, and a minor inversion layer at 1100 m (msl) reduced the plume rise, so the plume affected these stations located within 30 km from the Power Plant.

On the 23rd, the average height of the inversion layer was 1700 m (msl), so the plume was trapped below this layer. This fact, and the higher wind velocity – 25 m/s- causes a more reduced plume rise, so the impacts of the plume (Fig. 5) over the nearest F's

TABLE III

Automatic measurements for SO_2 emissions control at as Pontes power plant

Stack gas measurements (1/2 hour averages):	– SO_2 concentration. – Temperature. – Volumetric flow.
A Mourela meteorological station (1/2 hour averages):	– Wind velocity and direction at 10 m. – Wind velocity and direction at 80 m. – Temperature at 10 m. – Lapse rate, from 80 m to 30 m.
SO_2 glc stations (hourly averages):	– SO_2 concentration at 3 m over the terrain.

901

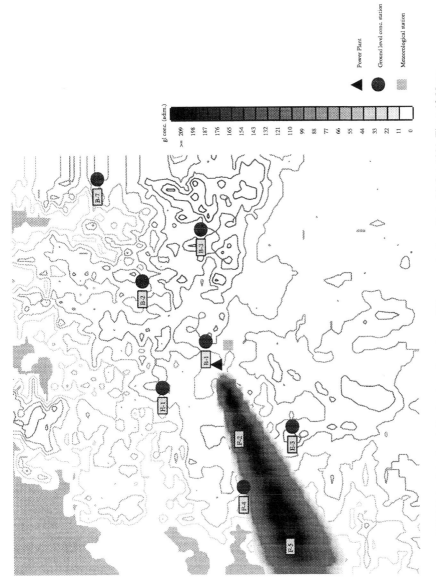

FIGURE 4   Calculated distribution of maximum SO$_2$ concentration. Date: 22-Oct-1991. Time: 13:00 am.

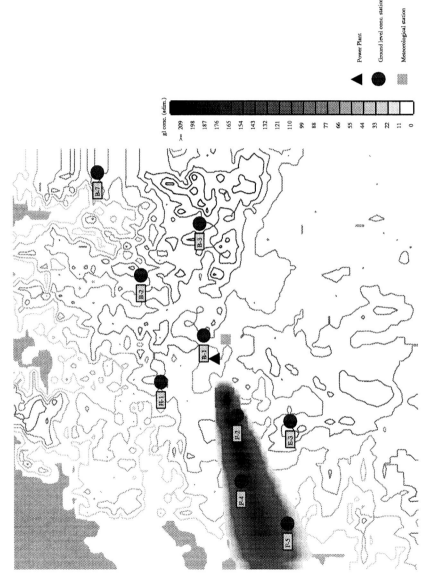

FIGURE 5   Calculated distribution of maximum SO$_2$ Concentration. Date: 23-Oct-1991 Time: 16:00 am.

TABLE IV

SO$_2$ observed and calculated impacts around As Pontes power plant

Date	Period	Remote stations affected	
		Model	Experimental
22/10/1991	9:30 — 20.00	F-2, F-4, F-5	F-2, F-4, F-5
23/10/1991	9:00 + 20.00	F-2, F-4	F-2, F-4
24/10/1991	9:00 — 20:00	None	None

stations (F-2 and F-4), were favored. The model and the observed values showed no SO$_2$ fumigation in F-5.

On the 24th, the inversion layer was below the 1300 m so, at first, the plume should be trapped; but the wind velocity was only 10 m/s, so the plume could rise over the inversion layer. In fact, no impacts are observed near the Power Plant, and the model suggested the same results.

The results obtained for the three days show that the simulation was always correct when no SO$_2$ glc was detected at the remote stations and that SO$_2$ glc tended to be affected mainly by the height of the temperature inversion and wind velocity.

On a qualitative basis it can be assured that the model is coherent with the experimental results and its phenomenological interpretation.

The system was further validated with the addition of a weather forecasting model to the atmospheric diffusion model. Also, a SODAR unit was installed next the Power Plant.

Figure 6 shows the predicted distribution of SO$_2$ concentration for May 13th, 1994, at 18:00 am. That day, only B-2 and B-7 stations detected SO$_2$ glc, in agreement to the system prediction. Since the system became operational, on June 1994, predicted and detected SO$_2$ glc on each remote station coincided over 89% of the time.

CONCLUSIONS

An emissions control system for a coal-fired power plant is presented, based in the use of models that allow the prediction of the impact of the plume on the ground, 24 hours in advance. The three models included in the system are coded in FORTRAN and can be applied independently to get estimates of SO$_2$ emissions, meteorological predictions and estimates of SO$_2$ glc.

A set of field experiments were conducted to validate the system developed. From the results obtained, when compared with the estimates predicted, it can be concluded that a control system of this kind will be a valid criterion for the definition of the mixtures of coals to be burnt at the power plant in order to avoid the occurrence of fumigation in the area around the plant. From the experience already gained, the quality of the estimates depends mostly in the availability of good meteorological predictions, that is to describe adequately the atmospheric flow.

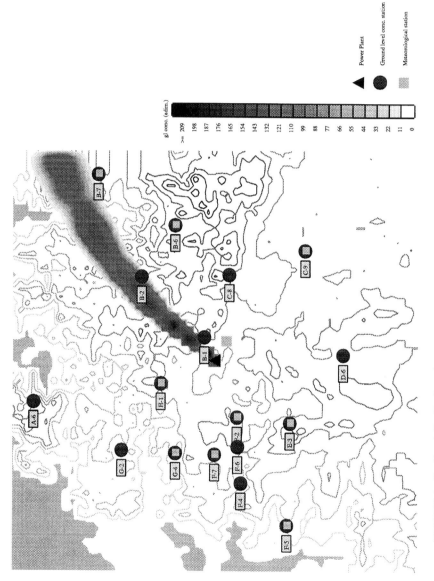

FIGURE 6   Predicted distribution of maximum SO_2 concentration. Date: 13-May-1994 Time: 18:00 am.

This system is being applied, on a trial basis, at As Pontes Power Plant both for environmental reasons and to optimize the consumption of imported sub-bituminous coal to be mixed with the local lignite. Until this moment, global results show a good qualitative agreement between predicted and observed SO_2, but the models are being continuously calibrated using the available automatic meteorological and SO_2 glc measurements, to optimize the response of the system.

ACKNOWLEDGEMENTS

The authors are grateful to the Meteorological Center of La Coruña (Spanish National Meteorological Institute) for providing wind and temperature sets of data, as well as for helpful discussions. The computational time assigned and technical support of the Centro de Supercomputación de Galicia are gratefully acknowledged. This work has been financially supported by OCIDE under research contract PIE 034-052.

NOMENCLATURE

a_y, b_y	Empirical parameters for calculation of traverse plume dispersion.
a_z, b_z	Empirical parameters for calculation of vertical plume dispersion.
B	Two-term virial equation global parameter.
B_{ij}	Two-term virial equation parameter between i and j gases.
c	Pollutant concentration.
D	Diameter of the stack.
F	Buoyancy.
f	Coriolis parameter.
g	Gravitational aceleration.
H_e	Stack height.
P	Pressure of stack gases.
Q_c	Total amount of material in a puff.
R	Ideal gases constant.
s	Top height in the meteorological model.
s_p	Stability Monin-Obukhov parameter.
T	Temperature.
T_e	Temperature of stack gases.
U	Average wind velocity: x component.
u	Fluctuation of wind velocity x component.
U_g	Wind velocity at the top level: x component.
V	Average wind velocity: y component.
v	Fluctuation of wind velocity y component.
v_e	Velocity of stack gases.
V_g	Wind velocity at the top level: y component.
v_H	Horizontal wind velocity $= (U^2 + V^2)^{1/2}$
v_I	Inlet stack gases velocity.
W	Average wind velocity: vertical component.
w_p	Plume rise contribution to vertical component.

X	Travel distance of the puff.
x	West-east coordinate.
y	South-north coordinate.
y_{gi}, y_{gj}	Molar fraction of stack gases.
z	Vertical coordinate.
z_g	Topographic height.
z_r	Eulerian receptor height, above the ground.
$\tilde{Z}$	Vertical puff scale in σ_z units.
Δq	Amount of material in a vertical segment of puff.
η	Following-terrain coordinate.
θ	Potential temperature.
Π	Atmospheric pressure.
ϕ	Latitude.
φ	Generic function.
ξ	Generic vertical coordinate in the puff.
σ_y	Standard deviation of horizontal Gaussian distribution in the puff.
σ_z	Standard deviation of vertical Gaussian distribution in the puff.

REFERENCES

Casares, J. J. and Souto, J. A. (1991) Estudio de modelos de difusión atmosférica para focos singulares. *V Encontro Galego-Portugués de Química-Medio Ambiente*, La Coruña.

Enger, L. (1990) Simulation of dispersion in moderately complex terrain. Part A. The fluid dynamic model. *Atmospheric Environment*, **24A**, 2431.

Enger, L., Koracin, D. and Yang, X. (1993) A numerical study of the boundary layer dynamics in a mountain valley. Part I. Model validation and sensitivity experiments. *Boundary-Layer Meteorology*, **66**, 357. Ludwig, F. L., Gasiorek, L. S. and Ruff, R. E. (1977) Simplification of a Gaussian puff model for real-time minicomputer use. *Atmospheric Environment*, **11**, 431.

Ludwig, F. L., Salvador, R. and Bornstein, R. (1989) An adaptive volume plume model. *Atmospheric Environment*, **23**, 1, 127.

Pooler, F. and Niemeyer, L. E. (1970) Dispersion from tall stacks: an evaluation. *2nd International Clean Air Congress*, New York, 1049.

Sheih, C. M. (1978) A puff pollutant dispersion model with wind shear and dynamic plume rise. *Atmospheric Environment*, **12**, 1933.

Yamada, T. and Mellor, G. (1975) A simulation of the Wagara atmospheric boundary layer data. *Journal of Atmospheric Sciences*, **31**, 2309.

Yamada, T. (1985) Numerical simulation of the night 2 data of the 1980 ASCOT experiments in the California Greysers Area. *Archives for Meteorology, Geophysics, and Bioclimatology*, Ser. A34, 233.

Dilute Oxygen Combustion:
A New Approach for Low NO$_x$ Combustion

H. KOBAYASHI and Z. DU   *PRAXAIR, Inc. Tarrytown, New York 10591 USA*

Abstract—An ultra low NO$_x$ combustion method based on the concept of reacting fuel with a very dilute oxygen stream has been developed recently. NO$_x$ emissions of 0.01 to 0.02 lb/10^6 Btu (0.02 to 0.04 mg/Kcal) were achieved in a test furnace operated at 2300 °F (1260°C) with various furnace nitrogen concentrations ranging from 34 to 68%.

Key Words: Combustion, oxygen, nitrogen oxides

INTRODUCTION

Many low NO$_x$ burners have been developed over the last two decades to meet the increasingly tightening emissions standards. It has been well established that the most effective design strategy for reduction of thermal NO$_x$ is to reduce the peak flame temperature. External and in-furnace flue-gas recirculation have been adopted in many burner designs for NO$_x$ reduction.

At Praxair Inc. (formerly Union Carbide Industrial Gases Inc.) a low NO$_x$ oxy-fuel burner using infurnace recirculation of flue-gas was developed in the late 1970's (see Anderson (1986)) and commercialized for high temperature industrial furnace applications (Type "A" Burner.) The burner demonstrated that the peak flame temperature of oxy-fuel flames could be reduced even below that of conventional air-fuel burners. NO$_x$ emissions as low as 0.001 lb as NO$_2$ per 10^6 Btu (0.002 mg/Kcal or equivalent to 0.72 ppm for CH$_4$-air combustion at 2% excess O$_2$ in flue-gas on a dry basis) was achieved in a test furnace with minimum air leakage. The source of nitrogen in the furnace was primarily molecular nitrogen contained in natural gas. Since it is not practical to totally eliminate nitrogen in most industrial furnaces and NO$_x$ formation is highly dependent on the flame temperature, control of the flame temperature is especially important for oxy-fuel combustion. Extensive studies have been conducted on the emissions of NO$_x$ from various industrial burners under oxygen enriched combustion conditions and published elsewhere (see Kobayashi, *et al.* (1989) or Kobayashi, *et al.* (1991)).

One of the fundamental technical questions that came out of these studies was how to achieve the lowest oxygen concentration of oxidant to react fuel in practical combustion systems. This question led to the development of the Dilute Oxygen Combustion (DOC) method described in this paper. The process has been patented and recently commercialized as Praxair Type L Burner.

DILUTE OXYGEN COMBUSTION

The ultimate design goal of the DOC method is to react fuel with an oxidant stream containing the lowest possible oxygen concentration. In Figures 1 and 2 the adiabatic

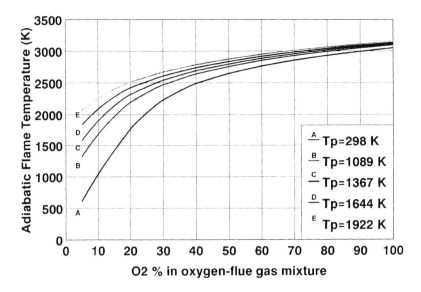

FIGURE 1

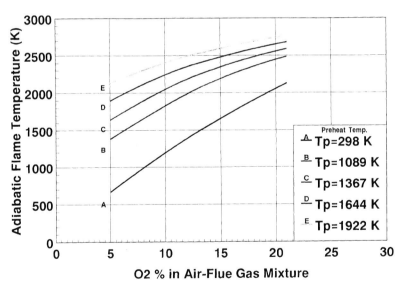

FIGURE 2

flame temperatures of methane-oxygen and methane-air combustion are shown as
a function of oxygen concentration 'and preheat temperature of oxidant at
a stoichiometric ratio corresponding to 2% O_2 in flue-gas on a wet basis. It is assumed
that the wet flue-gas is recirculated to dilute pure oxygen or air to provide the oxidant

with different oxygen concentrations. The lowest possible oxygen concentration in this case is 2% which is attained only with an infinite amount of flue-gas recirculation.

Without oxidant preheating, the calculated adiabatic flame temperature is 5030°F (2777°C) at 100% O_2 and decreases gradually at high oxygen concentrations. At low oxygen concentrations the adiabatic flame temperature decreases sharply from 3189°F (1754°C) at 25% O_2 to 2100°F (1149°C) at 15% O_2 and to 642°F (339°C) at 5% O_2 for the methane-oxygen combustion (Fig. 1). Similar temperature decreases are shown for the air case in Figure 2. It should be noted that the adiabatic flame temperature at 21% O_2 for the methane-oxygen case is approximately 2700°F (1482°C) as compared with 3369°F (1854°C) for the methane-air combustion with 2% excess O_2 in flue-gas. The difference is attributable to the higher heat capacities of CO_2 and H_2O as compared with N_2 and the endothermic water-gas shift reaction.

In order to establish a stable flame in a practical furnace a minimum adiabatic flame temperature of 1800 to 2000°F (982 to 1093°C) is considered to be required. Oxygen concentration of 12 to 13% may provide the lowest limit of DOC for the oxygen case and 10 to 11% for the air case based on this criterion. This limitation can be overcome by preheating of oxidant. The effects of oxidant preheating on adiabatic flame temperature are also shown in Figure 1 and 2 for oxidant preheat temperatures of 1500 to 3000°F (816 to 1649°C). At high oxygen concentrations, flame temperature increases only by 100 to 200°F (56 to 112°C) even with 3000°F (1649°C) preheat due to dissociation of various species and the small volume of oxidant available for preheating. At low oxygen concentrations, the increases in flame temperature become close to the increases in oxidant temperature. At 5% O_2, adiabatic flame temperatures are only 300 to 500°F (167 to 278°C) above the preheat temperatures. Therefore, the use of a preheated very dilute oxygen stream offers potential for stable low NO_x combustion.

Figure 3 shows the volume of oxidant required for combustion of methane-oxygen at 2% excess O_2 in the flue gas, which is approximately inversely proportional to the concentration of oxygen in oxidant. The key point is the huge amount of oxidant required in dilute oxygen combustion. For example, at 5% O_2 about 70,000 ft^3 (1980 m^3) of oxidant is required to burn 1×10^6 Btu (gross heating value) of methane (227×10^6 cal of net heating value), which is about seven times that required for the normal air combustion. It raises two technical problems for practical applications; (1) a large volume of flue-gas recirculation (FGR) is required and (2) fuel has to be mixed with a large volume of dilute oxidant for complete combustion.

Figure 4 depicts the concept of DOC methods developed by Praxair, Inc. An oxidant jet and a fuel jet are separately introduced into the furnace. The oxidant jet entrains furnace atmosphere by the aspiration effect of the turbulent jet and create an "oxidant mixing zone", in which oxygen concentration is reduced from the initial concentration to a value approaching that of the furnace gas. For example, if air is the oxidant, the initial oxygen concentration is about 21%. If the average oxygen concentration of furnace gas entrained into the oxygen stream is 2% and 20 moles of furnace gas is mixed per mole of oxidant in the oxidant mixing zone, oxygen concentration after mixing becomes about 2.9%. When pure oxygen or oxygen enriched air is used as the oxidant, higher entrainment ratios of the furnace gas to oxidant are required to reduce the oxygen concentration to the same level as in the air case. No combustion reaction takes

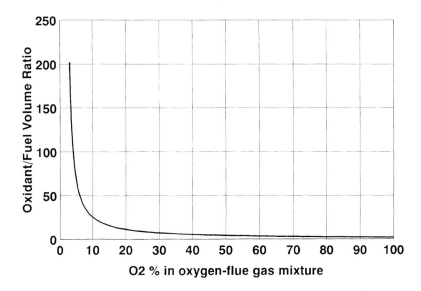

FIGURE 3

DILUTE OXYGEN COMBUSTION

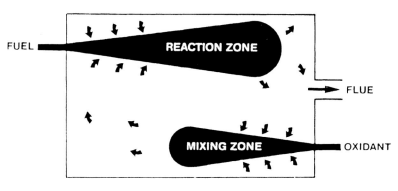

FIGURE 4

place in this zone since the furnace gas entrained into the oxidant jet is separated from the "fuel reaction zone".

Fuel is introduced separately into the furnace from the opposite wall in this particular arrangement to avoid direct mixing of the fuel and oxygen jets. Fuel reacts with oxygen and other species contained in the furnace atmosphere spontaneously as the temperature of furnace gas is substantially above the auto-ignition temperature of fuel and oxygen. The peak flame temperature is kept low due to the low oxygen concentration of "oxidant". By providing a good recirculation pattern within the furnace the oxygen concentration in the furnace atmosphere outside of the oxidant and fuel mixing zones can be maintained close to that of flue gas, which typically contains only about 2 to 3% O_2.

High momentum of the fuel jet is an important requirement for good combustion due to the large amount of oxidant that has to be mixed with fuel. For example, if the fuel is methane and the average concentration of oxygen in the furnace gas entrained into the fuel reaction zone is 3%, 67 volumes of furnace gas are required per volume of fuel to complete the combustion reactions. Small high velocity fuel jets can be used to entrain a large volume of furnace gas rapidly into the fuel jet. Thus the flame volume, or the fuel reaction zone within the furnace can be largely controlled by the turbulent jet mixing process.

Although the actual chemical reactions of the methane jet with the furnace gas containing a low concentration of oxygen and high concentrations of water and carbon dioxide are very complex and potential kinetic limitations may exist depending on the furnace gas temperature, the simple conceptual arrangement of the DOC method make it easier to characterize and scale the flames for practical applications than most of the commercial burners.

Figure 5 shows an alternate arrangement where the oxidant and fuel injection points are located close to each other on the same furnace wall. In this case it is particularly important to inject both oxidant and fuel streams at high velocities away from each other so that the oxidant mixing zone and the fuel reaction zone do not overlap. Since the dominant jet with a high momentum flux tends to overwhelm and entrain the weaker jet with a substantially lower momentum flux, a proper ratio of the fuel and oxidant momentum fluxes and sufficient spacing between any two jets must be maintained.

EXPERIMENTAL RESULTS

A series of tests were conducted in a refractory-lined cylindrical test furnace. The internal dimensions of the furnace are 3 feet (0.91m) in diameter and 7 feet 8 (2.34m)

**DILUTE OXYGEN COMBUSTION
(L- BURNER)**

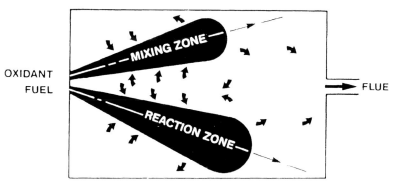

FIGURE 5

inches long. A primary burner port is located in the center of the end wall. A flue port is located on the axis of the furnace on the opposite end wall. Several access holes are placed on the flue end wall and the cylindrical side wall for injection of different gases and for insertion of heat sinks.

Tests were conducted at a constant firing rate of about 700 ft^3/hr (18.8 Nm^3/hr) of natural gas and at a constant furnace wall temperature of 2300 °F (1260°C), measured at the mid point of the cylindrical wall by a thermocouple. Two to four water-cooled heat sink pipes were inserted through the access holes to maintain the furnace temperature at the constant level. About 150 to 5000 ft^3/hr (4 to 134 Nm^3/hr) of nitrogen was introduced into the furnace through the three middle view ports to simulate an actual furnace with various nitrogen concentrations. Oxygen concentration in the flue-gas was monitored continuously by an in-situ sensor located at the flue port and kept at 2 to 2.5% on a wet basis by adjusting the flow rate of oxygen or natural gas. NO$_x$ was measured by a chemiluminescent type analyzer which was properly calibrated for the quenching effect of the background gases (N$_2$ and CO$_2$).

The result of the NO$_x$ measurements are plotted in Figure 6. As the baseline of NO$_x$ emissions for comparison a low NO$_x$ oxygen aspirator burner (Praxair Type A Burner) was used (see Anderson (1986)). This burner was optimized for NO$_x$ reduction with small fuel and oxygen nozzles. NO$_x$ emissions ranged from 0.0027 lb per 10^6 of gross Btu (0.0054 mg/Kcal net) for a nitrogen concentration of 6.8% to 0.0746 lb/ 10^6 Btu (0.149 mg/Kcal) for a 64.9% N$_2$. Since NO$_x$ emission was shown to increase approximately linearly with N$_2$ concentration for different burners tested in the initial series of the tests, only limited number of data with respect to nitrogen concentration were taken at eachconfiguration in the subsequent tests.

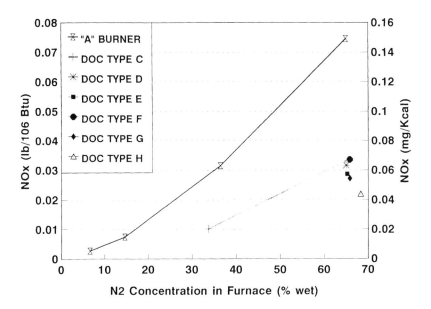

FIGURE 6

The results from the various DOC arrangements are presented in Figure 6 as C, D, E, F, G and H. In examples C, D and E, fuel was injected from the burner port and oxygen was injected from two oxygen lances through the lance holes located 180 degrees apart in the flue end wall. The oxygen jet velocities used for these tests were about 760 to 850 ft/sec (232 to 259 m/s). Different fuel and oxygen nozzles were used for each case. NO_x emission was reduced by about 50% compared to the base case. In examples F and G both fuel and oxygen were introduced from the burner, but special oxygen nozzles with 30 degree diverging angle were used to prevent the interaction of fuel and oxygen jets. Oxygen velocities were much higher. In case H the oxygen nozzles were the same as example E and fuel was injected through the flue port using a special fuel nozzle with sixteen 0.032 I.D. (0.813 mm) orifices. The lowest NO_x emission was achieved with this arrangement.

It is important to note that low NO_x emissions were achieved in examples C, D and E without using the very high velocity oxygen jets used in examples F, G and H. Since higher oxygen jet velocities require high oxygen supply pressures, the cost of oxygen becomes higher. It is desirable to achieve low NO_x emissions without using very high oxygen velocities. The stability of combustion was good for all of the above tests and CO emissions were typically 50 to 150 ppm on a dry basis.

CONCLUSIONS

An ultra low NO_x combustion method has been developed based on the concept of "Dilute Oxygen Combustion". DOC provides a practical method to burn different fuels with peak flame temperatures that are only 300 to 500 °F (167 to 278°C) above the furnace gas temperatures in many industrial furnaces. NO_x emissions of 0.01 to 0.03 lb/10^6 Btu (0.02 to 0.06 mg/Kcal or equivalent to 7 to 21 ppm for CH_4-air combustion at 2% excess O_2 in flue-gas on a dry basis) were achieved in a test furnace operated at 2300 °F (1260°C) with furnace nitrogen concentration ranging from 34 to 68%. The new method is applicable for both air and oxygen enriched combustion conditions.

REFERENCES

Anderson, J. E. "Oxygen Aspirator Burner And Process For Firing A Furnace", U.S. Patent 4,378,205, March 29, 1983.
Anderson, J. E. (1986) "A Low NO_x, Low Temperature Oxygen Fuel Burner", *Paper presented at the 1986 Symposium on Industrial Combustion Technologies*, Chicago, IL, April 29–30.
Kobayashi, H., Silver, L. S., Kwan, Y. and Chen, S. L. (1989) "NO_x Emission Characteristics of Industrial Burners and Control Methods under Oxygen Enriched Combustion Conditions", *Paper presented at the International Flame Research Foundation*, 9th Members Conference, Noordwijkerhout, May 24–26.
Kobayashi, H., Tuson, G. B. and Lauwers, E. J. (1991) "NO_x Emissions From Oxy-Fuel Fired Glass Melting Furnaces", *Paper presented at the European Society of Glass Science and Technology* Conference on Fundamentals of the Glass Manufacturing Process, Sheffield, England, September 9–11.
Kobayashi, H. (1991) "Segregated Zoning Combustion, U.S. Patent 5,076,779, Dec. 31.

Influence of a Hydrocarbon, C_3H_6 or C_3H_8, on the Reduction of NO by CO in the Absence and in the Presence of O_2, with a Pd/Al_2O_3 Catalyst

ANNE LEMAIRE, HÉLÉNE PRALIAUD, JEAN MASSARDIER
and MICHEL PRIGENT[a]   Institut de Recherches sur la catalyse,
C.N.R.S., 2 Avenue Albert Einstein, 69626 Villeurbanne,
France (Tel. 72.44.53.00 Fax 72.44.53.99); [a]Institut Français du
Pétrole, 1-4 avenue de Bois Préau, 92506 Rueil Malmaison, France

Abstract—The introduction of hydrocarbons inhibits the catalytic reduction of NO by CO in the presence and in the absence of O_2. This inhibition effect has been explained either by carbon poisoning of the active metal when the CO-NO reaction is carried out without oxygen (or with moderate amounts of O_2) or by oxygen poisoning in the presence of high oxygen amounts.

Key Words: Pd-based catalyst, NO reduction, CO-NO-O_2-hydrocarbons reactions, air pollution

INTRODUCTION

The catalytic control of the three major pollutants in automobile exhaust requires both reduction of nitrogen oxides and oxidation of hydrocarbons and carbon monoxide (Taylor and Schlatter, 1980; Summers *et al.*, 1988). Rhodium is active and selective for the reduction of NO and it is one of the components in automotive three-way catalysts. The standard three-way (TWC) formulations contain both platinum and rhodium. The Rh/Pt ratio in TWC is higher than mine ratio. It is desirable to reduce the use of rhodium because of its high cost and scarce abundance. Among the precious metals palladium is relatively abundant and less expensive than platinum and rhodium. So the use of Pd-based catalysts is explored as a substitute for Rh (Summers *et al.*, 1988; Muraki *et al.*, 1989; Duplan and Praliaud, 1991; Silver *et al.*, 1991; Massardier *et al.*, 1992)). From the comparison of catalytic behaviour their performances are similar to those of Rh. In particular they have been reported to carry out oxidation and reduction reactions in the same feedstream, for instance with CO-NO-O_2 mixtures. This paper reports on the influence of the addition of hydrocarbons, C_3H_6 or C_3H_8, to CO-NO and CO-NO-O_2 mixtures, on the reduction of NO by CO, with a Pd/Al_2O_3 solid.

EXPERIMENTAL

Palladium was deposited by wet impregnation of a gamma (delta) alumina (BET area around 100 m^2g^{-1}) coming from RHONE POULENC with Pd acetylacetonate dissolved in toluene (Vasudevan, 1983). The solid was then calcined under air flow at 723 K (programmed rate 5 K/mn, plateau 10 hours) and reduced under H_2 at the same temperature. After reduction the Pd content was 0.6 wt%.

Adsorption experiments were performed in a conventional volumetric apparatus. After reduction at 673 K and evacuation at 623 K the irreversible chemisorption uptake of H_2 at 298 K was measured by using the dual isotherm technique: after the determination of a first isotherm the sample was evacuated at 298 K for 20 minutes and a second isotherm was measured giving the part of reversibly adsorbed and absorbed H_2. A chemisorption at 348 K was also performed in order to avoid the hydride formation. Both methods gave the same results. The dispersion D was calculated by assuming an adsorption stoichiometry Hads/Pds = 1 where Pds represents one surface palladium atom. The mean diameter of metal particles was calculated from the well known relation $d_{nm} = 1.13/D$ (Wang, 1981). Such a particle size was also checked by direct observation using a JEOL 100CX electron microscope (Resolution $\leqslant 0.2$ nm).

Catalytic experiments were carried out at atmospheric pressure in a gas flow microreactor. The gases were diluted with helium. The gas concentrations (NO, CO, O_2, hydrocarbons) have been chosen to be representative of the exhaust gas mixtures. Analysis were performed by gas chromatography with a dual column (porapak and molecular sieve) and a TCD detector for O_2, N_2, CO, CO_2, N_2O, and a flame ionization detector for hydrocarbons. NO and N_2O were analyzed on-line by IR spectrometry (Rosemount analyzers).

The experiments were conducted as follows: the solid (10 mg), previously air-contacted, was loaded with 40 mg of diluent (inactive alpha Al_2O_3). A small amount of catalyst was used in order to prevent the mass and heat transfer limitations, at least for the low conversions. The catalyst was heated in a flow of N_2 up to 423 K and then contacted with the reactant gases (between 12 and 18 $1h^{-1}$). The analysis was performed at increasing and decreasing temperatures between 423 and 773 K with programmed rates of 2 K/mn. The stoichiometry of the feedstream was defined by the "s ratio" $= 2(O_2) + (NO)/(CO) + (2x + y/2)(C_xH_y)$.

Additional IR experiments have also been carried out on a Fourier transform spectrometer (I.F.S. 110 from BRUKER, resolution 4 cm^{-1}) in order to study the variations of the metal surface accessible to CO or NO after decomposition of the hydrocarbons.

RESULTS

First of all, the rough characterization of the Pd/Al_2O_3 sample gives the mean metallic particle size, about 3 nm from both electron microscopy and H_2 chemisorption results, leading to a dispersion D of 0.38.

Influence of the Addition of Hydrocarbons on the Reduction of
NO by CO in the Absence of O_2

The Figure 1 gives the CO and NO conversions as a function of the reaction temperature for the stoichiometric mixture CO-NO (s ratio = 1). The main reaction is

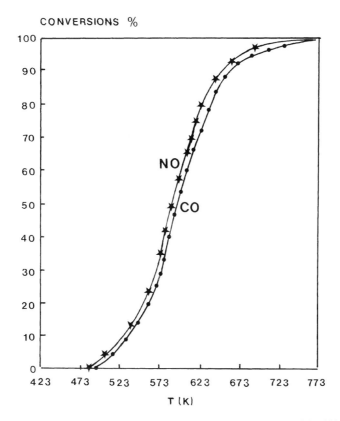

FIGURE 1   Reduction of NO by CO in the absence of O_2 (s ratio = 1) (1000 ppm CO, 1000 ppm NO, flow rate 16 l h^{-1}): NO and CO conversions.

the reaction $CO + NO \rightarrow CO_2 + 1/2\ N_2$. The slightly higher conversion of NO with respect to CO clearly shows that the N_2O formation (according to the reaction $CO + 2NO \rightarrow CO_2 + N_2O$) is weak and does not exceed 5% of the total NO converted, in our conditions.

On the Figure 2 is represented the influence of the addition of C_3H_6 and C_3H_8 to the initial CO-NO mixture on the NO conversion. The C_3H_6 and C_3H_8 contents have been choosen in order to obtain strong reducing mixtures. The "s ratio" is thus equal to 0.1 considering that the hydrocarbons react totally according to the reaction:

$$C_xH_y + (2x + y/2)\,NO \rightarrow x\ CO_2 + y/2\ H_2O + (x + y/4)N_2.$$

Whatever the nature of the hydrocarbon, the NO reduction is strongly inhibited in spite of the reducing character of the reagent mixture. The inhibition is stronger with C_3H_6 than with C_3H_8. The conversion of both hydrocarbons remains low and does not exceed 10–15 % near 773 K. Such a hydrocarbon inhibition has already mentioned on Pd catalysts by Muraki $et\ al.$ (1989).

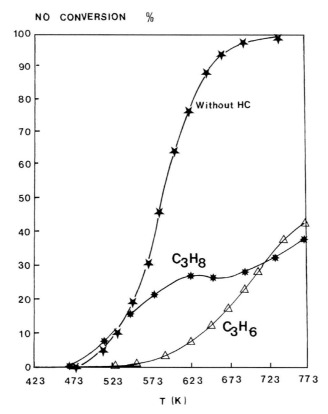

FIGURE 2   Influence of the addition of hydrocarbons on the NO reduction by CO in the absence of O_2: NO conversion:

– without hydrocarbon (HC)(s = 1)
– with 927 ppm C_3H_6 (s = 0.1)
– with 912 ppm C_3H_8 (s = 0.1).

Influence of the Addition of Hydrocarbons on the Reduction of NO by CO in The Presence of O_2

With O_2 in the reagent mixture, two main reactions occur almost simultaneously:

– the direct oxidation of CO by O_2: $CO + 1/2\ O_2 \rightarrow CO_2$
– the reduction of NO by CO: $CO + NO \rightarrow CO_2 + 1/2\ N_2$

The first reaction is followed by the O_2 consumption while the second is directly measured from the NO analysis.

The Figures 3 and 4 give the NO conversions in the absence and in the presence of the hydrocarbons and for various amounts of O_2.

When O_2 is added to the NO–CO mixture the NO reduction does not exceed 80 % (Figs. 3 and 4) whereas it reaches 100 % in the absence of O_2 (Fig. 1). The N_2O formation does not exceeds 10 % of the total NO conversion.

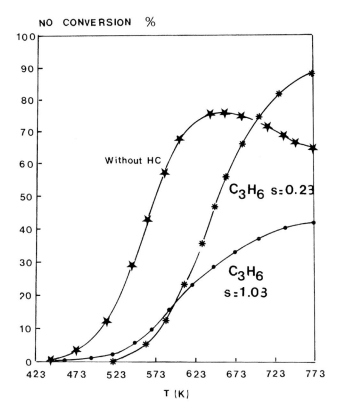

FIGURE 3   Influence of the addition of propene on the reduction of NO by CO in the presence of O_2

– without C_3H_6 (s = 1.03) (5930 ppm CO, 788 ppm NO, 2672 ppm O_2, flow rate 17.4 l h^{-1})
– adding C_3H_6 (s = 0.23)
– adding C_3H_6 and O_2 (s = 1.03).

Concerning the addition of the hydrocarbons, two situations have to be considered, depending on the nature of the hydrocarbon.

In the presence of unsaturated hydrocarbons, as illustrated by C_3H_6, whatever the "s ratio" of the reagent mixture, the NO reduction is inhibited. Nevertheless, for the reducing mixture (s = 0.23), the inhibition is not so pronounced than in the absence of O_2 (Figs. 2 and 3). Even, at high temperatures, the NO reduction increases with respect to the mixture without hydrocarbon. When the O_2 concentration increases, (s = 1.03), the inhibition drastically increases (Fig. 3).

In the presence of the saturated hydrocarbon C_3H_8 (Fig. 4), with the reducing mixture (s = 0.21) the inhibition is not observed at low temperatures (< 540 K) and, for high temperatures, the increase in the NO conversion is not observed, contrarily to the case of C_3H_6. The inhibition is pratically total with the oxidizing mixture (s = 1.03) (Fig. 4).

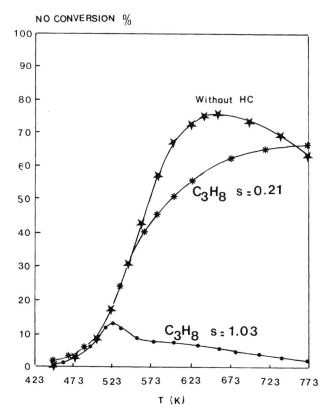

FIGURE 4   Influence of the addition of propane on the reduction of NO by CO in the presence of O_2 without C_3H_8 (s = 1.03)–adding C_3H_8 (s = 0.21)–adding C_3H_8 and O_2 (s = 1.03).

DISCUSSION

Depending on the oxidizing or reducing character of the reagent mixtures and also of the nature of the hydrocarbons, the addition of propane or propene modifies more or less drastically the NO reduction.

In the absence of O_2, the inhibition is more important than in the presence of low amounts of O_2 (s ≈ 0.2). Therefore, the decrease in the NO reduction by CO can be tentatively attributed to carbon deposits. As a matter of fact, the analysis of carbon species clearly evidences that the carbon content of the samples increases after reactions in the presence of hydrocarbons (Tab. I). At the same time the NO conversion decreases. Moreover, when the strongly poisoned solid is contacted with O_2 at 573 K, the initial activity is completely restored (Tab. I). Additional evidence of a decrease in the palladium accessibility due to the hydrocarbon is also given by the infrared spectra of CO or NO irreversibly adsorbed at 298 K, either on the initial catalyst, or on the sample treated by C_3H_6 at 473 K. On this last sample, no hydrocarbon species are

TABLE I

Influence of the hydrocarbons (C_3H_6, C_3H_8) on the amount of carbon deposits and on the NO conversion for the reactions performed in the absence of O_2

Treatment		"s ratio"	Carbon deposits (weight %)[a]	% NO conversion[b]
Catalyst alone [C]		–	0.1	–
[C] + (CO + NO)	(1)	1.0	0.1	≈ 40
[C] + (CO + NO + C_3H_6)	(2)	0.1	0.4	3
[C] + (CO + NO + C_3H_8)	(3)	0.1	0.2	20
(2) + O_2(573 K) then	(1)	1.0	–	≈ 33

[a] after reaction (1h) at 773K
[b] at 573 K

detected and the CO and NO spectrum are qualitatively unchanged with respect to those recorded on the initial catalyst. But the bands are less intense and the overall absorbance is decreased by a factor 2.6 for CO and 2.4 for NO.

It may be added that this inhibition by carbon deposits agrees well with the more important effect observed with C_3H_6 than with C_3H_8 (Tab. I, Fig. 2) since the formation of surface carbonaceous residues is easier with olefins than with saturated hydrocarbons (Barbier, 1986).

In the presence of O_2, two cases have to be considered:

i) Moderate amounts of O_2 leading to strongly reducing mixtures.
ii) Large amounts of O_2 leading to slightly oxidizing mixtures (s ≈ 1.03). This "s ratio" is obtained by addition of O_2 in order to balance the hydrocarbon concentration and is calculated assuming that the reagents are completely oxidized.

With moderate amounts of O_2 (s ≈ 0.2) and *for moderate temperatures*, the inhibition of the NO reduction by the addition of hydrocarbons is less pronounced than in the absence of O_2 (Figs. 2, 3, 4). Indeed, the carbon deposits are partially oxidized during the run. The inhibition is larger with propene than with propane, as observed without O_2. *At high temperatures* (> 723 K), when the NO conversion is enhanced by addition of propene (Fig. 3b) it can be supposed that: i) the carbon is eliminated from the surface, ii) the reduction of NO by the hydrocarbon or by CO formed by the incomplete oxidation of propene occurs. With propane, which is more difficult to oxidize than propene, such a phenomenon is not observed (Fig. 4).

In the presence of larger quantities of O_2, the NO reduction is strongly decreased and surprisingly in larger amounts with propane than with propene. In such conditions, i.e., with large O_2 amounts, the inhibition by carbon deposits can be reasonably ruled out. Moreover, it is obvious that the stronger effect with propane than with propene cannot be explained by higher hydrocarbon residues.

This carbon poisoning of the metal being set aside, it may be noticed that the "s ratio'(≈ 1.03) is calculated assuming that the hydrocarbon is completely oxidized. However, if this condition is not realized, the reagent mixture would show a strongly oxidizing character. Therefore, the NO reduction would be inhibited by the excess of unreacted O_2. The effect must be stronger when the hydrocarbon is more difficult to oxidize, *i.e.,* with propane, which is in fact observed.

To ascertain this assumption, *the oxidation by* O_2, in the *absence of* NO, *of the various reducing agents* has been undertaken. On the Figure 5, the conversions of CO, C_3H_6 and C_3H_8 on Pd/Al_2O_3 are compared. While the oxidation of C_3H_6 occurs about in the same temperature range as the CO one, C_3H_8 is only very weakly oxidized. At 773 K the C_3H_8 conversion does not exceed 20 %.

This order of oxidation by O_2 ($CO > C_3H_6 \gg C_3H_8$) qualitatively agrees with the stronger inhibiting effect observed with C_3H_8. Nevertheless, taking account of the small differences between the oxidations of CO and C_3H_6 by O_2, the inhibition by C_3H_6 would be very small. However, these oxidations by O_2 have been carried out in the absence of NO which is known to poison the oxidations (Oh and Carpenter, 1986). This poisoning effect could be more or less strong according to the reagent. The oxidation of C_3H_6, which is a large molecule, would be more disturbed by NO than the oxidation of CO.

To corroborate the *inhibition of the* NO *reduction by an* O_2 *excess* in the presence of hydrocarbons it has been verified that a similar effect is observed when the O_2 concentration is increased in the $CO-NO-O_2$ base mixture (Fig. 6). O_2 is already

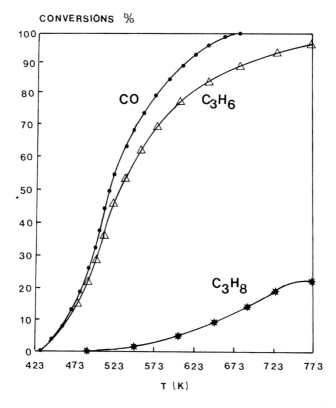

FIGURE 5   Oxidation of CO, C_3H_6 and C_3H_8 by O_2 (s ratios = 1).

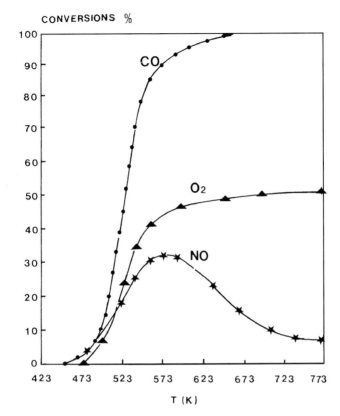

FIGURE 6   Reduction of NO by CO with O_2 in excess ($s = 2.13$) (5000 ppm CO, 667 ppm NO, 5000 ppm O_2, flow rate $12\,l\,h^{-1}$): conversions of CO, O_2, NO.

known to inhibit the NO reduction by competition with NO for adsorption sites (Egelhoff, 1982; Harrison et al., 1957; Shelef, 1975). In general the NO-CO reaction is found to proceed by a Langmuir-Hinshelwood mechanism between dissociated NO and molecular CO and the rate-limiting state would be the NO dissociation. The adsorbed nitrogen atoms desorb as N_2 and the adsorbed oxygen atoms either react with CO or the excess would form a surface oxide which inhibits the NO dissociation.

CONCLUSION

Whatever the hydrocarbons the reduction of NO by CO is inhibited when C_3H_6 or C_3H_8 are added to the two base mixtures CO-NO or CO-NO-O_2. Three cases have been considered: i) in the absence of O_2, ii) in the presence of moderate amounts of O_2, iii) in the presence of more oxidizing mixtures.

In the absence of O_2 the strong inhibition of the NO reduction is clearly explained by the formation of carbon deposits detected by chemical analysis. An additional treatment with O_2 restores the initial activity. Moreover, the inhibition is more important with the unsaturated than with the saturated hydrocarbon since the formation of carbon deposits is easier.

In the presence of moderate amounts of O_2 the addition of the hydrocarbon also inhibits the reduction of NO by CO but in a lower extent. At moderate temperatures, the inhibition is more pronounced with propene than with propane, as observed in the absence of O_2. Moreover, at high temperatures, when the direct oxidation of the propene by NO can occur, the NO conversion is enhanced.

In the presence of large quantities of O_2 the conversion of NO is strongly decreased, more specially with propane. The inhibition is mainly attributed to O_2 which is present in large excess. The more efficient reducing agent, CO, is completely oxidized. The oxidation of propane is weaker than the oxidation of propene, therefore the excess of O_2 is greater; as a consequence the inhibition is stronger with propane than with propene.

In conclusion, the inhibition of the NO reduction by hydrocarbons is explained according to the O_2 content of the mixtures and to the nature of the hydrocarbon: either by the formation of carbon deposits (for low O_2 contents and more specially with propene) or by the excess of unreacted O_2 (for high O_2 contents and more specially with propane). Therefore to resolve this inhibition problem the best catalysts would be those which are able to oxidize CO and the hydrocarbons in about the same temperature range and to avoid the formation of carbon deposits.

ACKNOWLEDGEMENTS

This work is carried out within the "Groupement de Recherches Catalyseurs d'épuration des gaz d'échappement automobile" funded by the "Centre National de la Recherche Scientifique", The "Institut Français du Pétrole" and the PIRSEM (Programme Interdisciplinaire de Recherches Scientifiques pour l'Energie et les Matières Premières).

REFERENCES

Barbier, J. (1986) Deactivation of reforming catalysts by coking. A review. *Appl. Catal.,* **23**, 225.

Duplan, J. L. and Praliaud, H. (1991) Pd/Al_2O_3 catalysts for the NO-CO-O_2 reaction: "in situ" determination of the palladium state under the reactant mixture. In Crucq, A. (ed.), *Studies in Surface Scienc and Catalysis,* Elsevier, **71**, 667–677.

Egelhoff, W. F. (1982) Nitric oxide reduction. In King D.A. and Woodruff D.P. (eds.), *The chemical Physics of Solid Surfaces and Heterogeneous Catalysis,* Elsevier, **4**, 397–426.

Harrison, B., Wyatt, M. and Gough, K. G. (1957) Catalysis of Reactions Involving the Reduction or Decomposition of nitrogen oxides. In Emmett P.H. (ed.), *Catalysis,* **5**, Reinhold New York, 127–171.

Massardier, J., El Hamadaoui, A., Bergeret, G. and Renouprez, A. (1992) Reduction of nitric oxide by carbon monoxide on palladium based bimetallic catalysts. In Guczi L. *et al.* (eds.), *Proceedings of the 10th Intern. Congress on Catalysis,* Elsevier, Part C, 2709–2712.

Muraki, J., Yokota, K. and Fujitani, Y. (1989) Nitric oxide reduction performance of automotive palladium catalysts. *Appl. Catal.,* **48**, 93.

Oh, S. H. and Carpenter, J. E. (1986) Role of NO inhibiting CO oxidation over alumina-supported rhodium. *J. Catal.,* **101**, 114.

Shelef, M. (1975) Nitric oxide: surface reactions and removal from auto exhaust. *Catal. Rev. Sci. Eng.,* **11**, 1–40.

Silver, R. G., Summers, J. C. and Williamson, W. B. (1991) Design and performance evaluation of automotive emission control catalysts. In Crucq A. (ed.), *Studies in Surface Science and Catalysis,* Elsevier, **71**, 167–180.

Summers, J. C., Williamson, W. B. and Henk, M. G. (1988) Uses of palladium in automotive emission control catalysts. Intern. Congress and Exposition, Detroit, Michigan, Feb. 29-March 4. SAE *Technical Paper Series*, 880281.

Taylor, K. C. and Schlatter, J. C. (1980) Selective reduction of nitric oxide over noble metals. *J. Catal.*, **63**, 53.

Vasudevan, S., Cosyns, E., Lesage, E., Freund, E. and Dexpert, H. (1983) The palladium alumina system: influence of the preparation procedures on the structure of the metallic phase. In Poncelet, G. *et al.* (eds.), *Scientific bases for the preparation of heterogeneous catalysts*, Elsevier, 463–470.

Wang, S. Y., Moon, S. H. and Vannice, M. A. (1981) The effect of SMSI (Strong metal-support interaction) behaviour on CO adsorption and hydrogenation on Pd catalysts. *J. Catal.*, **71**, 167.

Granulation of Filter Dusts from Combustion Plants

A. KAISER   *Maschinenfabrik Gustav Eirich, Walldürner Straße 50, D-74736 HARDHEIM, Germany*

Abstract—Increasing efforts to reduce the environmental pollution lead to increasing numbers of filter installations in all fields of industrial activities including combustion plants. Due to this development also increasing amounts of filter dusts are collected which have to be handled without causing further environmental pollutions. The filter dusts have various properties, but generally they are characterized by low bulk densities, small particle sizes and various contents of toxic compounds. Therefore they can cause secondary dust generation during transportation and handling and thus create new environmental problems or health hazards for employees. Due to their high specific surface area they also show a high leachability and therefore can cause water pollutions (Bonomo *et al.*, 1989). The reactivity of the dusts against moisture is mainly defined by their content of free CaO.

Depending on their characteristics filter dusts can be considered as valuable raw materials (Borgholm, 1992; Jiang and Roy, 1992; Sybertz, 1988), as more or less inert wastes or as hazardous wastes (Sattler and Ernberger, 1990). In almost all cases a particle size enlargement leads to positive effects like volume reduction, avoiding of secondary dust generation and also a remarkable reduction of leachability. Such pelletizing or agglomerating processes can be performed with different systems like disc pelletizer, pelletizing mixer or a mixer combined with a reaction drum.

Key Words: Fly ash, filter dust, granulation, pelletization, intensive mixer, disc pelletizer

INTRODUCTION

In order to reduce the environmental pollution in all fields of industrial production, increasing numbers of waste gas filtration systems have been installed in recent years and will be installed in the new future. Especially power stations and other combustion plants are more and more equipped with different types of flue gas treatment and filtration facilities. Due to this development also increasing amounts of filter dusts are collected from this flue gas treatment plants which have to be handled without causing further environmental pollutions. For the year 1990 only in West Germany about 3 million tons of pit coal fly ash and more than 6 million tons of lignite coal fly ash have been collected from power stations (Wagenknecht and Weiss, 1990). Additionally, 0.3–0.5 million tons of fly ash were produced by waste incineration plants. Furthermore, up to 80 million m^3 per year of municipal sewage sludges are expected to be produced in Germany at the end of this century, containing about 4 million tons per year dry material (Kassner, 1992). An increasing percentage of this sewage sludges will be incinerated creating remarkable additional amounts of fly ashes.

CHARACTERIZATION OF FILTER DUSTS
FROM COMBUSTION PLANTS

Filter dusts from combustion plants are characterized by strong variations of their properties, depending on the combustion system, the type of fuel, the flue gas treatment technology, the filter type etc. Generally they have low bulk densities < 0.8 t/m^3, typically in the range of 0.6 to 0.7 t/m^3. The particle size distribution normally is in the range of $< 100\,\mu$m (see also Fig. 1) and they have various contents of hazardous or toxic components like heavy metals or halogenated organic compounds. Their temperatures can be up to 200°C due to their very low thermal conductivity, even when they are stored in big silos. Because of their small particle sizes the dusts can cause secondary dust generation during transportation and handling and thus create new environmental problems or health hazards for employees, especially when they contain higher concentrations of toxic compounds. The small particle sizes lead also to high specific surface areas and therefore the dusts have high leachabilities and therefore can cause additional water pollutions after disposal. The reactivity of the fly ashes against moisture, their corrosiveness and other chemical properties are mostly defined by their

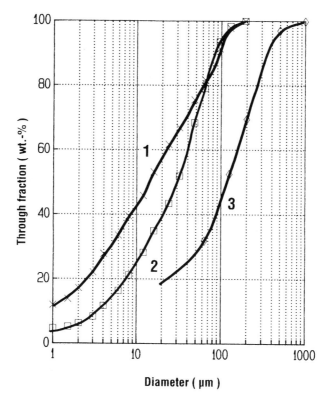

FIGURE 1   Typical particle size distributions of fly ashes from waste incineration plants (1 + 2) and from lignite power stations (3).

chemical composition, especially by their content of free, reactive lime (CaO) which can originate from the fuel or from the flue gas treatment processes and can reach amounts of 60 wt.-% of the fly ash or more. The great differences in reactivity against water can be seen in Figure 2 where quenching curves of lignite fly ashes with a high (about 65 %) and a low (< 20 %) content of CaO are compared with the quenching curve of pure quicklime. Also the chloride content of the ashes has to be taken into consideration to estimate their corrosion potential.

Depending on the characteristics of filter dusts they can be considered as valuable raw materials (e.g., fly ashes from pit coal fired power stations as fillers or cement substitutes in building materials), as more or less inert wastes which can be disposed of easily (e.g., lignite coal fly ashes) or as hazardous wastes (e.g., filter dusts from industrial waste incineration). Figure 3 shows treatment technologies and utilization possibilities of residual materials from different types of combustion plants.

AGGLOMERATION/PELLETIZATION SYSTEMS

In almost all cases a pelletization or agglomeration process for moistening the fly ashes and enlargement of the particle sizes leads to positive effects like avoiding secondary dust generation during transport and handling and volume reduction. Granulates with a sufficient strength and abrasion resistance, a good flowability and also a remarkably reduced leachability can be obtained. Depending on the characteristics of the individual dusts, especially their content of free CaO, and on the requirements which have to

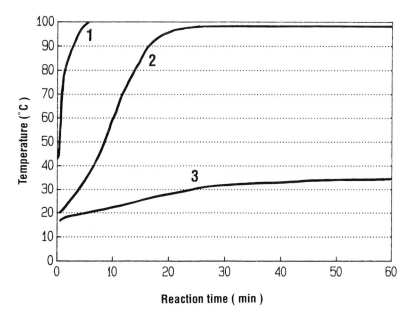

FIGURE 2   Quenching curves of pure quicklime (1) and lignite coal fly ashes with high (2) and low (3) CaO content.

930

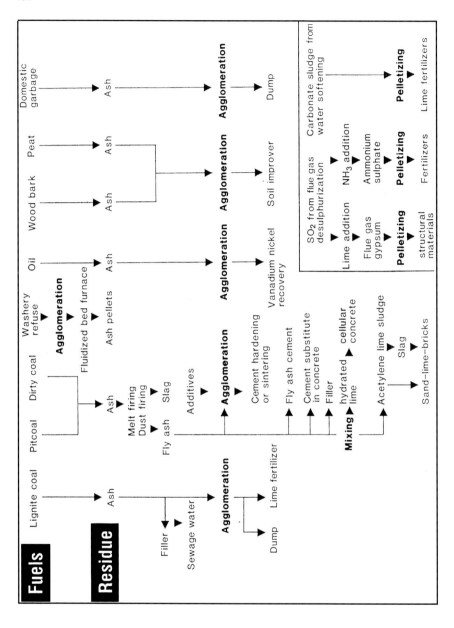

FIGURE 3 Treatment and utilization possibilities of residues from combustion plants.

be met (pellet size, strength etc.), different systems for agglomeration or pelletization can be used (Ries, 1979).

Disc Pelletizer

Fly ashes with low contents of reactive lime (< 10 %), a generally low reactivity and without hazardous components can be pelletized in a pelletizing disc. In this process the dry material is fed into an inclined rotating disc, water is sprayed through one or more nozzles onto the surface of the dust and pellets are formed by a rolling movement. A rotating scraper prevents sticking of the material to the bottom and a fixed hard metal plate keeps the wall of the disc clean. Figure 4 shows a picture of a pelletizing disc during operation.

FIGURE 4   Disc pelletizer.

This system has the advantages of relative low investment costs and low energy consumption. There are no incrustations and only few maintenance work ist necessary. However, it is difficult to operate the disc pelletizer automatically without supervision. The material has only a short residence time in the disc and there is only a limited mixing effect. An exhaustion system can be applied, but the disc is never completely closed and therefore a certain dust emission to the environment must be considered. At high ash temperatures steam generation can cause additional difficulties.

Pellets from a disc pelletizer are characterized by comparably high densities and a homogeneous size distribution. Typical pellet sizes of up to about 15 or 20 mm can be produced.

Intensive Granulating Mixer

Ashes with a medium lime content of up to about 20 % can be granulated in a countercurrent intensive granulating mixer. Also ashes with higher CaO contents can be granulated in such a system, if their reactivity is high enough and if therefore the highly exothermic reactions can be completed to a sufficient extent within a short time (normally < 5 min). The mixing principle of a countercurrent intensive mixer is schematically shown in Figure 5.

The mix container is rotating clockwise. An excentrically arranged high speed rotor which is running counterclockwise causes a very effective mixing effect. A fixed wall and bottom scraper keeps the pan clean and deflects the material into the zone of the high speed rotor.

If the particles are fine enough (which is normally the case with fly ashes) and there is a sufficient moisture content in the mix, an additional granulating effect can be obtained by choosing the right parameters (type of the rotor tools, rotation speed etc.). Granulates produced in a granulating mixer usually show a broader particle size distribution than pellets obtained from a disc pelletizer. Typical diameters are in the range of 0.5 to 8 mm. Depending on the residence time in the mixer normally the granulates also have a less perfect spherical shape than pellets from a disc. Figure 6 shows a comparison of the particle size distribution of granulated lignite fly ash from a countercurrent intensive granulating mixer and from a disc pelletizer.

Particle densities of granulates prepared in a granulating mixer and their crushing strength (which is mainly influenced by the CaO content of the ashes) are comparable to that of pellets from a disc. If the requirements are lower, e.g. if lignite dust should only be agglomerated for a dustfree transport back into the coal mine, very short residence times of less than 1 minute in the intensive mixer can be sufficient. In this case the resulting agglomerates are less densified and show an irregular shape as can be seen in Figure 7. Very short mixing times, however, are only applicable with fly ashes of very low reactivity. In the other case the exothermic hydration reaction continues after the agglomerates have been discharged from the mixer. The agglomerates are heated up and at least partially dried and dust is generated again during the subsequent handling and transport operations.

Countercurrent intensive mixers for continuous operation are available in sizes up to a volume of 7 m^3. Depending on the necessary mixing time a throughput of up to

FIGURE 5   Countercurrent intensive mixer.

several 100 m³/h can be reached with one mixing/agglomeration line. Figure 8 shows a photograph of a 5 m³ intensive mixer.

Main advantages of intensive granulating mixers are their excellent mixing efficiency due to their high mixing energy input, a very effective self cleaning effect (no growing and hardening incrustations inside the mixer or at the mixing tools and therefore only low maintenance work) and a controlled and adjustable residence time of the material in the mixer. The system ist completely closed and connected to an exhaustion system. In such mixing systems also utilization of waste water, e.g. from flue gas desulphurization plants, is possible without any problems.

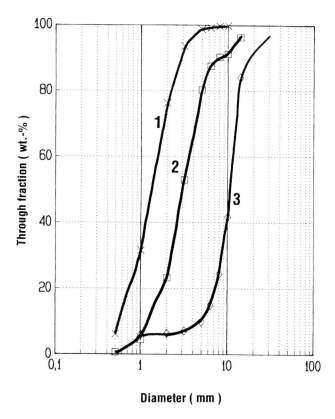

FIGURE 6   Particle size distribution of granulated lignite fly ash from a granulating mixer (1 + 2) and from a pelletizing disc (3).

Intensive Mixer and Reaction Drum

The mixing time in a granulating intensive mixer is limited for economical reasons, since with longer residence times the mixer volume becomes too large. For the granulation of fly ashes with high contents of CaO (> 20 % up to 70 % or higher) or for materials with a low reactivity (i.e., compounds with low hydration reactions) different systems have to be used. The basic problem of such systems is, that relative high amounts of water have to be added to the fly ash. The total amount of water necessary for the complete process is composed of:

a) water which is necessary for hydration reaction of CaO, $CaSO_4$ and other compounds (i.e., water which is chemically consumed and contributes to the formation of binding phases)

b) water which is evaporated in order to remove the hydration reaction heat from the mix (with higher CaO contents the temperature of the mix rapidly raises up to $\gg 100$ °C when no additional cooling is provided)

FIGURE 7   Lignite fly ash agglomerated in an intensive mixer for dustfree transport back to the coal mine.

c) water which is necessary to cool the dry fly ash to a temperature of about 80°C or lower in such cases, where the fly ash temperature is higher (this water is also evaporated)

d) water necessary as granulating liquid (the only water which remains as H_2O in the granulates after the process).

Normally it is impossible to add the total amount of necessary water (which can be remarkably higher than 50 % related to the weight of the dry fly ash) directly to the dust, since in this case a slurry-like slip would be formed which cannot be handled within a continuously operating mixer and cannot be granulated properly. Therefore it can become necessary to separate the hydration reaction physically into different steps.

One simple possibility for that is to agglomerate the fly ashes in a continuous intensive mixer using the main portion of the water and additionally to spray the second portion onto the agglomerates during the transport on a subsequent conveying

FIGURE 8   Countercurrent intensive mixer for continuous lignite fly ash agglomeration (5 m^3 useful volume, capacity 150 t/h dry ash, mixing time 1,5 min).

system. This is only useful, however, if the requirements for densification and strength of the agglomerates are low. Another possibility is to separate the complete reaction using 2 different mixers with an intermediate storage of the prereacted fly ash. However, there is a danger of hard lumps to be formed in the intermediate storage creating difficulties in the following discharging and mixing/granulating steps.

The most recommendable granulating system for such fly ashes with high CaO contents is a combination of an intensive mixer with a reaction drum, as it is shown schematically in Figure 9.

In this system the main part of the water is added into the mixer and an almost plastic mix is formed. Because of the chemical consumption of water and the reaction heat the system becomes partially dried until it is leaving the mixer. If there would be no further water addition, the agglomerates would burst and become a fine powder again during

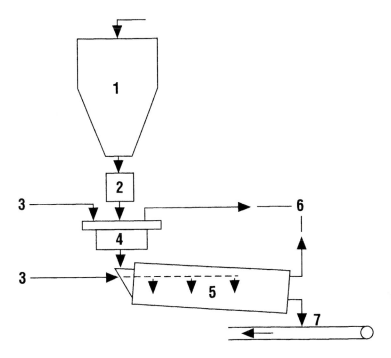

FIGURE 9   Combination of intensive mixer and reaction drum for granulation of ashes with high CaO contents or with long hydration times
(1 = fly ash surge hopper, 2 = fly ash flow control, 3 = process water, 4 = Eirich intensiv mixer, 5 = rotating drum reactor, 6 = scrubber system, 7 = discharge conveyor).

this drying process. Therefore the granulates are rolled for a certain time in a rotating drum, where additional water can be sprayed onto their surface. This process makes sure, that at each time of the reaction the necessary amount of water is available and on the other hand the granulates never become too wet. The residence time within the reaction drum can be adjusted according to the hydration behaviour of the material. During this rolling period the particles become strongly densified and get an almost perfect spherical shape. The size of this pellets can be adjusted mainly by the added moisture. Normally they are comparable to pellet sizes which can be obtained from a pelletizing disc. Figure 10 shows a photograph of lignite fly ash with a CaO content of about 65–70 wt.-% which has been pelletized in such a system.

General advantages of this combination of intensive mixer and rotating drum reactor are again the high mixing intensity and no building of incrustations since the reaction drum is also equipped with self cleaning facilities. The mixer is a completely closed system, the reaction drum as well as the mixer are connected to an exhaustion system, usually with an exhaust air scrubber. The system also needs only few maintenance. Its main advantage, however, is its high flexibility, since the amount of water and its distribution to the mixer and to the drum can be adjusted easily according to the demands of the properties of the dust. This is very important, since especially with poor qualities of lignite coal the characteristics of the fly ashes can change very rapidly.

FIGURE 10   Pellets of lignite fly ash with very high CaO content (65–70 wt.-%) prepared in an intensive mixer/reaction drum system.

Whereas the residence time in the intensive mixer should be kept at only a few minutes, the residence time in the drum can be much longer, e.g. 20 to 40 minutes. During that time the critical part of the exothermic reaction with the accompanying heat generation is finished. There are systems where the complete hydration reaction is lasting several hours, but the main exothermic reactions normally take place during the first 20 to 40 minutes. However, the final reaction can lead to a hardening of the material when it is stored in bulk. Therefore it can become necessary to let the pellets cool down in small quantities or during a transport process, e.g. on a belt conveyor.

PROPERTIES OF GRANULATED FLY ASHES

The sizes of granulates obtained from the different granulating systems normally are in the range of 0.5 mm up to about 20 mm, depending on the granulation system which has been used. The typical bulk densities of the granulates are about $0.9–1.2 \ t/m^3$. Starting from bulk densities of the fly ashes of about $0.6–0.7 \ t/m^3$, and regarding that remarkable amounts of water must be added to the ash, there is no big volume reduction. There are also systems, however, where volume reductions by a factor of $2–3$ can be obtained. e.g., a fly ash from an oil fired power station with high contents of vanadium and nickel with an initial bulk density of $0.25 \ t/m^3$ has been granulated to a resulting bulk density of $0.97 \ t/m^3$, which equals to a volume reduction by a factor of 3. This granulates have been considered to be used in a V and Ni recovery process.

The compressive strength of the granulates is strongly depending on the fly ash composition. For lignite fly ash with a high CaO content green strengths of 0.1–0.2 MPa have been obtained. After drying at ambient temperature in air the strength increased up to > 1 MPa. Pellets which have been rolled in a reaction drum for 20–30 minutes usually show a remarkably higher strength than granulates directly obtained from granulating mixers.

The reduction of the surface area which is caused by the granulating process leads also to a remarkable reduction of leachability of heavy metals from the ashes. Figure 11 shows a comparison of the leachability of different compounds from a lignite fly ash before and after granulation.

This effect can be increased dramatically by addition of different types of binders (Spicker et al., 1992). For fly ashes from waste incineration plants, e.g., with an addition of 15 wt.-% of special types of binders (modified cements) the following reductions of the leachability could be measured:

– lead reduced by a factor of 2–4,
– copper reduced by a factor of 2–10,
– zinc reduced by a factor of 20–200 and
– cadmium reduced by a factor of up to 1000.

These values, however, can not be generalized and must be proven separately with each system.

The reaction heat of the fly ashes in contact with water normally is sufficient to obtain a drying effect of the granulates at least at their surface. Due to this point the granulates generally show a good flowability and excellent handling and transporta-

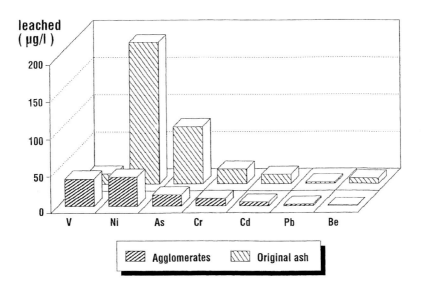

FIGURE 11   Comparison of the leachability of different heavy metals from original and granulated lignite fly ash.

tion properties compared to the original ashes. When the material is densified to a certain degree and the surface is smooth enough, normally a secondary dust generation caused by abrasion effects can be neglected.

SUMMARY

In many cases granulation or pelletization of fly ashes from combustion plants have positive effects in order to avoid secondary dust generation during transport and handling operations. These processes can also lead to a remarkable reduction of the leachability of hazardous components. Depending on the characteristics of fly ashes granulation can be performed with different systems. Pelletizing in a pelletizing disc is recommendable, when the dusts have almost no reactivity and do not contain toxic components. Higher requirements can be fulfilled with a countercurrent intensive granulating mixer, and in combination with a reaction drum almost each material can be handled. To choose the best system for a single application, it is important to discuss the properties of the fly ashes as well as the demands which are defined by the provided subsequent steps (further processing, utilization or disposal).

REFERENCES

Bonomo, L., Cernuschi, S., Giugliano, M. and de Paoli, I. (1989) Environmental behaviour of residues from coal and municipal solid waste combustion. In Thomé-Kozmiensky (ed.), *Müllverbrennung und Umwelt 3*, EF-Verlag für Energie- und Umwelttechnik, Berlin, Chap. 7, 571–582.
Borgholm, H. E. (1992) Reducing environmental pollution by using fly ash as a raw meal component. *Zement-Kalk-Gips*, **45**, 163–170.
Jiang, W. and Roy, D. M. (1992) Hydrothermal processing of new fly ash cement. *Ceram. Bull.*, **71**, 642–647.
Kassner, W. (1992) Alternative Verfahren zur Klärschlammentsorgung? Entsorgungspraxis **1–2/92**, 36–40.
Ries, H. B. (1979) Aufbaupelletierung, Verfahren und Anlagen. *Aufbereitungs-Technik*, **20**, 660–667.
Sattler, K. (ed.) and Emberger, J. (1990) Behandlung fester Abfälle. Vogel-Verlag, Würzburg.
Spicker, V., Oberste-Padtberg, R. and Roeder, A. (1992) Verfahren zur Verfestigung von staubförmigen Rückständen German Patent DE 41 20 911 C1 (05.11.92).
Sybertz, F. (1988) Puzzolanic activity of coal fly-ash. *Betonwerk + Fertigteil-Technik*, **1/1988**, 42–47.
Wagenknecht, P. and Weiss, V. (1990) Reststoffe aus Kraftwerken und Hausmüllverbrennungsanlagen. *Entsorgungspraxis*, **1–2/90**, 25–28.

Model-Based Study for Oxy-Fuel Furnaces for Low-NO$_x$ Melting Process

M. G. CARVALHO and M. NOGUEIRA   *Instituto Superior Técnico/ Technical University of Lisbon, Mechanical Engineering Department, Av. Rovisco Pais, 1096 Lisboa Codex, Portugal*

Abstract—A three-dimensional mathematical model describing the physical phenomena occurring in 100% oxy-fuel melting furnaces has been developed. This model is based on the solution of conservation equations for mass, momentum, energy, combustion related species, soot, nitric oxides, turbulent quantities. Sub-models for turbulence, combustion, soot formation/oxidation and NO formation/dissociation and radiative heat transfer in combustion chamber are considered. A finite difference/finite volume technique is used to solve the governing equations on a non-staggered non-orthogonal grid. The developed model is applied to study design strategies for more efficient and environmentally safer 100% oxy-fuel furnaces. The present work is motivated by the fact that 100% oxy-fuel burning systems are becoming one of the most promising combustion technologies, for a clean environment.

INTRODUCTION

Oxy-fuel systems are a strong alternative to the conventional air-fuel burning systems for reduced NO$_x$, CO$_2$, SO$_2$, CO and soot emissions and improved heat transfer rates. However, the viability of 100% oxy-fuel combustion systems is dependent of the thermal efficiency of these type of furnaces since the fuel savings (comparing with conventional air-fuel systems) have to balance the energy consumed in oxygen production. Thus, industrial combustion systems specially designed and optimized for 100% oxy-fuel burning are required.

The following reasons for the reduced emissions from 100% oxy-fuel can be pointed out:

- The reduced level of the N$_2$ present in the combustion chamber strongly limits the thermal NO and prompt NO formation in the combustion chamber.
- The improved fuel efficiency of these kind of furnaces (due to the absence of inert in the combustion products) allows a reduced fuel consumption with positive effects in the emissions of CO$_2$ and SO$_2$.
- The improved reactivity of the oxidant (100% oxygen) potentially limits the emission of CO.
- Soot emissions are limited because of the short character of the flame which reduces the time to the soot formation occurent under rich local mixture conditions.

The forementioned aspects give to the 100% oxy-fuel burning systems a significant potential as cleaner combustion technology. Furthermore, these firing systems are very efficient and easy to control. This allows improvements in quality (with reduction of product rejections on further stages of the production chain which also decrease the environmental impact of the industrial unit). The intensive heat transfer rates, possible with 100% oxy-fuel combustors, allow reduced equipment cost and furnace operation flexibility. Additionally, the drastic reduction in the flue gases mass flow rate, yields a considerable decrease in size and consequently in the cost of the equipment. These aspects turn the 100% oxy-fuel burning technologies a very attractive solution for **low cost pollution abatement**.

The oxy-fuel burning technologies are becoming economically attractive since the oxygen production cost has been reduced in the last years. However, this separation cost has to be balanced by the gains in energy efficiency of the oxy-fuel furnace. This aspect gives a crucial importance to the furnace thermal efficiency and therefore the optimization heat transfer process is a design priority in this kind of furnaces. There is reduced industrial and engineering experience of the design of these industrial furnaces. Thus, an industrial implementation, in short time, of efficient 100% oxy-fuel furnaces requires the use of sophisticated engineering design tools able to handle the particular features of these furnaces. The need of a rational use of oxygen and fuel demands the application, in the conception phase, of sophisticated predictive tools for a correct evaluation of the characteristics of the oxy-fuel combustion system, expressed in terms of thermal efficiency, pollutant emissions and effects on the industrial process.

Three-dimensional model-based tools, able to deal with the particular aspects involved in the oxy-fuel combustion, present significant advantages as an alternative to support the development of a new generation of low-NO_x furnaces based in the 100% oxy-fuel technology. The advantages of model based tools may be enumerated as follows:

– possibility to experiment a large number of design and operation scenarios;
– possibility to evaluate in an integrated way (heat transfer, combustion, turbulent flow) the effects on furnace performance of changes in operative and design parameters;
– possibility to experiment "exotic" solutions without technological risks.

Recently several researchers have directed their attention to the optimization of 100% oxy-fuel glass furnaces and to the technological problems related with the conversion of conventional air-fuel furnaces to oxy-fuel (Kobayashi and Richter (1990), Tuson *et al.* (1991), Kobayashi and Richter (1992), Eleazer *et al.* (1992), Farouk and Sidawi (1992)).

In the present paper, a three-dimensional mathematical model describing the physical phenomena occurring in 100% oxy-fuel industrial furnaces will be presented. This model is based on the solution of conservation equations for mass, momentum, energy and combustion related species and pollutant species. Sub-models for combustion, radiation, turbulent fluid flow, pollutants formation has been employed. The mathematical model is solved using a finite-difference/finite-volume method with a non-staggered, non-orthogonal grid.

The present paper demonstrates the capabilities of three-dimensional modelling for the optimization of the design of a furnace. This study also shows the strong influence of several geometrical parameters on the furnace efficiency and presents a methodology of optimization of those parameters through the use of a three-dimensional model.

Oxygen burning systems are applicable to high temperature directly heated continuous furnaces. Melting furnaces with a high heat absorption at the maximum tempera-

ture and furnaces in which the load has to be heated and kept at constant temperature
are clear examples of the applicability of the 100% oxy-fuel systems. The developed
model was applied to study industrial size melting furnaces.

An extensive parametric study was performed varying the parameters around
a reference value for an oxy-fuel furnace geometry arrangement for three power rates.
The studied parameters are:

– burner momentum
– width/length ratio
– combustion chamber height
– burner height
– staggered/non-staggered burner arrangement

These design parameters are an example of the several combinations of geometrical
values to be considered in the conception of such furnace.

The design solutions obtained from the above referred parameters are evaluated in
the point of view of thermal efficiency and NO_x emissions. The working conditions
considered for the parametric study represent standard conditions under which
melting furnaces have to be operated. Thus, practical design suggestions may be
extracted from the present study. Detailed results for a base case are presented through
plots representing predictions of temperature, NO_x concentration and flow pattern.

Physically-Based Modelling Procedure

In this section a comprehensive model able to simulate turbulent fluid flow, heat
transfer, combustion and pollutants formation in 100% oxy-fuel industrial combustors
is described. This model has been developed at IST–Technical University of Lisbon as
an extension of several developments reported in (see e.g., Carvalho *et al.* (1988),
Carvalho and Nogueira (1991), Azevedo *et al.* (1993)).

The thermal fluid behaviour of the gas mixture inside the combustion chamber was
predicted by solving the governing partial differential equations set in its steady-state
time-averaged using the finite difference/finite volume method. The transport equations considered may be casted into:

$$\text{div}(\rho \vec{u} \phi) = (\Gamma_\phi \,\text{grad}\, \phi) + S_\phi \tag{1}$$

where ϕ represents the transported property, $\vec{u}$ the velocity vector, Γ_ϕ the diffusion
coefficient and S_ϕ a source term.

The discretized form of the transport equation (1) was solved over a domain
represented by a numerical non-orthogonal grid. The hybrid central/up-wind scheme
was used to discretize the convective term of (1). The developed model follows
a nonstaggered grid approach for the calculation of the velocity components. An ADI
solution method is used to solve the discretized algebraic equations set.

The following global assumptions were followed in the present model: the Bous-
sinesq approximation is considered valid in this incompressible flow which, naturally,
presents a Newtonian behaviour; the transient effects due to fluctuations in operating
conditions, are assumed to be negligible; therefore limit stationary operating condi-

tions are considered. A time average description of the turbulent fluctuations was adopted; the ideal gas assumption was followed to describe the gas mixture behaviour.

Turbulence Model

The well established "k-ε" model eddy viscosity/diffusivity turbulence model (Launder and Spalding (1972)) was used. In this model, transport equations for two turbulent scales k (kinetic turbulent energy) and ε (dissipation of k) are solved, and effective (turbulent plus laminar) viscosity, μ_{eff}, and diffusivity, $\Gamma_{\phi,\text{eff}}$, are considered.

Combustion Model

The combustion model is based on the ideal fast single step reaction between the fuel and the oxidant (Pun and Spalding (1967)). This model is based on the assumption that the reaction rates associated with the fuel oxidation have very small time scales when compared with those characteristics of the transport phenomena. Chemical reactions take place instantaneously as soon as the reactants are brought together. Under this assumption, the instantaneous thermochemical state of the gaseous mixture can be determined as a function of strictly conserved scalars. Similarity between the mass diffusion coefficients of all chemical species and the thermal diffusion is also assumed. If, in addition, the system is adiabatic then all the strictly conserved variables are linearly related. Hence, knowledge of one of them is sufficient to define the instantaneous composition of the gaseous mixture. Consequently, a fuel/oxidant mixture fraction, f, has been defined as a passive scalar. A transport equation for f is solved, allowing to predict the mass concentration of the combustion related chemical species.

The short flame typical of a firing system requires particular attention on the grid distribution and refinement in the near burner region.

Turbulent Fluctuations Model

The fluctuating nature of the turbulent reaction is accommodated through a modelled equation for the variance of the mixture fraction fluctuations (g). A statistical approach to describe the temporal nature of the mixture fraction fluctuations was assumed together with a clipped normal density function (Lockwood and Naguib (1975)). The local definition of the Probability Density Function requires the knowledge of the mixture fraction, f, and its variance, g, for which transport equation is solved.

The time-averaged value of every dependent scalar, ϕ, on the local turbulent fluctuations is calculated through the following expression:

$$\phi = \int_0^1 \phi(\hat{f}) \, \text{PDF}(\hat{f}) \, d\hat{f} \tag{2}$$

where f represents the mixture fraction instantaneous value and PDF a Gaussian like probability density fraction.

In the present work particular attention was given to the numerical integration of (2). The integration was performed for each control volume for which f and g are known. A large band adaptive procedure was used to handle this numerical integration with a reduced computational load. The expression (2) is applied to the following set of variables:

- Chemical species concentration
- Soot formation/oxidation
- Radiative properties of the gas media
- Thermo-physical properties as temperature, density and specific heat

Typically, in oxy-fuel burning systems, there is a strong scale disparity between the burner geometrical details and the overall furnace size. In spite of the non-premixed flame, if only a single grid node is considered to represent the O_2 and fuel injection, a constant value of f is kept over all domain. In this case the PDF shape remains based in the local value of g. A special source term in the g transport equation was used to simulate the effect, of the non-premixed injection of O_2 and fuel, even when a very coarse representation of the near burner region is adopted. This source term is linearly dependent on the distance between the O_2 and fuel injection nozzles. This approach allows considerable savings in CPU time turning the model more easily applicable for design purposes.

Soot Formation/Oxidation Model

Soot is a determinant radiative participating specie which strongly influences the radiative flux distribution in the combustion. In the present model transport equation for the soot mass concentration was solved which includes in the source terms the formation/oxidation rates (Khan and Greeves (1974), Abbas et al. (1981) and Magnussen and Hjertager (1976)).

The soot formation process is modelled through an Arrenhius like formulation which may be expressed as:

$$\dot{S}_{S^+} = \int_0^1 C_{S^+} . \hat{m}_{\text{fu}} . \exp(-E/R\hat{T}_g) . \text{PDF}(\hat{f}) d\hat{f} \tag{3}$$

where C_{S^+} is a formation model constant, $\hat{m}_{\text{fu}}$ the local instantaneous fuel mass fraction, E an activation energy, $\hat{T}_g$ the instantaneous local gas temperature. S_{S^+} is the formation source term to be used in the soot concentration transport equation.

The soot oxidation is assumed to be limited by the instantaneous concentration of oxygen and controlled by the turbulent mixing rate. The time averaged soot oxidation source term, S_{S^+}, may be expressed as follows:

$$\dot{S}_{s^-} = C_{s^-} - m_s \left(\frac{\varepsilon}{k}\right) \int_0^1 \min\left\{1, \left(\frac{\hat{m}_{\text{ox}}}{S_s m_s + S_{\text{fu}} \hat{m}_{\text{fu}}}\right)\right\} \text{PDF}(\hat{f}) d\hat{f} \tag{4}$$

where C_{s^-} is an oxidation model constant, $\hat{m}_{\text{ox}}$ the instantaneous oxygen mass fraction and S the stoichiometric ratio.

Nitric Oxides Formation/Dissociation Model

In the present work the NO formation-oxidation-transport process is modelled based on the Zeldovich mechanism assuming local chemical equilibrium (Zeldovich et al. (1949)). A transport equation for the thermal NO mass fraction was solved.

The major portion of NO_x in practical systems has been found to be NO which may be formed from different mechanisms: thermal NO yield from the oxidation of molecular nitrogen; fuel; NO from oxidation of nitrogen combined in the fuel. NO from reactions between nitrogen and CH radicals followed by oxidation.

In the present work only thermal NO was considered since no nitrogen is bounded in the fuel, and due to the rapid oxidation of CH radicals in oxy-fuel flame. In the case of oxy-fuel burning systems, the NO formation is due to the non negligible presence of N_2 mixed in the fuel (example: some kinds of natural gas) or in the oxidant stream (due to non-complete separation of the air) and due to the incoming air leakage occurring when the furnace is not pressurized.

At high temperature (above 1850 K) the thermal NO formation/dissociaiton reactions are slow relative to those of combustion. Thus the modelling of NO concentration may presume a finite rate reaction.

The kinetic path of NO formation cannot be represented only through the direct reaction between oxygen and nitrogen molecules ($N_2 + O_2 \leftrightarrow 2NO$). The presence of atoms and radicals have to be considered. In the present work the Zeldovich mechanism is used to present the thermal NO formation/dissociation process. From this mechanism a source term driving the NO concentration up to equilibrium conditions is coupled with a NO mass fraction transport equation (Semião (1991)). An Arrhenius formulation is used to model the reaction rate of each step considered in the Zeldovich mechanism.

Temperature and Density Calculation

The temperature calculation follows the assumption that the local instantaneous value may be calculated in an adiabatic basis as far as the time averaged local enthalpy value is known. The enthalpy is calculated considering the local net radiation source. Thus the temperature may be calculated as

$$\hat{T} = (\hat{h} - \hat{m}_{fu} H_{fu})/\hat{C}_{p.mix} \tag{6}$$

where $\hat{h}$ is the instantaneous enthalpy, H_{fu} the fuel heating power and $\hat{C}_{p.mix}$ the mixture specific heat.

In the present application the linear formulation for the $C_{p,i}$ temperature dependence was adopted. A transport equation was solved for the gas mixture enthalpy in which the local radiative source is considered as source term. The instantaneous value of the enthalpy was calculated through the following linear relations conventionally used for modelling of air-fuel combustors:

$$\hat{f} < \bar{f} \Rightarrow \hat{h} = h_0 + (\bar{h} - h_0)\hat{f}/\bar{f}, \quad \hat{f} > \bar{f} \Rightarrow \hat{h} = \bar{h} + (h_1 - \bar{h})/(\bar{f} - 1) \tag{7}$$

where h_0 and h_1 are enthalpies obtained for $f = 0$ and $f = 1$ respectively.

The time averaged temperature and density were calculated using the expression (2).

Thermal Radiation Model

The "discrete transfer" radiation prediction procedure of Lockwood and Shah (1980) was applied in this study to model the radiative fluxes inside the combustion chamber enclosure. Following the discrete transfer assumptions the enclosure may be considered as divided in:

– Surface zones in which boundary conditions are specified;
– Discrete solid angles around a pre specified direction from a number of locations on the walls;
– Volume zones in which volumetric net radiative flux is accounted.

The "discrete transfer" method is based on a direct solution of the radiation intensity, I, along a pre-specified direction represented by a chosen solid angle from known conditions over a surrounding surface, up to a point of impingement on an opposite wall. Fluxes on a given surface zone are calculated by integrating over the whole visible hemisphere the radiative intensity. The energy exchanged by radiation in volume zone (expressed as source in the enthalpy transport equation) is calculated through the summation of the intensity variation each time a "ray" crosses a zone. The (gas + soot) absorption coefficient spectral dependence, due to the presence of CO_2, SO_2, H_2O and C_xH_y is considered in basis of the assumption due to Truelove (1972).

As boundary condition, a linear heat transfer rate between the walls internal surface temperature and the outside temperature was considered. The internal surface temperature and the ouside temperature was considered. The internal temperature is calculated based on of the equilibrium between the flux calculated by the discrete transfer method and the referred linear heat losses rate.

Typically, the oxy-fuel furnaces flames are confined to a small region of the domain which yields the presence of very high temperature gradients. Special care is required when the Discrete Transfer is used for very fine and non uniform grids due to the so-called "ray-effect" (Carvalho et al. (1993)). In the present work, a coarser grid than the one used for the flow and scalar transport equations was adopted for the radiative transfer calculation. The radiative source term calculated for the coarser grid through the radiation model, has to be transformed into the source terms for the finer grid of the enthalpy transport equation. This transformation takes into account the volume of each fine cell, the emissivity of the medium in the cell and the forth power of temperature of the cell. This procedure allows significant gains in the accuracy of the radiative transfer calculation (overcoming the ray effect) as well as significant savings in the computational load.

In the present work the discrete transfer procedure was extended to be applicable to non-orthogonal grids covering an enclosure bounded by non-orthogonal boundaries. This is an essential aspect in the calculation of oxy-fuel burning systems since the character highly participating medium (almost without the presence of the transparent N_2) requires the rigorous calculation of the radiative path lengths. Therefore, the commonly used approximation of considering the real combusting system as a rectangular enclosure is not applicable to oxy-fuel systems where all the components of the combustion products are non-transparent.

MODEL RESULTS – APPLICATION TO AN INDUSTRIAL SIZE MELTING FURNACE

The three-dimensional model was applied to the optimization of the design (an operating condition) of industrial oxy-fuel furnace. The working conditions listed in

Table I are considered as standards. The model was applied to predict the performance of the furnace for the standard conditions and a parametric study was performed in order to optimize the design and operating conditions of the furnace.

In Figure 1 results of the prediction of the three-dimensional velocity field are plotted. In this figure it is possible to observe the upwards flow in front of each burner. The three-dimensional character of the present flow is well apparent in the figure. In Figure 2 a detail of this flow pattern is presented through a horizontal plane crossing the burners. The non symmetrical character of the present flow is well apparent in the

TABLE I

Working conditions of the considered base-case

Model Input Parameter	Applied value
Combustion chamber length	6.93 m
Combustion chamber width	4.15 m
Combution chamber height	1.16 m
Crown radius	4.59 m
Number of oxy-fuel burners	6
Relative location of the burners	frontal
Location of burners row (height)	0.37 m
Burner injection velocity	20 m/s
Type of fuel	Algerian natural gas
Fuel mass flow rate	0.0234 kg/s
Oxygen/fuel mass fraction	3.2 kg_{O_2}/kg_{N_2}
Oxygen inlet temperature	310 K
Fuel inlet temperature	310 K
Outside temperature	298 K
Walls and crown emissivity	0.6
Walls and crown heat transfer rate	0.710 $W/(K m^2)$
Load emissivity	0.8
Load melting temperature	1823 K

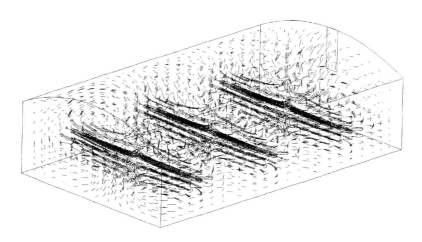

FIGURE 1   Predicted flow pattern in an oxy-fuel glass-melting furnace – 3D view (time step of marker particules: 0.2 s).

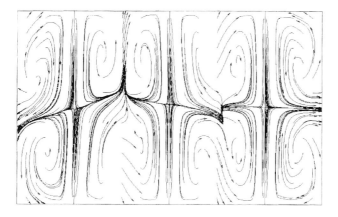

FIGURE 2   Predicted flow pattern in an oxy-fuel glass-melting-2D view in a plane at the burners level.

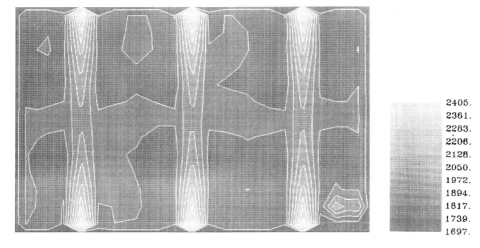

2405.
2361.
2283.
2206.
2128.
2050.
1972.
1894.
1817.
1739.
1697.

FIGURE 3   Predicted temperature (K) field in an oxy-fuel glass-melting furnace-2D view in a plane at the burners level.

figure. The temperature field is shown in Figure 3. The high temperature may be easily observed in this figure. The NO mass fraction is presented in Figure 4. The highest values of NO concentration are found in the back region of the tank, near the outlet.

The present geometry and operating conditions are typical of industrial furnaces used for the melting of glass or non-ferrous metals.

MODEL BASED PARAMETRIC STUDY FOR IMPROVEMENT OF THERMAL EFFICIENCY AND REDUCTION OF NO_x EMISSIONS

In the present section, a parametric study aimed to predict the effect of several design parameters on the furnace energy efficiency and specific NO_x emission is presented. In this parametric study the following geometrical parameters are considered:

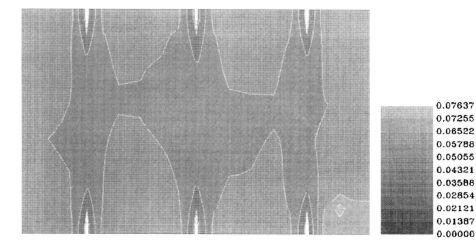

| | 0.07637 |
| 0.07255 |
| 0.06522 |
| 0.05788 |
| 0.05055 |
| 0.04321 |
| 0.03588 |
| 0.02854 |
| 0.02121 |
| 0.01387 |
| 0.00000 |

FIGURE 4   predicted NO mass fraction (gNO/kg$_{mix}$) distribution in an oxy-fuel glass-melting furnace-2D view in a plane at the burners level.

- furnace aspect ratio (furnace width/furnace height)
- height of the combustion chamber enclosure
- distance between the burners and the melting load surface
- burner diameter (i.e. gas and O_2 injection mean velocity)
- burners arrangement (staggered or frontal)

Three different furnace sizes, corresponding to three heating levels (0.351MW; 1.053MW; 1.775MW), are considered. These three heating levels indicate three possible pull rates since, in the present study, the load surface is kept constant and equal to 36.7 kW/m². Thus, these three heating levels or melting pull correspond to three values of the area of the heat transfer of the load surface: 9.6m², 28.8m², 48m².
The furnace energy efficiency is given by the following ratio:

Efficiency = total heat transfer to the load/fuel input energy

In the case of oxy-fuel furnaces, this efficiency is expected to be significantly higher than in the case of conventional furnaces. This is an important requirement for oxy-fuel burning systems, since the saving in energy consumption should overcome the econ-omic cost of the oxygen production.

The predicted emission of NO_x is given according to the following ratio:

Specific NO_x emission = emission of NO_x (kg/s)/heat flux to the load (TJ/s)

The considered index allows the comparison of the emissions of different furnace sizes and several load pull rates.

The values of specific NO_x emissions are expected to be significantly lower than the produced by conventional air fuel furnaces. This fact, which is due to the limited character of the N_2 concentration in the furnace enclosure constitutes one of the strong motivations for the use of 100% oxy-fuel industrial burning systems.

Influence of the Aspect Ratio on the Thermal Efficiency

From Table II, it may be observed an almost continued increase of the efficiency with the value of the aspect ratio defined as follows:

Aspect ratio = furnace width/furnace length

for all the levels of heating power.

Important differences are found when the aspect ratio falls from 0.4 to 0.3 for the smallest considered furnace (heating power of 0.351 MW). This may be due to the reduced distance between the side walls where burners are located. This effect is attenuated for higher furnace sizes (higher heating power values) since the distance between side walls is consequently increased.

The highest efficiency value is found to be obtained for an aspect ratio of 0.3 for the highest heating power level. For the lowest one, the maximal efficiency is found to be obtained for an aspect ratio of 0.5.

From the obtained result it should be pointed out that the furnace aspect ratio is a design parameter with strong influence on the thermal efficiency.

In the present study improvements up to 5% in the thermal efficiency were achieved varying the aspect ratio when compared with the values used in the reference case.

Influence of the Combustion Chamber Height on the Thermal Efficiency

The furnace energy efficiency increases with the level of pull, i.e., with the level of the heating power of the considered equipment. From Table III it may be observed that the energy efficiency of the smaller furnace (heating power of 0.351 MW) decreases with the combustion chamber height. First, this may be due to the increase of the side and front walls area and consequently the increase of heating losses. However, the gas in the

TABLE II

Influence of the combustion chamber aspect ratio on the furnace energy efficiency

Heating Power (MW)	Aspect Ratio (width/length)	Efficiency (W_{fuel}/W_{load})
0.351	0.3	0.48
1.053	0.3	0.53
1.755	0.3	0.60
0.351	0.4	0.55
1.053	0.4	0.57
1.775	0.4	0.58
0.351	0.5	0.56
1.053	0.5	0.56
1.775	0.5	0.57
0.351	0.6	0.55
1.053	0.6	0.56
1.775	0.6	0.57

TABLE III

Influence of the combustion chamber height on the furnace energy efficiency

Heating Power (MW)	Chamber Height (m)	Efficiency (W_{fuel}/W_{load})
0.351	0.84	0.56
1.053	0.84	0.56
1.755	0.84	0.56
0.351	1.00	0.56
1.053	1.00	0.57
1.775	1.00	0.57
0.351	1.16	0.55
1.053	1.16	0.56
1.775	1.16	0.57
0.351	1.32	0.51
1.053	1.32	0.54
1.775	1.32	0.57

regions above the burners row, acts as a thermal insulator causing the reduction of the radiative transfer from the combustion region to the crown yielding an increase of the temperature in that region with consequent improvement of the effectiveness of the heat transfer to load. This second effect becomes more important when the crown surface (through which most part of the heat losses takes place) is predominant over the surface of the side walls which were varied in the present study. This is the case of the largest furnace considered.

Thus, it may be inferred, that the furnace height upper limit is constrained by the heat losses through the side walls and the lower limit is determined by the confinement of the flame region.

In the present study improvements up to 1% in the thermal efficiency were found varying the furnace height when compared with the values used in the reference case.

Influence of the Distance between Burners and the Melting Load Surface on the Thermal Efficiency

From Table IV, it may be observed that the energy efficiency diminishes when the burner height increases. This trend is reinforced for the smaller furnace case. This is an expected result, since the proximity of the flame to the melting load surface is favourable to direct radiate transfer from the flame region to the load. However, the distance between the burners and the load surface has a lower limit imposed by the chemical and mechanical interaction between the flame gases and load. A secondary effect has to be considered in the present analysis: if the total furnace height is kept constant the insulating effect due to the presence of the combustion products in the region above the flames is reduced when the burner distance to the load increases. This will cause a reduction of the furnace efficiency due to the processes referred in the previous chapter. This effect acts in the same direction as the primary effect above referred. In the larger studied furnaces the secondary effect is less important than in the small ones since the heat losses through the crown and walls are not so important as in

TABLE IV

Influence of the distance between the level of the burners and the level of melting load on the furnace energy efficiency

Heating Power (MW)	Burner Height (m)	Efficiency (W_{fuel}/W_{load})
0.351	0.23	0.57
1.053	0.23	0.57
1.755	0.23	0.57
0.351	0.37	0.55
1.053	0.37	0.56
1.775	0.37	0.57
0.351	0.51	0.52
1.053	0.51	0.54
1.775	0.51	0.56

the smaller ones. This fact explains the strong decay in the energy efficiency when the burners height is augmented, for the smaller studied furnaces.

In the present study, improvements up to 2% in the thermal efficiency were found varying the furnace height when compared with the values used in the reference case.

Influence of the Fuel/Oxygen Injection Velocity on the Thermal Efficiency

In the present parametric study, the furnace energy efficiency increases with the level of pull for the lower injection velocities.

From the results listed in Table V, it may be found that the efficiency decreases for the injection velocities below 20m/s. Above this value, a constant efficiency was found. This

TABLE V

Influence of the gas and oxygen injection mean velocity on the furnace energy efficiency

Heating Power (MW)	Injection Mean Velocity (m)	Efficiency (W_{fuel}/W_{load})
0.351	5	0.50
1.053	5	0.52
1.775	5	0.53
0.351	10	0.53
1.053	10	0.55
1.775	10	0.57
0.351	20	0.55
1.053	20	0.56
1.775	20	0.57
0.351	30	0.56
1.053	30	0.56
1.775	30	0.56
0.351	40	0.56
1.053	40	0.56
1.775	40	0.56
0.351	50	0.56
1.053	50	0.56
1.775	50	0.56

is mainly due to the small length of the flames produced by the very low momentum burners. In these conditions, very non-uniform heat flux fields are produced with damaging effects to the radiative heat transfer efficiency. Above 20m/s this effect is no more evident since the frontal location of the burners limits the flame region.

In the present study, the maximal thermal efficiency is found for the reference case.

Influence of the Burner Arrangement (Frontal or Staggered) on the Thermal Efficiency

The furnace energy efficiency increases with the level of pull, i.e., with the level of the heating power of the considered equipment. From Table VI, it may be observed that, in general, the energy efficiency increases with staggered burner arrangement. This effect is due to the more uniform heat flux transferred to the load in the case of the staggered arrangements. This effect may be expected since in the frontal arrangement, the flame region is very limited–large regions without combustion were left between burners.

In the present study improvements up to 1% in the thermal efficiency were found changing the burners arrangement from the frontal position to the staggered one.

Influence of the Furnace Size on the Thermal Efficiency

From the analysis presented, above it may be concluded that, in general, the thermal efficiency is improved for larger furnaces. This is an expected result since the heat losses through the walls are reduced when the size of the furnace is increased.

Influence of the Aspect Ratio on the Specific Emission of NO_x

From the results presented in Table VII, it may be observed that the specific emission of NO_x (per amount of energy transferred for the furnace load) strongly augments for the smaller values of the furnace aspect ratio. This is due to the increase of the time during which the gases are flowing at high temperature levels. In fact, the proximity of the front walls in the case of the aspect ratio 0.3, yields a reduction of the flame region and the formation of an upwards hot gas flow in which the NO_x formation occurs at significant rates. In Table VII it may be also observed that smaller NO_x emissions are obtained for the lower pull furnaces with the exception of the cases in which the aspect ratio

TABLE VI

Influence of the burner arrangement on the furnace energy efficiency

Heating Power (MW)	Burner Arrangement	Efficiency (W_{fuel}/W_{load})
0.351		0.56
1.053	Staggered	0.57
1.775		0.57
0.351		0.55
1.055	Frontal	0.56
1.775		0.57

TABLE VII

Influence of the combustion chamber aspect ratio on the furnace
NO_x specific emission (per energy transferred to the melting load)

Heating Power (MW)	Aspect Ratio (width/length)	Specific NO_x emission (Kg_{NO_x}/TJ)
0.351	0.3	10.1
1.053	0.3	11.4
1.755	0.3	7.81
0.351	0.4	6.30
1.053	0.4	7.23
1.775	0.4	8.34
0.351	0.5	5.24
1.053	0.5	7.08
1.775	0.5	7.01
0.351	0.6	4.06
1.053	0.6	5.82
1.775	0.6	5.83

decreases substantially. In these situations, the proximity of the front walls becomes critical and the flame confinement effects become predominant. The smaller values of specific NO_x emission are obtained for the reference case.

Influence of the Combustion Chamber Height on the Specific Emission of NO_x

The upwards hot flow produced in the central plane of the furnace, due to the frontal location of the burners, is the reason that explains the increase of the NO_x specific emission for the larger furnace height. From Table VIII, it may be observed that the higher values of NO_x emission are obtained for the highest furnace height considered in the present study. This result is due to the larger residence time of the gases inside the referred upwards hot flow with consequent increase of the thermal NO formation. The reduction of the furnace height, tested in the present study, allows a reduction in the NO_x emissions up to 15%.

TABLE VIII

Influence of the combustion chamber height on the furnace NO_x
specific emission (per energy transferred to the melting load)

Heating Power (MW)	Chamber Height (m)	Specific NO_x emission (Kg_{NO_x}/TJ)
0.351	1.00	3.42
1.053	1.00	5.61
1.775	1.00	5.64
0.351	1.16	4.06
1.053	1.16	5.82
1.775	1.16	5.83
0.351	1.32	7.12
1.053	1.32	7.21
1.775	1.32	7.06

Influence of the Distance between Burners and the Melting Load Surface
on the Specific Emission of NO_x

From the results listed in Table IX, it may found that the higher values of the specific NO_x emission are obtained for the lower burner height in the larger furnace size. This result may be explained by the higher temperatures obtained in this situation which are caused by the confined character of the flow field in this case with consequent slower mixing of the combustion products. However, this effect is balanced by the higher heat transfer rates produced in these more confined flames. This second effect will cause a reduction of the gases temperature in the flame region yielding a reduction of the NO_x formation rate, effect which is predominant for the smaller furnaces. This is the reason for the lower NO_x specific emissions obtained for the lower burner location, for the case of the smaller furnaces.

Influence of the Fuel/Oxygen Injection Velocity on the Specific Emission of NO_x

From the Table X, it may be observed that the specific emission of NO_x increases with the size of the furnace for all the considered burner velocities. The results, listed in Table X, may also indicate that the specific NO_x emission is lower for the higher injection velocities. This fact may be explained by the reduction of the time during which the gases are present in the hot part of the flame region.

An increase of the burner velocity from 20m/s (reference case) to 50m/s may cause a reduction of the specific NO_x emission up to 45% when compared with the value obtained for the reference case.

Influence of the Burner Arrangement (Frontal or Staggered) on the
Specific Emission of NO_x

The results, listed in Table XI, may also indicate that the specific NO_x emission is lower for the burners staggered arrangement than for the frontal arrangement. This fact is e xplained by the absence of the upwards hot flow referred before. A faster mixing of the

TABLE IX

Influence of the distance between the burners and the melting load on the furnace NO_x specific emission (per energy transferred to the melting load)

Heating Power (MW)	Burner Height (m)	Specific NO_x emission (Kg_{NO_x}/TJ)
0.351	0.23	2.97
1.053	0.23	6.69
1.775	0.23	7.51
0.351	0.37	4.06
1.053	0.37	5.82
1.775	0.37	5.87
0.351	0.51	4.03
1.053	0.51	5.69
1.775	0.51	5.72

TABLE X

Influence of the gas/oxygen burner velocity on the furnace NO_x specific emission (per energy transferred to the melting load)

Heating Power (MW)	Injection Mean Velocity (m)	Specific NO_x emission (Kg_{NO_x}/TJ)
0.351	5	6.62
1.053	5	7.54
1.775	5	8.60
0.351	10	5.28
1.053	10	7.14
1.775	10	7.35
0.351	20	4.06
1.053	20	5.82
1.775	20	5.83
0.351	30	2.36
1.053	30	5.18
1.775	30	5.87
0.351	40	1.87
1.053	40	4.68
1.775	40	5.24
0.351	50	1.83
1.053	50	4.26
1.775	50	4.48

TABLE XI

Influence of the burner arrangement on the furnace NO_x specific emission (per energy transferred to the melting load)

Heating Power (MW)	Bunrer Arrangement	Specific NO_x emission (Kg_{NO_x}/TJ)
0.351		3.03
1.053	Staggered	3.89
1.775		4.08
0.351		4.06
1.053	Frontal	5.82
1.775		5.83

combustion products, ocurrent in the present situation, may be positive in terms of NO formed in the combustion chamber.

Similarly to the observed in previous parametric studies, from the Table XI, it may be observed that the specific emission of NO_x increases with the size of the furnace.

CONCLUDING REMARKS

In the present paper a numerical procedure able to predict the thermal behaviour of industrial oxy-fuel furnaces was presented. This three-dimensional mathematical model solves the fluid flow, radiative heat transfer, combustion and pollutant formation in this type of furnaces. The numerical solution of this model is based in the domain discretization through an non-orthogonal non-staggered grid. The radiative heat transfer solution applies the discrete transfer method, also over a non-orthogonal grid.

The model was applied to the industrial requirements involved in the high temperature melting process such as that of glass melting and non-ferrous metals melting. A parametric study, in which several design parameters were varied, was performed in order to demonstrate the applicability of the proposed model to the optimization of the furnace design.

From the present application, the following main conclusions may be withdrawn:

- the NO_x emission of 100% oxy-fuel melting furnaces is found to be $1/50$–$1/100$ times lower than the NO_x specific emission of conventional air-fuel melting furnace;
- the thermal efficiency of 100% oxy-fuel melting furnaces nearly duplicates the thermal efficiency of conventional melting furnaces;
- the furnace thermal efficiency is strongly dependent on several design parameters - variations of 10% may be achieved for different geometrical arrangements;
- the design of 100% oxy-fuel furnaces is a completely multi-dimensional problem - an 100% oxy-fuel furnace presents much larger design freedom degree than a conventional air-fuel furnace;
- in the design of 100% oxy-fuel furnaces the thermal efficiency is a particularly important requirement since the gains in the thermal efficiency should balance the oxygen production cost.

In the case of 100% oxy-fuel furnaces the development, test and application of advanced design tools design is a priority in order to overcome the lack of technological experience in such equipment, the strong requirements for energy efficiency and the urgent character of the adoption of cleaner combustion technologies. From the above considerations, it may be inferred that three-dimensional modelling is a powerful tool for multidimensional optimization of oxy-fuel furnaces design and operation.

ACKNOWLEDGEMENTS

This work has been performed within the JOULE project 0051-C (SMA) 'Energy Saving and Pollution Abatement in Glass-making Furnaces, Cement Kilns and Baking Ovens. Also the Scholarship of JNICT, CIENCIA/BOLSAS, contract BD/39/90-IB, is acknowledged.

REFERENCES

Abbas, A.S., Koussa, S. S. and Lockwood, F. C. (1981) The Prediciton of a Variety of Heavy Oil Flames. *Proc. ASME Winter Combustion Annual Meeting, Spec. Sess. on Two-Phase Combustion Liquid Fuels*, Washington, Nov.

Azevedo, J. L. T., Carvalho, M. G., Coelho, P. J., Coimbra, C. F. M. and Nogueira, M. (1993) Modelling of Combustion and NO_x Emissions in Industrial Equipment. *Pure and Appl. Chem.*, **65**(2) 345–354.

Carvalho, M. G. and Nogueira, M. (1991) Evaluation of Glass Quality via 3-D Mathematical Modelling of Glass Melting Furnaces. Fundamentals of Glass Manufacturing Process, European Society of Glass Science and Technology, pp. 169–177.

Carvalho, M. G., Farias, T. and Fontes, P. (1993) Multidimensional Modelling of Radiative Heat Transfer in Scattering Media. *Journal of Heat Transfer*, **115**, 486.

Carvalho, M. G., Oliveira, P. and Semião, V. (1988) A Three-dimensional Modelling of an Industrial Glass Furnace. *Journal of the Institute of Energy*, **448**, 143–156.

Eleazer, P. B., Slavejkov, A. G., Baxter, A. and Neff, G. (1992) Optimization of Oxy-Fuel Fired Glass Melting. *European Seminar on Improved Technologies for the Rational Use of Energy in the Glass Industry,* Wiesbaden.

Farouk, B. and Sidawi, M. (1992) Modelling Oxygen-natural Gas Combustion in an Industrial Furnace. Nat. Heat Transfer Conference, San Diego

Khan, I. M. and Greeves, G. (1974) A Method for Calculating the Formation and Combustion of Soot in Diesel Engines. *Heat Transfer in Flames.* Ed. Afgan and Beer, pp. 391–402.

Kobayashi, H. and Richter, W. Design Considerations and Modelling of the Glass Melter Combustion Space for Oxy-Fuel Firing. *Int. Symp. on NO_x Control, Waste Incineration and Oxygen Enriched Combustion,* San Francisco (1990).

Kobayashi, H. and Richter, W. (1992) Combustion Space Modelling of Oxy-Fuel Fired Glass Melter. *XVI Int. Congress on Glass,* Madrid.

Launder, B. E. and Spalding, D. B. (1972) *Mathematical Models of Turbulence.* Academic Press, New York.

Lockwood, F. C. and Naguib, A. S. (1975) The Prediction of the Fluctuations in the Properties of Free, Round Jet, Turbulent Diffusion Flame. *Combustion and Flame,* **24**(1) 109.

Lockwood, F. C. and Shah, N. G. (1980) A New Radiation Solution Method for Incorporation in General Combustion Prediciton Procedurs. *18th Symp. (Int.) on Combustion,* The Combustion Institute, p. 1405.

Magnussen, B. F. and Hjertager, B. H. (1976). On Mathematical Modelling of Turbulent Combustion with Special Emphasis on Soot Formation and Combustion. *16th Symp. (Int.) on Combustion,* Combustion Institute.

Pun, W. M. and Spalding, D. B. (1967) A Procedure for Predicting the Velocity and Temperature Distributions in a Confined Steady, Turbulent, Gaseous Diffusion Flame. *Proc. Int. Astronautical Federaiton Meeting,* Belgrade.

Semião, V. (1990) *PhD Thesis,* Technical University of Lisbon.

Truelove, J. S. (1972) Mathematical Modelling of Radiant Heat Transfer in Furnaces. Heat Transfer and Fluid Service, *Chemical Engineering Division, Aere Harwell Rept.,* No. HL76/3448/KE.

Tuson, G., Higdon, R. and Moore, D. (1991) 100% Firing of Regenerative Container Glass Melters. *52nd Conf. on Glass Problems,* University of Illinois.

Zeldovich, Ya. B., Sadvnikov, P. Ya. and Frand-Kamenetskii, D. A. (1949) Oxidation of Nitrogen in Combustion. *Academy of Sciences of USSR,* Institute of Chemical Physics.

A Numerical and Experimental Study of the Turbulent Combustion of Natural Gas with Air and Oxygen in an Industrial Furnace

B. FAROUK, M. M. SIDAWI and C. E. BAUKAL*   *Department of Mechanical Engineering, Drexel University Philadelphia, Pa 19104; *Applied Research Division, Air Products and Chemicals Inc. Allentown, Pa 18195*

Abstract— A three-dimensional computational model has been developed for the simulation of the reactive turbulent flows within the combustion space of an industrial furnace fired by natural gas (containing a small amount of nitrogen). Simulations were carried out with both air-fuel and with oxygen-fuel burners. A two-step approach is used to model the formation of NO in the furnace. The first step involves the main exothermal reaction of natural gas in air/oxygen. Local instantaneous equilibrium is considered where the combustion process is assumed to be micro-mixing limited. The second step involves the solution of the Zeldovich reaction scheme for the generation of NO. Wall and gas phase radiation is treated by a gray six-flux model. The predicted NO emissions are compared with experimental data obtained in a geometrically similar furnace.

Key Words: Mathematical model, industrial furnace, NO_x emission, oxygen-natural gas combustion

INTRODUCTION

Increasing environmental awareness and tightening clean air legislation are forcing operators of industrial furnaces to reduce NO_x emissions and other pollutants (Zabielski and Woody, 1987). Several competing technologies for control of NO_x emissions treat the flue-gas after it leaves the furnace (post processing). In one such method, ammonia is injected into the flue-gas stream which converts most of the NO_x to N_2 and H_2O. Another method suggests the use of catalysts to perform this conversion. Removal of atmospheric nitrogen from the combustion environment (use of oxygen instead of air in the burners) has also been suggested as an alternative method to substantially reduce NO_x emissions from industrial furnaces. Due to the intensity of the combustion process of natural gas with oxygen (without any diluent like nitrogen), and the resulting changes in the flow rate and composition of flue gases, careful considerations must be given in the design of industrial furnaces fired by oxygen-fuel burners. Fluid dynamic effects on the combustor performance due to the low flow rate in oxygen-fuel burners need to be properly understood before industrial usage. The effects of the oxygen-fuel burners on the flame structure, furnace ceiling temperature distribution, pollutant formation, and heat transfer effectiveness to the load also need to be evaluated.

Three dimensional simulation of industrial furnaces (including utility boilers, glass furnaces etc.) for studying combustion characteristics is appearing as a successful technique to support design improvements and optimization efforts (Gosman *et al.*,

1980; Carvalho and Lockwood, 1985; Carvalho and Coelho, 1990). A three-dimensional computational model is presented here that takes into account the chemical and physical phenomena occurring within furnaces fired by either air-fuel or oxygen-fuel burners. The comprehensive model (under development) will serve several valuable purposes. Simulations of an industrial furnace can be made to quantify the energy savings, cost savings, and evaluate reduced environmental impact of conversion to oxygen-fuel firing from air-fuel firing. The model will also serve as a design tool to optimize burner location and firing rate and burner design.

In this paper, we present results from the computational model for the simulation of the reactive turbulent flows within the combustion space of an industrial furnace fired by a single burner located at the center of a side wall. A two-step approach is used to model the formation of NO in the furnace. The first step involves the main exothermal reaction of natural gas in air/oxygen. Local instantaneous equilibrium is considered where the combustion process is assumed to be micro-mixing limited. The second step involves the solution of the Zeldovich reaction scheme for the generation of NO and N. The calculated values of NO emission are compared to measurements carried out to determine the quantity of NO emitted from the flue of a pilot scale furnace under similar conditions. For air-fuel operation, the flue NO is primarily due to atmospheric nitrogen. For oxygen-fuel cases, the flue NO results from small amounts of nitrogen found in the fuel stream.

COMBUSTION SPACE GEOMETRY

The experiments were conducted in a pilot scale furnace which is shown schematically in Figure 1. The inside of the 28 m^3 rectangular furnace is 5.38 m long, 2.44 m wide and 2.13 m high. The upper halves of the side walls in the furnace are water cooled to simulate a load and prevent the furnace from overheating. The roof and lower side walls have boiler plate on the outside with 15.2 cm thick ceramic fiber blanket, and firebrick going from outside to inside. Temperatures are measured along the floor, walls and roof by using R type thermocouples. The gas flow rates are measured by the temperature corrected pressure drop across orifice plates. A more complete description of the instrumentation is provided by Joshi *et al.* (1986).

The circular exhaust flue has an inside diameter of 0.61 m . The flue is equipped with a damper to control the furnace pressure. The flue damper was adjusted to ensure the pressure was positive to exclude air infiltration. The effect of air infiltration on the NO production could thus be avoided. Oxygen in the flue was measured with a type OA.244 Taylor Servomax Oxygen Purity Analyzer. Nitric oxide and carbon monoxide were measured with Beckman Model 865 NDIR Analyzers. The NO data was corrected to 0% oxygen in the flue.

The North American 4425-8A nozzle mix burner was used for the air-fuel firings. The fuel was introduced through a 18.3 mm ID central nozzle. The air was supplied through 8 concentric 9.5 mm ID nozzle symmetrically located around the core nozzle. An Air Products KT-3 nozzle mix burner was used for the oxygen-fuel cases. This burner basically consisted of three concentric tubes. Fuel gas was supplied through the 15.9

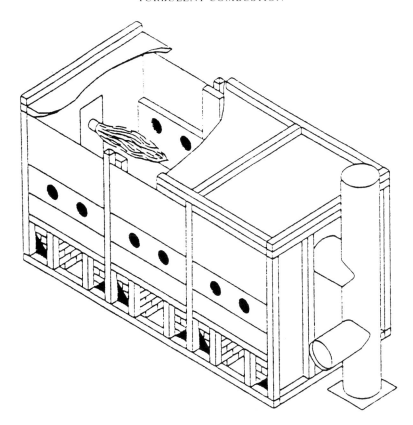

FIGURE 1   Schematic of the pilot furnace.

mm ID inner core and through the outer annulus. The outer annulus has a 41.2 mm ID and a 50.8 mm OD. The oxidizer is supplied through the inner annulus between the two fuel passages. This annulus has a 34.8 mm ID and a 38.2 mm ID. The natural gas used in the experimental studies had the following molar composition: 96.6% CH_4, 1.7% C_2H_6, 0.7% CO_2, 0.4% N_2, 0.4% C_3H_8 and trace amounts of other hydrocarbons. The gross heating value was 38.5 MJ/Nm^3.

For the computational model, a geometrically similar furnace (with the same dimensions in length, width and height) was considered. Instead of a circular opening, a square exhaust port was considered. The exhaust port has sides 0.6 m long and the center of the opening is located along the symmetry axis of the furnace, 1.33 m from the bottom. Due to the very small burner openings (compared to the overall dimensions of the furnace walls), a simplified burner geometry was considered for the computations. This was done to avoid an excessively large computational grid and severe grid nonuniformity. For the present computations, natural gas enters through a central hole (0.0092 m^2) in the burner while the oxidant (air or pure oxygen) enters the furnace through four openings (each having an area of 0.0077 m^2) located around the natural gas port. The oxidant ports are located along the periphery of a circle, 90° apart. For the present calculations, the walls of the three-dimensional domain are all taken to be black (unity emissivity). The top half of the two side walls (along the y-z plane) are assumed to be cooled to model the pilot furnace condition. The rest of the walls in the furnace (including the ceiling and the floor) are considered to be insulated.

MATHEMATICAL FORMULATION

The reactive flow field is obtained from the solution of the Favre averaged conservation equations of mass, momentum and energy along with models for turbulent combustion and radiation. Using Cartesian tensor notation, governing equations for turbulent steady flow are given as:

$$\frac{\partial}{\partial x_j}(\rho u_j) = 0 \tag{1}$$

$$\frac{\partial}{\partial x_j}(\rho u_i u_j) = -\frac{\partial p}{\partial x_j} + \frac{\partial}{\partial x_i}\left[\mu_{\text{eff}}\left(\frac{\partial u}{\partial x_j} + \frac{\partial u_j}{\partial x_i}\right) - \delta_{ij}\frac{2}{3}\frac{\partial u_j}{\partial x_j}\right] + \rho g_j \tag{2}$$

where $\mu_{\text{eff}} = \mu + \mu_t$. All dependent variables in the above equation (except density) are Favre averaged. Density is averaged in the conventional (Reynolds) sense. The turbulent viscosity, μ_t is evaluated in conjunction with the k-ε-g turbulence model for which the transport equations for the turbulent kinetic energy, k the dissipation rate of turbulent kinetic energy, ε and the square of the fluctuation of species concentration, g are solved.

The energy equation, expressed in terms of mixture enthalpy is given by:

$$\frac{\partial}{\partial x_j}(\rho u_j h) = \frac{\partial}{\partial x_j}\left(\left(\frac{\lambda}{C_p} + \frac{\mu_t}{\sigma_h}\right)\frac{\partial h}{\partial x_j}\right) + S_{\text{rad}} \tag{3}$$

For each individual species i, the enthalpy, h_i includes the standard-state enthalpy of formation and the sensible enthalpy:

$$h_i = h^o_{f,i_{298K}} + \int_{298K}^{T} c_{p,i}dT$$

where $c_{p,i}$ is the constant-pressure specific heat capacity for each individual specie as a function of temperature. Three-dimensional $(x$-y-$z)$ version of the above equations are considered in the present work.

Radiation Model

In the energy equation, S_{rad} represents the radiation source term for enthalpy h. In the present analysis the radiation is accounted for through the gray six-flux model derived from the work of Schuster (1905) and Hamaker (1947). It is noted here that the gray gas approximation is used to obtain preliminary results only as the radiation behavior of the gases is non-gray. For a three-dimensional situation, the six-flux model calculates the radiation fluxes in six mutually perpendicular directions. The six differential equations describing the variation of the radiation fluxes are combined resulting into three governing equations for the composite fluxes R_x, R_y and R_z in the three co-ordinate directions (x, y, and z):

$$\frac{d}{dx}\left(\frac{1}{a}\frac{dR_x}{dx}\right) = -\{a(R_x - E_b)\} \tag{4}$$

$$\frac{d}{dy}\left(\frac{1}{a}\frac{dR_y}{dy}\right) = -\{a(R_y - E_b)\} \tag{5}$$

$$\frac{d}{dz}\left(\frac{1}{a}\frac{dR_z}{dz}\right) = -\{a(R_z - E_b)\} \tag{6}$$

The radiation source term, S_{rad} is expressed in terms of those fluxes, the emissive power E_b and the absorption coefficient, 'a' as follows:

$$S_{rad} = 2a\{(R_x - E_b) + (R_y - E_b) + (R_z - E_b)\} \tag{7}$$

Once R_x, R_y and R_z are determined, the net radiative heat fluxes Q_{rx}, Q_{ry} and Q_{rz} in the x, y and z directions, respectively can be calculated from:

$$Q_{rx} = -\frac{2}{a}\frac{dR_x}{dx} \tag{8}$$

$$Q_{ry} = -\frac{2}{a}\frac{dR_y}{dy} \tag{9}$$

$$Q_{rz} = -\frac{2}{a}\frac{dR_z}{dz} \tag{10}$$

The present model incorporates a constant absorption coefficient of 0.2 m^{-1} for the oxygen-fuel calculations and a value of 0.1 m^{-1} for air-fuel case. These values were estimated using the expected flue-gas composition and furnace geometries. Since atmospheric nitrogen is not present for the oxygen-fuel case, a higher absorption coefficient for the combustion products is to be expected. Some combustion products (e.g. CO, CO_2, OH, H_2O etc.) are radiatively participating where the respective absorption coefficients are non-gray and temperature dependent. Hence, improved radiation modeling must be done in future to improve the accuracy of the predictions. A more rigorous evaluation of the local absorption coefficient within the furnace needs to be considered where the local non-gray absorption coefficient is dependent on the local mixture composition and local temperature.

Turbulent Combustion Model

It is important to properly account for the turbulent interactions with the combustion chemistry in diffusion flames. In the present burner geometry, the fuel and oxidizer are separated in different streams. They enter the furnace in separate eddies which must be intimately contacted on a molecular level before reaction can occur. For the combustion model employed, the assumption is made that this micro-mixing process is rate-limiting, not the kinetic process. The assumption has been found to be valid for end-fired furnaces (Smoot and Pratt, 1979). This also allows the chemistry to be computed from equilibrium considerations. Departure from chemical equilibrium appear in reacting flows where the characteristic time of the hydrodynamics becomes comparable with the characteristic time of the chemical reaction. Significant departures from chemical equilibrium can occur in turbulent flows by intense mixing (not expected in the furnace considered). However, not all the chemical species in a reacting flow are necessarily in equilibrium. For example, in laminar premixed flames and presumably in turbulent non-premixed flames radical species such as hydroxyl molecules and oxygen atoms may exist at several times their equilibrium values. Super-equilibrium radical concentrations in a turbulent jet diffusion flames were computed using a two scalar pdf model (Correa *et al.*, 1984). For non-equilibrium reactions that can occur in methane flames, partial equilibrium can be used to describe the mixing limited reactions (Colson, 1988; Chen and Kollmann,1988). Such improved models, however, have not yet been applied for predicting transport in large non-adiabatic furnaces.

To describe the degree of "mixedness" or "unmixedness" a differential equation for the 'mixture fraction' f is solved. It is defined as

$$f = \frac{m_{fu}}{m_{fu} + m_{ox}} \tag{11}$$

where

m_{fu}:   mass originated as fuel

m_{ox}:   mass originated as oxidizer

The value of any conserved scalar, s can then be calculated from the local value of 'f':

$$s = f \cdot s_{fu} + (1 - f) \cdot s_{ox} \tag{12}$$

where again

s_{fu}:   value originated as fuel

s_{ox}:   value originated as oxidizer

The effects of fluctuations in mixture fraction, which arise in all turbulent diffusion flames are modeled by introducing the probability density function (pdf) for the mixture fraction. The transport equations for the (Favre averaged) mean, f and variance, g of mixture fraction are given as

$$\frac{\partial}{\partial x_j}(\rho u_j \bar{f}) = \frac{\partial}{\partial x_j}\left[\left(\frac{\mu_t}{\sigma_t}\right)\frac{\partial \bar{f}}{\partial x_j}\right] \tag{13}$$

$$\frac{\partial}{\partial x_j}(\rho u_j g) = \frac{\partial}{\partial x_j}\left[\left(\frac{\mu_t}{\sigma_t}\right)\frac{\partial g}{\partial x_j}\right] + C_{g1}\mu_t\left(\frac{\partial \bar{f}}{\partial x_j}\right)^2 - C_{g2}\rho g\varepsilon/k \tag{14}$$

where

$$\bar{f} = \int_0^1 f P(f)\,df$$

$$g = \int_0^1 (f - \bar{f})^2 P(f)\,df$$

C_{g1} and C_{g2} are model constants taken to be 2.8 and 2.0 respectively, Spalding (1971). Besides f and g, the pdf shape is also needed. In regions of small g, the shape of the pdf makes little difference. The time-mean (Favre averaged) value of any scaler, f which is a unique function of mixture fraction such as density, temperature and species concentration, can be obtained by convolution over the probability density function:

$$\bar{\phi} = \int_0^1 \phi(f)P(f)\,df$$

In a furnace which is assumed locally adiabatic (no gas radiation or heat loss from walls), the mixture enthalpy is a conserved scalar. Thus with the assumption of equal diffusivities, the instantaneous local enthalpy may be directly calculated from f. The pdf for f then also applies for h. However for the present problem with significant radiation, the above approach is not valid. Favre averaged properties in this case need to be obtained by convolution over a joint pdf. Following Pratt (1979), the problem of obtaining the joint pdf is simplified by partitioning the enthalpy into:

(a) the energy (h_a) which would be convected if there were no radiation and
(b) the residual energy (h_r) due to radiation and wall heat loss effects:

$$h = h_a + h_r$$

$$= (f.h_{fu} + (1-f).h_{ox}) + h_r$$

It is then assumed that the effects of fluctuations of h_r are small compared with the effects of fluctuations of h_a.

For a given mixture fraction, equilibrium density, temperature and species concentrations are obtained by employing a Gibbs energy minimization scheme. Seventeen species (CH_4, C_2H_6, C_3H_8, CO_2, CO, O, O_2, OH, H, H_2, H_2O, HO_2, N, N_2, NO, NO_2 and N_2O) are considered in the Gibbs minimization scheme. The thermochemical package CREK, Pratt and Wormeck (1976), is adapted for the calculations of chemical equilibrium. Temperature dependent thermodynamic properties are considered for each individual specie. The turbulent flow mean (Favre averaged) compositions and other properties are obtained by weighting with the *pdf* of f. The shape for the $P(f)$ distribution is assumed to be a two-Dirac delta functions for the present calculations. In regions of small g, the pdf shape makes little difference. In the study of turbulent diffusion flames, Kent and Bilger (1977) used a 'clipped' Gaussian distribution which accounts for intermittency, and noted significant sensitivity of the predictions to the form of the pdf shape used, particularly with respect to nitric oxide concentrations.

Boundary Conditions

Due to symmetry along the z axis, only one half of the combustion space was considered for the computations. Zero velocities are specified along solid boundaries of the combustion space. Along the symmetry plane, zero gradient boundary condition was considered for all the dependent variables. Zero axial gradient was considered for the variables along the outlet. Most of the other boundary conditions are self-evident but a few require some discussion.

The top half of the vertical side walls (in the y-z plane) were considered water cooled. An over all heat transfer coefficient boundary condition was considered for the cooled surface where

$$Q_W = U(T_w - 373)$$

A value of $U = 10$ W/m^{2}K was considered for the oxygen-natural gas and the air-natural gas cases. For the radiation flux variables R_x, R_y and R_z boundary conditions need to be specified along the x, y, and z directions respectively. For the insulated walls, zero gradient boundary condition is used for R_x, R_y and R_z along the appropriate walls. For the water cooled walls (in the y-z plane), the following boundary condition is used for R_x:

$$\frac{dR_x}{dx} = \pm \frac{a.\varepsilon}{2-\varepsilon}(E_{bw} - R_x)$$

where the positive sign is used when the radiative flux leaving the wall is in the negative x direction.

NO$_x$ Calculations

The term NO$_x$ usually implies two major oxides, nitrogen oxide (NO) and nitrogen dioxide (NO$_2$). In combustion, NO is the dominant of the two components, NO$_2$ being mainly derived from NO. It is the prediction of NO we are concerned with in this paper. There are three accepted methods for NO$_x$ formation. Thermal NO$_x$ is formed by the reaction of nitrogen with oxygen at high temperature. 'Prompt' NO$_x$ is formed by relatively fast reaction between nitrogen, oxygen and hydrocarbon radicals. Fuel NO$_x$ is formed by the direct oxidation of organo-nitrogen compounds contained in the fuel. It is the first of these three mechanisms we are concerned with here since it is the one that can be most affected by the burner design.

The model employed here incorporates one of the simplest and most widely used mechanisms for calculations involving nitric oxide formation, viz. the Zeldovich mechanism:

$$N_2 + O \Leftrightarrow NO + N \tag{15}$$

$$O_2 + N \Leftrightarrow NO + O \tag{16}$$

$$OH + N \Leftrightarrow NO + H \tag{17}$$

As mentioned earlier, the basic assumption employed here is that the above reactions can be considered as decoupled from the main fuel burning reaction. The species O, O$_2$, OH, H and N$_2$ are assumed to exist at chemical equilibrium and remain invariant. This is with justification that they are formed by reactions which are considerably faster than those involving NO formation. The mass fractions of N and NO are determined by solving conservation equations with convection, diffusion and source/sink terms. The source/sink terms for the two conservation equation come from a summation of the reaction rate for all of the Zeldovich reactions. The forward and

TABLE I

Constants used for the forward and reverse reactions

Reaction	Forward			Reverse		
	10^B	T_{act}	N	10^B	T_{act}	N
$N_2 + O = NO + N$	1.8×10^8	3.83×10^4	0	3.8×10^7	4.25×10^2	0
$O_2 + N = NO + O$	1.8×10^4	4.68×10^3	1	3.8×10^3	2.08×10^4	1
$OH + H = NO + H$	7.1×10^7	4.5×10^2	0	1.7×10^8	2.45×10^8	0

backward reactions are given by Arrhenius expressions of the form:

$$R\left(\frac{\text{kg-mol}}{\text{m}^3.\text{s}}\right) = K\rho^2 C_1 C_2$$

where C_1 and C_2 (kg-mol/kg) are the concentrations of participating species and K the reaction rate coefficient, given by

$$K(\text{m}^3/\text{kg-mol.s}) = 10^B T^N \exp(-T_{act}/T)$$

with $10^B T^N$ being the pre-exponential factor, and T_{act} the activation temperature. The values of these quantities (Hanson and Saliaman, 1984) are given above:

NUMERICAL METHOD

The governing equations for mass, momentum and energy along with the equations for the mean mixture fraction, k, ε and g were solved by a variant of the SIMPLE algorithm (Spalding, 1980). The radiative flux equations and the NO and N transport equations (for NO_x prediction) were also solved by the above method. The thermochemical package CREK, was integrated with the above solution algorithm for the calculations of chemical equilibrium. As stated earlier, due to symmetry, only one half of the furnace was considered for the computations. A mesh size of $10 \times 31 \times 25$ ($x \times y \times z$) was considered for the above calculations. All computations were done in an IBM 3191 computer at Drexel University.

RESULTS AND DISCUSSIONS

The computations and the experiments were performed for air-natural gas and oxygen natural gas firings. Predictions were obtained for an air-fuel case where natural gas enters the furnace through the central opening (see Fig. 1) at 300 K with an inlet velocity of 10.0 m/sec while air enters through the four openings around the natural gas port at 300 K and inlet velocity of 29.0 m/sec. This corresponds to a firing rate of approximately 3.54 MW. In the second case, the air streams are replaced by oxygen streams (nitrogen is withdrawn from the inlet streams). The oxygen streams enter the furnace at 300 K and with inlet velocity of 20.0 m/sec. For both cases, the flow rates are adjusted for an equivalence ratio of near unity. For the oxygen-fuel case, the firing rate is found to be 7.1 MW.

Figure 2 shows the velocity vectors in the *y-z* plane of the computational domain for the air-fuel case along the mid-plane ($x = 0.0$ m). Figure 3 displays the velocity vectors for above plane for the oxygen-fuel case.

Due to the geometry of the furnace, recirculating regions are found near the upper and lower corners in both cases. The bottom recirculation region is found to be small for both the air-fuel and oxygen-fuel cases. However, due to the larger mass flow rate, the top recirculation region is much larger for the air-fuel case. The convective heat transfer to the top and bottom walls will be much higher for the air-fuel case, if a heat loss boundary is considered for the furnace walls. The effect of buoyancy is significant in the flow field for the oxygen-fuel case.

Temperature contours in the *y-z* plane of the combustion space are shown in Figure 4 for the air-fuel case along $x = 0.0$ m, the plane passing through the fuel inlet port. Similar results are shown in Figure 5 for the oxygen-fuel case. The maximum flame temperature for the air-fuel case is close to 1800 K while that for the oxygen-fuel

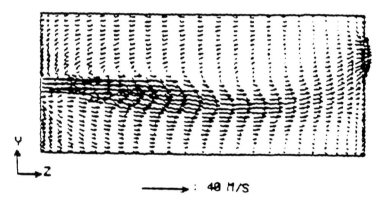

FIGURE 2   Velocity vectors for the air-fuel case along the *y-z* plane at $x = 0.0$ m ($u_{in,air} = 29.0$ m/s; $u_{in,fuel} = 10.0$ m/s.).

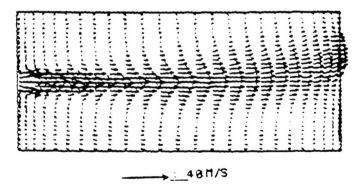

FIGURE 3   Velocity vectors for the oxygen-fuel case along the *y-z* plane at $x = 0.0$ m ($u_{in,ox} = 11.4$ m/s; $u_{in,fuel} = 20.0$ m/s.).

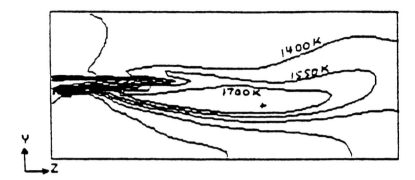

FIGURE 4   Temperature contours for the air-fuel case along the y-z plane at $x = 0.0$ m ($T_{in, air} = 300$ K; $T_{in, fuel} = 300$ K).

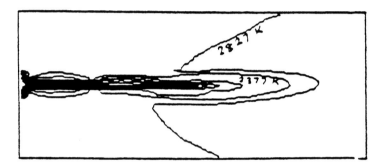

FIGURE 5   Temperature contours for the oxygen-fuel case along the y-z plane at $x = 0.0$ m ($T_{in, ox} = 300$ K; $T_{in, fuel} = 300$ K).

case is 2700 K. The flame shapes are very different for the two cases shown. While a broad flame shape is observed for the air-fuel case, the effect of lack of mixing is evident for the oxygen-fuel case. It is noted here that to enhance mixing, a special type of burner was used for the oxygen-fuel cases in the experiments. For the air-fuel case, temperature along the center increase to a value near 1600 K and remains fairly constant thereafter. For the oxygen-fuel case, the temperature change along the center line is not monotonic and is strongly influenced by dissociation of CO_2 and H_2O and subsequent oxidation of CO and H_2.

Figures 6(a), 6(b), and 6(c) show the isotherms along the x-z plane at three locations for the air-fuel case. Even though, a symmetry plane was assumed in the calculations, the results are shown for the entire x-z plane of the combustion space by imaging. In Figure 6(a), the isotherms near the furnace floor ($y = 0.236$ m) are shown where the adjoining walls are all considered to be insulated and no inlet or outlet ports are located. Temperatures in this plane are found to be high and strong inversions of the isotherms are displayed which are caused by the recirculating flow. The temperature contours along the plane passing through the fuel inlet port ($y = 1.01$ m) is shown in Figure 6(b). A broad flame shape is observed. The temperature field near the ceiling ($y = 1.7$ m) is shown in Figure 6(c). The effect of the water cooled side walls is evident

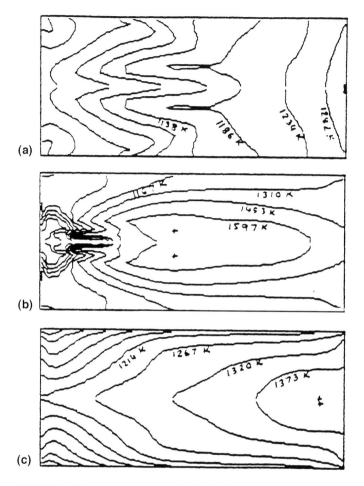

FIGURE 6   Temperature contours for the air-fuel case along the *x-z* plane.
(a) *y* = 0.0.236 m
(b) *y* = 1.01 m
(c) *y* = 1.7 m.

from the sharp temperature gradients observed along the edges. Similar results for the oxygen-fuel case are shown in Figures 7(a), 7(b) and 7(c). Qualitatively, similar results are observed but significantly higher temperatures are observed compared to the air-fuel case. This is partly due to the higher firing rate in this case. Comparing Figures 6(b) and 7(b), it is observed that a thinner flame exists for the oxygen-fuel case. The effect of recirculating flow field is more pronounced in Figure 7(b) compared to Figure 6(b). The effect of the water cooled sidewalls is again evident from Figure 7(c) where sharp gradients of temperature are found near the edges.

Very little CO is produced for the air-fuel case downstream in the furnace. For the oxygen-fuel case, much larger values of CO_2 and CO mass fractions are found throughout the combustion space due to the absence of ballast nitrogen. Significant dissociation of CO_2 occurs due to the high temperature in the furnace for the oxygen-fuel case. Similar dissociation of H_2O also occurs in the oxygen-fuel case.

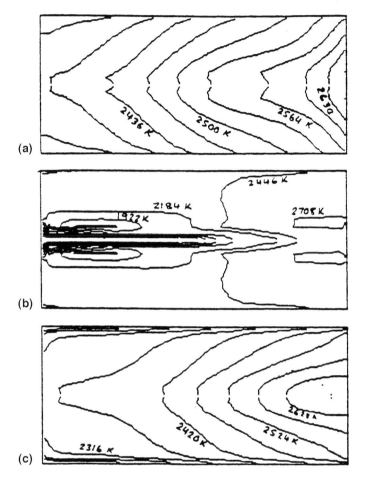

FIGURE 7   Temperature contours for the oxygen-fuel case along the x-z plane.
(a) $y = 0.0.236$ m
(b) $y = 1.01$ m
(c) $y = 1.7$ m.

NO_x Measurements and Predictions

For air-fuel cases, the burner was fired between 0.44 and 0.73 MW in the experiments. For an equivalence ratio of one with air as the oxidizer, Figure 8 shows that normalized flue NO is constant (43 ng/J) at these rates which are in the middle portion of the recommended range. The equivalence ratio and percent oxygen in the air stream were varied in another series of tests. Flue NO went up significantly as more oxygen (and therefore less nitrogen) was present in the air stream. Flue NO also went up when the equivalence ratio was reduced for the air-fuel case.

The predicted mean equilibrium NO emission was found to be 140 ng/J for the air-fuel case (3.54 MW) computed. With the Zeldovich mechanism, the predicted NO

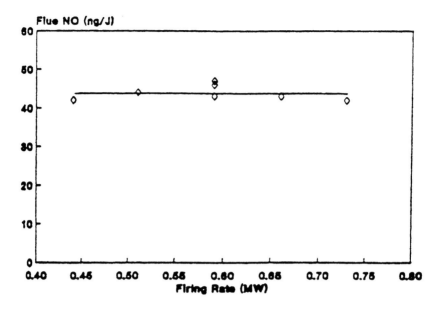

FIGURE 8   Flue NO vs. firing rate for 4425-8A burner at stoichiometric conditions with air.

emission level reduced significantly to a value of 60 ng/J. This compares favorably to the measured values reported earlier. The predicted NO mass fraction contours are shown in Figure 9 in the *y-z* plane of the combustion space for the air-fuel case along *x* = 0.0 m, the plane passing through the fuel inlet port . The contour plot reveals the dependence of NO formation on temperature and residence time.

For the experimental oxygen-fuel firings, the KT-3 burner was tested at firing rates ranging from 0.69 to 1.10 MW. For stoichiometric flames with oxygen as the oxidizer, Figure 10 shows that normalized flue NO (5.2 ng/J) was not substantially affected by the firing rate at the lower end of the burner's designed range. As in the air-fuel cases, the flue NO was found to be strongly affected by the equivalence ratio. Flue NO was found to increase significantly as equivalence ratio decreases.

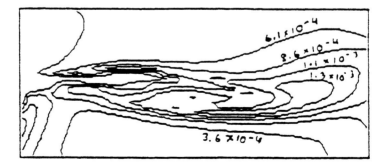

FIGURE 9   Predicted NO mass fraction contours for the air-fuel case along the *y-z* plane at *x* = 0.0 m.

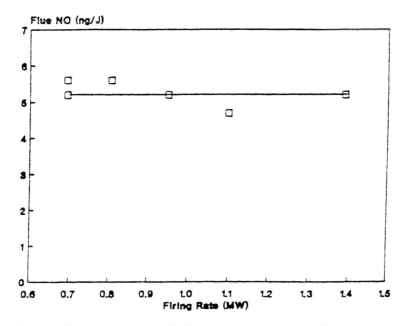

FIGURE 10 Flue NO vs. firing rate for the KT-3 burner at stoichiometric conditions with pure oxygen.

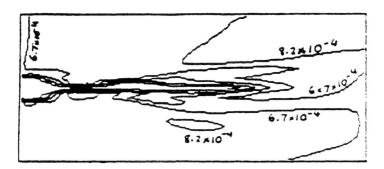

FIGURE 11 Predicted NO mass fraction contours for the oxygen-fuel case along the y-z plane at $x = 0.0$ m.

The predicted mean equilibrium NO emission was found to be 25 ng/J for the oxygen-fuel case (7.1 MW) computed. With the Zeldovich mechanism, the predicted NO emission level again reduced significantly to a value of 10 ng/J. This also compares favorably to the measured values reported earlier. The predicted NO mass fraction contours are shown in Figure 11 in the y-z plane of the combustion space for the oxygen-fuel case along $x = 0.0$ m, the plane passing through the fuel inlet port. The NO contours show steep gradients along the axis of the furnace due to the narrow flame shape observed in the oxygen-fuel case.

CONCLUSIONS

Comparing the results obtained for the two cases (air-fuel and oxygen-fuel firings) studied, the significant effects of withdrawing atmospheric nitrogen from the oxidizer stream can be realized. The withdrawal of atmospheric nitrogen from the oxidizer stream changes the flame temperature, the composition, and volume of the combustion products. The present results also illustrate that pure oxygen combustion characteristics can not be predicted by direct extrapolation of air combustion conditions.

The air-fuel burner tests demonstrated that the normalized flue NO was independent of the firing rate. The comparison of the measured NO emission and the predicted value was good, realizing the inadequacy in the present calculations to resolve the burner near-field. The tests also demonstrated that flue NO was directly proportional to equivalence ratio. For the oxygen-fuel cases, the burner tests again showed that normalized flue NO was not a function of firing rate. The flue NO in this case was caused by the nitrogen in the natural gas. In general, normalized flue NO was much lower for the oxygen-fuel burner than for the air-fuel burner. This suggests that NO emissions can be significantly reduced by replacing air-fuel burners with oxygen-fuel burners in industrial furnaces running under positive pressure.

ACKNOWLEDGEMENTS

The support provided by Air Products and Chemicals Inc. and the CertainTeed Corporation is gratefully acknowledged.

NOMENCLATURE

a	absorption coefficient
c_P	specific heat at constant pressure
C_{g1}, C_{g2}	constants in combustion model
E_b	black body emissive power, $\sigma.T^4$
f	mixture fraction
g	square of the fluctuation of concentration; also acceleration due to gravity
h	mixture enthalpy
h_f	enthalpy of formation
k	kinetic energy of turbulence
m	mass fraction
p	pressure
$P(f)$	the pdf of 'f'
Q_x, Q_y, Q_z	radiative heat flux components
R	universal gas constant
R_x, R_y, R_z	composite radiation fluxes
S_{rad}	radiation source of enthalpy

s	a conserved scalar
T	absolute temperature
U	over all heat transfer coefficient
u_i, u_j	velocity component in the i and j directions
x	coordinate direction
y	coordinate direction
z	coordinate direction

Greek symbols

μ	viscosity
ρ	density
σ_ϕ	Schmidt or Prandtl number for any variable ϕ
σ_B	Stefan-Boltzmann constant
ε	dissipation of turbulent kinetic energy
ε	emissivity of wall

Subscripts

eff	effective (including the effects of turbulence)
fu	fuel
i	i-th species; also used as index for tensor notation
j	index for tensor notation
ox	oxidant
t	turbulent

REFERENCES

Carvalho, M. G. and Lockwood, F. C. (1985) Mathematical Simulation of an End-port Regenerative Glass Furnace. *Proc. Inst. Mech. Engrs.*, **199**, (c2), 113.

Carvalho, M. G. and Coelho, P. J. (1990). Numerical Prediction of an Oil-Fired Water Tube Boiler. *Eng. Computations*, **7**, 227.

Chen, J-Y, and Kollmann, W. (1988) pdf Modeling of Chemical Nonequilibrium Effects in Turbulent Nonpremixed Hydrocarbon Flames". *Twenty second Symposium (International) on Combustion*, 645.

Colson, R. W. (1988) *Thesis, M. S.* Chemical Engineering Department, Brigham Young University.

Correa, S. M., Drake, M. C., Pitz, R. W. and Shyy, W. (1984) Prediction and Measurement of a Non-Equilibrium Turbulent Diffusion Flame. *Twentieth Symposium (International) on Combustion*, 337.

Hamaker, H. C. (1947) Radiation and Heat Conduction in Light-Scattering Material. *Philips Research Report*, **2**, 55.

Hanson, R. K. and Saliaman, S. (1984) Survey of Rate Constants in H/N/O System *Combustion Chemistry* (Ed. Gardiner, W. C.), 361.

Gosman, A. D., Lockwood, F. C., Megahed, I. E. A. and Shah, N. G. (1980) The Prediction of the Flow, Reactions and Heat Transfer in the Combustion Chamber of a Glass Furnace. AIAA *18th Aerospace Sciences Meeting*, January, 14.

Joshi, S. V., Becker, J. S. and Lytle, G. C. (1986) Effects of Oxygen Enrichment of Air-Fuel Burners. *Proceedings of the Industrial Combustion Technologies Symposium, Chicago*, II., 165.

Kent,J. H. and Bilger, R. W. (1977) The Prediction of Turbulent Diffusion Flame Fields and Nitric Oxide Formation. *Sixteenth Symposium (international) on Combustion*, 1643.

Pratt, D. T. and Wormeck, J. J. (1976) CREK: A Computer Program for Calculation of Combustion Reaction Equilibrium and Kinetics in Laminar or Turbulent Flow. *Washington State University* Report WSU-ME-TEL-76-1, Seattle, Washington.

Schuster, A. (1905) Astrophysical Journal, **21**, 1.

Smoot, L. D. and Pratt, D. T. (1979) *Pulverized Coal Combustion and Gasification*, Plenum Press, New York, NY.

Spalding, D. B. (1971) Concentration Fluctuation in a Round Turbulent *Jet. Chemical Engineering Science*, **26**, 95.

Spalding, D. B. (1980) Mathematical Modeling of Fluid-Mechanics, Heat-Transfer and Chemical-Reaction Processes. A lecture course, *CFDU Report*, HTS/80/1.

Zabielski, M. F. and Woody, B. A. (1987) NO_x Chemistry in Natural Gas Flames. *GRI Report* 87-02-00.

Development of Burners and Combustion Facilities in Kakogawa Works

AKIRA YAMAMOTO, HIDEO TATEMICHI, RYUICHI ODAWARA,
and TAKAO MINE   *Kakogawa Works, Kobe Steel, LTD.*
1 Kanazawacho, Kakogawa, Hyogo, Japan

Abstract—In steelmaking industry, it has become necessary to reduce air pollutants caused by combustion, at the same time, to develop the combustion and heating technologies which contribute to energy saving and improvement in product quality such as uniform material quality and improved dimensional accuracy. To achieve these purposes, it is important to develop burners for individual processes suited to heating purposes.

In Kakogawa Works, various developments and improvements of burners have been made at the in-plant Combustion Test Center. Up to the now, we have applied many burners for the commercial production processes suitably and got the good results.

This paper describes examples of techniques to reduce air pollutants (NO_x and CO_2) generated by combustion at the heating equipments using byproduct gas–Coke Oven Gas (COG), Blast Furnace Gas (BFG), LD Converter Gas (LDG) - in the integrated steelmaking plant, together with the viewpoint of burner development and equipment design.

Key Words: Low NO_x burner, energy saving, integrated steel works, ladle, reheating furnace ceramic tube, radiant tube

1. INTRODUCTION

In recent years, from the viewpoint of environmental protection, the reduction of air pollutants has become a critical problem in the world. In the steelmaking industry, the reduction of NO_x, SO_x, and CO_2 generating from combustion has been a priority subject. What is important in implementing these countermeasures is to achieve the reduction of air pollutants while realizing more effective production process, that is, energy saving and improvement of product quality such as higher uniformity of product quality and higher dimensional accuracy.

To achieve these purposes, the heating and combustion techniques must be developed. At Kakogawa Works, large integrated steelmaking works, we have carried out various approaches positively on developing burners as well as on designing heating equipments including the combustion control systems.

Experimental attempts using full-scale testing devices, combined with numerical simulations of temperature and flow, effectively enforced at the stage of designing heating equipments. As a result, various heating equipments and burners which is used with the byproduct gases have been developed and improved, and we have been reducing air pollutants caused by combustion.

We report a technique of CO_2 reduction by energy saving in drying equipment for ladle in steelmaking process and the combustion techniques of low-NO_x burners for plate reheating furnace and for cold-rolled coil batch annealing furnace in rolling process.

2. HIGH-SPEED LADLE DRYING TECHNIQUE IN STEELMAKING PROCESS

This equipment dries and preheats the refractory of ladles, which is containers to transfer molten steel from converter to casting equipment, after periodical repairing. Figure 1 is a schematic drawing of the ladle drying equipment. Table I shows specifications of the heating equipment.

In order to increase the efficiency of the refractory repairing work, we have made attempts to replace the conventional brick lining construction inside the ladle with that using a monolithic lining. To achieve this purpose, for the heating equipment, the performances different from conventional ones must be provided; that is, in the drying process (the initial heating stage), to prevent explosion of the monolithic refractory, which contains 6–7% moisture and provides insufficient strength, first, it is suitable to heated to a specified temperature slowly. And then, in the preheating process (the second heating stage), rapid heating should be realized.

To achieve these requirements, the flexible drying and heating equipment was developed. We provided the function of combustion with high turndown ratio and of strong agitating force as a slow and uniform heating means during drying. And we adopted high-load heating during the preheating period thereafter.

To this equipment, the diffusion burners which divide combustion air to primary and secondary air were adopted. The burner produced stable flames at the burner section using primary air. By the flow of the outer secondary air, combustion gas was agitated in the ladle and effectively added convectional effects. It was intended to achieve low-temperature low-speed uniform heating. The two flow rate measuring systems

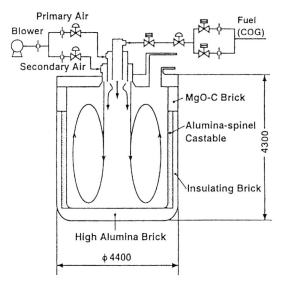

FIGURE 1   Schematic illustration of improved drying and heating equipment.

TABLE I

Specification of Berner

Item	Specification
Capacity	9.4 GJ/hr.
Turn Down Ratio	8% (Min./Max. FLOW)
Fuel	COG (19MJ/Nm^3), 20°C
Air Temperature	300°C

separately provided for fuel line and combustion air line according to the flow rate level have enabled stable combustion control even at high turndown ratio.

Figure 2 shows the air ratio required for uniform heating in correlation with the fuel flow rate. These characteristics are determined by experimental approaches using on-line production facilities and indicate the conditions which can achieve a 50°C difference between the upper surface temperature and the lower surface temperature of the refractory at the reference position.

The application of this combustion equipment and control system has greatly improved the delays of heating at the ladle bottom during drying (at low-combustion load) which has been, in particular, a problem caused by short flame with conventional equipment.

Figures 3, 4 show ladle drying characteristics of the conventional equipment and the developed one respectively. The refractory temperatures are represented the range of

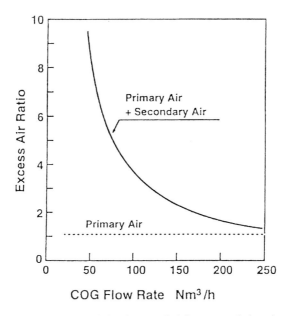

FIGURE 2   Relation between fuel flow rate and air ratio.

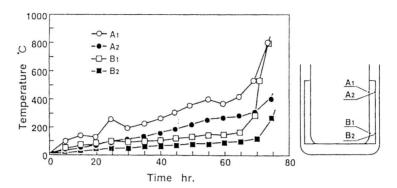

FIGURE 3   Transition of refractory temperature at conventional equipment.

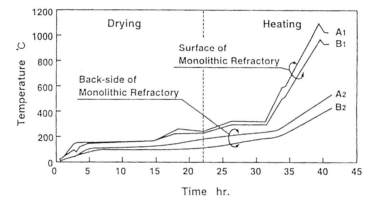

FIGURE 4   Transition of refractory temperature at improved equipment.

temperature measured (a total of 18 points of surface and internal temperatures in the circumferential and vertical directions of the ladle side wall). In the drying process, surface temperature difference is within $30°C$ in both circumferential and vertical directions, which indicates uniformly heating. The time required for the lowest internal temperature to reach at $110°C$ is 22 hours, extremely shorter than that of the conventional equipment (63 hr.).

As a result, the fuel consumption per cycle decrease to 5200 Nm^3 from 13600 Nm^3 at the conventional equipment, achieving a 62% reduction in consumed energy (energy saving). With this improvement, an amount of CO_2 generation by combustion from this process has reduced nearly similar to the reduction ratio of the fuel. Presently, it is assumed that energy saving is the most effective countermeasure to reduce combustion products, including CO_2, and the improvement of heating performance achieved by this equipment is significant.

3. UNIFORM HEATING TECHNIQUE AND LOW NO$_x$ TECHNIQUE IN ROLLING PROCESS

(1) *Reheating Furnace*

In Kakogawa Works, a box-type reheating furnace has recently been adopted to improve heating efficiency, and as a result, side burners and roof burners are installed in place of conventional axial burners. Figure 5 shows schematic drawings of a new plate reheating furnace and the skid configuration.

In this furnace, reduction of temperature unevenness (skid mark) in slab generated by the skids which supports and transports slabs has been achieved by using the local heating equipment positively. It was installed for the low-temperature slab portion in the soaking zone, together with optimizing the skid button height and performing skid shift.

Figure 6 shows the local heating equipment using the ceramic tube, and Figure 7 shows the burner. This is a diffusion type gas burner utilizing the side burner and is used with MIX gas (COG, BFG, LDG). It is operated under the preheating air temperature of 600 °C, and the maximum combustion capacity of the burner is 2.5 GJ/hr.

Figure 8 shows the temperature distribution of the slab lower surface at the lateral direction of the conventional furnace. The slab temperature in the figure indicates the numerical simulation, which is obtained by taking the view factor into account, and measurement at the on-line production facilities. In the plate reheating furnace, the lowest temperature arises at nearly middle between the walking beam and the stationary beam in the slab longitudinal direction because of shadow effect caused by

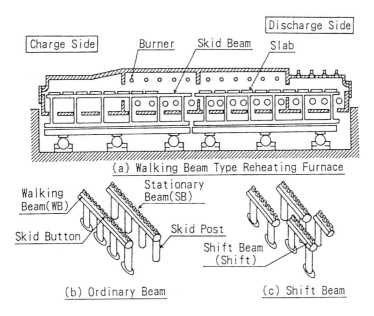

FIGURE 5   Outline of walking beam type reheating furnace and the arrangement of skid beams.

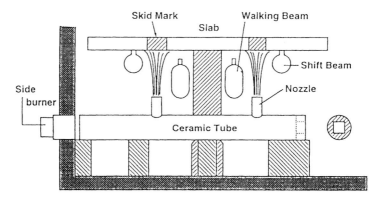

FIGURE 6   Local heating equipment of soaking zone.

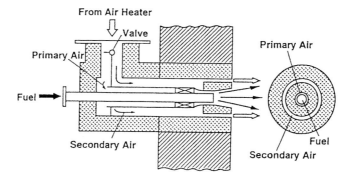

FIGURE 7   Side burner applied to ceramic tube.

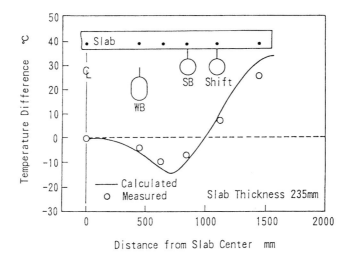

FIGURE 8   Temperature distribution of slab lower surface.

the skid placed under the slab. Figure 9 shows the measurement results of the slab temperature in the case of the local heating equipment in the furnace. Heating the slab for 20 minutes promotes the rise in temperature of the slab lower surface about 30 °C as compared without local heating.

We have carried out the combustion test of the ceramic tube local heating equipment, before they were installed on actual furnace, and investigated the characteristics of the combustion and the heating. Generally, in combustion characteristics at narrow space in tubes, flame temperature rises higher than that of open diffused flames such as side burners and roof burners under similar combustion conditions (air temperature, combustion load, furnace temperature, etc.). And finally NO_x generation is promoted. In Kakogawa Works, various experimental studies for the secondary air ratio and tube arrangement varied were carried out. As a result, NO_x generation was able to be controlled to as low as 105 ppm under the maximum combustion load even in the case of using with the ceramic tube. In the overall exhaust gas from the reheating furnace containing the exhaust gas of side burners and roof burners, the NO_x concentration in exhaust gas is kept below 80 ppm, and is suitable to the Japanese environmental standard value of 90 ppm for this reheating furnace.

Figure 10 shows the relationship between NO_x emission, tube arrangement, and the rate of primary air flow. As the clearance between the burner and the ceramic tube is wider, more the circulation rate of exhaust gas which flows into the tube increase, and flame temperature decreases. And as the secondary air flow rate increases, to become slow combustion, the area rate of high temperature decrease and NO_x generation is controlled. However, in the former event, the jet temperature at the ceramic tube nozzle decreases, it provides a disadvantage which is low heat transfer rate to the slab. Finally, in order to achieve both low NO_x concentration and efficient heating of the slab in a well-balanced means, we adopted the primary air ratio 10% with optimized clearance between the burner and the ceramic tube, and applied to the actual furnace.

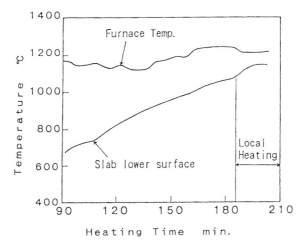

FIGURE 9   Promoted heating for skid mark by local heating.

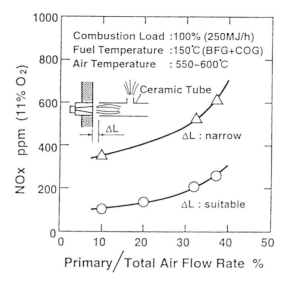

FIGURE 10   NO$_x$ emissions from local heating equipment.

(2)   *Heat Treatment Furnace*

In Kakogawa Works, there are more than 2200 pieces of radiant tubes, which are used in continuous sheet annealing and batch type heat treatment processes including continuous galvanizing facilities. For these heating equipments, the countermeasure of effective energy savings is preheating of combustion air by recovering the exhaust heat from radiant tube, but a high-concentration of NO$_x$ is likely to occur because of high-load high-temperature heating at a small and narrow space in the tube. As a result, combustion air temperature is restricted from the viewpoint of the environmental standard, and even if various combustion methods were adopted, practically it is almost operated in 400–450°C as the highest limit in Japan.

In the development of radiant tube burner, it is important to control the NO$_x$ generation in high combustion load and to decrease the unburned fuel (fuel gas, soot, cracking carbon) in low combustion load together with uniform heating for longitudinal direction in the radiant tube.

Figure 11 shows the developed burner (**KOBELCO RT BURNER**) which is based on an original combustion idea. Table II shows the burner specification. The burner is basically a partial premixing type, and the remarkable characteristic is that the axises of fuel and primary air nozzle including the pilot burner are offset to be slightly lower than that of the radiant tube.

Figure 12 show a conceptual drawing which is compared the combustion process of the new burner with that of the typical concentric burner.

In the case of high-load combustion, in the conventional burners, rapid mixing of the fuel and the air takes place in the first straight tube because of large momentum in the flow. Therefore, high-temperature area widely generates in the vertical and longitudinal directions. As a result, the concentration of NO$_x$ extremely increases because of thermal NO$_x$ generated at the high-temperature area in the tube.

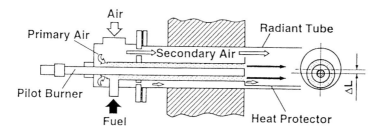

FIGURE 11   Eccentric axial nozzle burner.

TABLE II

Specification of Radiant Tube and Burner

Item	Specification
Radiant Tube	
Size	Diameter: 6 inches
	Length: 2500 mm (straight tube)
Shape	U-Type
Burner	
Capacity	500 MJ/hr.
Fuel	COG (19 MJ/Nm^3), 20°C
Air Temperature	550°C (max)

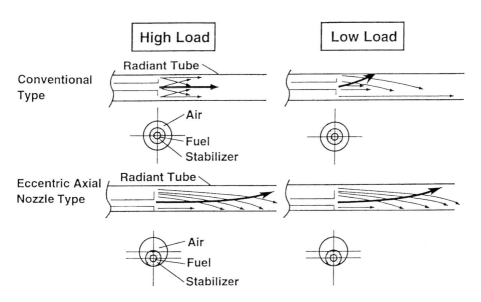

FIGURE 12   Mixing process between air and fuel.

As against this, in the developed burner, by designing the fuel and secondary air nozzle axis to be not concentric and by feeding the combustion air to the upper layer in the tube more plenty than that of lower layer forcibly, the mixing rate of fuel and combustion air in the tube is controlled to enable slow combustion to take place. At the initial combustion stage, the air flow rate which contributes to combustion reaction is restricted and the generation of the high-temperature area nearby the fuel nozzle is controlled. The fuel which does not burn at this stage slowly rises by its buoyancy and diffusion as it goes downstream, mixes with the air, and completes the combustion. That is, because of the low flame temperature caused by decreasing of the partial pressure in oxygen at the flame front, the NO_x generation in the radiant tube is reduced in compared with former's.

In the case of low-load combustion, in the conventional burner, as the velocities at the point of the nozzles decrease, the momentums become smaller and the mixing performance lowers. In addition, on the downstream side, the flame containing unburned fuel floats up at the position comparatively near the fuel nozzle. As a result, a two-phase flow composing of flame (fuel) and combustion air occurs, and it makes partly the fuel change into solid free-carbon, which is discharged from the radiant tube with keeping incomplete combustion. In this condition the NO_x concentration becomes lower, but the unburned fuel makes troubles and damages on the equipments installed at the downstream side of the radiant tube. As against this, in the new burner, a relatively large volume of air existing in the upper layer at the point of the nozzle tip end and fuel in the lower layer crosses and mixes adequately to complete the combustion. As a result, unburned fuel concentration containing soot becomes lower, and it achieves more clean and stabilized combustion. Consequently, even in the case where high turndown ratio is required, it is possible to apply this burner.

Figure 13 shows the distribution of O_2 concentration and temperature in the first straight tube at the off-line test equipment. In the case of eccentric axial burner, the flame temperature in the initial combustion stage is lower than that of the conventional burner, and the area of the highest flame temperature shifts downstream. In other words, slow combustion occurs, and this means that it would be possible to control the NO_x generation. In the case of low combustion load at the new burner, the area existing with more than 5% O_2 is narrower than that of the conventional type and the combustion reaction is promoted. This means that it is caused by rapid mixing of fuel and air.

Figure 14 shows the effect of combustion load on NO_x and soot emission. At the highest load, even when 550 °C of combustion air is used, the NO_x concentration is controlled to 145 ppm. And it is indicated that soot concentration at the turndown ratio 1/5 is low. It is possible to achieve further cleaner and stabler application, if the combustion control system, by properly increasing the air ratio at the low combustion load, is adopted.

Finally, this new type of burner was applied to six units (120 sets) of batch type cold-rolled coil annealing equipment. Furthermore, in conventional furnaces, this combustion idea has been applied to conventional partial pre-mixing type burners (1372 sets) when the equipment was improved to use high-temperature combustion air for energy saving. And burners were improved to those that can maintain low-NO_x

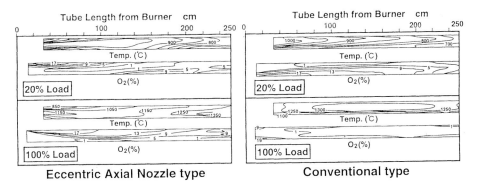

Eccentric Axial Nozzle type   Conventional type

FIGURE 13   Temperature and O_2 distribution in radiant tube.

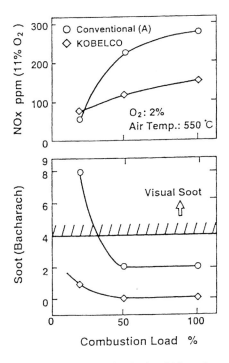

FIGURE 14   Effect of combustion load on NO_x and soot emission.

stable combustion performance even when high-temperature preheated combustion air is used at 500–550 °C.

Figure 15 shows the improved burner and Figure 16 shows the test results at the off-line combustion test equipment. In improving the burner, we optimized the secondary air nozzle design by adopting the technique of the eccentric combustion, where by the secondary air flows at the upper layer in the tube are more rich than these at the lower layer, together with further improving the distribution rate of combustion air (primary vs. secondary). That is to say, we optimized the flow condition of the fuel

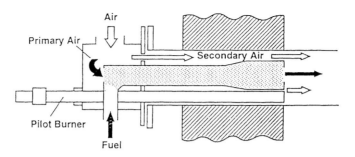

FIGURE 15   Improved burner (partial premixing type).

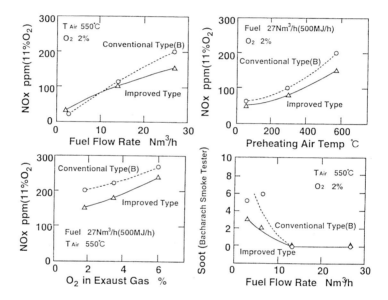

FIGURE 16   NO$_x$ and soot emission from improved burner.

and the secondary air by the off-line combustion test. Consequently, we have achieved the improvement of the fuel/air concentric flow burners, and achieved 10–20% reduction in NO$_x$ generation at high air temperature on high combustion load.

As described above, by adopting this burner or the basic combustion idea in this burner, it is able to control the NO$_x$ generation lower. And then, it is obtained the more large energy saving effects in the production facilities achieving the application of higher combustion air temperature. Finally, this will contribute to reduction of air pollutants caused by combustion.

CONCLUSION

In the steelmaking industry, it becomes more important to carry out well-balanced equipment designs and developments of the burner with environmental protection taken into account, in addition to production activities such as improvement in quality and productivity.

Presently, to reduce pollutants resulting from combustion in the steelmaking process, we believe that the promotion of energy saving, which contributes to the reduction of the fuel consumption and the NO_x, CO_2, and the application of the low NO_x burner for heating equipments are the leading countermeasures. The developments and the improvements of these technologies have been positively carried out.

In the future, combining the optimization of design achieved by progress and improvement of accuracy in theoretical approaches, such as numerical fluid dynamics and simulations of combustion reaction, with off-line and on-line experimental approaches, the present production processes as described above will be further improved. And at the same time, these progressive combustion and heating technologies will be applied to designing new heating equipments and will bring about good results.

We are positively going to carry out the energy saving, the developments of burner and heating equipment, continuously, with the largest concerns about the harmonization between industry and environment.

REFERENCE

Odawara, R. *et al.* (1993) The Drying and Heating Equipment of Monolithic Ladle Refractory, ISIJ, Japan, **79-TI3**.
Tatemichi, H., Mine, T. and Suzuki, T. Heat Transfer Technology in Reheating Furnace, *29th National Heat Transfer Symposium of Japan*, (1992) D136 (P185, 186).
Hasegawa, T. and Hoshino, T. (1992) Interpretation on Fuel Staged Combustion Mechanism for High Performance Burner Design, JFRC, Japan.
Odawara, R. *et al.* (1985) Development of Radiant Tube Burner for High Air Temperature, JFRC, Japan.

Ultra Low NO$_x$ Burners for Industrial Process Heaters

R. T. WAIBEL *John Zink Company Tulsa, Oklahoma, USA*

Abstract— This paper describes the factors that affect NO$_x$ emissions from industrial burners and the methods employed to control NO$_x$ emissions. The basics of NO$_x$ emissions are briefly discussed as well as the furnace operating parameters that influence NO$_x$ emissions. Emissions data for burners employing NO$_x$ control technologies such as combustion staging and flue-gas recirculation are presented. Finally data are presented for a new burner design employing a unique combination of staged fuel combustion, lean premix combustion, lean-rich-lean reburn technology and self-induced flue-gas recirculation. Data showing emissions of less than 40 mg/NM^3 (20 PPM) are presented from full-scale burners in a test furnace as well as field data from an industrial process heater in a refinery.

NO$_x$ EMISSIONS

Emissions of oxides of nitrogen, commonly referred to collectively as NO$_x$, are regulated because of their adverse effects on health and the environment. They play an important role in acid rain, the formation of harmful ozone and photochemical smog in the lower atmosphere and the depletion of the beneficial ozone in the upper atmosphere. Figure 1, derived from Dignon and Hameed (1989), shows data for the historical growth in NO$_x$ emissions.

The most environmentally important oxides of nitrogen are:

$$NO,$$

$$NO_2,$$

and, more recently,

$$N_2O.$$

N$_2$O is a recent concern because it is a "greenhouse" gas which contributes to global warming and because it can aid in destroying the upper atmosphere ozone layer which protects us from ultraviolet radiation. Fortunately, very little N$_2$O is emitted from the flame of a typical burner.

Over 90% of the NO$_x$ from a typical flame is in the form of NO and the remainder is NO$_2$. However, since NO is eventually converted to NO$_2$ in the atmosphere, most regulations treat all of the NO$_x$ as NO$_2$.

NO$_x$ emissions from combustion sources are due to the oxidation of atmospheric N$_2$ and the oxidation of nitrogen chemically bound in fuel molecules. Mechanisms for forming NO include the fuel NO$_x$, prompt NO$_x$ and thermal NO$_x$ mechanisms. Some of these mechanisms are well understood, while others are still under investigation. The following discussion of NO$_x$ formation mechanisms is derived from Miller and Bowman (1989).

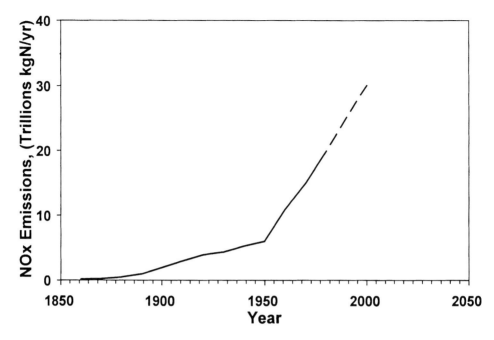

FIGURE 1   Global NO_x emissions.

Fuel NO_x and Prompt NO_x

Fuel NO_x is produced if nitrogen is chemically bound in the fuel molecule and is primarily a concern with heavy oils and solid fuels. Some gaseous fuels, however, can contain NH_3, HCN or amine carry-over as potential sources of fuel bound nitrogen. For fuels with organically-bound nitrogen, the fuel NO_x mechanism begins with the decomposition of the organic molecule in the flame zone:

$$C_xH_yN \rightarrow C_{x-1}H_{y-1} + HCN$$

or

$$C_xH_yN \rightarrow C_xH_{y-1} + NH,$$

depending on the nature of the carbon/hydrogen/nitrogen bonds. The HCN or NH reacts further and may be oxidized to NO.

Prompt NO_x is the NO_x formed from N_2 in the very early portion of the flame zone where the fuel and air are first reacting. It is formed in a part of the flame where little, if any thermal NO_x should be formed and, by definition, is that NO_x formed from molecular nitrogen which is in excess of the NO_x predicted by the thermal NO_x mechanism. There are several reaction paths postulated for forming prompt NO_x. One of the more important involves the reaction of molecular nitrogen with hydrocarbon radicals formed during the decomposition of the fuel in the initial reaction zone. The major reactions are:

$$CH + N_2 \leftrightarrow HCN + N,$$

and

$$C + N_2 \leftrightarrow CN + N.$$

The fuel NO$_x$ and prompt NO$_x$ mechanisms proceed identically after these initiation reactions. The major steps in the mechanisms are shown in Figure 2.

The fuel NO$_x$ contribution to the total NO$_x$ emissions depends on the concentration of nitrogen bound in the fuel. This is shown in Figure 3. It can range from a few PPM to several hundred PPM of NO$_x$.

The fuel NO$_x$ emissions do not increase linearly with the fuel nitrogen content of the fuel because the fraction of bound nitrogen converted to NO$_x$ is a function of the concentration of nitrogen in the fuel. It can range from nearly 100% at trace concentrations to only a few percent converted at very high concentrations. The exact emission values for a specific level of bound nitrogen are affected by other parameters such as burner design and excess oxygen.

Prompt NO$_x$ typically accounts for only a few PPM of NO$_x$. However, as NO$_x$ control techniques for thermal NO$_x$ improve, prompt NO$_x$ is becoming a large fraction of the remaining NO$_x$ and may prove to be a limiting factor.

Thermal NO$_x$

For most gaseous fuels the major concern is thermal NO$_x$. Thermal NO$_x$ forms via the Extended Zeldovich Mechanism:

$$O + N_2 \leftrightarrow NO + N,$$

$$N + O_2 \leftrightarrow NO + O$$

and

$$N + OH \leftrightarrow NO + H,$$

and is the primary source of uncontrolled NO$_x$ emissions from fuels that do not contain organically bound nitrogen.

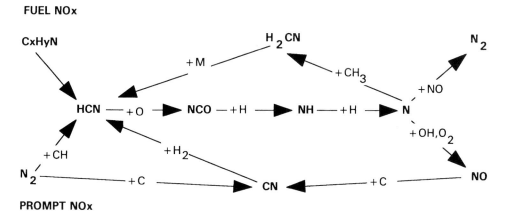

FIGURE 2   Major steps in the fuel and prompt NO$_x$ mechanism.

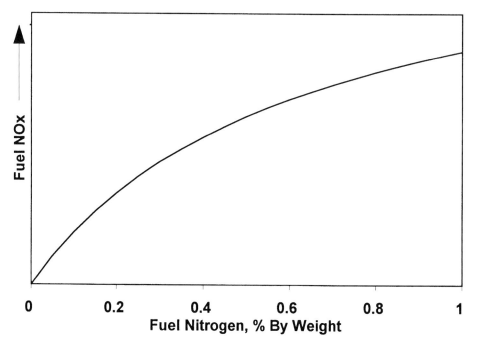

FIGURE 3   Fuel NO$_x$ emissions vs fuel nitrogen concentration.

Fortunately the NO$_x$ formation reactions have high activation energies and, therefore, the formation rate is kinetically controlled. This means that NO$_x$ levels produced in flames are far below the levels predicted by equilibrium calculations. Formation rates, however, are sensitive to temperature and increase exponentially with the reaction temperature.

The thermal NO$_x$ formation rate can be expressed in the form:

$$d[NO]/dt = A \cdot [N_2] \cdot [O_2]^{1/2} \cdot \exp(- E_A/RT).$$

Figure 4 is a plot of this equation showing the exponential dependence of the NO formation rate on temperature and illustrating the importance of reaction temperature on thermal NO$_x$ emissions.

NO$_2$ Formation

NO$_2$ is primarily formed in the lower temperature regions of the flame from the NO produced in the high temperature regions of the flame. The major formation reaction is:

$$NO + HO_2 \leftrightarrow NO_2 + OH.$$

The major reaction removing NO$_2$ from the flame zone is:

$$NO_2 + H \leftrightarrow NO + OH.$$

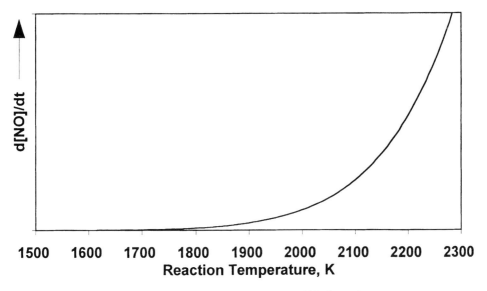

FIGURE 4   Effect of reaction temperature on NO$_x$ formation rate.

The removal step is favored in the flame zone and only a small portion of the NO$_x$ exits the flame zone in the form of NO$_2$.

N$_2$O Formation

Another oxide of nitrogen of concern is N$_2$O. It is formed by the following mechanism:

$$O + N_2 + M \leftrightarrow N_2O + M,$$

$$NCO + NO \leftrightarrow N_2O + CO,$$

and

$$NH + NO \leftrightarrow N_2O + H.$$

The N$_2$O removal mechanism includes:

$$H + N_2O \leftrightarrow N_2 + OH,$$

$$O + N_2O \leftrightarrow N_2 + O_2,$$

and

$$O + N_2O \leftrightarrow NO + NO.$$

Although NO and NO$_2$ molecules last only a matter of days in the atmosphere, N$_2$O is a very stable species that can last 100 to 200 years in the lower atmosphere. Because of its long life span some N$_2$O eventually reaches the upper atmosphere where it decomposes under ultra violet light and its reaction products efficiently remove ozone from the upper atmosphere.

Fortunately very little, if any, N_2O is produced in the flames of process heater burners because the removal reactions are very efficient at high temperatures. The major combustion source appears to be fluidized bed furnaces. Another source, paradoxically, is selective catalytic NO_x reduction units with noble metal catalysts. It has been shown that at least some of the NO_x thought to be destroyed by these units is actually converted to N_2O (de Soete, 1990).

OPERATING PARAMETERS AFFECTING NO_x

Thermal NO_x is the major source of NO_x from the combustion of gaseous fuels. The parameters that influence the oxygen concentration in the flame zone or the temperatures achieved in the flame zone will affect thermal NO_x emissions. The most important parameters are:

- Excess Air
- Fuel Composition
- Air Preheat Temperature
- Furnace Temperature

Excess Air

Excess air provides for additional oxygen beyond the stoichiometric air requirement and is generally required to minimize the emissions of CO and unburned hydrocarbons. It accomplishes this, however, by increasing the concentration of oxygen in the flame zone, which tends to increase NO_x. Excess air also decreases the overall flame temperature and contributes to a loss in thermal efficiency. Figure 5 shows the effect of excess air, expressed as percent excess oxygen, on NO_x emissions.

As the excess air is steadily increased the reduction in NO_x due to the reduction in flame temperature finally overcomes the increase in NO_x due to oxygen concentration and the NO_x emissions peak. Further increases in excess air then reduce NO_x emissions.

Fuel Composition

Fuel composition influences thermal NO_x because of its direct effect on flame temperature. Different fuels are capable of achieving different flame temperatures and the maximum potential flame temperature for a fuel is best defined by the adiabatic flame temperature. The adiabatic flame temperature is the theoretical temperature attained when a fuel/air mixture is burned to completion and all of the sensible and chemical energy of the reactants is transferred to the products of combustion.

Table I provides a list of the calculated adiabatic flame temperatures for a group of fuel gases. The variation in flame temperature with composition is apparent from the table, ranging from 2107 K (3334°F) for methane to 2284 K (3652°F) for hydrogen. Although practical flames transfer heat away from the flame zone, the adiabatic flame temperature provides a good method for evaluating the potential effect of fuel gas

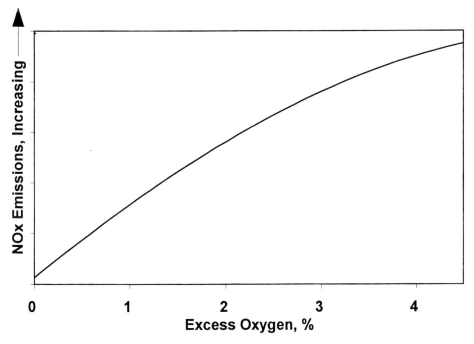

FIGURE 5   NO$_x$ vs excess O$_2$ for a typical diffusion flame burner.

TABLE I

Adiabatic flame temperature for selected fuel components

Component	Adiabatic Flame Temperature	
	K	°F
CH$_4$	2107	3334
C$_2$H$_6$	2128	3371
C$_3$H$_8$	2131	3377
C$_4$H$_{10}$	2131	3377
C$_4$H$_8$	2176	3458
C$_3$H$_6$	2189	3481
C$_2$H$_4$	2225	3546
H$_2$	2284	3652

composition on flame temperatures and, therefore, the potential effect on thermal NO$_x$ emissions.

Air Preheat Temperature

Air preheat affects thermal NO$_x$ by its direct effect on flame temperature. Preheating the combustion air adds sensible heat to the flame reactants which increases the heat in the products of combustion and, thus, increases the flame temperature. Figure 6 shows the effect of air preheat temperature on NO$_x$. Note that the NO$_x$ essentially follows an exponential increase with increasing air preheat temperature. A reasonably good rule

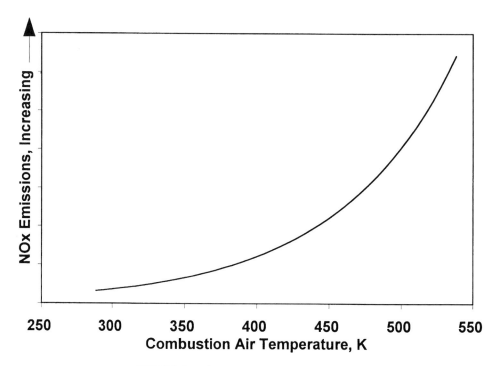

FIGURE 6   NO$_x$ vs combustion air temperature.

of thumb from industrial experience is that the thermal NO$_x$ emissions will double as the combustion air temperature is increased from ambient to about 530 to 590 K (500 to 600°F).

Furnace Temperature

The effect of furnace temperature on NO$_x$ is shown in Figure 7. Furnace temperature affects thermal NO$_x$ emissions by its effect on the rate of heat transfer from the flame and, as a result, it influences the actual temperatures attained within the flame zone. The lower the furnace temperature, the higher the heat transfer rate from the flame and the lower the actual peak flame temperatures within the flame zone. Lower peak flame temperatures mean lower thermal NO$_x$ emissions.

NO$_x$ CONTROL

The major contributors to NO$_x$ emissions are thermal NO$_x$ and, if fuel-bound nitrogen is present, fuel NO$_x$. Most refinery process heaters in the US are fueled by refinery fuel gas and, thus, thermal NO$_x$ is the primary concern. As noted previously, thermal NO$_x$ is strongly influenced by peak flame temperatures and the key to controlling thermal NO$_x$ is to moderate peak flame temperatures.

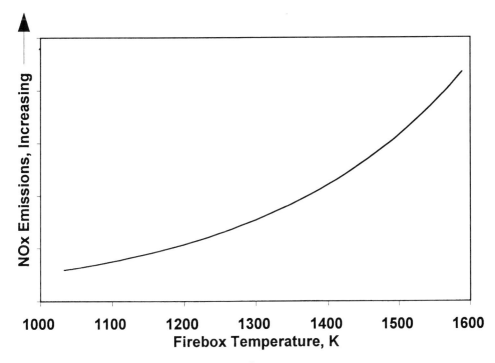

FIGURE 7   NO$_x$ vs furnace temperature.

Historically, thermal NO$_x$ control techniques have included excess air control, air or fuel combustion staging and flue-gas recirculation. Low excess air operation provides only limited benefit. However, it is compatible with, and can be used together with, most of the other NO$_x$ control techniques. Combustion staging and flue-gas recirculation have proven to be more beneficial. The combination of fuel staging and flue-gas recirculation has proven to be the most beneficial combination until recent developments. Low NO$_x$ industrial burners have been developed utilizing these NO$_x$ control techniques and their performance results have been reported (Waibel *et al.*, 1986, 1990 & 1991; Napier *et al.*, 1989; Waibel and Civardi, 1990).

Staged Combustion

NO$_x$ emissions can be reduced by introducing either the combustion air or the fuel into the flame in stages. With air staging a portion of the combustion air, typically about 50% to 75%, is supplied to a primary combustion zone with all of the fuel. This produces a fuel-rich flame zone. NO$_x$ emissions in this flame zone are reduced due to the substoichiometric combustion conditions. The remainder of the air is injected downstream, forming a secondary flame zone where combustion is completed. NO$_x$ formation in this secondary flame zone is reduced because the inerts from the primary flame zone reduce flame temperatures.

Fuel staging is the reverse of air-staging. Generally 30 to 50% of the fuel is injected into the combustion air to form a lean primary flame zone. Although excess oxygen is available, NO_x is minimized by the low flame temperatures that are generated due to the lean combustion conditions. The remainder of the fuel is then injected downstream forming a secondary flame zone where combustion is completed. NO_x formation rates in this zone are low because the inerts from the primary flame zone lower the flame temperatures and local oxygen concentrations.

Figure 8 shows a comparison of NO_x emissions from a conventional, non-low NO_x burner, a staged air burner and a staged fuel burner under similar operating conditions. This data is for burners firing natural-gas with 15% excess air in a 1144 K (1600° F) firebox.

Flue-Gas Recirculation

Flue-gas recirculation (FGR) can be utilized alone or in combination with staged combustion to reduce thermal NO_x emissions. Flue-gas recirculation involves the addition of inert products of combustion into the air or fuel prior to combustion. These products of combustion can be obtained from the furnace exhaust via fans and duct work or can be induced directly into the burner from the furnace chamber using the fuel or air momentum. FGR reduces NO_x by reducing flame temperatures and local oxygen concentrations in the flame zone. FGR combined with fuel staging produces very low NO_x emissions.

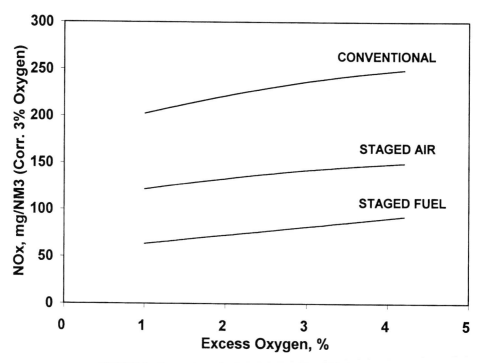

FIGURE 8   Comparison of conventional and low NO_x burners.

Figure 9 shows a plot of NO$_x$ for different levels of flue-gas recirculation with a staged fuel burner. The FGR rate is given as a function of the normal flue-gas flow rate. An FGR rate of 10% is a quantity of flue-gas equal to 10% of the normal products of combustion.

The effect of FGR on NO$_x$ is the mirror image of the effect of air preheat temperature on NO$_x$. As FGR is added to the flame, it reduces the flame temperature and NO$_x$ initially drops rapidly. As the FGR rate is increased, the effectiveness of the additional FGR decreases. Eventually, above 25% to 30% FGR, flame stability diminishes and further addition of FGR is not effective.

INFURNO$_x^{TM}$ Technology

The bold lines in Figure 10 show how NO$_x$ emissions vary with stoichiometry for a diffusion flame compared with a premix flame. Under substoichiometric or fuel-rich conditions, both types of flames show a rapid drop in NO$_x$ as the flames burn under richer and richer conditions. On the other hand, the premix flame shows a peak in NO$_x$ emissions very near to stoichiometric conditions and shows a reduction in NO$_x$ as the flame is operated under increasingly lean conditions. For a diffusion flame the peak NO$_x$ level is greater than that for the premix flame and the peak occurs well into the lean operating regime.

Introducing flue-gas recirculation (FGR) reduces NO$_x$ emissions for both types of flames. This is shown by the narrow lines in Figure 10. The FGR dilutes local oxygen

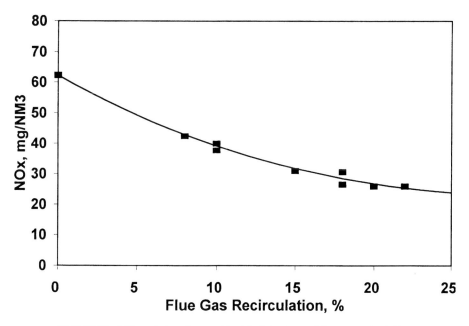

FIGURE 9   NO$_x$ emissions for combined fuel staging and flue gas recirculation.

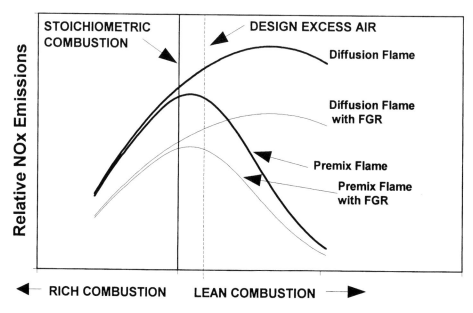

FIGURE 10   Relative NO$_x$ emissions for premix and diffusion flames.

concentrations and, more importantly, it reduces flame temperatures. These two factors combine to reduce NO$_x$ emissions.

On this basis, if one were only interested in limiting NO$_x$, it would be advisable to operate either very lean with a premix flame or very rich with either type of flame and to include FGR.

Unfortunately, operating very lean or very rich would be very inefficient and would lead to increases in other pollutants. However, if we burn part of the fuel in a lean premix flame and the remainder in a rich flame, then combine the products of combustion from these two flames in a final burnout zone, we can achieve very low emissions and an acceptable overall excess air level. This combination is depicted schematically in Figure 11.

If the lean and rich combustion zones are organized properly the burner can employ an additional NO$_x$ control technique which is often referred to as reburning. It is possible to arrange for the products of combustion from the lean primary flame to pass through the rich secondary flame prior to the final burnout zone. This "lean-rich-lean" combustion scheme allows for the NO$_x$ in the primary zone combustion products to either react with the N atoms in the rich secondary flame to form N$_2$ or to react with the hydrocarbon radicals and thus to recycle back to HCN as shown in the Prompt/Fuel NO$_x$ Mechanism. This permits the N molecules that were once in the form of NO to have another chance to end up as N$_2$ rather than remain as NO. At the same time they are providing a source of NO molecules to help the N molecules in the fuel rich region to form N$_2$. John Zink Company has termed this unique combination of NO$_x$ control techniques "INFURNO$_x^{TM}$" technology.

The development work was done utilizing a full-scale burner with a nominal capacity of 2.5 to 2.9 MWt (8 to 10 MMBtu/hr). Figure 12 shows test furnace data for the burner

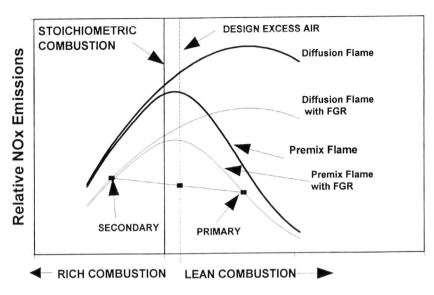

FIGURE 11   Comparison of INFURNO$_x^{TM}$ burner with other burner types.

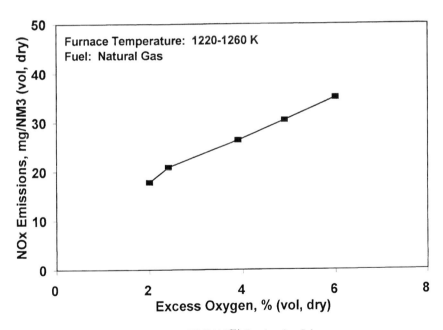

FIGURE 12   INFURNO$_x^{TM}$ Combustion Scheme.

firing natural-gas over a wide operating range under simulated industrial conditions. The data show NO$_x$ emissions well below 40 mg/NM^3 (20 PPM) under these typical industrial furnace conditions. Carbon monoxide emissions were less than 4 mg/NM^3 (5 PPM) for all test points.

Performance data for six INFURNO$_x^{TM}$ burners installed in a process heater firing refinery fuel gas is shown in Figure 13. This fuel gas typically produces higher NO$_x$ emissions than natural gas. In this case the NO$_x$ emissions are higher than seen for

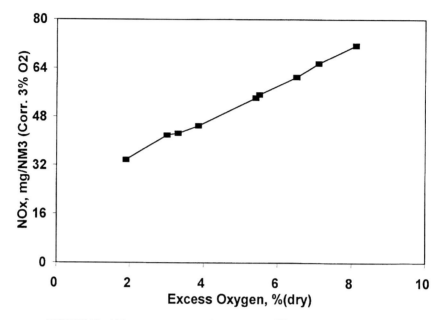

FIGURE 13   NO$_x$ vs excess oxygen for INFURNO$_x^{TM}$ burner firing natural gas.

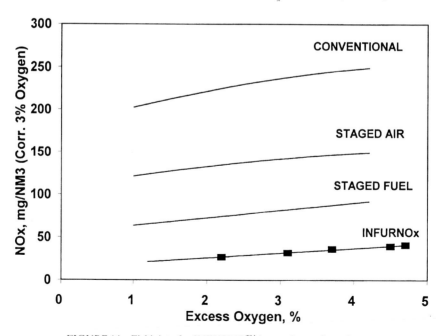

FIGURE 14   Field data for INFURNO$_x^{TM}$ burner in a refinery heater.

natural gas. However they are well below the 60 mg/NM^3 (30 PPM) emission limit for this furnace with up to 6% excess oxygen. At 2% excess oxygen (10% excess air), NO$_x$ emissions are below 40 mg/NM^3 (20PM).

Figure 14 shows a comparison of NO$_x$ emissions among conventional, staged air, staged fuel and INFURNO$_x^{TM}$ burners firing natural gas with 15% excess air in a 1144 K (1600° F) furnace.

As NO$_x$ regulations have tightened it has become more and more difficult to design burners to meet the more stringent requirements. The INFURNO$_x^{TM}$ burner design has achieved significant reductions over previous burner designs by providing a unique combination of NO$_x$ control technologies.

REFERENCES

1. Dignon, J. and Hameed, S. (1989). *J. Air Pollution Control Assoc.*, No. 39, pg. 180.
2. Miller, J. A. and Bowman, C. T. (1989). *Prog. Energy Combustion Science*, **15**, 287–338.
3. Napier, S., Claxton, M. and Waibel, R. T. *Low NO$_x$ Burners for Stringent Emission Regulations, American Flame Research Committee International Symposium on Combustion in Industrial Furnaces and Boilers*, (1989). Short Hills, NJ.
4. de Soete, G. G. (1990). Nitrous Oxide Formation and Destruction by Industrial NO Abatement Techniques, Including SCR, *American Flame Research Committee Spring Meeting*, Tucson.
5. Waibel, R. T. and Civardi, D. (1990). *Bruciatori A Bassa Emissione Di NO$_x$ Per Applicazioni Industriali*, Energia Ed Ambiente, Prospettive Per Gli Anni 90, Capri.
6. Waibel, R. T., Napier, S., Arledge, W. and Method, W. *Meeting SCAQMD Rules 1109 and 1146 with Low NO$_x$ Burners, American Flame Research Committee International Symposium*, (1990). San Francisco.
7. Waibel, R. T., Nickeson, D., Radak, L. and Boyd, W. Fuel Staging Burners for NO$_x$ Control, *ASM Symposium on Industrial Combustion Technologies*, (1986). Chicago.
8. Waibel, R. T., Price, D. N., Tish, P. S. and Halprin, M. L. (1990). Advanced Burner Technology for Stringent NO$_x$ Regulations, *American Petroleum Institute Mid Year Refining Meeting*, Orlando.

The Prediction of the Performance Effects of Powdered Coal Injection into the Tuyere of a Blast Furnace

F. C. LOCKWOOD[a] and K. TAKEDA[b]   [a]*Imperial College of Science Technology and Medicine London, United Kingdom,* [b]*Kawasaki Steel Corporation, Chiba, Japan*

Abstract—This paper concerns the mathematical modelling of the consequences of pulverised coal injection in the blow pipe of a blast furnace. The process reduces the coke requirements and increases the overall efficiency of the furnace so reducing the carbon emissions to the atmosphere. The mathematical model which has been constructed encompasses the flows in the blow pipe, tuyere, raceway and in the surrounding packed bed.

Sensitivities to operational parameters which are not known with precision are explored. Some results of computer runs to find optimised design conditions are reported.

INTRODUCTION

Background

This paper concerns the prediction of the effect of powdered coal (PF) injection in the blow pipe of a blast furnace on the performance of the furnace. Coal injection is done primarily in order to reduce the cost of operation since the requirement for the more expensive coke fuel is correspondingly reduced. The interest to this conference of PF injection in blast furnaces is that the total energy demands for pig iron production, and so the CO_2 burden, are reduced. Moreover, the energy requirements of the coke ovens and power plant are correspondingly reduced. Coke consumption of 500 kg per ton of hot metal (tHM) can be reduced to 320 kg/tHM with a PF injection rate of 200 kg/tHM. Additionally, a lowered demand on coke production allows greater flexibility of coke oven operation and the opportunity to further reduce emissions.

A brief description of the blast furnace and its operation is merited, refer to Figure 1. It is essentially a high temperature moving-bed chemical reactor for the production of molten iron. The charge, iron oxides, coke and flux are introduced at the top of the furnace which is typically 25m high and 10m in diameter. The coke supplies the heat for the process, serves as a reducing agent, provides mechanical support for the descending material and ensures permeability for the ascending gases. Preheated air for reaction, and sometimes auxiliary fuels, are introduced into the furnace through 'tuyeres' via 'blowpipes'. The velocities of these streams, $150-250$ ms^{-1}, form voids called 'raceways' which extend to the order of 1m. Reactions between the air and coke particles entrained in the raceway flow produce a hot reducing gas composed of carbon monoxide, hydrogen and nitrogen which permeates the surrounding charge, termed 'burden material', to which it transfers heat as it ascends through the furnace.

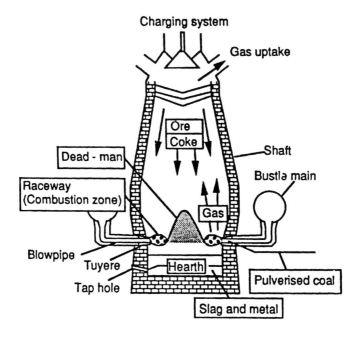

FIGURE 1   Schematic diagram of a blast furnace showing its major features.

A stagnant zone of coke, termed the 'dead man' forms in front of the raceway and the low permeability of this zone can cause serious operational difficulties.

Coal injection was originally tried as an alternative to oil during the 1970 oil crisis. In the event it was found to offer potential for higher levels of coke replacement and improved furnace control. Coal injection is expected to become increasingly attractive over the next decade as coke production declines. PF injection rates have increased from negligible values at the end of the 1970's to values as high as 200 kg/tHM in some installations in Western Europe. In a joint venture involving British Steel, Hoogovens and ILVA, trials with rates as high as 300 kg/tHM have been reported (Campbell et al., 1991).

COMMENTS ON PREVIOUS MODELLING EFFORTS

Broadly, the work has proceeded on three types of models: (i) one-dimensional models of the raceway with and without pulverised coal, or other auxiliary fuel, injections (Kuwabara et al., 1981), (Hatano et al., 1981), (Suzuki et al., 1986), (Jamuluddin et al., 1986), (Ohno et al., 1992), and (Yamagata et al., 1992); (ii) Two and three dimensional models of raceways of presumed configurations (Suzuki et al., 1984), (Jamaluddin et al., 1986), (Apte et al., 1988) and (Sugeyama et al., 1990); (iii) Two-dimensional models for the simultaneous prediction of the reactions and the raceway configurations (Hatano et al., 1980) and (Nogami et al., 1992).

Although a good deal of chemistry can be embodied by the one-dimensional models, the important phenomena of the mixing of the PF stream with the hot enveloping air is not well simulated. The most comprehensive two-dimensional models are due to (Hatano *et al.*, 1981) and (Nogami *et al.*, 1992). These models approximate the coke packed bed with a 'quasi-fluid' where the motion of the coke particles was obtained from momentum transport equations similar to those of the gas phase. None of the previous models considers both the turbulence of the raceway flow and that in the surrounding packed bed. The turbulence is relevant to the important phenomena of: the dispersion of the particulate phase, the mixing of the gas species and the combustion of the volatile material, and to a lesser extend it, to all the other features of the injected flows.

PRESENT PAPER

This paper is a summary only of an extensive PhD thesis, (Takeda, 1993). Many details of interest to the specialist reader have of necessity been omitted and such readers are invited to refer to the thesis or contact the authors directly.

A few preliminary remarks on the basic code used and its application are merited. It bears the name FAFNIR and predicts the flow, heat transfer and combustion for two dimensional geometries. It was originally developed for the prediction of pulverised coal-fired furnaces, and in particular, cement kilns, see for example Lockwood *et al.*, (1988). It is applied here to the turbulent flow encompassing the blowpipe, tuyere and the raceway. It has been further developed to embody modelling for the interaction of the injected air and PF with the coke particles in the raceway and for the flow through the surrouding packed bed. A schematic view of the relevant phenomena modelled by the modified FAFNIR code is presented by Figure 2.

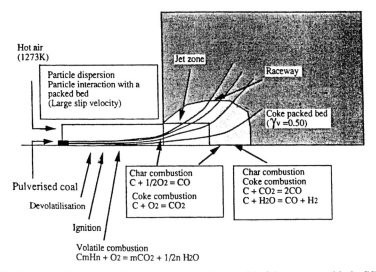

Hot air
(1273K)

Jet zone

Raceway

Particle dispersion
Particle interaction with a
packed bed
(Large slip velocity)

Coke packed bed
($\gamma_v = 0.50$)

Pulverised coal

Devolatilisation

Char combustion
$C + 1/2O_2 = CO$

Coke combustion
$C + O_2 = CO_2$

Char combustion
Coke combustion
$C + CO_2 = 2CO$
$C + H_2O = CO + H_2$

Ignition

Volatile combustion
$C_mH_n + O_2 = mCO_2 + 1/2n\ H_2O$

FIGURE 2   Raceway phenomena relevant to the simulation model of the raceway with the PF injection.

THE MATHEMATICAL MODEL

The Basic FAFNIR Code

FAFNIR solves the governing 'Navier Stokes' partial differential equations for the gas phase using a finite volume treatment. It embodies options for various differencing schemes and both the $k \sim \varepsilon$ and Reynolds stress models of turbulence. The calculations presented herein are based on the former turbulence model, the latter having been found to offer little benefit for the present application. The particulate phase is handled in lagrangian fashion with the flights of the particles being predicted through a stepwise solution of their vectorial momentum equation incorporating information about the local value of the drag coefficient.

The simultaneous solution of the enthalpy equation enables the particle temperatures to be determined. Modelling is included for devolatilisation, volatiles combustion and char burnout. The thermal radiation transfer is handled by a non equilibrium diffusion model. All of this is standard and is well documented in many references. (Lockwood et al., 1988) and (Lockwood and Malalasekera, 1988.)

Application of FAFNIR to the Blast Furnace

Eleven transport equations are solved for the gas phase variables. In addition to those for the momenta in the two co-ordinates directions these comprise the nine scalar variables of Table I. The mass fractions of carbon, oxygen and hydrogen are selected as independent variables because they are conserved with respect to the gaseous reactions: $CO + 1/2O_2 = CO_2$ and $H_2 + O_2 = H_2O$. The complete reaction scheme adopted is summarised by Figure 3.

As indicated, separate heterogeneous reaction schemes are assumed for the char and coke reactions. The char carbon is assumed to react simultaneously with oxygen and carbon dioxide to form carbon monoxide with the final product being carbon dioxide. Coke particles having temperatures in excess of 80% of the local gas temperature burn to form carbon dioxide. The primary product of the coke combustion is presumed to be carbon dioxide because of its relatively slow diffusional rate through the boundary layer enveloping the relatively large coke particles the typical size of which is 25mm.

TABLE I

Scalar variables in the model

Symbol of scalar variables	Meaning
k	Turbulent kinetic energy
ε	Eddy dissipation rate
f_C	Mass fraction of carbon
f_H	Mass fraction of hydrogen
f_O	Mass fraction of oxygen
f_{vol}	Mass fraction of volatiles
H	Enthalpy of gas
f_{CO_2}	Mass fraction of carbon dioxide
f_{H_2O}	Mass fraction of water

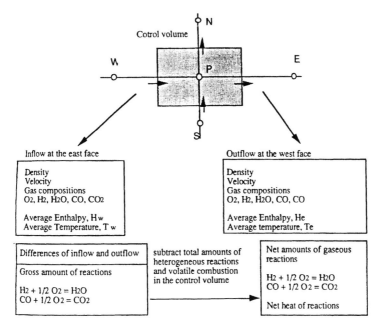

FIGURE 3   Calculation of the net gaseous reactions in a control volume.

The secondary reactions with carbon dioxide and water occur simultaneously. The rate constants of the heterogeneous reactions have been selected from values reported in the literature. Equilibrium of the major gas species O_2. CO, CO_2, H_2 and H_2O is assumed.

A conventional stochastic model has been incorporated in FAFNIR to account for the turbulence dispersion of the char particles. The treatment of the interaction of the flow with the coke particles in the raceway presumes that the raceway can be divided into two regions: a jet region where the voidage is so large (>0.98) that the interaction is insignificant; and a transition zone between the jet region and the packed bed containing coke particles presumed to be stationary with respect to the gas. This transition zone is handled in the same manner as the packed which is the subject of the following paragraphs, being distinguished only by greater voidage. The voidage is presumed to vary linearly across the transition region from zero in the jet region to an assumed prevailing value for the packed bed. The calculation of the flow in the packed bed is based on the distribued resistance concept originally introduced by Patankar and Spalding (1978) and subsequently developed and applied by Sha *et al.* (1982) and Lybaert *et al.* (1991). The use of the concept in this application is distinguished by the calculation of the important turbulent properties of the flow.

The governing partial differential equations are solved. The r-directed momentum equation, for example, reads:

$$\frac{1}{r}\frac{\partial}{\partial r}\left[r\rho_g\gamma_v\bar{v}\bar{v}\right] + \frac{1}{r}\frac{\partial}{\partial z}\left[r\rho_g\gamma_v\bar{u}\bar{v}\right]$$

$$=\frac{1}{r}\left[\frac{\partial}{\partial r}\left(r\mu_{\text{eff}_i}v\frac{\partial\bar{v}}{\partial r}\right) + \frac{\partial}{\partial r}\left(r\mu_{\text{eff}_i}v\frac{\partial\bar{v}}{\partial r}\right)\right] - \gamma v\frac{\partial\bar{P}}{\partial r} + \gamma v S_v + S_{p.v} - R_{fs,r} \qquad (A)$$

where, the source term S_v is:

$$S_v = \frac{1}{r}\frac{\partial}{\partial r}\left(r\mu_{\text{eff}}\frac{\partial \bar{v}}{\partial r}\right) + \frac{\partial}{\partial z}\left(\mu_{\text{eff}}\frac{\partial \bar{u}}{\partial r}\right) + \frac{\rho_g \bar{w}^2}{r} - 2\mu_{\text{eff}}\frac{\bar{v}}{r^2} - \frac{\partial}{\partial r}\left[\frac{2}{3}\rho_g k\right]$$

Here r_g is the gas density, γ_v is the void fraction of the packed bed, v is the r-directed velocity (the overbars indicating time averaged quantities), μ_{eff} is the effective turbulent viscosity, P is pressure. $S_{p,v}$ is the source term due to interaction of the gas with the packed bed. $R_{fs,r} = \gamma_V R_o$ V where R_o is the flow resistance coefficient derived from Ergun's equation:

$$R_o = 150\,\mu_g\left[\frac{1-\gamma_v}{d_p\gamma_v}\right]^2 + 1.75\,\rho_g\left[\frac{1-\gamma_v}{d_p\gamma_v}\right]|\vec{u}| \tag{B}$$

where d_p is the diameter of the packed solid material and μ_g is the laminar viscosity of the gas. The z-directed momentum equation is analogous.

The turbulence dissipation rate, ε, may be assumed to be dominated by the frictional loss with the packed bed in which case the full governing partial differential equation for ε reduces to:

$$\rho_g \varepsilon = C_k R_o \bar{u}^2 \tag{C}$$

The hydraulic diameter of the packed bed is used to obtain a turbulent mixing length:

$$1_m = C_{lm}\left[\frac{\gamma_v}{1-\gamma_v}d_p\right] \tag{D}$$

where C_{lm} is an empirical constant. Dimensional considerations leads to:

$$\mu_t = C_\mu \rho_g \varepsilon^{1/3} l_m^{3/4} \tag{E}$$

where C_μ is a further empirical constant. The preceeding three equations allow the determination of the turbulent viscosity in the packed bed.

Effective diffusivities, D_{eff}, for heat and mass are obtained from:

$$\rho_g D_{\text{eff}} = \left(\frac{\mu_1}{\sigma_1} + \frac{\mu_t}{\sigma_t}\right) \tag{F}$$

where σ_l and σ_t are laminar and turbulent Prandt/Schmidt numbers. All of the empirical parameters: C_{lm}, C_k, σ_l, and σ_t have been determined with reference to measured data for packed beds.

VALIDATION STUDIES

Predictions have been obtained for the conditions cited in Tables II and III for the Chiba no. 1 blast furnace of Kawasaki Steel.

Validation of the mathematical model is hampered since for the most part the raceway is not accessible for measurement. Also the diameter of the coke, its reactivity and concentrations in the raceway will all be functions of the operating conditions in ways which can neither be predicted nor measured. In order to test the sensitivities to

TABLE II

Blast furnace data (Kawasaki Steel Chiba No. 1 blast furnace)

Hearth diameter	7.2m
Inner volume	966 m^3
Number of tuyere	14
Diameter of tuyere	0.13 m
Length of tuyere	– m
Blowpipe inner diameter	– m
Top pressure	– Pa (abs.)
Productivity	1600 ton hot metal/day
Raceway depth	1.5 m

TABLE III

Operating conditions on Chiba No. 1 blast furnace of Kawasaki Steel Corp. during the measurements

Blast temperature	970–1000°C
Blast pressure	$1.08–1.2 \times 10^5$ Pa (abs)
Oxygen concentration in blast	0.233 kg/kg
Moisture concentration in blast	18–24 g/Nm^3
Flow rate of blast	1482 Nm^3/hr/tuyere

TABLE IV

Conditions of sensitivity tests

Sensitivity tests	Conditions
1) Geometry of raceway	Length of $\gamma_r = 0$ zone
2) Coke particle size	dp = 0.010, 0.020, 0.050
3) Chemical reaction rate of coke	K_0 = 0.00737, 0.00369
4) Slip velocity between gas and coke	C_{sf} = 0.9, 0.45

these uncertainties predictions have been obtained for the range of conditions stated in Table IV.

As an example, predictions of the O_2 and CO concentrations with distance from the tuyere nose, showing the influence of the assumed diameters of the coke particles, are displayed in Figure 4. The range of diameters shown extends to the extremes expected in real furnace operation. Not surprisingly the smallest particles result in more prominent CO profiles and a faster depletion of oxygen. Less expected is the early fall in O_2 concentration for the largest particles, a consequence of the enhanced turbulent mixing which they produce. On the whole though the influence of the coke particle size does not seem to be a dominant factor. The complete sensitivity studies suggest that uncertainties about raceway conditions do not preclude the useful application of the model.

Some predictions are shown in Figure 5. for the axial variations in O_2, CO and CO_2 concentrations in the raceway when a relatively high volatile coal, 38.7%, is injected.

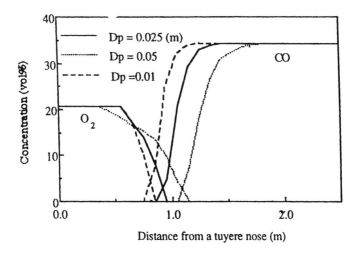

FIGURE 4   Influences of coke particle diameter in a packed bed on the combustion process.

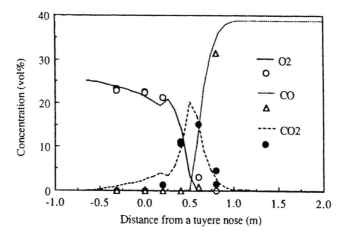

FIGURE 5   Predictions of axial profiles of gas compositions in the raceway with the injection of high volatile coal, VM = 38.7%.

This is one of the rare cases where measured data are available. The agreement between the predictions and measurements is remarkably good. The results over the entire range of sensitivity and validation studies performed were sufficiently encouraging to justify a general exploration of design variables.

SEARCH FOR DESIGN IMPROVEMENTS

Ways of enabling ever higher PF injection rates are sought. Certainly, injection rates significantly in excess of 200 kg/tHM are targeted by the forerunning industries. The most important limiting factor is that good burnout of the PF in the raceway must be achieved if satisfactory furnace operation is to be maintained.

In Figure 6 predictions of the O_2 concentrations and temperature along the axis of the blow pipe and raceway for the Chiba no. 1 furnace are presented for injection rates in the range 0 to 250 kg/tHM. The effect of even small amounts of PF injection appears to be significant; but once PF is introduced the variations consequent of increasing the injection rate are relatively small. Expectedly, the O_2 concentration decreases with increasing injection rate. The temperatures are suppressed in the blow pipe by the addition of PF due to particle heating and the latent heat of devolatilisation. The

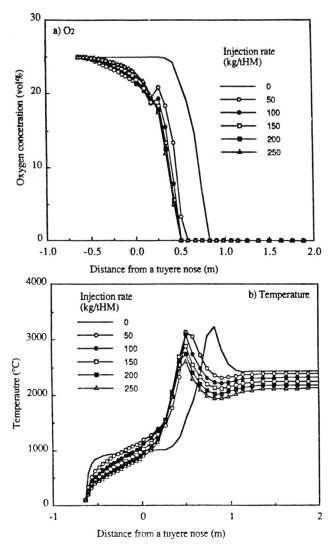

FIGURE 6   Predictions of axial profiles of oxygen temperatures for injection rates from 0 to 200 kg/t.

subsequent rapid temperature rise is the obvious consequence of ignition. The temperature rise is largest for the smallest injection rates since the excess air factor is decreasing with injection rate.

Axial profiles of burnout and temperature along the blow pipe axis are presented in Figure 7 for an injection rate of 200 kg/tHM. Burnout is here defined as the cumulative mass release from the coal divided by its initial combustible mass. The initial rapid rise in the burnout so defined is due to the combined effects of devolatilisation and char combustion. The rate of change of burnout flattens in the vicinity of the packed bed boundary (1.5m) illustrating the need to complete combustion of the coal well within the raceway region. Temperatures considerably in excess of those in the packed bed are achieved in the raceway encouraging high coal combustion rates. Once the raceway boundary is reached the temperature is caused to descend to the bulk value of the material of the packed bed.

A number of methods for improving the burnout of the PF in the raceway have been explored. These included: a 2% oxygen enrichment of the primary air, the secondary air and of both streams; variations in burner design (injection pipe diameter, pipe wall thickness and injection with swirl); alterations to the raceway conditions (raceway depth, coke particle diameter and coke reactivity, and alteration of the fineness of grind of the PF.

The results of all of these runs are conveniently summarised by Figure 8 where the values achieved for the overall and char burnouts are reported relative to a base case having none of the modifications. The overall burnout is here the ratio of the burned mass to the initial combustible mass of coal while the char burnout is the mass of burned char divided by that of the initial char. Notable improvements in burnout are

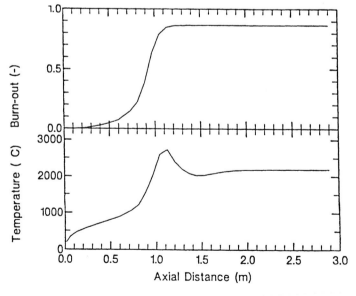

FIGURE 7   Axial profiles of the PF overall burnout and temperatures at the center.

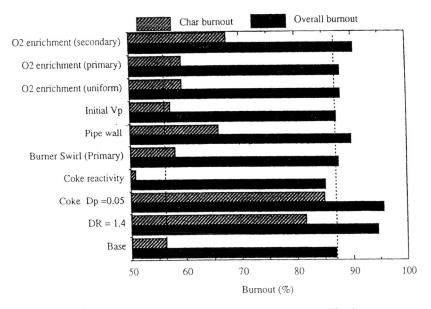

FIGURE 8   Comparisons of the burnout for the various modification.

achieved by: oxygen enrichment to the secondary or blast air stream; an increase in the coke particle diameter; increasing the outer diameter of the injection lance and extending the raceway depth (DR).

Oxygen enrichment to the same level, 2%, of the secondary stream is more effective than enrichment of the primary flow simply because the secondary flow rate is much greater. Increasing the coke particle diameter increases the raceway turbulence and so the mixing. Increasing the outer diameter of the injection lance creates a bluff body which has the same effect. Increasing the raceway depth simply increase the available residence time for PF particles to burn out. In practice the raceway depth can be increased by increasing the blast velocity, but this will have the converse effect on residence time.

CONCLUSIONS

The mathematical model described herein certainly appears to be a useful tool for assisting the engineer to find ways to maximise coal injection rates in blast furnaces. In particular it inspires the user to think about the physical phenomena occurring in the blow pipe, raceway and adjacent packed bed and such thought processes must eventually lead to design improvements. Some areas for possible design improvements have been identified. These need to be validated by equipment-based studies. And the computer based design explorations so far performed lead naturally to further potentially promising areas for exploitation. There is much still to be done.

REFERENCES

Abbas, T., Costen, P. G., Lockwood, F. C. and Romo-Millares, C. (1993) The effect of particle size on NO formation in a large-scale pulverised coal-fired laboratory furnace: measurements and modelling. *Combustion and Flame*, **93**, 316–326.

Apte, V. B., Wall, T. F. and Truelove, J. S. (1988) Gas flows in raceways fromed by high velority jets in a two dimensional packed bed. *Chem Eng. Res. and Des.*, **66**, 357–362.

Badzioch, S. and Hawksley, P. G. W. (1970) Kinetics of thermal decomposition of pulverised coal particles. *Ind. Eng. Chem. Process. Des. Develop.*, **9**, 521–530.

Bortz, S. and Flament, G. (1983) Experiments on pulverised coal combustion under conditions simulating blast furnace environments. *Ironmaking and Steelmaking*, **10**, 222–229.

Burgess, J. M. (1985) Fuel combustion in the blast furnace raceway zone. *Progress of Energy and Combustion Science*, **11**, 61–82.

Burgess, J. M., Jamaluddin, A. S., McCarthy, M. J., Mathieson, J. G., Nomur, S., Truelove, J. S. and Wall, T. F. Pulverised coal ignition and combustion in the blast furnace tuyere zone. *Proceedings of the Joint Symposium of ISIJ and AIMM*, Tokyo, Japan, pp. 129–141 (1983).

Campbell, D. A., Piennan, G., Malgarinio, G. and Smith, R. B. Oxy-coal injection at Cleveland Ironworks. *Proceedings of Second European Ironmaking Congress*, Glasgow, pp. 263–274 (1991).

Clixby, G. and Willmers, R. R. (1991) The blast furnace raceway. *2nd European Iron Making Congress*, pp. 434–437.

Hatano, M., Kurita, K. and Tanaka, T. Aerodynamic study on raceway in blast furnace, *International Blast Furnace Hearth and Raceway Symposium*, Newcastle, Australia, Symposium Series, Australasian Institute of Mining and Metallurgy, **26**, 4–1–4–10 (1981).

Jamaluddin, A. S., Wall, T. F. and Truelove, J. S. (1986) Mathematical modelling of combustion in blast furnace raceways, including injection of pulverised coal. *Ironmaking and Steelmaking*, **13**, 91–99.

Kuwabara, M., Hsieh, Y. S. Isobe, K. and Muchi, I. Mathematical modelling of the tuyere combustion zone of the blast furnace, *International Blast Furnace Hearth and Raceway Symposium*, Newcastle, Australia, Symposium Series, Australasian Institute of Mining and Metallurgy, **26**, (1981).

Lockwood, F. C. and Mahmud, T. The prediction of swirl burner pulverised coal flames, *Proceedings of 22nd Symposium (International) on Combustion*, The Combustion Institute, pp. 165–173 (1988).

Nogami, H., Miura, T. and Furukawa, T. (1992) Simulation of transport phenomena around raceway zone in the lower part of blast furnace, *Tetsu to Hagane*, **78**, 1222–1129.

Ohno, Y., Furukawa, T. and Matuura, M. (1992) Combustion behaviour of pulverised coal in a raceway cavity of blast furnace and a large amount injection technology, *Tetsu to Hagane* (in Japanese), **78**, 50–57.

Sugiyama, T., Matsuzaki, S. and Nakagawa, T. Analysis of flow and reaction in lower part of blast furnace. *Proceedings of the Sixth International Iron Steel Congress*, Nagoya, ISIJ, pp. 414–421 (1990).

Suzuki, T., Smoot, L. D., Fletcher, T. H. and Smith, P. J. Prediction of high intensity pulverised coal combustion. *Combustion Science and Technology*, **45**, 167–183 (1986).

Takeda, K. (1993) The mathematical modelling of pulverised coal combustion in a blast furnace, *Ph D. Thesis*, University of London.

Yamagata, C., Suyama, S., Horisaka, S., Kajiwara, Y., Nishizawa, R., Komatsu, S., Shibata, H. and Aminaga, Y. Fundamental study on the internal state of the blast furnace at high pulverised coal injection. *McMaster Symposium*, Toronto, Canada (1991).

A Detailed Kinetic Study of Ammonia Oxidation

R. P. LINDSTEDT, F. C. LOCKWOOD and M. A. SELIM   *Department of Mechanical Engineering, Imperial College of Science, Technology and Medicine, Exhibition Road, London SW7 2BX, UK*

Abstract—A comprehensive study of ammonia oxidation in laminar flames and flow reactors using detailed chemical kinetic modelling is reported. A flat premixed NH_3/NO flame and counterflow $CO/O_2/N_2$ and $CH_4/O_2/N_2$ diffusion flames doped with ammonia have been studied along with $NH_3/NO/O_2$ and $NH_3/NO/O_2/C_2H_6$ flow reactors. Available kinetic data has been reviewed and a detailed reaction mechanism for the C/N system is proposed and results from computations are compared with experimental data. The destruction and formation of NO is found to be dominated by NH_2, NH and N radicals in most of the cases studied. It is shown that the relative significance of the different NH_i radicals in the various NO formation channels depends entirely on the flame conditions and that the role of the amidogen is crucial in nitric oxide reduction.

Key Words: Ammonia flame chemistry, SNCR, NO_x formation

INTRODUCTION

The ability to predict emissions of nitrogen oxides (NO_x) from combustion applications has become an important part of flame modelling. Many studies (MacLean and Wagner, 1967; Fenimore and Jones, 1961; Lyon and Benn, 1979; Bian *et al.*, 1986; Bian *et al.*, 1990; Branch *et al.*, 1982; Miller *et al.*, 1983; Miller and Bowman, 1989; Lindstedt *et al.*, 1994) have focused their interest on the understanding of the mechanisms of the ammonia oxidation process. Ammonia is important not only for its role in nitric oxide reduction, but also for the role of NH_i fragments in the process of conversion of nitrogen containing fuels. However, the understanding of the ammonia oxidation mechanism is not complete. There is still uncertainty with respect to the products formed in some elementary reactions as well as a lack of direct measurement of kinetic data for many important reactions (e.g., Lindstedt *et al.*, 1994).

Nevertheless, different NO_x reduction techniques have been employed world-wide in order to meet the NO_x emission standards established by legislation. Among the techniques frequently used in practical systems for post flame nitric oxide reduction are Selective Catalytic Reduction (SCR) and Selective Non-Catalytic Reduction (SNCR). In both techniques ammonia is injected in the post flame zone of combustors. The difference between the two techniques is that in SCR ammonia is injected in presence of a catalyst and at temperature typically between 700 and 750 K while SNCR acts without a catalyst and within a temperature window between 1100 and 1400 K. The latter method is popular for economic reasons and has been shown to be successful in reducing NO emissions through homogeneous gas phase reaction (Lyon, 1973; Gebel *et al.*, 1989).

1021

The present study is aimed to make a contribution towards an established detailed reaction mechanism for the successful kinetic modelling of ammonia oxidation for an extended range of combustion conditions. The study is complementary to that reported by Lindstedt *et al.* (1994) who considered a wide range of flat premixed laminar flames with molecular oxygen acting as the prime oxidant. The focus of the present study is the investigation of ammonia oxidation under (i) conditions where nitric oxide is the prime oxidant (ii) in counterflow diffusion flames of methane and carbon monoxide doped with ammonia and (iii) in flow reactors in the presence of molecular oxygen, nitric oxide and active hydrocarbon fragments. All of these conditions add new aspects to the previously reported study (Lindstedt *et al.*, 1994). The first flame is of direct relevance to nitric oxide reduction technologies and the diffusion flames extend the understanding of the ammonia oxidation process to combustion environments where diffusion is predominant and hydrocarbon fragments are present. Finally the flow reactor studies permit a direct assessment of the ability of the reaction mechanism to predict the SNCR temperature window. Model predictions are validated by comparing with experimental data from literature (Hahn and Wendt, 1981; Hahn *et al.*, 1981; Muris, 1993; Vandooren *et al.*, 1994).

BASIC EQUATIONS AND COMPUTATIONAL PROCEDURE

To construct a detailed scheme for ammonia oxidation which gives good agreement for flames under a wide range of combustion condition is difficult. In the present study a number of different cases for which experimental data is available have been computed to evaluate proposed reaction channels. The configurations are listed in Table I. The premixed flame is approximated as a burner stabilised flame and the pressure is assumed to be constant and thermal diffusion is ignored. The computational technique is described in detail elsewhere (Jones and Lindstedt, 1988a). The counterflow diffusion flames were computed using the customary similarity transformed co-ordinate system (Jones and Lindstedt, 1988b) and the basic equations and the computational procedures are therefore not repeated here. The flow reactors were computed using well established techniques (e.g., Muris, 1993). The transport coefficients for viscosities and binary diffusivities have been evaluated using the theory by Chapman and Enskog (see Reid and Sherwood, 1960). Thermal conductivities were evaluated using the Mason and Monchick (1962) approximation. Mixture properties were evaluated using the semi-empirical Wilkes formula (see Reid and Sherwood, 1960) and thermodynamic data was computed using the CHEMKIN thermodynamic data base (Kee *et al.*, 1987).

CHEMICAL KINETICS

The starting point of the present work is the N/H/O mechanism by Lindstedt *et al.* (1994) which comprises 21 species and 95 reactions. For reasons of compactness only a subset of this reaction mechanism is listed in Table II. For the methane and carbon monoxide diffusion flames and the ethane doped flow reactor a detailed C/N/H/O

TABLE I

Gas compositions for premixed and diffusion flames given in mole fractions

Flame	NH$_3$	NO	Ar	ϕ	P(kPa)
[a]F1AP	0.461	0.472	0.067	1.46	7.20

Flame	P(kPa)	$a(s^{-1})$	Fuel side					Oxidizer side			
			CH$_4$	CO	N$_2$	H$_2$O	NH$_3$	O$_2$	N$_2$	NH$_3$
[b]F1AD	101	3.62	0.1	–	0.899	–	0.001	0.52	0.48	0.0
[b]F1BD	101	3.62	0.1	–	0.90	–	0.0	0.52	0.4778	0.0022
[c]F2AD	101	3.62	–	0.362	0.599	0.03	0.0088	0.191	0.809	0.0
[c]F2BD	101	3.62	–	0.362	0.608	0.03	0.0	0.191	0.8071	0.0191

Initial Gas Composition and Residence Times (τ) for Flow Reactors. Concentrations in ppm

	NH$_3$	NO	C$_2$H$_6$	O$_2$	τ(s)	P(kPa)
[d]FR1	750	500	–	20000	525/T(K)	101
[d]FR2	750	500	100	20000	525/T(K)	101

[a]Vandooren et al. (1994)
[b]Hahn and Wendt (1981)
[c]Hahn et al. (1981)
[d]Muris (1993)

mechanism comprising 419 reactions and 67 species was used. For the latter mechanism the C$_1$/C$_2$ hydrocarbon mechanism by Leung and Lindstedt (1994) was adopted along with a newly developed C/N sub-mechanism. The latter sub-mechanism was initially extracted from the work by Miller and Bowman (1989) and Drake and Blint (1991) and subsequently updated by the use of more recent kinetic data. The reactions from the C$_1$/C$_2$ mechanism mentioned in the path analyses can be found in Table III and a complete listing of the C/N sub-mechanism can be found in Table IV.

AMMONIA OXIDATION IN PREMIXED LAMINAR FLAMES

Vandooren et al. (1994) investigated a low pressure laminar premixed flat NH$_3$/NO flame diluted with argon both computationally and experimentally. The present reaction scheme, formulated following an extensive validation process (Lindstedt et al., 1994), differs from that used by Vandooren et al. (1994) in a number of aspects such as in the product channels of the important NO removal reactions by amidogen (NH$_2$). A further study of the present flame is therefore of considerable interest as NO reactions with NH$_2$, NH and N radicals play a crucial role in determining the flame structure. Furthermore, this flame is directly related to the NO$_x$ reduction technologies using ammonia. A brief analysis of the main reaction paths in

TABLE II

Part of N/H/O sub-mechanism from Lindstedt *et al.* (1994) Rates in the form $k = AT^n \exp(-E/RT)$ Units are kmole, cubic meters, seconds, degrees Kelvin and kJ/mole

No.	Reactions		A	n	E	References
2.1	$NH_3 + M = NH_2 + H + M$		$0.140\,E+03$	0.06	379.07	Miller and Bowman (1989)
2.2	$NH_3 + H = NH_2 + H_2$		$0.636\,E+03$	2.39	42.56	Miller and Bowman (1989)
2.3	$NH_3 + OH = NH_2 + H_2O$		$0.204\,E+04$	2.04	2.37	Miller and Bowman (1989)
2.4	$NH_3 + O = NH_2 + OH$		$0.210\,E+11$	0.00	37.65	Hanson and Salimian (1984)
2.5	$NH_2 + H = NH + H_2$		$0.568\,E+07$	1.14	8.26	Lindstedt *et al.* (1994)
2.6	$NH_2 + OH = NH + H_2O$		$0.900\,E+05$	1.50	-1.91	Cohen and Westberg (1991)
2.8	$NH_2 + O = HNO + H$		$0.990\,E+12$	-0.50	0.00	Lindstedt *et al.* (1994)
2.11	$NH_2 + NO = N_2 + H_2O$		$0.300\,E+18$	-2.60	3.87	Lindstedt *et al.* (1994)
2.12	$NH_2 + NO = NNH + OH$		$0.140\,E+10$	0.00	0.00	Lindstedt *et al.* (1994)
2.13	$NH_2 + NO = N_2O + H_2$		$0.500\,E+11$	0.00	102.43	He *et al.* (1992)
2.16	$NH_2 + HO_2 = NH_3 + O_2$		$0.452\,E+11$	0.00	0.00	Sarkisov *et al.* (1984)
2.20	$NH + H = N + H_2$		$0.100\,E+11$	0.00	0.00	Bian *et al.* (1990)
2.21	$NH + O = NO + H$		$0.700\,E+11$	0.00	0.00	Cohen and Westberg (1991)
2.23	$NH + OH = N + H_2O$		$0.200\,E+07$	1.20	0.03	Cohen and Westberg (1991)
2.24	$NH + OH = HNO + H$		$0.400\,E+11$	0.00	0.00	Lindstedt *et al.* (1994)
2.25	$NH + O_2 = NO + OH$		$0.100\,E+11$	-0.20	20.04	Lindstedt *et al.* (1994)
2.28	$NH + NO = N_2O + H$		$0.294\,E+12$	-0.40	0.00	Miller and Melius (1992)
2.29	$NH + NO = N_2 + OH$		$0.216\,E+11$	-0.23	0.00	Miller and Melius (1992)
2.30	$NH + NO = NNH + O$		$0.560\,E+10$	0.21	45.47	Bozzelli *et al.* (1994)
2.32	$N + O_2 = NO + O$		$0.640\,E+07$	1.00	26.28	Miller and Bowman (1989)
2.33	$N + OH = NO + H$		$0.380\,E+11$	0.00	0.00	Miller and Bowman (1989)
2.34	$N + NO = N_2 + O$		$0.330\,E+10$	0.30	0.00	Miller and Bowman (1989)
2.43	$NNH + M = N_2 + H + M$		$0.170\,E+10$	0.00	59.86	Lindstedt *et al.* (1994)
2.45	$NNH + NH_2 = N_2 + NH_3$		$0.500\,E+11$	0.00	0.00	Miller and Bowman (1989)
2.46	$NNH + NO = N_2 + HNO$		$0.910\,E+09$	0.00	0.00	He *et al.* (1992)
2.50	$N_2O + H = N_2 + OH$		$0.223\,E+12$	0.00	70.08	Marshall *et al.* (1989)
2.53	$N_2O = N_2 + O^a$	$k_\infty =$ $0.130\,E+12$	0.00	249.42	Tsang and Herron (1991)
		$k_0 =$ $0.722\,E+15$	-0.73	262.72	
2.55	$HNO + M = H + NO + M$		$0.236\,E+14$	0.00	203.68	He *et al.* (1992)
2.56	$HNO + H = H_2 + NO$		$0.603\,E+10$	0.00	0.00	Bulewicz and Sugden (1964)
2.57	$HNO + OH = H_2O + NO$		$0.130\,E+05$	1.88	-4.00	Lindstedt *et al.* (1994)
2.62	$HNO + O_2 = NO + HO_2$		$0.316\,E+10$	0.00	12.55	Fujii *et al.* (1981)

$^a F_c = 1.167 - 1.25\,E - 04\,T$

TABLE III

Part of C_1–C_2 mechanism from Leung and Lindstedt (1994). Rates in the form $k = AT^n \exp(-E/RT)$ Units are kmole, cubic meters, seconds, degrees Kelvin and kJ/mole

No.	Reactions	A	n	E	References
1.23	$CO + OH = CO_2 + H$	$6.320E+03$	1.50	-2.08	Baulch *et al.* (1992)
1.76	$CH_2O + H = CHO + H_2$	$2.288E+07$	1.05	13.70	Baulch *et al.* (1992)
1.78	$CH_2O + OH = CHO + H_2O$	$3.400E+06$	1.18	-1.87	Baulch *et al.* (1992)
1.85	$CH_3 + O = CH_2O + H$	$8.430E+10$	0.00	0.00	Baulch *et al.* (1992)
1.87	$CH_3 + OH = {}^1CH_2 + H_2O$	$4.000E+10$	0.00	20.47	Grotheer *et al.* (1992)
1.91	$CH_3 + O_2 = CH_2O + OH$	$3.300E+08$	0.00	37.40	Baulch *et al.* (1992)
1.119	$CH_4 + H = CH_3 + H_2$	$1.325E+01$	3.00	33.63	Baulch *et al.* (1992)
1.121	$CH_4 + OH = CH_3 + H_2O$	$1.560E+04$	1.83	11.60	Baulch *et al.* (1992)

TABLE IV

C/N/H/O sub-mechanism. Rates in the form $k = AT^n \exp(-E/RT)$ Units are kmole, cubic meters, seconds, degrees Kelvin and kJ/mole

No.	Reactions	A	n	E	References
3.1	$NO + C = N + CO$	$0.280E+11$	0.00	0.00	Dean et al. (1991)
3.2	$NO + C = CN + O$	$0.200E+11$	0.00	0.00	Dean et al. (1991)
3.3	$NO + CH = HCN + O$	$0.110E+12$	0.00	0.00	Dean et al. (1991)
3.4	$NO + CH = CO + NH$	$0.110E+11$	0.00	0.00	Dean et al. (1991)
3.5	$NO + {}^3CH_2 = HCN + OH$	$0.346E+10$	0.00	-1.58	Atakan et al. (1992)
3.6	$NO + {}^1CH_2 = HCN + OH$	$0.200E+11$	0.00	0.00	Miller and Bowman (1989)
3.7	$NO + CH_3 = HCN + H_2O$	$0.120E+12$	0.00	121.30	Henning and Wagner (1994)
3.8	$NO + CHO = HNO + CO$	$0.723E+10$	0.00	0.00	Tsang and Herron (1991)
3.9	$NO + CH_2O = HNO + CHO$	$0.102E+11$	0.00	170.77	Tsang and Herron (1991)
3.10	$NO + CH_2O = CH_3O + HNO$	$0.130E+12$	-0.70	0.00	Atkinson et al. (1992)
3.11	$NO + HCCO = HNCO + CO$	$0.480E+11$	0.00	2.90	Boullart et al. (1994)
3.12	$NO + HCCO = HCN + CO_2$	$0.120E+11$	0.00	2.90	Boullart et al. (1994)
3.13	$NO_2 + CH = CHO + NO$	$0.101E+12$	0.00	0.00	Wagal et al. (1982)
3.14	$NO_2 + CH_3 = CH_3O + NO$	$0.130E+11$	0.00	0.00	Glaenzer and Troe (1974)
3.15	$NO_2 + CH_4 = CH_3 + HNO_2$	$0.120E+11$	0.00	12.55	Slack and Grillo (1981)
3.17	$NO_2 + CO = NO + CO_2$	$0.903E+11$	0.00	141.34	Tsang and Herron (1991)
3.18	$N_2O + C = CN + NO$	$0.100E+11$	0.00	0.00	Drake and Blint (1991)
3.19	$N_2O + CO = N_2 + CO_2$	$0.320E+09$	0.00	85.00	Fujii et al. (1985)
3.20	$N_2 + CH = HCN + N$	$0.610E+09$	0.00	58.20	Lindackers et al. (1990)
3.21	$N_2 + {}^3CH_2 = HCN + NH$	$0.482E+10$	0.00	149.66	Sanders et al. (1987)
3.22	$N + CH = CN + H$	$0.126E+11$	0.00	0.00	Messing et al. (1981)
3.23	$N + {}^3CH_2 = HCN + H$	$0.500E+11$	0.00	0.00	Millers and Bowman (1989)
3.24	$N + CH_3 = HCN + H_2$	$0.700E+10$	0.00	0.00	Marston et al. (1989)
3.25	$N + CH_4 = NH + CH_3$	$0.100E+11$	0.00	100.42	Drake and Blint (1991)
3.26	$N + HCCO = HCN + CO$	$0.500E+11$	0.00	0.00	Drake and Blint (1991)
3.27	$N + C_2H_3 = HCN + {}^3CH_2$	$0.200E+11$	0.00	0.00	Miller and Bowman (1989)
3.28	$N + C_2H_4 = HCN + CH_3$	$0.121E+08$	0.00	0.00	Levy and Winkler (1989)
3.29	$N + C_3H_3 = HCN + C_2H_2$	$0.100E+11$	0.00	0.00	Miller and Bowman (1989)
3.30	$NH + CH = HCN + H$	$0.500E+11$	0.00	0.00	Drake and Blint (1991)
3.31	$NH + {}^3CH_2 = H_2CN + H$	$0.300E+11$	0.00	0.00	Drake and Blint (1991)
3.32	$NH_2 + CH = H_2CN + H$	$0.300E+11$	0.00	0.00	Drake and Blint (1991)
3.33	$NH_2 + CH_4 = NH_3 + CH_3$	$0.579E+10$	0.00	55.10	Hack et al. (1988)
3.34	$NH_2 + C_2H_4 = NH_3 + C_2H_3$	$0.206E+08$	0.00	10.96	Bosco et al. (1984)
3.35	$NH_2 + C_2H_6 = NH_3 + C_2H_5$	$0.971E+10$	0.00	44.40	Enbrecht et al. (1987)
3.36	$NH_2 + HNCO = NH_3 + NCO$	$0.400E+09$	0.00	0.00	Mertens et al. (1991)
3.37	$HCN + O = CN + OH$	$0.270E+07$	1.58	111.29	Perry and Melius (1985)
3.38	$HCN + O = NH + CO$	$0.350E+01$	2.64	20.84	Miller and Bowman (1989)
3.39	$HCN + O = NCO + H$	$0.140E+02$	2.64	20.84	Miller and Bowman (1989)
3.40	$HCN + OH = CN + H_2O$	$0.904E+10$	0.00	44.89	Baulch et al. (1992)
3.41	$HCN + OH = HNCO + H$	$0.480E+09$	0.00	46.02	Drake and Blint (1991)
3.42	$HCN + OH = HOCN + H$	$0.920E+10$	0.00	62.76	Drake and Blint (1991)
3.43	$HCN + OH = NH_2 + CO$	$0.783E-06$	4.00	16.74	Drake and Blint (1991)
3.44	$CN + O = CO + N$	$0.102E+11$	0.00	0.00	Baulch et al. (1992)
3.45	$CN + H_2 = HCN + H$	$0.310E+03$	2.45	9.30	Atakan et al. (1989a)
3.46	$CN + OH = NCO + H$	$0.603E+11$	0.00	0.00	Baulch et al. (1992)
3.47	$CN + O_2 = NCO + O$	$0.662E+10$	0.00	-1.70	Baulch et al. (1992)
3.48	$CN + NO_2 = NCO + NO$	$0.300E+11$	0.00	0.00	Miller and Bowman (1989)
3.49	$CN + N_2O = NCO + N_2$	$0.100E+11$	0.00	0.00	Miller and Bowman (1989)
3.50	$CN + CH_4 = HCN + CH_3$	$0.217E+11$	0.00	5.12	Miller and Bowman (1989)
3.51	$CN + N = C + N_2$	$0.104E+13$	-0.50	0.00	Miller and Bowman (1989)
3.52	$CN + NO = N_2 + CO$	$0.108E+12$	0.00	33.59	Mulvihill and Phillips (1975)
3.53	$CN + NO = N + NCO$	$0.964E+11$	0.00	176.26	Natarajan and Roth (1988)
3.54	$NCO + M = N + CO + M$	$0.102E+13$	0.00	195.39	Baulch et al. (1992)

TABLE IV

(Continued)

No.	Reactions	A	n	E	References
3.55	$NCO + H = NH + CO$	$0.500E + 11$	0.00	0.00	Miller *et al.* (1984)
3.56	$NCO + O = NO + CO$	$0.562E + 11$	0.00	0.00	Lauge and Hanson (1984)
3.57	$NCO + OH = NO + CO + H$	$0.100E + 11$	0.00	0.00	Miller *et al.* (1984)
3.58	$NCO + NO = N_2O + CO$	$0.340E + 10$	0.00	-1.69	Cooper and Hershberger (1992)
3.59	$NCO + NO = N_2 + CO_2$	$0.450E + 10$	0.00	-1.69	Cooper and Hershberger (1992)
3.60	$NCO + NO = N_2 + CO + O$	$0.235E + 10$	0.00	-1.69	Cooper and Hershberger (1992)
3.61	$NCO + N = N_2 + CO$	$0.200E + 11$	0.00	0.00	Miller *et al.* (1984)
3.62	$NCO + H_2 = HNCO + H$	$0.207E + 04$	2.00	25.19	Tsang (1992)
3.63	$HNCO + H = NH_2 + CO$	$0.300E + 11$	0.00	19.12	Tsang (1992)
3.64	$HNCO + O = NH + CO_2$	$0.960E + 05$	1.41	35.67	Tsang (1992)
3.65	$HNCO + O = NCO + OH$	$0.223E + 04$	2.11	47.80	Tsang (1992)
3.66	$HNCO + OH = NCO + H_2O$	$0.638E + 03$	2.00	10.72	Tsang (1992)
3.67	$HNCO + CN = HCN + NCO$	$0.150E + 11$	0.00	0.00	Tsang (1992)
3.68	$HNCO + NH_2 = NH + NCO$	$0.200E + 11$	0.00	49.87	He *et al.* (1992)
3.69	$HOCN + H = CN + H_2O$	$0.100E + 10$	0.00	0.00	Szekely *et al.* (1984)
3.70	$HOCN + H = NCO + H_2$	$0.100E + 10$	0.00	0.00	Szekely *et al.* (1984)
3.71	$N + CH_3 = H_2CN + H$	$0.710E + 11$	0.00	0.00	Marston *et al.* (1989)
3.72	$NO + CH_3 = H_2CN + OH$	$0.520E + 10$	0.00	101.43	Henning and Wagner (1994)
3.73	$H_2CN + H = HCN + H_2$	$0.100E + 12$	0.00	0.00	Gardiner *et al.* (1993)
3.74	$H_2CN + OH = HCN + H_2O$	$0.100E + 12$	0.00	0.00	Gardiner *et al.* (1993)
3.75	$H_2CN + M = HCN + H + M$	$0.300E + 12$	0.00	92.05	Miller and Bowman (1989)
3.76	$H_2CN + N = {}^3CH_2 + N_2$	$0.200E + 11$	0.00	0.00	Miller and Bowman (1989)
3.77	$CH_3 + NO = CH_3NO$	$0.400E + 10$	0.00	-1.00	Henning and Wagner (1994)
3.78	$CH_3NO + M = CH_2NOH + M$	$0.100E + 12$	0.00	231.00	Estimated, Fitzer *et al.* (1994)
3.79	$CH_2NOH + M = HCN + H_2O$	$0.100E + 12$	0.00	0.00	Estimated, Fitzer *et al.* (1994)

the ammonia destruction process and in the subsequent removal of nitric oxide by NH_2, NH and N radicals is presented below. Comparisons with flames reported in the earlier study by Lindstedt *et al.* (1994) are also made. A schematic representation of the principal reaction paths for ammonia oxidation to molecular nitrogen and nitrous oxide is outlined in Figure 1.

Ammonia Breakdown

Ammonia consumption in the present nitric oxide supported ammonia flames occurs predominantly via reaction with the H radical (R2.2), whereas in ammonia-oxygen flames it is dominated by OH (R2.3) radical attack (Lindstedt *et al.*, 1994). However, the reaction with the OH radical remains important in the present flame and accounts for about 35% of the total NH_3 consumption.

$$NH_3 + H = NH_2 + H_2 \tag{R2.2}$$

$$NH_3 + OH = NH_2 + H_2O \tag{R2.3}$$

Amidogen predominantly reacts with nitric oxide- the primary oxidant in the present flame. However, H radical attack (R2.5) also contributes significantly towards NH_2

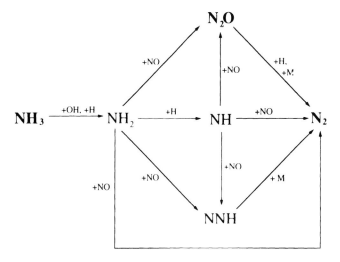

FIGURE 1   Schematic representation of major ammonia destruction paths for flame F1AP.

consumption ($\sim$ 20% of the total NH_2 destruction).

$$NH_2 + H = NH + H_2 \tag{R2.5}$$

$$NH_2 + NO = N_2 + H_2O \tag{R2.11}$$

$$NH_2 + NO = NNH + OH \tag{R2.12}$$

$$NH_2 + NO = N_2O + H_2 \tag{R2.13}$$

The specific contributions of the individual reaction channels for the amidogen reactions with nitric oxide (R2.11, R2.12 and R2.13) are around 20, 45 and 5%, respectively. In ammonia flames the process of NH_2 destruction is dominated by the H radical in fuel-rich and stoichiometric flames and by the O radical in lean flames. The contribution of reactions involving NO is much less significant and amounts to between 10 and 15%.

The imidogen destruction process is also dominated by reactions with nitric oxide and yields predominantly nitrous oxide and molecular nitrogen.

$$NH + NO = N_2O + H \tag{R2.28}$$

$$NH + NO = N_2 + OH \tag{R2.29}$$

$$NH + NO = NNH + O \tag{R2.30}$$

Among the above reactions R2.28 is the principal NH consumption channel and responsible for about 47% of the total consumption. The reminder of the imidogen passes through reactions R2.29 and R2.30 which contribute around 27 and 17% respectively. In ammonia-oxygen flames the imidogen consumption is dominated by the OH radical, while in lean ammonia and nitric oxide doped hydrogen-oxygen flames the O radical is of primary importance. However, nitrous oxide formation through

R2.28 is always the most important secondary channel for NH destruction in pure ammonia flames (Lindstedt *et al.*, 1994). In ammonia doped hydrogen flames and rich ammonia flames, the formation of atomic nitrogen (N) via the reaction of imidogen with the hydrogen radical is dominant.

The NNH radical mainly undergoes thermal decomposition to produce molecular nitrogen and a hydrogen radical.

$$NNH + M = N_2 + H + M \qquad \text{(R2.43)}$$

The primary H radical formation channel is the OH radical attack on molecular hydrogen and the above reaction is the most important secondary reaction. Nitrosyl hydride (HNO) is primarily produced from the reaction of NNH with nitric oxide (R2.46) and is rapidly recycled back to NO through thermal decomposition (R2.55).

$$NNH + NO = N_2 + HNO \qquad \text{(R2.46)}$$

$$HNO + M = NO + H + M \qquad \text{(R2.55)}$$

In ammonia flames the formation path for HNO is very different and passes via the reaction of amidogen with the oxygen radical (Lindstedt *et al.*, 1994).

Nitric Oxide Destruction

Nitric oxide is consumed by reactions with NH_2 and NH radicals. Amidogen plays the dominant role in this process as imidogen formation is a secondary path among the parallel channels for amidogen consumption as outlined above. The total contribution of amidogen (R2.11, R2.12 and R2.13) to the nitric oxide removal process in the present mechanism is estimated to be more than 75%.

$$NH_2 + NO = N_2 + H_2O \qquad \text{(R2.11)}$$

$$NH_2 + NO = NNH + OH \qquad \text{(R2.12)}$$

$$NH_2 + NO = N_2O + H_2 \qquad \text{(R2.13)}$$

The individual net contributions of the above reactions (R2.11, R2.12 and R2.13) to nitric oxide consumption is around 21, 50 and 5% respectively. Nitric oxide consumption by imidogen is, however, still significant ($\sim 23\%$).

$$NH + NO = N_2O + H \qquad \text{(R2.28)}$$

$$NH + NO = N_2 + OH \qquad \text{(R2.29)}$$

$$NH + NO = NNH + O \qquad \text{(R2.30)}$$

The individual contributions of the above reactions (R2.28, R2.29 and R2.30) to nitric oxide consumption are around 12, 7 and 4% respectively. Since atomic nitrogen formation is insignificant in this flame its contribution to the NO consumption process is negligible ($< 2\%$).

Amidogen and imidogen also play a significant role in the nitric oxide removal process in pure ammonia and ammonia/nitric oxide doped hydrogen flames (Lindstedt *et al.,* 1994). However, the relative importance of NH_2 and NH depends entirely on combustion conditions such as the stoichiometry. It should also be mentioned that for ammonia doped hydrogen flames the contribution of atomic nitrogen (N) to the nitric oxide removal process is not negligible and indeed becomes dominant in rich flames. The NNH radicals formed during the process of NO removal by amidogen and imidogen also reacts with NO. However, this process has again no net effect on NO consumption as HNO is subsequently recycled back to NO via thermal decomposition.

Nitrous Oxide Formation and Destruction

Nitrous oxide is an important member of the nitrogen oxide family - e.g., due to its important role in stratospheric ozone layer destruction. As outlined above this species is formed during the process of NO reduction by amidogen and imidogen (see also Miller and Bowman, 1989; Vandooren, 1992; Lindstedt *et al.,* 1994) and is controlled ($\sim 70\%$) by the reaction of imidogen with nitric oxide (R2.28). The same also holds true for ammonia-oxygen and ammonia-nitric oxide doped hydrogen-oxygen flames as discussed by Lindstedt *et al.* (1994). The destruction of nitrous oxide is dominated by the H radical. In the present flame the contribution of reaction R2.50 is about 65% with the thermal decomposition responsible for the reminder.

$$N_2O + H = N_2 + OH \tag{R2.50}$$

$$N_2O + M = N_2 + O + M \tag{R2.53}$$

This situation is similar to that encountered in ammonia and nitric oxide doped hydrogen-oxygen flames (Lindstedt *et al.,* 1994).

Molecular Nitrogen Formation

Molecular nitrogen is the major nitrogenous product of ammonia oxidation and is principally formed through reactions R2.11 ($\sim 25\%$) and R2.43 ($\sim 27\%$) as outlined above. However, nitrous oxide destruction channels (R2.50) and (R2.53) also play an important role in molecular nitrogen formation and their respective contributions are around 12 and 7%. Imidogen reactions with nitric oxide (R2.29) also produces molecular nitrogen and accounts for about 8% of the total N_2 formation. The additional NNH destruction channels (R2.45, R.246) also make significant contributions ($\sim 7\%$ each) towards molecular nitrogen formation.

$$NNH + NH_2 = N_2 + NH_3 \tag{R2.45}$$

$$NNH + NO = N_2 + HNO \tag{R2.46}$$

The reaction of NH_2 with NO is also a major contributor to N_2 formation in ammonia-oxygen flames. However, in these flames reaction R2.12 is dominant and hence N_2 is formed predominantly through the NNH intermediate. The dominance of NNH formation over direct N_2 formation (R2.11) can be attributed to the higher flame temperature (> 2000 K). In stoichiometric and rich ammonia doped hydrogen flames, the ammonia destruction process mainly passes via hydrogen abstraction reactions to produce atomic nitrogen and eventually molecular nitrogen through the Zel'dovich mechanism (Zel'dovich *et al.*, 1947).

Comparison with Experimental Data

The present reaction scheme predicts the experimental profiles for all reactants and major products well. Comparisons of predictd and experimental profiles of ammonia nitric oxide are shown in Figure 2 and the predicted decay rates for both are in good agreement with the measurements. Predicted and experimental profiles for molecular nitrogen can be seen in Figure 3 and the agreement is excellent. Similar agreement is also obtained for molecular hydrogen and water. Predicted concentration profiles of important intermediates, N_2O and NH_2, also agree well with measured profiles as shown in Figure 4 though an under-prediction of the peak NH_2 concentration by about 25% is noted. The latter can partly be attributed to the rate adopted for the NO removal reaction "$NH_2 + NO$ = products". However, the present rate and branching ratio for this channel has been found to give arguably the best possible agreement of predicted and experimental NO profiles for a wide range of ammonia-oxygen and ammonia/nitric oxide doped hydrogen-oxygen-argon flames (Lindstedt *et al.*, 1994). The agreement obtained by Vandooren *et al.*, (1994) was similar though the current predictions for N_2O are much improved.

An under-prediction of the H radical by about a factor 2 is, however, observed. To investigate this further a calculation adopting the $N_2 + H + OH$ product distribution for reaction R2.12— as proposed by Vandooren *et al.* (1994)— was made. Some

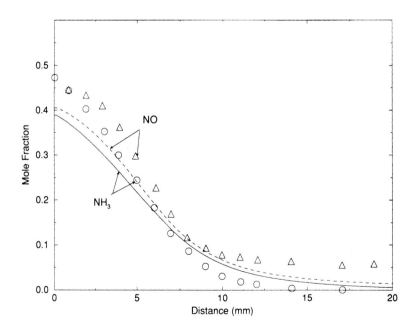

FIGURE 2   Comparison of predicted ammonia and nitric oxide profiles with experimental data from Vandooren *et al.* (1994) for flame F1AP.

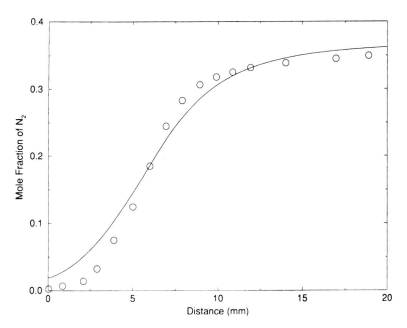

FIGURE 3   Comparison of the predicted molecular nitrogen profile with experimental data from Vandooren *et al.* (1994) for flame F1AP.

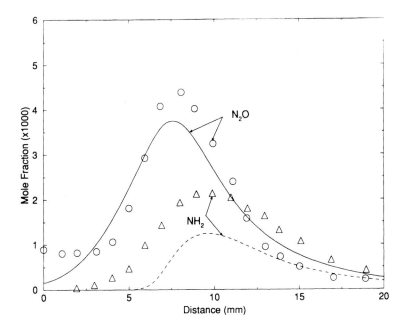

FIGURE 4   Comparison of predicted nitrous oxide and amidogen profiles with experimental data from Vandooren *et al.* (1994) for flame F1AP.

improvement was noted, though the H radical profile remained under-predicted by about 35%. It is thus unlikely that this reaction is the sole reason for the observed discrepancy. Also, the latter product distribution for R2.12 results in over-predictions of burning velocities for ammonia-oxygen flames by around 30% (Lindstedt *et al.*, 1994). While the formation of NNH and OH has also been preferred over the product distribution $N_2 + OH + H$ by numerous kinetic studies (Andersen *et al.*, 1983; Silver and Kolb, 1987; Dolson, 1986; Atakan *et al.*, 1989b; Bulatov, 1989; Miller and Bowman, 1989; Stephens *et al.*, 1993 and Diau *et al.*, 1994) it is quite possible that the $N_2 + OH + H$ channel is important and consequently further work is highly desirable. In this context is should also be noted that present mechanism accurately predicts the experimental NNH concentrations for ammonia-oxygen flame and NH_3/NO doped hydrogen-oxygen-argon flames as shown by Lindstedt *et al.* (1994).

NITRIC OXIDE REDUCTION BY AMMONIA IN FLOW REACTORS

It has been experimentally observed that NO removal through SNCR works efficiently in oxygen containing flue-gases within a narrow temperature window centred around 1200 K (Gebel *et al.*, 1989). The location of the SNCR temperature window is sensitive to the flue-gas composition and it is therefore essential to know its dependence on chemical species in order to determine the ammonia injection point in a practical combustor. A series of flow reactor calculations for a wide range of temperatures (800 end as 1400 K) have been made to investigate the ability of the present N/H/O reaction mechanism to adequately model the SNCR temperature window. The model predictions are compared with experimental data obtained in a flow reactor by Muris (1993). The starting gas mixture composition contained N_2 doped with 500 ppm of NO, 750ppm of NH_3 and 20000ppm of O_2. The mechanism responsible for the NO reduction and its temperature dependence can be understood by a study of the ammonia destruction process at different temperatures.

Ammonia breakdown does not occur at lower temperatures as the formation of radicals is initiated through the thermal decomposition of ammonia itself (R2.1). This process commences close to a temperature of 1100 K and results in fast amidogen production accompanied by a large NO reduction. Subsequent ammonia destruction is found to be mainly through reactions with OH and O radicals

$$NH_3 + M = NH_2 + H + M \qquad (R2.1)$$

$$NH_3 + OH = NH_2 + H_2O \qquad (R2.3)$$

$$NH_3 + O = NH_2 + OH \qquad (R2.4)$$

The dominant channel is through reaction R2.3 which contributes about 85% of the total ammonia consumption at 1100 K.

Amidogen acts as the principal NO reduction element over the studied temperature range. At the cold side of the temperature window (1100 K) about 96% of the amidogen is consumed through reactions with NO resulting in a corresponding reduction in nitric oxide.

$$NH_2 + NO = N_2 + H_2O \qquad (R2.11)$$

$$NH_2 + NO = NNH + OH \qquad (R2.12)$$

The contributions of reactions R2.11 and R2.12 to NH_2 removal are around 61 and 35% respectively and formation of imidogen is negligible. However, reactions of O and HO_2 do contribute to amidogen destruction.

$$NH_2 + O = HNO + H \qquad (R2.8)$$

$$NH_2 + HO_2 = NH_3 + O_2 \qquad (R2.16)$$

The NNH radical formed through reaction R2.12 mainly decomposes to molecular nitrogen and the hydrogen radical.

$$NNH + M = N_2 + H + M \qquad (R2.43)$$

From the above discussion it is clear that NO depletion principally occurs through reactions with amidogen. At higher temperatures reaction R2.11 becomes slower due to its negative temperature dependency and at the same time NO formation through the HNO intermediate (R2.8) becomes more active. This results in a lowering of the NO removal rate and consequently the NO reduction efficiency starts to decrease with an increase in temperature. The present reaction mechanism can predict the SNCR temperature window well as shown in Figure 5. However, the predicted maximum

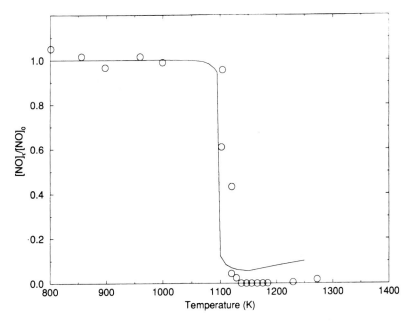

FIGURE 5   Comparison of the predicted SNCR temperature window with experimental results obtained in a flow reactor (FR1) by Muris (1993).

reduction efficiency is somewhat lower than that measured. The latter discrepancy can, at least partly, be attributed to the uncertainties associated with the temperature dependencies of the individual reaction channels for the reaction "$NH_2 + NO$ = products".

NITRIC OXIDE REDUCTION BY AMMONIA IN FLOW REACTORS IN THE PRESENCE OF HYDROCARBONS

The effects on the SNCR chemistry of hydrocarbon addition to the flue-gas is of obvious interest since most practical combustors use conventional fossil fuels. The present validation has therefore been extended to the C/N/H/O system by adding 100 ppm of ethane to the former gas mixture. This addition of ethane affects the SNCR temperature window significantly by shifting the cold boundary of the window towards lower temperatures by about 50 K (1050 K) and narrowing down the width of the window (Muris, 1993). The shift of the temperature window can be attributed to the earlier initiation of radical formation through the thermal decomposition of ethane and the agreement between the predicted and measured shift is excellent as can be seen in Figure 6. The presence of ethane results in significantly increased radical concentrations, e.g., factors of three for H and O and a factor of six for OH at 1100 K, when compared to the undoped case.

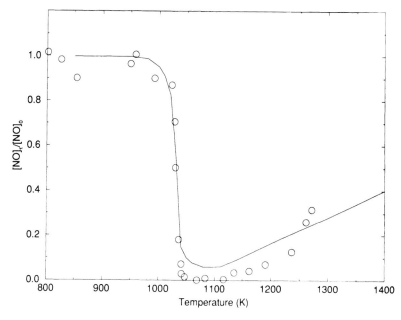

FIGURE 6   Comparison of predicted SNCR temperature window with experimental results obtained in a flow reactor (FR2) by Muris (1993).

AMMONIA OXIDATION IN COUNTERFLOW DIFFUSION FLAMES

A study of counterflow diffusion flames of methane and carbon monoxide doped with ammonia is of particular interest as it permits an extension of the present work to include the effects of carbon containing elements on ammonia oxidation in flames. Calculations have been conducted for both fuels with ammonia blended with the fuel and oxidiser streams- giving a total of four flames (Flame F1AD-1BD and F2AD-2BD). A schematic representation of major reactions in an ammonia doped counterflow methane-oxygen diffusion flame can be found in Figure 7. The corresponding experimental data was obtained from Hahn nad Wendt (1981) and Hahn *et al.* (1981) respectively.

Methane Breakdown

For methane flames (F1AD and F1BD) the fuel consumption is dominated by H and OH radical attack accounting for around 55 and 40% of methyl formation respectively.

$$CH_4 + H = CH_3 + H_2 \tag{R1.119}$$

$$CH_4 + OH = CH_3 + H_2O \tag{R1.121}$$

Subsequent methyl radical consumption proceeds predominantly through reaction with the O radical (R1.85) while reactions with OH (R1.87) and O_2 (R1.91) form secondary channels.

$$CH_3 + O = CH_2O + H \tag{R1.85}$$

$$CH_3 + OH = {}^1CH_2 + H_2O \tag{R1.87}$$

$$CH_3 + O_2 = CH_2O + OH \tag{R1.91}$$

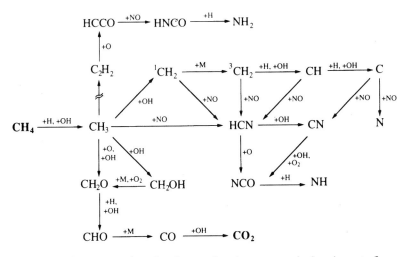

FIGURE 7 Schematic representation of major reactions in an ammonia doped counterflow methane-oxygen diffusion flame.

Formaldehyde consumption principally proceeds through reaction with H and OH radicals to produce the formyl radical.

$$CH_2O + H = CHO + H_2 \tag{R1.76}$$

$$CH_2O + OH = CHO + H_2O \tag{R1.78}$$

The formyl radical subsequently decomposes to carbon monoxide. The main carbon monoxide conversion reaction is via OH radical attack (R1.23)

$$CO + OH = CO_2 + H \tag{R1.23}$$

The latter also applies to the carbon monoxide flames (F2AD and F2BD).

Ammonia Breakdown and Nitric Oxide formation

In the present diffusion flames ammonia destruction leads solely to NH_2- mainly through OH radical attack.

$$NH_3 + H = NH_2 + H_2 \tag{R2.2}$$

$$NH_3 + OH = NH_2 + H_2O \tag{R2.3}$$

The exception is the carbon monoxide flame with ammonia doped in the fuel stream (Flame F2AD) where H radical attack (R2.2) becomes the dominant path. Amidogen destruction paths lead to NH, HNO, NNH and N_2.

$$NH_2 + H = NH + H_2 \tag{R2.5}$$

$$NH_2 + OH = NH + H_2O \tag{R2.6}$$

$$NH_2 + O = HNO + H \tag{R2.8}$$

$$NH_2 + NO = N_2 + H_2O \tag{R2.11}$$

$$NH_2 + NO = NNH + OH \tag{R2.12}$$

Among the above reactions the H radical attack (R2.5) is dominant when ammonia is injected from the fuel side- irrespective of whether the fuel is carbon monoxide or methane. For the methane flame with an ammonia doped oxidiser stream R2.6 becomes the dominant channel, while for the corresponding carbon monoxide flame R2.8 becomes the principal NH_2 consumption path. For both of the latter flames R2.11 becomes comparable.

Imidogen (NH) consumption leads to N, NO and HNO mainly through the following reactions.

$$NH + H = N + H_2 \tag{R2.20}$$

$$NH + OH = N + H_2O \tag{R2.23}$$

$$NH + OH = HNO + H \tag{R2.24}$$

$$NH + O_2 = NO + OH \tag{R2.25}$$

For the methane flame with the ammonia doped fuel stream (F1AD) the formation of HNO through R2.24 is found to be the dominant channel for NH destruction while NO formation through the reaction with O_2 is found to be dominant in flames with ammonia doped oxidant streams (F1BD, F2BD). However, for the carbon monoxide flames with the ammonia doped fuel stream (F2AD) hydrogen radical attack (R2.20) becomes the principal NH destruction path.

Atomic nitrogen is predominantly converted to NO and N_2 via the Zel'dovich mechanism. Atomic nitrogen (N) conversion to nitric oxide occurs predominantly through reaction with OH for all the flames with the exception of the carbon monoxide flame with ammonia injected in the oxidiser stream where reaction with molecular oxygen becomes dominant. The importance of the Zel'docich mechanism is greatest in Flame F2AD as a consequence of the dominance of H radical attacks in the NH_3 break-up sequence (R2.2, R2.5 and R2.20). Thus ammonia consumption leads to atomic nitrogen before it is converted to NO or N_2. Reaction path analyses for all four flames show that direct reactions of NH_i fragments with hydrocarbon species are not important.

Nitrosyl hydride (HNO) leads to NO through reactions with OH, O_2 and H

$$HNO + H = NO + H_2 \qquad\qquad (R2.56)$$

$$HNO + OH = NO + H_2O \qquad\qquad (R2.57)$$

$$HNO + O_2 = NO + HO_2 \qquad\qquad (R2.62)$$

The reaction with H dominates the HNO break-up process in the carbon monoxide flame with the ammonia doped fuel stream. The reaction with O_2 becomes dominant with ammonia doped oxidiser streams and the reaction with the OH radical (R2.57) dominates the HNO consumption process in methane flames with ammonia injected in the fuel stream.

Nitric Oxide Removal and Molecular Nitrogen Formation

Nitric oxide removal occurs through reactions with NH_i and hydrocarbon fragments. For both carbon monoxide flames and the methane flame with ammonia doped in the oxidiser stream the nitric oxide removal process is dominated by the NH_i radicals.

$$NH_2 + NO = N_2 + H_2O \qquad\qquad (R2.11)$$

$$NH_2 + NO = NNH + OH \qquad\qquad (R2.12)$$

$$NH + NO = N_2O + H \qquad\qquad (R2.28)$$

$$NH + NO = N_2 + OH \qquad\qquad (R2.29)$$

$$N + NO = N_2 + O \qquad\qquad (R2.34)$$

For the carbon monoxide flame with the fuel stream doped with ammonia the dominant NO removal reactions are R2.28 (15%) and R2.34 (64%). For the case of the ammonia doped oxidiser stream the situation is quite different. In this case the major contributions are through R2.11 (44%) and R2.34 (30%). A similar trend was also observed for the methane flame with ammonia doped in the oxidiser stream. In this flame reactions R2.11 and R2.12 are the major contributors and account for 40 and 25% respectively and the reminder passes via reactions with hydrocarbon fragments. However, for the methane flame with ammonia dopant in the fuel stream the relative importance of NO consumption reactions is different. For this flame only 20% of the total NO removal passes through NH_i fragments will R2.28 and R2.34 being the major contributors with 6 and 7% respectively. It is rather interesting to observe this significant change in the NO consumption channels.

Hydrocarbon fragments clearly play an important role in the NO conversion process in methane flames. The overall contribution of hydrocarbon fragments to NO removal is about 80 and 35% respectively for methane flames with ammonia injected in the fuel and oxidiser streams. The reactions making significant contributions are outlined below.

$$C + NO = N + CO \qquad (R3.1)$$

$$C + NO = CN + O \qquad (R3.2)$$

$$CH + NO = HCN + O \qquad (R3.3)$$

$$CH_3 + NO = HCN + H_2O \qquad (R3.7)$$

$$CHO + NO = HNO + CO \qquad (R3.8)$$

$$HCCO + NO = HNCO + CO \qquad (R3.11)$$

$$HCCO + NO = HCN + CO_2 \qquad (R3.12)$$

For the case with ammonia in the fuel stream the reactions with HCCO, CH, C, CH_3 and CHO contribute around 26, 20, 12, 5 and 5% to the total NO removal respectively. Clearly, the reactions involving the ketenyl radical constitute a major path, though the contribution of these reactions drops to around 10% if ammonia is injected with the oxidizer stream. The rates and product channels for the reaction of HCCO with NO have recently been investigated by Boullart *et al.* (1994) and Nguyen *et al.* (1994) and the present study has adopted the results of these investigations which have important implications for reburn chemistry.

HNCO produced through the NO removal process is predominantly consumed through H radical attack (R3.63) and hydrogen cyanide consumption is dominated by OH (53%) and O radical (34%) attack.

$$HNCO + H = NH_2 + CO \qquad (R3.63)$$

$$HCN + O = NCO + H \qquad (R3.39)$$

$$HCN + OH = CN + H_2O \qquad (R3.40)$$

The cyano radical is oxidised to NCO via the following reactions.

$$CN + OH = NCO + H \qquad (R3.46)$$

$$CN + O_2 = NCO + O \qquad (R3.47)$$

The isocyanato radical (NCO) principally proceeds to NO formation (85%) and the NH radical (14%).

$$NCO + H = NH + CO \qquad\qquad (R3.55)$$

$$NCO + O = NO + CO \qquad\qquad (R3.56)$$

$$NCO + OH = NO + CO + H \qquad\qquad (R3.57)$$

Reaction R3.56 dominates the NCO consumption and contributes about 47% of the total consumption rate. From the above discussion it is clear that most of the NO consumed by hydrocarbon fragments is in fact recycled back to NO.

Methane Flames

Calculations of the methane flames were initially made only by using the rate of strain reported by Wendt and co-workers ($3.62\,s^{-1}$). However, these computations resulted in significantly wider temperature profiles compared to those obtained experimentally. Calculations were therefore in addition performed at a higher rate of strain to match the temperature profile more closely. Predicted results corresponding to a rate of strain of $10\,s^{-1}$ were found to be suitable for all cases-including the carbon monoxide flames. The agreement obtained for the major species profiles was also found to be somewhat improved with the higher rate of strain. Comparisons of predicted temperature and major species profiles (CH_4, O_2, CO_2, H_2O) for Flame F1AD show good agreements with experimental results. For example, a comparison of predicted CH_4 and O_2 profiles is shown in Figure 8. The agreement of predicted results with experimental data for NO

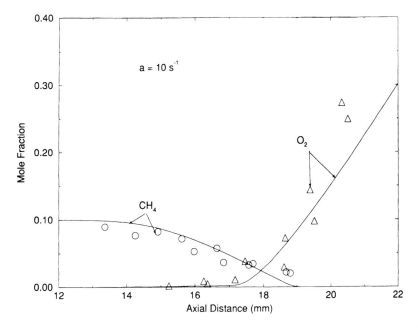

FIGURE 8   Comparison of predicted CH_4 and O_2 profiles with experimental results of Hahn and Wendt (1981) for flame F1AD.

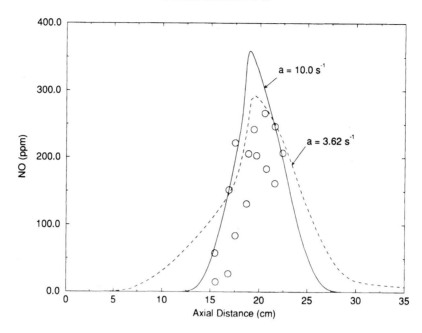

FIGURE 9   Comparison of predicted NO profiles with experimental results of Hahn and Wendt (1981) for flame F1AD.

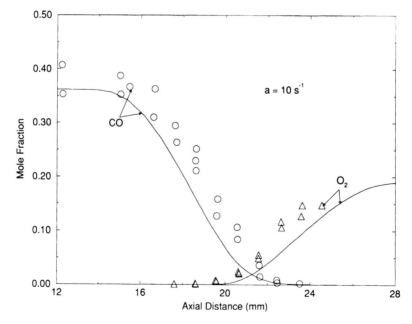

FIGURE 10   Comparison of predicted CO and O_2 profiles with experimental results of Hahn *et al.* (1981) for flame F2BD.

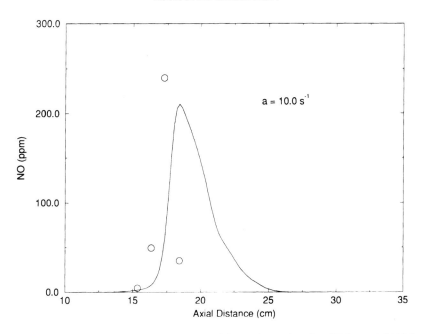

FIGURE 11   Comparison of predicted NO profiles with experimental results of Hahn *et al.* (1981) for flame F2AD.

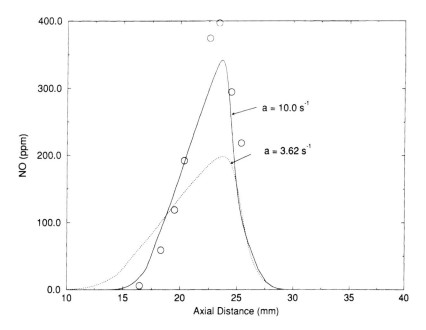

FIGURE 12   Comparison of predicted NO profiles with experimental results of Hahn *et al.* (1981) for flame F2BD.

is good as can be seen in Figure 9. In this Figure the NO profile obtained with a rate of strain $3.62 s^{-1}$ is also shown to indicate the width of reaction zone with the experimentally reported strain rate. Similar agreement of predicted results with the experimental data is also obtained for Flame F1BD.

Carbon Monoxide Flames

Comparisons of predicted CO and O_2 decay profiles for flame F2BD with experimental profiles show good agreement as can be seen in Figure 10. The agreement between measured and predicted peak NO concentrations for flames F2AD and F2BD is excellent as shown in Figures 11 and 12. Wendt and co-workers also predicted these flames and their agreement was similar except for Flame F2AD where an underprediction of the peak NO concentration by an order of magnitude was observed.

CONCLUSIONS

In the present study an extensive amount of information on kinetic data for elementary reactions related to SNCR and reburn chemistry has been assessed and a detailed reaction mechanism constructed and validated. It has been found that while hydrocarbon radicals have only a minor direct influence on the ammonia destruction process, the indirect influence via the H, O and OH radical pool can be substantial to the point where the relative importance of the NH_i radicals to the nitric oxide reduction process changes.

The overall agreement between predicted results and experimental data is very satisfactory. Generally it was found that reactions involving amidogen have a dominant role in the NO reduction process. However, for diffusion flames with ammonia injected with the oxidiser stream the contributions of imidogen and the nitrogen radical also become important, while for methane diffusion flames with ammonia injected with the fuel stream, the role of hydrocarbon fragments in the NO conversion process becomes dominant. It has also been shown that the ketenyl radical has a primary role in nitric oxide reduction in the presence of hydrocarbons.

During the present work some areas where additional kinetic information is needed to have identified. Of particular importance are rates and product distributions for reactions of nitric oxide with amidogen and the methyl radical. Similarly, the reaction of HCCO with NO has been shown to be of major influence and further work, particularly at higher temperatures, is highly desirable.

ACKNOWLEDGEMENTS

M. A. Selim gratefully acknowledges the financial support of the Association of Commonwealth Universities and the British Council through the Commonwealth Scholarship scheme.

REFERENCES

Andersen, P., Jacobs, A., Kleinermanns, C. and Wolfrum, J. *Nineteenth Symp. (Int.) on Combustion*, (1982) The Combustion Institute, 11.

Atakan, B., Jacobs, A., Wahl, M. Weller, R. and Wolfrum, J. (1989a) *Chem. Phys. Lett.*, **154**, 449.

Atakan, B., Jacobs, A., Wahl, M. Weller, R. and Wolfrum, J. (1989b) *Chem. Phys. Lett.*, **155**, 609.

Atakan, B., Kocis, D., Wolfrum, J. and Nelson, P. *Twenty-Fourth Symp. (Int.) on Combustion*, (1992) The Combustion Institute, 691.

Atkinson, R., Baulch, D. L., Cox, R. A., Hampson, R. F., Jr., Kerr, J. A. and Troe, J. (1992) *J. Phys. Chem. Ref. Data*, **21**, 1125.

Baulch, D. L., Cobos, C. J., Cox, R. A., Esser, C., Frank, P., Just, Th., Kerr, J. A., Pilling, M. J., Troe, J., Walker, R. W. and Warnatz, J. (1992) *J. Phys. Chem. Ref. Data*, **21**, 411.

Bian, J., Vandooren, J. and Van Tiggelen, P. J. *Twenty-first Symp. (Int.) on Combustion*, (1986) The Combustion Institute, 953.

Bian, J., Vandooren, J. and Van Tiggelen, P. J. *Twenty-third Symp. (Int.) on Combustion*, (1990) The Combustion Institute, 379.

Bosco, S. R., Nava, D. F., Brobst, W. D. and Stief, L. J. (1984) *J. Chem. Phys.*, **81**, 3505.

Boullart, W., Nguyen, M. T. and Peeters, J. (1994) *J. Phys. Chem.*, **98**, 8036.

Bozzelli, J. W., Cheng, A. YY. and Dean, A. M. Presented in *Twenty Fifth Symp. (Int.) on Combustion*, (1994) The Combustion Institute.

Branch, M. C., Kee, R. J. and Miller, J. A. (1982) *Combustion, Science and Technology*, **29**, 147.

Bulatov, V. P., Ioffe, A. A., Lozovsky, V. A. and Sarkisov, O. M. (1989) *Chem., Phys. Lett.*, **161**, 141.

Bulewicz, E. M. and Sugden, T. M. (1964) *Proc. Roy. Soc. London*, **A277**, 143.

Cohen, N. and Westberg, R. K. (1991) *J. Phys. Chem. Ref. Data*, **20**, 1211.

Cooper, W. F. and Hershberger, J. F. (1992) *J. Phys. Chem.*, **96**, 771.

Dean, A. J., Hanson, R. K. and Bowman, C. T. (1991) *J. Phys. Chem.*, **95**, 3180.

Diau, E. W., Yu, T., Wagner, M. A. G. and Lin, M. C. (1994) *J. Phys. Chem.*, **98**, 4034.

Dolson, D. A. (1986) *J. Phys. Chem.*, **90**, 6714.

Drake, M. C. and Blint, R. J. (1991) *Combustion, Science and Technology*, **75**, 261.

Ehbrecht, J., Hack, W., Rouveirolles, P. and Wagner, H. Gg (1987) *Ber. Bunsenges Phys. Chem.*, **91**, 700.

Fennimore, C. P. and Jones, G. W. (1961) *J. Phys. Chem.*, **65**, 298.

Fitzer, J., Lindstedt, R. P., Muris, S., Sick, V. and Wolfrum, J. Work in Progress Poster 93. Twenty-fifth Symp. (Int.) on Combustion (1994).

Fujii, N., Kakuda, T., Sugiyama, T. and Miyama, H. (1985) *Chem. Phys. Lett.*, **112**, 489.

Fujii, N., Miyama, H., Koshi, M. and Asaba, T. *Eighteenth Symp. (Int.) on Combustion*, (1981) The Combustion Institute, 873.

Gardiner, W., Lissianski, V., Okoroanyanwu, U., Yang, H. and Zhao, M. *Proc. Second International Conference on Combustion Technologies for A Clean Environment*, (1993) II, 16.

Gebel, K., Madlsperger, G., Hein, K. and Bokenbrink, K. D. (1989) *Proc. VGB Conference Powerplants and Environment*, Essen, Federal Republic of Germany, 173.

Glaenzer, K. and Troe, J. (1974) *Ber. Bunsenges Phys. Chem.*, **78**, 182.

Grotheer, H. H., Kelm, S., Driver, H. S.T., Hutcheon, R. J., Lockett, R. D. and Robertson, G. N. (1992) *Ber. Bunsenges Phys. Chem.*, **96**, 1360.

Hack, W., Kurzke, H., Rouveirolles, P. and Wagner, H. Gg. *Twenty-first Symp. (Int.) on Combustion*, (1988) The Combustion Institute, 905.

Hahn, W. A. and Wendt, J. O. L. *Eighteenth Symp. (Int.) on Combustion*, (1981) The Combustion Institute, 121.

Hahn, W. A., Wendt, J. O. L. and Tyson, T. J. (1981) *Combustion, Science and Technology*, **27**, 1.

Hanson, R. K. and Salimian, S. (1984) In *Combustion Chemistry*, W. C. Gradiner, Jr. (Ed) Springer-Verlag, NY.

He, Y., Lin, M. C., Wu, C. H. and Melius, C. F. *Twenty-fourth Symp. (Int.) on Combustion*, (1992) The Combustion Institute, 711.

Henning, G. and Wagner, H. Gg. (1994) *Ber. Bunsenges. Phys. Chem.*, **98**, 749.

Jones, W. P. and Lindstedt, R. P. (1988a) *Combustion and Flame*, **73**, 233.

Jones, W. P. and Lindstedt, R. P. (1988b) *Combustion, Science and Technology*, **61**, 31.

Kee, R. J., Rupley, F. R. and Miller, J. A. (1987) *The CHEMKIN Thermodynamic Data Base*, Sandia Rep. SAND 87–8215.

Leung, K. M. and Lindstedt, R. P. (1995) *Combustion and Flame*, **102**, 129.

Levy, E. M. and Winkler, C. A. (1962) *Can. J. Chem.*, **40**, 686.

Lindackers, D., Burmeister, M. and Roth, P. *Twenty-third Symp. (Int.) on Combustion*, (1990) The Combustion, Institute, 257.

Lindstedt, R. P., Lockwood, F. C. and Selim, M. A. (1994) *Combustion, Science and Technology*, **99**, 253.

Louge, M. Y. and Hanson, R. K. *Twentieth Symp (Int.) on Combustion*, (1984) Institute, 665.

Lyon, R. K. (1973) Method for the Reduction of Concentration of NO in Combustion Effluents using Ammonia, U.S. Patent 3,900,544.

Lyon, R. K. and Benn, D. J. *Seventh Symp. (Int.) on Combustion*, (1979) The Combustion Institute, 601.

MacLean, D. I. and Wagner, H. Gg. (1967) *Eleventh Symp. (Int.) on Combustion*, The Combustion Institute, 871.

Marshall, P., Ko, T and Fontijn, A. (1989) *J. Chem. Phys.*, **93**, 1922.

Marston, G., Nesbitt, F. L. and Stief, L. J. (1989) *J. Chem. Phys.*, **91**, 3483.

Mason, E. A. and Monchick, L. (1962) *J. Chem. Phys.*, **36**, 1622.

Mertens, J. D. Kohse-Hoinghaus, C. Hanson, R. K. and Bowman, C. T. (1991) *Int. J. Chem. Kinet.*, **23**, 655.

Messing, T. Filseth, S. V. Sadowski, C. M. and Carrington, T. (1981) *J. Chem. Phys.*, **74**, 3874.

Miller, J. A. and Bowman, C. T. (1989) *Prog. Energy. Combustion and Science*, **15**, 287.

Miller, J. A. and Melius, C. F. *Twenty-fourth Symp. (Int.) on Combustion*, (1992) The Combustion Institute, 719.

Miller, J. A. Branch, M. C. and Kee, R. J. (1981) *Combustion and Flame*, **43**, 81.

Miller, J. A., Branch, M. C., MacLean, W. J., Chandler, D. W., Smooke, M. D. and Kee, R. J. (1984) *Twentieth Symp (Int) on Combustion.*, The Combustion Institute, 673.

Miller, J. A. Smooke, M. D. Green, R. M. and Kee, R. J. (1983) *Combustion, Science and Technology*, **34**, 149.

Mulvihill, J. N. and Phillips, L. F. *Fifteenth Symp (Int.) on Combustion*, (1975) The Combustion Institute, 1113.

Muris, S. (1993) Ph.D. Thesis. Ruprecht Karls Universitaet Heidelberg, Germany.

Natarajan, K. and Roth. P. *Twenty-first Symp. (Int.) on Combustion*, (1988) The Combustion Institute, 729.

Nguyen, M. T., Boullart, W. and Peeters, J. (1994) *J. Phys. Chem.*, **98**, 8030.

Perry, R. A. and Melius, C. F. *Twentieth Symp.(Int.) on Combustion*, (1985) The Combustion Institute, 639.

Reid, R. C. and Sherwood, T. K. (1960) McGraw Hill, New York.

Sanders, W. A., Lin, C. Y. and Lin, M. C. (1987) *Combustion. Science and Technology*, **51**, 103.

Sarkisov, O. M., Cheskis, S. G., Nadtochenko, V. A., Sviridenkov, E. A. and Vendeneev, V. I. (1984) *Arch. Combust.*, **4**, 111.

Silver, J. A. and Kolb, C. E. (1987) *J. Phys. Chem.*, **91**, 3713.

Slack, M. W. and Grillo, A. R. (1981) *Combustion and Flame*, **40**, 155.

Stephens, J. W. Morter, C. L., Farhat, S. K. Glass, G. P. and Curl, R. F. (1993) *J. Phys. Chem.*, **97**, 8944.

Szekely, A., Hanson, R. K. and Bowman, C. T. (1984) *Int. J. Chem. Kinet.*, **16**, 1609.

Tsang, W. (1992) *J. Phys. Chem. Ref. Data*, **21**, 753.

Tsang, W. and Herron. J. (1991) *J. Phys. Chem. Ref. Data*, **20**, 609.

Vandooren, J. (1992) *Combustion, Science and Technology*, **84**, 335.

Vandooren, J., Bian, J. and Van Tiggelen, P. J. (1994) *Combustion and Flame*, **98**, 402.

Wagal, S. S., Carrington, T., Filseth, S. V. and Sadowski, C. M. (1982) *J. Chem. Phys.*, **69**, 61.

Zel'dovich, Ya. B., Sadovnikov, P. Ya. and Frank-Kamenetskii, D. A. (1947) *Oxidation of Nitrogen in Combustion*, Translated by M. Shelef, Academy of Science of the USSR, Moscow.

A Route to NO$_x$ Reduction via Flame Stretch

D. PROCTOR[a], I. G. PEARSON and S. A. BRUMALE
[a]*CSIRO Div. of Building, Construction and Eng., Highett VIC 3190 Australia*

Abstract—This paper describes how using acoustic techniques can substantially reduce NO$_x$ emissions from some types of burner. The increased turbulence and consequent reduction of both flame height and particle residence time as well as increased flame stretch, leads to lower flame temperatures and lower NO$_x$ production.

Some data are presented from the novel pulse combustion facility at the CSIRO, Division of Building, Construction and Engineering which show noteworthy NO$_x$ emission reduction. The distributions of NO$_x$ emissions in the flame are presented and whilst the local concentrations may be higher with larger flame stretch parameters, it is shown that the total global emission of NO$_x$ is generally reduced markedly. It is clear that the NO$_x$ production and destruction chemistry is being modified considerably by flame stretching and acoustic parameters.

Key Words: NO$_x$ chemistry, fluid dynamics, flame stretch, oscillations

INTRODUCTION

Various substitution strategies have been put forward as ways of ameliorating the global temperature increase which is likely to be making its presence felt early in the 21st centuary. LPG(propane) and natural-gas (methane) have been discussed as possible substitutes for other fuels with carbon to hydrogen ratios, such as coal, to reduce the amount of CO$_2$ emitted to the atmosphere from combustion. The trend has been to produce natural-gas burners that are very efficient and emit substantially lower levels of pollutants, such as nitrogen oxides N$_2$O, NO and NO$_2$, which in themselves contribute to global warming either directly in the case of N$_2$O, or through photochemical reactions involving unburnt hydrocarbons, NO$_x$ and UV light from solar radiation to produce O$_3$. Clearly NO$_x$ emissions from combustion systems are important.

The current research activity around the world into producing efficient low NO$_x$ burners has taken several different routes to try and achieve this aim. These include staged combustion, internal and external combustion gas recirculation, catalytic burners, metal and ceramic matrix burners and O$_2$ controlled combustion. This paper concentrates on another method that of changing the fluid dynamics to influence the chemistry of combustion, which in some instances feeds back to modify the fluid dynamics. The mechanism by which this occurs is via flame stretch which produces a cooler radical deficient flame. Two types of burner fall into this category – the pulse combustion burner and the precessing jet burner (Proctor *et al.* (1993)). Pulsed combustion devices appear to offer enormous potential from both aspects. This paper looks at the possible mechanisms as to how it is brought about in pulse combustion burners.

The effectiveness of most combustion systems depends on the ability to maintain a stable flame at reasonably high combustion efficiency as well as low pollutant emission. All aspects are closely related to the flow characteristics inside the combustor and the degree of mixing obtained of the injected fuel and the oxidant. The mixing involves several important processes. Large-scale structures bring into the mixing layer large amounts of reacting components from the two separated streams. The fine-scale eddies enhance the mixing at the molecular level, between the reactants, which is a necessary condition to initiate the chemical reactions. Unfortunately, intense fine-scale mixing produces intense, high temperature burning which results in high emissions of thermally generated NO_x. When a flame propagates in a non-uniform flow it experiences strain and curvature effects. The fractional rate of change of the flame area constitutes the flame stretch (Candel and Poinsot, 1990). The flame stretch parameter determines the available flame surface density, and consequently the reactivity of the fuel and oxidants.

Stretching of the flame and the body of the flame also controls some of the mixing and consequently the emission of pollutants. Large flame stretching parameters generally lead to lower temperatures and modified residence times and hence reduced emissions of thermally generated NO_x. The modification of the flame stretch can be achieved by modification of the fluid- dynamics. Besides changing the flame- stretch of the flame, the modified flow field can also lead to a change in the chemistry taking place by altering the chemical equlibrium. The understanding of these complex gas dynamic processes and their interactions requires analysis of the interactions between the fluid-dynamics, chemical reactions, acoustic waves, flame stretch and heat release of the reactive system (Broadwell and Dimotakis, 1986; Ballal, 1986).

EXPERIMENTAL APPARATUS AND FACILITIES

Experiments were carried out in a combustion research facility which allowed optical and other measurements to be made. The facility, shown diagramatically in Figure 1, comprised a vertical combustion tunnel around which was mounted an optical table capable of traversing in three mutually orthogonal directions. A two mirror schlieren system was used to visualize the gas flames and the resulting schlieren and Laser induced fluorescence images were captured electronically via a Charge Injection Device (CID) video camera and digitizing frame grabber (Pearson *et al.* (1991)).

The facility enabled us to examine independently the various factors which govern pulse combustion. In a pulse combustor, the frequency of operation and firing rate are essentially fixed, once the combustor and tail-pipe geometry have been fixed. In the experimental facility both the firing rate and the frequency can be varied independently over a wide range of operating conditions. This has enable us to look at what happens in a flame as it undergoes progressive stretching due to increasing the amplitude of the pulsations whilst maintaining the pulsation frequency and the firing rate constant. The facility is described in detail elsewhere (Proctor and Pearson (1991)). Has the following features:

1. a 248 mm square duct/combustion chamber with transparent quartz windows on three sides over a 500 mm length, coupled to a variable speed exhaust fan, which could vary the air flow rate from 0.04 to 8.0 m/s in the tunnel test section, in which was centrally mounted a concentric of tubes:

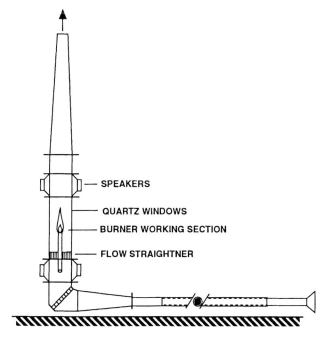

FIGURE 1   Experimental pulse combustion research facility.

(a) an acoustically stabilised jet of CH_4 fuel gas issuing from a 10 mm ID tube and 12.5 mm OD.

(b) acoustically stabilised co-flowing air supply in a 29.1 mm ID and 31.8 mm OD tube.

2. two sets of loud speakers above and below the movable air/fuel jet, with two physically opposed 205 mm diameter speakers in each set.

The speaker sets were operated in anti-phase such that the acoustic mode generated in the duct gave a pressure node, with a resultant velocity anti-node at the burner nozzle tip. By providing the speaker amplifiers with a preset waveform and frequency, the pulsations were made repetative and the process outcomes could be interogated over any desirable time frame. All the results presented here were made at a frequency of 58 Hz, since this was one of the natural resonant frequencies of the ductwork. The firing rate is constant at 27 kW in all images and figures.

When oscillatory flows are applied to a methane/air diffusion flame, as the amplitude of the pulsation is increased, the following pictures shown in Figure 2 emerge, starting off with zero amplitude or no forcing on the left-hand side. It is clear from these flame images that there is a considerable degree of stretching taking place as the amplitude of the oscillations is increased. If we now turn to the corresponding NO_x emissions. These have been aquired using conventional chemi-luminescent gas analysis equipment. Species analysis measurements were made using an Analgas brand unit with Horiba flame ionization, infrared gas analysers and polarographic oxygen analyser, and Thermo Electron chemi-luminescence detectors. The gas analysis equipment

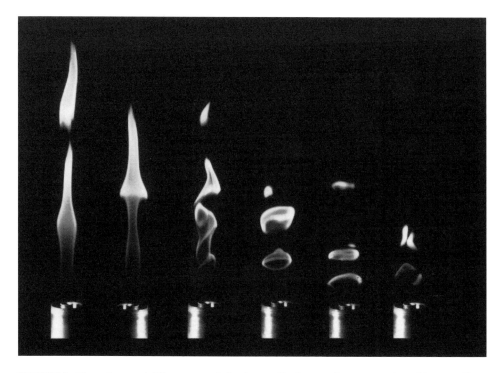

FIGURE 2   Flame images at different acoustic forcing amplitude, same frequency and gas firing rate. (See Color Plate II).

comprised of individual components dedicated to determine the concentrations of specific species in the sampled gases. The instrument was capable of monitoring levels of carbon monoxide (CO), carbon dioxide (CO_2), oxygen (O_2), total unreacted hydrocarbons (THC), nitric oxide (NO), and total oxides of nitrogen (NO_x). NO_2 concentrations were derived by subtracting NO concentrations from total NO_x concentrations. This method assumes, of course, that the total oxides of nitrogen are composed entirely of NO and NO_2, completely ignoring N_2O and other minor nitrogen species. Table I outlines the principal method of operation for each instrument and indicates the accuracy of the measurement. Representative samples of constituent gases were drawn from the flame area with a stainless steel suction probe. A water cooled condenser, coupled with a bed of calcium chloride dessicant, was used to remove water vapour from the sampled gases. Because of the nature of the gas analysis system, no attempts were made to phase resolve the measurements. All samples were made over many periods of the forcing function. The results consist of phase averaged local NO, NO_2 and NO_x concentrations.

RESULTS AND DISCUSSION

Contour maps of the temporal average of NO, NO_2 and total NO_x concentrations measured are presented in Figure 3. The top set are for the natural flame and the lower

TABLE I

Instrumentation for species analysis

Species	Description	Accuracy
CO	Non-dispersive infra-red	+ 1.0%
CO_2	Non-dispersive infra-red	+ 1.0%
O_2	Polarographic oxygen probe	+ 0.1%
THC	Flame ionization detector	+ 1.0%
NO	Ozone chemi-luminescence	+ 1.0%
NO_x	Ozone chemi-luminescence	+ 1.0%

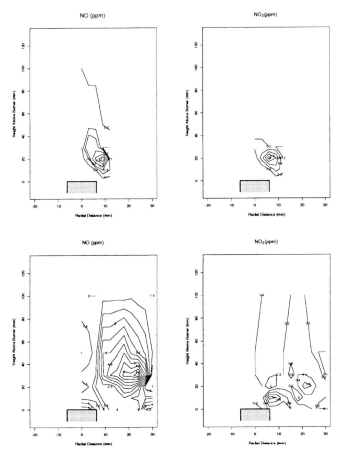

FIGURE 3 Unforced flame and flame forced at 20volts NO and NO_2 emissions.

set for the flame forced at 20 volts at 58 Hz. The left column are for NO and the right column are for NO_2. 58 Hertz was used for forcing as it coincided with one of the duct's primary modes, thus enabling high forcing amplitudes from the loudspeakers because of the low system impedance at that frequency. The flow rate of fuel was 7.05×10^{-4}

m^3/s, which corresponds to a Reynolds number (based on jet exit diameter) of 5960. Global emission concentrations, taken well downstream from the exhaust system, as a function of acoustic pulsation amplitude are presented in Table II. The global concentrations measured all show a marked decrease (up to 47%) with increasing amplitude of pulsation. The local figures show a localized increase in value, thus demonstrating the altered production and destruction mechanisms.

The applied forcing simulates the acoustic field in a self sustaining pulse combustor (Proctor and Pearson (1991)). The very large flame stretch rates thus obtained are clearly affecting the NO_x chemistry, both locally and globally. Drake (1985) showed that thermal NO_x levels in flames maximize at positions of maximum temperature and are linearly proportional to reaction times in high temperature zones, i.e., thermal NO_x levels decrease with flame stretch.

NO production and subsequent removal is rather complex but there are distinct sets of reactions producing NO interspersed with some other removal steps. The first of these is the Zeldovich thermal NO_x mechanism, (Zeldovich *et al.* (1985)).

$$N_2 + O \rightleftharpoons NO + N \tag{1}$$

$$O_2 + N \rightleftharpoons NO + O. \tag{2}$$

This Zeldovich mechanism is of most significance in the unforced flame. As the forcing amplitude increases, this mechanism will be of less consequence and the subsequent reaction rates will also be less because of lower temperature and lack of available radicals.

The main "Prompt" route to NO starts with either the formation of CN and N radicals via

$$C + N_2 \rightleftharpoons CN + N, \tag{3}$$

or reaction (4) according to Fenimore (1971)

$$CH + N_2 \rightleftharpoons HCN + N, \tag{4}$$

followed by a series of rapid chain reactions, Miller *et al.* (1985), terminating with reaction (8) below:

$$HCN + O \rightleftharpoons NCO + H \tag{5}$$

TABLE II

Global oxides of nitrogen emissions as a function of pulsation amplitude for a fixed firing rate and frequency at 3% oxygen

Amplitude driving voltage	NO (ppm)	NO_x (ppm)	NO_2 (ppm)
0.0	1.09	1.82	0.73
7.5	1.09	1.82	0.73
9.0	1.09	1.76	0.67
12.5	1.06	1.64	0.52
20.0	1.03	1.45	0.42
27.5	0.97	1.21	0.24

$$NCO + H \rightleftharpoons NH + CO \tag{6}$$

$$NH + H \rightleftharpoons N + H_2 \tag{7}$$

$$N + OH \rightleftharpoons NO + H. \tag{8}$$

Coupled with the last five reactions are a set of recycling reactions involving C, CH and the singlet and triplet states $^1CH_2$ and $^3CH_2$ in the removal of NO, favoured under fuel rich conditions of the unforced flame.

Removal of NO starts with the reverse of some of the above reactions. These then provide the necessary reactants for the next removal mechanism for NO, this being another of the "Prompt" recycle reactions.

$$CH + NO \rightleftharpoons HCN + O \tag{9}$$

$$H_2 + NO \rightleftharpoons HNO + H. \tag{10}$$

The last removal reactions are the reverse of reaction (8) above and the formation N_2O.

$$H + NO \rightleftharpoons N + OH \tag{12}$$

$$NH + NO \rightleftharpoons N_2O + H \tag{13}$$

$$NO + NO \rightleftharpoons N_2O + O. \tag{14}$$

Reactions (9) to (14) will be biased most strongly in the unforced case. This "prompt" recycle mechanism will occur less as the pulsation amplitude increases due to the lack of necessary radicals. The N_2O route to NO, the reverse of reactions (13) and (14), becomes more important as the temperature decreases, and the pressure increases. This source of NO is most important under conditions where the total NO formation is relatively low (Bowman (1992)).

Detailed chemical kinetic modelling of NO_2 formation in laminar flames (Sano (1982)) explains theoretically the formation and disappearance of NO_2 early in the preflame zones. Sano's model calculations of hot burnt gas mixing with cool air (1984) demonstrate that NO_2 is formed by:

$$NO + HO_2 \rightleftharpoons NO_2 + H \tag{15}$$

and destroyed through reactions:

$$NO_2 + H \rightleftharpoons NO + OH \tag{16}$$

$$NO_2 + OH \rightleftharpoons NO + HO_2 \tag{17}$$

There are two NO_x mechanisms occuring in flames of this character: Zeldovich "thermal" and "prompt". By pulsing the flame we have biased the production from "thermal" towards "prompt" as well as lowering the absolute values and rates. In the unforced flames the "prompt" NO mechanism is occuring before the "thermal" mechanism, but occurs after the "thermal" in the stretched flames.

To gain some understanding of how mixing parameters and structure formation affect flame stretch and flame chemistry, schlieren studies were made. Figure 4 a representative series of schlieren images taken at random times in the cycle. The sequence shows a series of vortical rings ("doughnuts") being emitted from the nozzle exit.

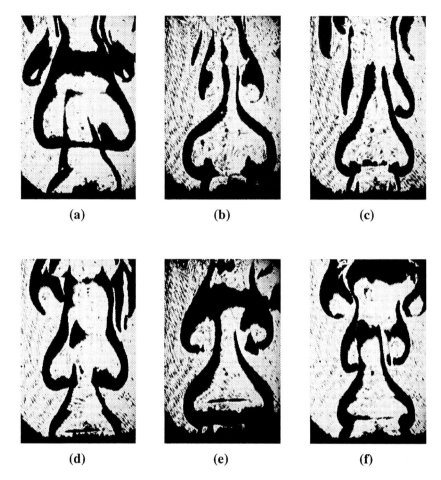

FIGURE 4   Schlieren images of the pulsed combustion flame.

Detailed study of the flame shape, both via direct viewing and recorded video, indicates a series of rings joined by thin filamental tubes. There is great stretching of the flame front between pulsations forming the cone shaped, inner flow. Each vortical ring emission corresponds to the peak amplitude of the gas exit plane acoustic particle velocity. The filamental flame front joining each vortical ring in a thin tube formation corresponds to the peak negative velocity at the nozzle exit and allows re-ignition of each subsequent fresh charge or pulse. Close study of several of the schlieren images shows a flat, horizontal, plate-like structure in the centre of the flow. It is believed that this corresponds to a "slug" of fuel being emitted before the shear layer instability causes it to roll up into a vortical ring. As Cabelli *et al.* (1988) showed, the acoustics of the system, both upstream and downstream, control very strongly the shape and form of structures in flames of this nature. Clearly, the chemical kinetics are also being modified by the acoustics, and hence the flame stretch, of the system.

Pindera and Talbot (1986) present a theoretical treatise suggesting, in general, that the velocity field in the burning regions is rotational, that is, flames produce vorticity. The circulation of the vortical flow depends linearly upon the flame stretch. It is not surprising, therefore, to find that increasing flame stretch by imposing pulsations leads to increased vorticity and hence better mixing. The changes in mixing and kinetics account for the modifications to the NO_x chemistry.

The rate of formation of NO_x's is normally given by equations of the form:

$$\frac{d[NO]}{dt} = \mathscr{A} \exp(E_0/\mathbf{R}T)[A]^a[B]^b\ldots - \mathscr{A}' \exp(E_0'/\mathbf{R}T)[NO]^{no}[\ldots]. \qquad (18)$$

with the assumption that $[NO]$ or $[\ldots]$, etc. ~ 0, (where A and B are those reactants on the LHS of reactions (1), (2) or (8) for example). There are a number of situations where this appears not to be true. Increasing the amplitude of the pulsations causes the flow field to trap the NO_x's at the point where they would form. Consequently, the effect of this is to negate the assumption of negligible $[NO]$ and suppress the formation of further NO_x's. It might almost be considered as a case of very local exhaust gas recirculation.

CONCLUSIONS

The following conclusions can be drawn from this work:

1. Local concentration of NO_x's increases while global concentration decreases as the flame is stretched more and more. One possible explanation is that the higher local concentration of NO and NO_2 are trapped and as a consequence supress the forward rate of the formation reactions of NO and NO_2 in the highly forced case. Whereas in the unforced case, the NO and NO_2 "leak" away to the next formation cell towards the tip of the flame.
2. Thermal NO_x decreases as the "prompt" NO_x increases as the flame is progressively stretched because the radical pool has been reduced as well as the flame temperature.
3. The "prompt" NO_x moves from a pre-flame to a post-flame production zone. The evidence for this is that the NO_x's tend to reside within the flame in the unforced case and on the air side in the highly forced flames.

REFERENCES

Ballal, D. R. (1986) Studies of turbulent flow-flame interactions. *AIAA Journal*, **24**(7), 1148–1154.

Bowman, C. T. Control of combustion–generated nitrogen oxide emissions, technology driven by regulation. 24th *Int. Symp. on Combustion, The Combustion Institute*, Sydney, 3rd–7th July (1992).

Broadwell, J. E. and Dimotakis, P. E. (1986) Implications of recent experimental results for modelling reaction in turbulent flows. *AIAA Journal*, **24**(6), 875–889.

Cabelli, A., Pearson, I.G., Shepherd, I. C., and Hamilton, N. B. Fluid dynamic structures in jet diffusion flames: acoustic effects. Proc. 1st World Conference on Experimental Heat Transfer, Fluid Mechanics and Thermodynamics, Dubrovnik, Yugoslavia, 4–9 September 1988, pp. 627–631 (1988).

Candel, S. M. and Poinsot, T. J. (1990) Flame stretch and the balance equation for the flame area. *Combustion Science and Technology*, **70**, 1–15.

Drake, M. C. (1985) Kinetics of nitric oxide formation in laminar and turbulent methane combustion. Final Report for Gas Research Institute contract No. 5081-263-0600. GRI Chicago, Illinois 1985.

Fenimore, C. P. *13th International Symposium on Combustion* Combustion Institute, Pittsburgh pp. 373–379 (1971).

Miller, J. A. and Bowman, C. T. (1989) *Prg. Energy Combust. Sci.* **15**, 287–338

Miller, J. A., Branch, M. C., McLean, W. J., Chandler, D. W., Smooke, M. D. and Kee, R. J. *20th International Symposium on Combustion*, Combustion Institute, Pittsburgh pp. 673–684 (1985).

Pearson, I. G. and Proctor, D. A Novel Experimental Pulsed Combustion Natural-Gas Burner. International Symposium on Pulsating Combustion, Monterey, California, August 5–8 (1991).

Proctor, D., Nathan, G. J., Luxton, R. E., Pearson, I. G., Brumale, S. A., Mann, B. A., Schnieder, G. M. and Newbold, G. J. R. (The efficient low NO_x burning of gas for large scale industrial applications. *2nd Int. Conf. on Combustion Technologies for a Clean Environment*, Lisbon, Portugal, July) (1993).

Pindera, M. Z. and Talbot, L. Flame Induced Vorticity: Effects of Stretch. *21st International Symposium on Combustion*, Combustion Institute, Pittsburgh pp. 1357–1366 (1986).

Sano, T. (1982) *Combust. Science and Tech.*, **29**, 261.

Sano, T. (1984) *Combust. Science and Tech.*, **31**, 129.

Vandsburger, U., Lewis, G., Seitzman, J. M., Allan, M. G., Bowman, C. T. and Hanson, R. K. Flame-flow structure in an acoustically-driven jet flame. Western States Section Combustion Institute Meeting, Univ. Arizona, Paper No. 86-19, October (1986).

Zeldovich, Y. B., Barenblatt, G. I., Librovich, V. B. and Makhviladze, G. M. (1985) The Mathematical Theory of Combustion and Explosions, Consultants Bureau, New York pp. 30–36.

Soot Morphology and Optical Properties in Nonpremixed Turbulent Flame Environments

G. M. FAETH and Ü. Ö. KÖYLÜ   *Department of Aerospace Engineering
The University of Michigan, Ann Arbor, MI 48109-2118, U.S.A.*

Abstract— Motivated by the importance of soot to the emission of particulates and other pollutants from combustion processes, current understanding of soot morphology and optical properties is reviewed, emphasizing nonpremixed flame environments. The understanding of soot morphology in flames has grown rapidly in recent years due to the development of methods of thermophoretic sampling and analysis by transmission electron microscopy (TEM). The results show that soot consists of nearly spherical primary particles having diameters generally less than 60 nm, which collect into open structured aggregates that are mass fractal objects. Aggregates grow by cluster/cluster aggregation to yield broad aggregate size distributions with the largest aggregates containing thousands of primary particles and reaching dimensions of several μm. The optical properties of soot aggregates generally are not suited for the Rayleigh and Mie scattering approximations which has led to the development of approximate Rayleigh-Debye-Gans (RDG) scattering models for polydisperse fractal aggregate populations of soot. Evaluation of RDG models for conditions where both soot structure and scattering properties are known indicates encouraging agreement between predictions and measurements at both visible and infrared wavelengths, as well as reasonable accuracy for Rayleigh scattering theories in the infrared. Thus, there is potential for nonintrusive measurements of both soot concentrations and structure in flame environments, which should be helpful for diagnosing problems of particulate and pollution emissions from combustion processes. However, additional work is needed to realize this potential, including: reliable solutions of the inverse problem, to yield soot properties from scattering and extinction measurements, must be developed; existing uncertainties about soot refractive indices, including effects of fuel type and flame conditions, must be resolved; and more definitive assessment of the limitations of existing approximate theories for soot optical properties must be obtained.

Key Words: Soot structure, soot scattering, optical diagnostics, soot refractive indices

INTRODUCTION

Practical hydrocarbon-fueled flames generally contain and emit soot, which affects their structure, and their radiation, pollutant and particulate emission properties. Thus, numerous studies of soot processes in flame environments have been reported. In spite of this interest, however, processes of soot nucleation, growth and oxidation in flames are unusually complex and their current understanding is very incomplete. This is unfortuante because anticipated regulations will impose new and more stringent limitations on the emissions of particulates (largely soot) and pollutants from combustion processes, which will be difficult to satisfy without an improved technology base concerning combustion-generated soot. Enhanced experimental capabilities are required study soot processes in flames, with optical diagnostics being favored because they are convenient and nonintrusive. Thus, the objective of the present paper is to review current understanding of soot optical properties needed to develop reliable optical diagnostics for soot concentrations and structure.

Most practical hydrocarbon-fueled flames contain soot, which motivates study of soot phenomena as a necessary step toward the development of detailed numerical simulations of combustion processes, i.e., computational combustion. In particular, the significance of soot chemistry for practical flames is highlighted by the impact of soot on the emission of other pollutants. For example, while soot particles themselves constitute an obvious, and generally visible, emission of unburned hydrocarbons, they also can serve as a carrier for other undesirable pollutants like PAH (Lahaye and Prado, 1981). Another interaction is the well-known trade-off between NO_x emissions and the emission of soot particulates and carbon monoxide which limits capabilities to meet requirements for all regulated emissions using existing combustion modification technologies (Bowman, 1992). The correlation between soot and carbon monoxide emissions is well known (Tewarson, 1988), and it is not surprising because both emissions result from incomplete combustion of fuel carbon, e.g., carbon monoxide is the main product of the oxidation of soot and is likely to be present as long as soot is present. This relationship is quantified in Figure 1 for turbulent diffusion flames burning in air at long residence times where both soot and carbon monoxide generation factors (kg of soot and carbon monoxide emitted per kg of fuel carbon burned) are independent of the residence time. It is evident that there is a strong correlation between soot and carbon monoxide emissions, relatively independent of fuel type, with heavily sooting fuels emitting roughly 0.37 kg of carbon monoxide per kg of fuel

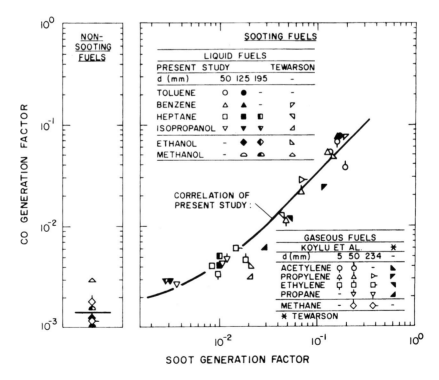

FIGURE 1   Carbon monoxide generation factors as a function of soot generation factors for turbulent diffusion flames in air. Measurements from Köylü and Faeth (1991), Köylü *et al.* (1991) and Tewarson (1988).

carbon burned (Köylü and Faeth, 1991). Thus, progress toward understanding pollutant emissions from practical combustion devices clearly will require an improved understanding of soot processes in flames.

Studies of soot nucleation, growth and oxidation in flame environments depend on measurements to a significant degree, due to the complexity of these phenomena. The use of intrusive probes for measurements of soot properties has not been popular, however, due to concern about the disturbances caused by intrusive probes on flame structure. In particular, problems of probes are severe in soot-containing flame environments due to thermophoretic deposition of soot particles on surfaces. This causes probes to clog and to grow soot deposits to sizes that clearly disturb test flames. Thus, nonintrusive optical methods have been emphasized generally using laser extinction and scattering in the visible to infer soot concentrations and surface areas relevant to heterogeneous soot chemistry. These measurements generally are interpreted based on either the Rayleigh scattering approximation at the small particle limit or the Mie scattering approximation for an equivalent spherical particle. Nevertheless, even early studies raised questions about the general validity of these methods. For example, transmission electron microscopy (TEM) measurements for premixed acetylene, benzene and propane flames showed that soot consisted of small spherical primary particles collected into open structured aggregates having a broad distribution of sizes; additionally, many aggregates were too large for reasonable application of the Rayleigh scattering approximation, and too open structured for proper representation as equivalent compact spheres using the Mie scattering approximation (Erickson et al., 1964; Dalzell et al., 1970; Wersborg et al., 1972). Furthermore, direct measurements at these conditions confirmed these concerns about soot scattering properties: strong forward scattering was observed which is not representative of Rayleigh scattering behavior, while use of the Mie scattering approximation for an equivalent sphere still did not provide an adequate fit of the data (Erickson et al., 1964; Dalzell et al., 1970; Wersborg et al., 1972; Magnussen, 1974). Subsequently, the absence of reliable nonintrusive and intrusive diagnostics for soot has remained as a major impediment to definitive measurements needed to study soot phenomena in flame environments.

Recently, more effective theories of soot optical properties have been developed which appear to have the potential to remove past limitations and provide reliable nonintrusive optical diagnostic for soot properties in flames. These methods are based on the Rayleigh-Debye-Gans scattering approximation while treating soot aggregates as polydisperse collections of mass fractal aggregates (denoted the RDG-PFA scattering approximation in the following), see Jullien and Botet (1987), Martin and Hurd (1987), Dobbins and Megaridis (1991) and Köylü and Faeth, (1992b,c). The objective of this review is to describe the RDG-PFA approach and its evaluation, and to highlight areas where research is needed so that RDG-PFA theory can be exploited to establish reliable nonintrusive optical diagnostics for soot properties. The paper begins with consideration of the structure of soot aggregates and existing evidence for their characterization as mass fractal objects. Soot optical properties are then considered, treating RDG-PFA scattering theory, its evaluation based on scattering measurements, the implications of the theory for measuring soot properties, and issues that still must be resolved to achieved reliable measurements of soot properties.

SOOT PHYSICAL PROPERTIES

General Properties

The physical properties of soot aggregates must be addressed in order to properly model their optical properties; therefore, they will be considered first. Numerous TEM photographs of soot aggregates at various flame conditions have appeared in the literature, all indicating a reasonably generic morphology or structure, see Dalzell *et al.* (1970), Dobbins and Megaridis (1987), Erickson *et al.* (1964), Köylü and Faeth (1992, 1993c), Medalia and Heckman (1969), Nelson (1989), Samson *et al.* (1987) and Wersborg *et al.* (1973) for examples. A typical TEM photograph of soot aggregates appears in Figure 2. These conditions involve soot emitted from long residence time turbulent acetylene/air diffusion flames where emitted soot properties are independent of residence time (Köylü and Faeth, 1992). The aggregates in the photograph are somewhat large, due to the long residence time available for aggregation, but otherwise they are representative of soot found in flame environments.

Quantitative examination of soot aggregates generally indicates that the aggregates consist of nearly spherical primary particles having relatively uniform diameters at

FIGURE 2   TEM photograph of typical soot aggregates emitted from turbulent acetylene-air diffusion flames. From Köylü and Faeth (1992).

a given flame condition. The primary particles tend to be some what merged, rather than just touching at points, due to soot growth subsequent to the joining of adjacent primary particles. Nevertheless, models of soot aggregates as monodisperse spherical primary particles that just touch one another are a reasonable approximation of TEM observations, and this approximation generally has been adopted for studies of soot optical properties.

As seen in Figure 2, soot aggregates have branched and open structures with a large variation of the number of primary particles per aggregate even at a given flame condition. Thus, the open structure of soot aggregates does not suggest that their optical properties can be modeled effectively by treating them as equivalent spherical objects having the same volume. Additionally the largest aggregates have dimensions of several μm, exceeding the wavelength of light in the visible portion of the spectrum, which raises questions about the use of the small particle (Rayleigh) scattering limit to treat their optical properties. As discussed earlier these concerns are amply justified by observations of soot scattering properties that were not consistent with either of these scattering approximations.

Other measurements of the physical properties of soot aggregates relate to their density, porosity and composition. Not surprisingly, these properties generally are found to be similar to carbon blacks with two major exceptions: soot during the last stages of oxidation has significant porosity, even extending to the presence of hollow cenospheres, and soot from some internal engine combustion processes (like diesel engines at heavily sooting conditions) contain high levels of volatile matter (Lahaye and Prado, 1981). Otherwise, soot densities are typical of carbon blacks, e.g., values in the range 1820–2050 kg/m^3, see Dobbins et al. (1993) and references cited therein. Soot aggregates also appear to be relatively nonporous with measurements of surface area compatible with shapes observed on TEM photographs except as noted earlier. Finally, soot mainly consists of carbon. For example, soot emitted from long residence time turbulent diffusion flames (involving the combustion of toluene, benzene, acetylene, propylene and propane burning in air) had the following elemental mole ratio ranges: C:H of 8.3–18.3, C:O of 58–109, C:N of 292–976. However, the presence of volatiles, as well as annealing processes at high temperatures, can affect concentrations of noncarbon substances, and possibly optical properties like soot refractive indices as well. Finally, TEM microstructure studies showing somewhat different structure near the core and at the surface of primary particles, as well as the formation of pores and cenospheres at high levels of soot oxidation, raise questions about the uniformity of primary particle properties (Lahaye and Prado, 1981). Thus, while models of soot optical properties generally adopt approximations of uniform primary particle properties, and ignore potential variations of refractive indices with fuel type and flame conditions, these approximations clearly merit additional scrutiny.

Primary Particle Size

Recent measurements of soot aggregate properties in flame environments have employed thermophoretic sampling and analysis by TEM. This is reasonably reliable because effects of aggregate size bias are fortuitously rather small for thermophoretic sampling, assuring representative statistics (Rosner et al., 1991), while TEM analysis is not dependent upon questionable approximations about the optical properties of soot. This work has included both the fuel-rich and fuel-lean regions of laminar diffusion flames (Dobbins and Megaridis, 1987; Megaridis and Dobbins, 1990; Köylü and Faeth, 1993c; Samson et al., 1987) as well as soot emitted from long residence time turbulent diffusion flames (Köylü and Faeth, 1992). Some typical structure properties of soot aggregates, including primary particle diameters, are summarized in Table I for soot emitted from long residence time turbulent flames of various fuels burning in air. This soot is representative of soot emitted from combustion processes, and also is of interest for evaluation of predictions of soot optical properties to be discussed later.

Primary particle diameters at a given flame condition are nearly monodisperse. For example, the standard deviations of the primary particles listed in Table I were in the range 17–25% of the mean primary particle diameter. Primary particle size distributions can be fitted accurately to log normal distribution functions but due to their small standard deviations are approximated reasonably well by Gaussian functions (Köylü and Faeth, 1992). Mean primary particle diameters vary with flame condition and fuel type but diameters less than 60 nm generally are observed in flame environments, with the largest diameters associated with heavily sooting fuels, see Table I. This implies primary particle size parameters, $x_p < 0.4$ for $\lambda > 500$ nm, so that it is reasonable to assume that individual primary particles behave like Rayleigh scattering particles, i.e., total scattering and absorption cross sections are within 1 and 5%, respectively, of estimates based on the Rayleigh scattering approximation for individual primary particles (Köylü and Faeth, 1993a).

Aggregate Properties

Soot aggregates are small near the soot inception point but they aggregate rapidly with $\bar{N}$ in the range 200–600 for soot emitted from turbulent diffusion flames, see Table I. Unlike primary particle diameters, however, aggregate size distributions are quite broad, with standard deviations comparable to $\bar{N}$ and large values of the second moment of the distribution. $\bar{N}^2/(\bar{N})^2$, e.g., values up to 3.3 have been observed for soot emissions. Nevertheless, aggregate size distributions are reasonably represented by

TABLE I

Structure of soot emitted from long residence time turbulent diffusion flames[a]

Fuel	$\bar{d}_p$ (nm)	$\bar{N}$	$\bar{N}^2/(\bar{N})^2$	D_f	$(\bar{R}_g^2)^{1/2}$ (nm)
Toluene	51	526	3.23	1.73	672
Acetylene	47	417	3.20	1.79	481
Benzene	50	552	2.71	1.71	686
Propylene	41	460	2.61	1.75	431
Ethylene	32	467	2.42	1.73	376
n-Heptane	35	260	2.05	1.73	299
Propane	30	364	2.56	1.74	305
Isopropanol	31	255	2.61	1.70	275

[a]From Köylü and Faeth (1992), r.m.s. value of R_g found assuming $D_f = 1.81$ and $k_f = 8.1$ from Köylü and Faeth (1993b).

a log normal distributions function (Dobbins and Megaridis, 1987; Köylü and Faeth, 1992; Lahaye and prado, 1981). Based on the properties of this distribution function, 95% of the soot aggregates summarized in Table I would contain 30–1800 primary particles. This means that the optical properties of aggregates are complex in the visible – ranging from Rayleigh scattering for small aggregates, where each primary particle essentially scatters light independently, to significant interactions between scattering from the individual primary particles of large aggregates.

Numerous recent evaluations have shown that flame-generated soot aggregates exhibit mass fractal-like behavior with a Hausdorf or fractal dimensions, $D_f < 2$, even when the number of primary particles in an aggregate is small (Jullien and Botet, 1987; Köylü and Faeth, 1992, 1993c; Megaridis and Dobbins, 1990; Samson et al., 1987). Mountain and Mulholland (1988) have carried out stochastic simulations of the aggregation process finding that cluster/cluster aggregation yields properties similar to soot aggregtates, with the progressive buildup of aggregate size leading naturally to scaling relationships representative of fractal behavior. The fractal dimension has important implications for soot optical properties because the scattering per primary particle continues to grow as the size of the aggregate increases for $D_f > 2$ but reaches a constant saturated value for $D_f < 2$ (Berry and Percival, 1986; Dobbins and Megaridis, 1991; Nelson, 1989).

The mass fractal approximation implies the following relationship between the primary particle diameter, the number of primary particles, and the radius of gyration of the aggregate (Julhen and Botet, 1987):

$$N = k_f(R_g/d_p)^{D_f} \qquad (1)$$

where k_f is the fractal prefactor and the aggregates are assumed to consist of monodisperse nonoverlapping primary particles. Results summarized in Table I indicate that the fractal dimensions of soot emitted from flames are relatively independent of fuel type. Recent studies of soot within laminar diffusion flames, where aggregates are still relatively small, yield fractal dimensions in the range 1.7–1.8 as well (Köylü and Faeth, 1993c; Megaridis and Dobbins, 1990). Fractal prefactors have received less attention, however, recent determinations based on both TEM and scattering measurements for various flame conditions suggest values in the range 7.0–9.2 (Köylü and Faeth, 1992, 1993b; Puri et al., 1993). Thus, unlike other aggregate structure properties, D_f and k_f appear to be relatively durable properties of soot in flame environments.

SOOT OPTICAL PROPERTIES

General Description

In view of the difficulties encountered when using the Rayleigh and Mie scattering approximations for soot optical properties, recent work has focused on the Rayleigh-Debye-Gans (RDG) scattering approximation. The RDG approximation implies that effects of multiple- and self-scattering are ignored so that the electromagnetic fields within each primary particle are the same as the incident field, and differences between the phase shift of light scattered from various points within a particular primary particle are ignored. This requires that both $|m - 1| \ll 1$ and $2x_p|m - 1| \ll 1$ (Kerker, 1969; van de Hulst, 1957; Bohren and Huffman, 1983), which is questionable for soot aggregates due to the relatively large refractive indices of soot. In addition, recent computational studies suggest significant effects of multiple scattering for large soot aggregates, similar to those summarized in Table I, see Berry and Percival (1986), Chen et al. (1990), Ku and Shim (1992) and Nelson (1989). Thus, the RDG scattering approximation requires evaluation with experiments as discussed in the following.

Major assumptions concerning soot physical properties, used in RDG scattering theories, can be summarized as follows: spherical primary particles having constant diameter, primary particles just touch one another, uniform refractive indices, log normal aggregate size distributions and the aggregates are mass fractal-like objects that satisfy Equation (1) with constant values of D_f and k_f.

RDG-PFA Theory

The RDG-PFA scattering theory is based on methods described by Freltoft et al. (1986), Jullien and Botet (1987), Lin et al. (1989), Martin and Hurd (1987) and Dobbins and Megaridis (1991) for a single aggregate, as extended by Köylü and Faeth (1993b) for polydisperse aggregate populations. Only the main results of RDG-PFA scattering theory will be summarized in the following, original sources should be consulted for details.

Under the present assumptions, primary particles satisfy the Rayleigh scattering approximation, yielding the following expressions for their optical properties (Bohren and Huffman, 1983; Kerker, 1969):

$$C_a^p = 4\pi x_p^3 E(m)/k^2, \quad C_s^p = 8\pi x_p^6 F(m)/(3k^2), \quad C_{vv}^p = x_p^6 F(m)/k^2 \quad (2)$$

In Equation (2) and the following equations, subscripts for differential scattering cross sections denote the direction of polarization vectors with respect to the scattering plane defined by the light source, the soot aggregate and the observer, v and h denote polarization vectors normal and parallel to this plane, respectively; and the first and second subscripts refer to incident and scattered light, respectively.

The treatment of aggregate optical properties begins with the scattering cross sections for a single fractal aggregate under the RDG approximation (Kerker, 1969):

$$C_{vv}^a(\theta) = C_{hh}^a(\theta)/\cos^2\theta = N^2 C_{vv}^p f(qR_g) \quad (3)$$

The form factor, $f(qR_g)$, is expressed as follows in the small angle (Guinier) and large angle (power-law) regimes (Freltoft et al., 1986; Jullien and Botet, 1987; Lin et al., 1989; Martin and Hurd, 1987):

$$f(qR_g) = \exp(-(qR_g)^2/3), \quad \text{Guinier regime} \quad (4)$$

$$f(qR_g) = (qR_g)^{-D_f}, \quad \text{power-law regime} \quad (5)$$

Following Dobbins and Megaridis (1991), the boundary between the Guinier and power-law regimes is taken to be $(qR_g)^2 = 3D_f/2$, chosen to match the value and derivative of $f(qR_g)$ where the two regimes meet. The total scattering cross section then becomes:

$$C_s^a = N^2 C_s^p g(\lambda, R_g, D_f) \tag{6}$$

where $g(\lambda, R_g, D_f)$ has different forms if the power-law regime is reached for $\theta \leqslant 180$ or not, see Köylü and Faeth (1993b) for these expressions.

The mean optical cross sections of populations of randomly oriented polydisperse aggregates are found by integrating over all aggregate sizes, as follows:

$$\bar{C}_j^a = \int_{N=1}^{\infty} C_j^a(N) p(N) \, dN; \quad j = pp,s,a \tag{7}$$

where $p(N)$ is the aggregate size distribution function. Equation (7) must be integrated numerically when scattering extends over both the Guinier and power-law regimes. However, simple closed form expressions are possible when all aggregates are either in the Guinier ($p(N) \ll 1$ for $N \geqslant N_c$) or power-law ($p(N) \ll 1$ for $N \leqslant N_c$), where

$$N_c = k_f (3D_f/(2q^2 d_p^2))^{D_f/2} \mathrm{m} \tag{8}$$

This yields:

$$\bar{C}_{vv}^a(0)/C_{vv}^p = \bar{N}^2 \exp(-q^2 \bar{R}_{gG}^2/3), \quad \text{Guinier regime} \tag{9}$$

$$\bar{C}_{vv}^a(0)/C_{vv}^p = \bar{N} k_f (q d_p)^{-D_f} = \bar{N}^2 (q^2 \bar{R}_{gG}^2)^{-D_f/2} \text{ power-law regime} \tag{10}$$

where expressions for $\bar{R}_{gG}^2$ and $\bar{R}_{gL}^2$ can be found in Köylü and Faeth (1993c). The general expression for the total scattering cross section of a polydisperse aggregate population is

$$\bar{C}_s^a = C_s^p \int_{n=1}^{\infty} N^2 g(\lambda, R_g, D_f) p(N) dN \tag{11}$$

which must be numerically integrated in both scattering regimes. The absorption cross section is evaluated quite simply, as follows:

$$\bar{C}_a^a = \bar{N} C_a^p \tag{12}$$

Evaluation of RDG-PFA Predictions

Existing computer simulations of the optical properties of soot do not provide an adequate basis for evaluating approximate scattering theories because they involve either fundamentally accurate solutions for small aggregatees where effects of multiple- and self-induced scattering are small, or approximate solutions having uncertain accuracy for the large soot aggregates of interest for practical flames (Köylü and Faeth, 1993a).

The predictions of RDG-PFA theory have been evaluated for conditions where both soot structure and scattering properties were measured. These evaluations included the large soot aggregates emitted from long residence time turbulent diffusion flames (see Table I for the properties of this soot), as well as relatively small soot aggregates found at fuel-rich conditions within laminar ethylene and acetylene/air diffusion flames

(Köylü and Faeth, 1993b, c). This range of conditions was studied in order to adequately cover the Guinier (for small aggregates) and power-law (for large aggregates) regimes. Measured volumetric optical cross sections were converted to optical cross sections, and vice versa, assuming that absorption is not affected by aggregation, which follows from Equation (12) based on RDG theory, but also is accurate within experimental uncertainties for scattering properties (ca. 20%) based on more complete scattering theories allowing for effects of multiple- and self-induced scattering (Chen *et al.*, 1990; Farias and Carvalho, 1993; Iskander *et al.*, 1989; Köylü and Faeth, 1993a). This approximation yields:

$$\bar{C}_j^a(\theta) = \pi \bar{N} d_p^3 \bar{Q}_j^a(\theta)/(6f_s), \quad j = pp, s, a \tag{13}$$

where $\bar{N}$ and d_p were found from TEM measurements and f_s from extinction and scattering measurements. Finally, all computations employed the refractive indices of Dalzell and Sarofim (1969) in order to be consistent with earlier work (Köylü and Faeth, 1992, 1993a), and because they were preferred based on scattering measurements to be discussed later.

Measured and predicted values of $\bar{Q}_{vv}^a(\theta)$ are plotted as a function of qd_p in Figure 3 for soot emitted from long residence time flames fueled with acetylene, propylene, ethylene and propane. The substantial departure from Rayleigh scattering behavior (where $\bar{Q}_{vv}^a(\theta)$ would be independent of qd_p) is evident, with forward scattering roughly 100 times larger than back scattering for all the fuels. In fact, the large size of the aggregates prevented fully reaching the Guinier regime even though scattering angles as small as $5°$ were considered, however, the measurements provide an extended range within the power-law regime. Within this regime Equations (10) and (13) yield D_f and k_f from the slope and magnitude of plots of $\bar{Q}_{vv}^a(\theta)/\bar{Q}_{vv}^p$ as a function of qd_p from Figure 3. These values are summarized in Table II for the four fuels, along with values found from TEM measurements of soot structure from Köylü and Faeth (1992) and Puri *et al.* (1993). The scattering measurements are seen to be in good agreement with the TEM measurements.

Two sets of predictions based on RDG-PFA theory are illustrated in Figure 3, one set entirely based on TEM measurements of soot structure and the other set based on refitting the higher moments, $\bar{N}^2/(\bar{N})^2$, of the aggregate size distribution functions in order to best fit the scattering predictions and measurments, keeping all other structure properties the same. This was done because the higher moment was not found very accurately during the TEM structure measurments (experimental uncertainties were only less than 40–90%) due to the broad size distribution of aggregates (Köylü and Faeth, 1992). Predictions based on refitted structure properties are seen to be in good agreement with the measurements, while the required changes of the higher moments of the size distribution functions were within experimental uncertainties. Evaluation of predictions of extinction at 632.8 and 1152 nm also was reasonably satisfactory with differences between predictions and measurements less than 18%.

An obvious weakness of the evaluation of predictions using large soot aggregates is that the aggregates were to large to reach the Guinier regime. This was rectified during subsequent work considering relatively small soot aggregates observed at fuel-rich conditions in laminar flames (Köylü and Faeth, 1993c). In this case, the predictions based on TEM structure measurements agreed with the scattering measurements

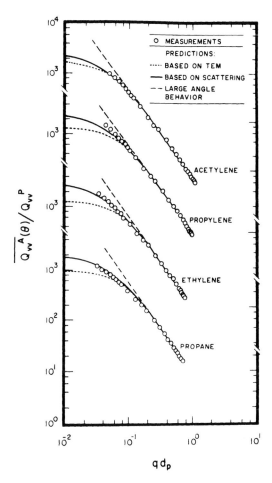

FIGURE 3 Measured and predicted volumetric vv cross sections as a function of the modulus of the scattering vector for soot emitted from turbulent diffusion flames in air. From Köylü and Faeth (1992).

within 37%, while refitting the higher moments of the aggregate size distribution function, within their experimental uncertainties, yielded results similar to the fitted predictions illustrated in Figure 3.

A final test of the RDG-PFA predictions involved the consideration of scattering patterns. Typical results are illustrated in Figures 4 and 5 for the smallest and largest soot aggregates studied, in order to indicate the range of behavior observed. Predictions on these figures are seen to be in reasonably good agreement with the measurements. Comparing the $\bar{C}_{vv}^{a}(\theta)$ plotted in Figures 4 and 5 illustrates the effect of aggregate size on scattering patterns, with the very strong forward scattering observed for large aggregates evolving toward a uniform distribution of scattering, typical of Rayleigh scattering (see Equation (2)), as the aggregates become smaller. The variation of the slope of $\bar{C}_{vv}^{a}(\theta)$ at angles in the range 45–135° often is used for nonintrusive measurements of aggregate size (Dobbins et al., 1990), e.g., dissymmetry ratios, $\bar{C}_{vv}^{a}(45°)/\bar{C}_{vv}^{a}(135)$,

TABLE II

Soot aggregate fractal properties from thermophoretic sampling and light scattering measurements[a]

Fuel	Acetylene	Propylene	Ethylene	Propane
Fractal Dimension, D_f:				
Thermophoretic sampling	1.79	1.75	1.73	1.74
Light scattering	1.85	1.84	1.83	1.77
Fractal Prefactor, k_f:				
Thermophoretic sampling[b]	9.2	—	8.6	
Light scattering[c]	7.0	8.6	8.8	8.0

[a]For soot emitted from long residence time turbulent diffusion flames, from Köylü and Faeth (1993b).
[b]From Puri et al. (1993).
[c]Based on the soot refractive indices of Dalzell and Sarofim (1969).

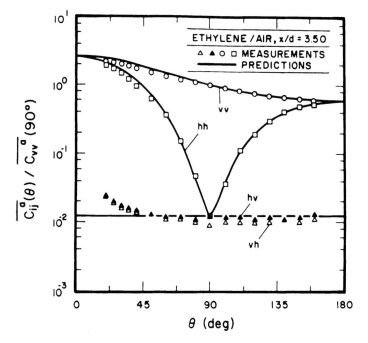

FIGURE 4   Measured and predicted angular scattering patterns for small soot aggregates in ethylene/air diffusion flames. From Köylü and Faeth (1993c).

are seen to be larger for the large soot aggregates of Figure 5 than the small soot aggregates of Figure 4. However, this approach must be used with caution: dissymmetry ratios for angles in the range 45–135° are not very sensitive to aggregate size for $\bar{N} > 100$ (Köylü and Faeth, 1993a), and become insensitive to aggregate size once again for small aggregates approaching the small particle (Rayleigh) scattering limit. In contrast, effects of aggregate size are most evident at small scattering angles ($\theta < 45°$), which provides a more reliable regime for nonintrusive measurements of aggregate size.

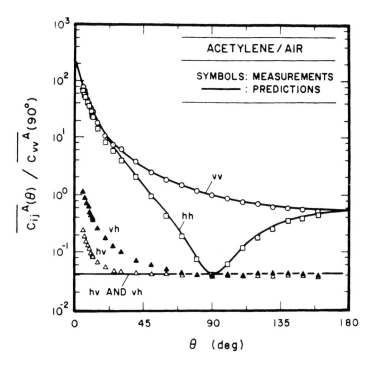

FIGURE 5 Measured and predicted angular scattering patterns for large soot aggregates in acetylene air diffusion flames. From Köylü and Faeth (1993b).

Results at other test conditions were similar to the findings illustrated in Figures 3–5. Thus, RDG-PFA theory yielded predictions that agreed with mesurements, within experimental uncertainties, over the following ranges: x_p of 0.128–0.330, $\bar{N}$ of 31–467, $\bar{N}^2/(\bar{N})^2$ of 1.82–3.23, $\bar{C}_{vv}^a(0°)/\bar{C}_{vv}^a(180°)$ of 10–1000 and values of the albedo of 0.06–0.29.

Implications of Soot Scattering Properties

Further evaluation of the RDG-PFA theory for soot optical properties would be desirable. In particular, the studies of Köylü and Faeth (1993a, b) involved a limited range of soot aggregate properties and problems of adequately measuring the higher moments of the aggregate size distribution using TEM compromised the evaluation to some extent. Uncertainties also remain about effects of primary particles, nonuniformity of primary particle physical properties, and variations of fractal properties and refractive indices as a function of fuel type and flame condition. Some of these issues can be addressed by computer simulations using more advanced theories than the RDG approximation , e.g., the methods developed by Borghese *et al.* (1984) and Iskander *et al.* (1989). Nevertheless, the evaluation of RDG-PFA scattering theory has been reasonably promising thus far; therefore, it is worthwhile to exploit the theory to gain insight about potential nonintrusive optical diagnostics for soot properties.

Laser extinction often is used to measure soot concentrations in flame environments, with the measurements interpreted based on the small-particle (Rayleigh) scattering approximation, see Faeth *et al.* (1989), Köylü and Faeth (1993a), Santoro *et al.* (1983), Tien and Lee (1982), and references cited there in. Under the Rayleigh scattering approximation, scattering is negligible so that extinction and absorption are essentially identical and extinction measurements are proportional to the volume fraction of soot in the optical path through Equation (1). Nevertheless, the differential scattering cross sections illustrated in Figures 3–5 indicate significant scattering levels for large soot aggregates so that attributing all the extinction to absorption can overestimate soot volume fractions significantly.

The effect of the departure of soot aggregate scattering from the Rayleigh scattering limit on laser extinction measurements of soot volume fractions is illustrated in Figure 6. This is a plot of the ratio of the total scattering to absorption cross sections, $\bar{\rho}_{sa}$, based on RDG-PFA theory, as a function of mean aggregate size (Köylü and Faeth, 1992). Results are illustrated for two different wavelengths (note the different scales of the ordinates of the two plots): 632.8 nm, which is representative of wavelengths used to find soot volume fractions from laser extinction measurements, and 2000 nm which is representative of wavelengths where radiant emission from flames is relatively large. The results are presented as $\bar{\rho}_{sa}E(m)/F(m)$ to avoid complications of the uncertainties of soot refractive indices, this parameter also is proportional to the fractional error incurred by using the Rayleigh scattering approximation to find soot volume fractions from laser extinction measurements. These plots have been constructed using the soot properties of Table I, except for varying $\bar{N}$, with the data symbols on each plot indicating behavior for the actual $\bar{N}$ of each fuel.

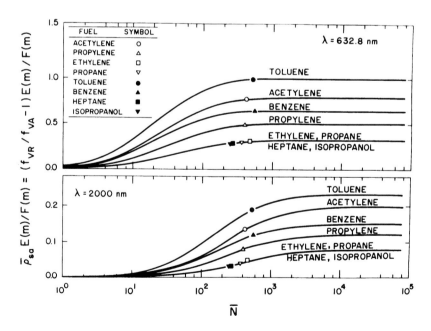

FIGURE 6   Ratios of scattering to absorption cross sections as a function of aggregate size for soot aggregates. From Köylü and Faeth (1992).

Results illustrated in Figure 6 show the transition between nearly Rayleigh scattering behavior for small aggregates, $\bar{\rho}_{sa} \approx 0$, to conditions where $\bar{\rho}_{sa}$ becomes independent of aggregate size (or saturated scattering) for large aggregates. Saturated scattering is characteristic of scattering from fractal objects having $D_f < 2$; for $D_f > 2$. $\bar{\rho}_{sa}$ would continue to increase as $\bar{N}$ increases (Berry and Percival, 1986; Dobbins and Megaridis 1991). The variation of $\bar{\rho}_{sa}$ with fuel type and wavelength seen in Figure 6 is caused by variations of primary particle size parameters, e.g., $\bar{\rho}_{sa} \sim x_p^{3-D_f}$ within the saturated scattering regime (Dobbins and Megaridis, 1991). The large soot aggregates summarized in Table I approach saturated scattering at 632.8 nm, and are intermediate between Rayleigh and saturated aggregate scattering at 2000 nm. The additional scattering from aggregates causes soot volume fractions for these large aggregates to be significantly overestimated when laser extinction measurements are analyzed using the Rayleigh scattering approximation. For example, noting that $F(m)/E(m)$ is of order unity, as discussed later, such measurements at 632.8 nm overestimate soot volume fractions by nearly a factor of two for a heavily sooting material like toluene and by roughly 30% for lightly sooting materials like ethylene, propane, n-heptane and isopropanol. Thus, it appears that some consideration of scattering is required for laser extinction measurements of soot volume fractions in the visible unless evidence is available that both primary particle diameters and aggregate sizes are small.

Effects of scattering on laser extinction measurements of soot volume fractions become smaller at longer wavelengths, as seen from the results at a wavelength of 2000 nm in Figure 6. This is illustrated more directly in Figure 7, where $\bar{\rho}_{sa}$ is plotted as a function of wavelength for the large soot aggregates summarized in Table I. It is seen

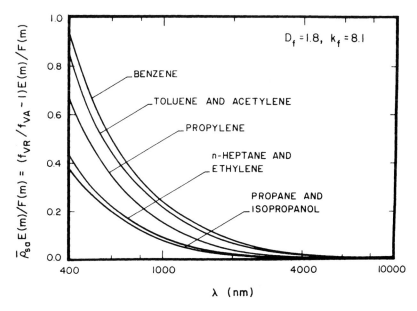

FIGURE 7   Mean ratios of scattering to absorption cross sections as a function of wavelength for soot aggregates from turbulent diffusion flames in air. From Köylü and Faeth (1993b).

that $\bar{\rho}_{sa}$ progressively decreases with increasing wavelength, reaching values less than 20% for wavelengths greater than 2000 nm, even for very large toluene soot, aggregates. Thus, extinction measurements in the infrared offer a simple means of measuring soot volume fractions without the complications of simultaneous measurements of scattering or soot structure, as long as wavelengths are chosen to avoid effects of absorption from infrared gas bands. Additionally, results of Figures 6 and 7 suggest relatively small effects of scattering on the radiation properties of soot aggregates in flame environments, where the significant wavelengh range is in the infrared.

Nonintrusive optical measurements of soot structure and concentration properties impose somewhat different requirements. For example, it has just been noted that soot concentrations are most easily measured at the Rayleigh scattering conditions, however, these same conditions provide relatively little information about soot structure. In particular, primary particles scatter essentially independently in the Rayleigh scattering regime; therefore, no information about aggregate properties can be found at tehse conditions. Additionally, C_{vv}^p is independenet of scattering angle from Equation (2) so that dissymmetry ratios are unity and independent of primary particle diameter. Only the primary particle diameter is available, based on absolute scattering measurements to find $\bar{\rho}_{sa}$ and then using Equation (2), e.g., $x_p = (3\bar{\rho}_{sa} E(m)/(2F(m))^{1/3}$.

Similar to the Rayleigh scattering regime, the saturated scattering regime, which is dominated by power-law scattering behavior, has limited capabilities for inferring soot structure properties. At these conditions, the slope of plots of $\bar{Q}_{vv}^a(\theta)$ as a function of qd_p yields the fractal dimension but this property is relatively universal for soot aggregates in any event. A corollary of this behavior is that dissymmetry ratios are relatively independent of aggregate size, as discussed in connection with Figures 4 and 5. Thus, direct measurements of $\bar{C}_{vv}^a(\theta)$ and $\bar{C}_a^a$ can yield d_p through Equations (10) and (12) if the fractal properties D_f and k_f can be reliably prescribed, implying structure information and absolute scattering measurement requirements somewhat analogous to the Rayleigh scattering regime.

Scattering in the small-angle (Guinier) regime provides conditions where soot aggregate structure can be found using nonintrusive optical methods. Dobbins et al. (1990) have considered this inverse problem using RDG-PFA theory, however, their polydisperse aggregate considerations were limited to conditions where all aggregates were in the Guinier regime, see Köylü and Faeth (1993b, c) for a discussion of this issue. These conditions probably were not met for the ethylene soot aggregates that were measured by Dobbins et al. (1990). In particular, they found values of $\bar{\rho}_{sa}$ up to 30% at a wavelength of 632.8 nm, which would imply nearly saturated behavior, with scattering dominated by the power-law regime, based on the results for ethylene soot illustrated in Figure 6. Thus, while Dobbins et al. (1990) provide an interesting example of inferring soot structure properties from scattering measurements based on RDG-PFA theory, their conclusions should be considered with caution until the impact of power-law scattering on their results has been assessed.

Köylü and Faeth (1993b, c) also report results where soot structure properties were inferred from soot scattering properties, as discussed earlier. However, this was a mixed approach where scattering and TEM measurements were combined in order to optimize the higher-order moments of the aggregate size distribution function. Thus, algorithms to infer the properties of the soot aggregate size distribution function from extinction and scattering measurements in the Guinier regime must still be developed and evaluated. As discussed earlier, primary particle diameters can be found from scattering measurements in the Rayleigh and power-law regimes; an approach to yield primary particle diameters in the Guinier regime clearly is of interest as well.

Refractive Indices

Computations of soot optical properties require accurate refractive indices for their wavelength range of interest, typically $0.1-10\mu m$, in order to treat optical diagnostics and continuum radiation for soot. In particular, $E(m)$ is needed for estimates of absorption, $F(m)$ is needed for estimates of scattering and the ratio $F(m)/E(m)$ is needed for estimates of the importance of scattering as well as for nonintrusive determinations of primary particle diameters using RDG-PFA and Rayleigh scattering theories. Thus, current understanding of soot refractive indices is briefly reviewed in the following see Chang and Charalampopoulos (1990), Charalampopoulos (1992), Tien and Lee (1982) and Viskanta and Mengüc (1987), and references cited therein for a more complete discussion.

Early determinations of soot volume fractions involved *ex situ* reflection coefficient measurements from compressed soot samples removed from flames, see Dalzell and Sarofim (1969), Felske *et al.* (1984) and references cited therein. These approaches were questioned, however, based on potential modification of soot properties by sampling and compression, as well as effects of surface roughness and voids on reflectance properties that are difficult to evaluate (Felske *et al.*, 1984; Lee and Tien, 1980; Charalampopoulos, 1992). Thus, most recent work has involved *in situ* measurements of extinction and scattering within both premixed and nonpremixed (diffusion) flames. see Chang and Charalampopoulos (1990), Charalampopoulos (1992), Lee and Tien (1980), Roessler and Faxvog (1980) and Vaglieco *et al.* (1990) for examples. Unfortunately, these determinations did not involve characterization of soot structure and employed Rayleigh and Mie scattering approximations whose accuracy is questionable based on results discussed here. Additionally, the extinction and scattering measurements must be supplemented by other information in order to find refractive indices. Three methods for providing auxiliary information have been used: Drude-Lorentz dispersion models for the wavelength dependence of soot refractive indices (Dalzell and Sarofim, 1969; Lee and Tien, 1980), Kramers-Krönig relationships based on the casuality requirements of inputs and outputs of passive physical systems (Chang and Charalampopoulos, 1990; Charalampopoulos, 1992), and results based on other measurements that ultimately rest on either dispersion models or Kramers-krönig relationships (Vaglieco *et al.*, 1990). Unfortunately, there are considerable differences of empirical dispersion model constants among the various investigations while use of Kramers-Krönig relationships invariably requires uncertain extrapolations of soot refractive indices to ranges that can not be measured because the wavelength range, $0-\infty$, is involved in these computations. Thus, there are significant uncertainties about all existing measurements of soot refractive indices, which constitutes a major impediment to proper characterization of soot optical properties.

Results concerning the effects of fuel type and flame condition on soot refractive indices are somewhat contradictory. Early work suggested that soot refractive indices were relatively independent of fuel type and did not find significant effects of flame condition, including effects of H/C ratio and temperature (Tien and Lee, 1992). However, later work reported significantly lower refractive indices when the soot contained high levels of volatile matter, as well as effects of annealing of soot in the high-temperature post-flame region of premixed flames where it was hypothesized that refractive indices increased due to removal of hydrogen near the surface of the soot particles by reactive effects (Charalampopoulos, 1992). These effects are difficult to resolve, however, due to the fundamental limitations of the methods used to measure soot volume fractions, discussed earlier. For example, continued aggregation of soot in post flame gases can easily be interpreted as an increase in soot refractive indices when using questionable Rayleigh and Mie scattering approximations to estimate soot optical properties.

Typical measurements of the real and imaginary parts of the refractive indices of soot, extending over the wavelength range $0.1 - 20\,\mu m$, are illustrated in Figure 8. The following measurements are shown on the figure: acetylene soot based on the reflectance of acetylene soot pellets and a dispersion model (Dalzell and Sarofim, 1969); polystyrene and plexiglass soot based on *in situ* extinction measurements in diffusion flames and a dispersion model (Lee and Tien, 1980); and propane soot based on *in situ* extinction measurements in premixed flames and the use of Kramers-Krönig relationships (Chang and Charalampopoulos, 1990). It is evident that there are significant differences among these evaluations of soot refractive indices, including other measurements in the literature would yield even wider variations (Charampopoulos, 1992). These differences can be attributed to approximations used to interpret the optical measurements, effects of fuel type, and effects of flame conditions–particularly the temperature for the near ultraviolet region (Lee and Tien, 1980). However, concerns about the approximate theories used to interpret the data are an overriding consider-

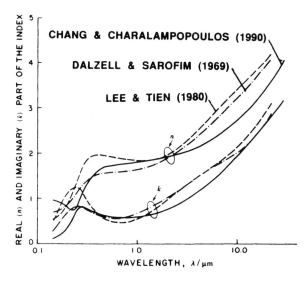

FIGURE 8 Measurements of soot refractive indices as a function of wavelength. From Chang and Charalampopoulos (1990).

ation and it is difficult to see how the other issues can be resolved until a sound method for measuring soot refractive indices has been established. Careful characterization of soot structure properties coupled with the use of RDG-PFA scattering theory certainly merits consideration as a means of measuring soot refractive indices more reliably.

Variations of soot refractive indices within the range illustrated in Figure 8 have a substantial effect on estimations of soot optical properties. This can be seen from results in Table III, where the refractive indices, $E(m)$, $F(m)$ and the ratio $F(m)/E(m)$ are summarized for wavelengths of 514.5 and 1000 nm. Results in the table are based on the measurements of Dalzell and Sarofim (1969), Lee and Tien (1980) and Chang and Charalampopoulos (1990). The variations of refractive index properties over these three studies are substantial roughly 40% for $E(m)$, 80% for $F(m)$ and 90% for $F(m)/E(m)$. From Equations (2), (3) and (6), it is evident that uncertainties in $E(m)$, $F(m)$ and $F(m)/E(m)$ yield corresponding uncertainties in absorption, scattering and soot volume fractions. Thus, effects of uncertainties of refractive index properties are comparable to effects of aggregate scattering in the visible, illustrated in Figure 6 and persist into the infrared as well.

Köylü and Faeth (1993b) have reported a very limited independent assessment of soot refractive indices at 514.5 nm, based on RDG-PFA scattering predictions and measurements in the power-law regime for large soot aggregates emitted from long residence time turbulent diffusion flames. For these conditions, Equations (2) and (10) can be rearranged to yield:

$$F(m)/E(m) = 4\pi(q\,d_p)^{D_f}(\bar{Q}^a_{vv}/(k_f x^3_p Q^{\bar{a}}_a) \tag{14}$$

Then using measured values of d_p, k_f, $\bar{Q}^a_{vv}$ and $\bar{Q}^a_a$, it was found that $F(m)/E(m) = 0.74$ with a standard deviation of 0.09 over all the conditions illustrated in Figure 3. Referring to Table III, this result is in best agreement with the measurements of Dalzell

TABLE III

Refractive index properties of soot

Source	n	k	$E(m)$	$F(m)$	$F(m)/E(m)$
$\lambda = 5.14.6$ nm:					
Dalzell and Sarofim (1969)[a]	1.57	0.56	0.26	0.22	0.84
Lee and Tien (180)[b]	1.90	0.55	0.19	0.30	1.58
Chang and Charalampopoulos (1990)[c]	1.63	0.48	0.21	0.20	0.95
$\lambda = 1000$ cm:					
Dalzell and Sarofim (1969)[a]	1.65	0.72	0.31	0.30	0.97
Lee and Tien (1980)[b]	1.90	0.80	0.27	0.39	1.44
Chang and Charalampopoulos (1990)[c]	1.65	0.50	0.22	0.22	1.00

[a]For acetylene soot pellets at room temperature.
[b]For polystyrene and plexiglass diffusion flames.
[c]For the post flame region of fuel-rich premixed propane oxygen flames.

and Sarofim (1969), suggesting that the soot aggregate scattering approximation used by Tien and Lee (1980) and Chang and Charalampopoulos (1990) may have compromised their results. Nevertheless, more study clearly is needed to resolve the current unacceptably large uncertainties of soot refractive indices.

CONCLUSIONS

Existing information concerning the structure and optical properties of soot within diffusion flame environments was reviewed. The main conclusions from the review are as follows:

1. Soot consists of nearly spherical and monodisperse primary particles collected into mass fractal aggregates exhibiting a log normal distribution function for the number of primary particles per aggregate. While primary particle diameters and aggregate sizes vary with fuel type and flame condition, the fractal dimension and prefactor appear to be relatively robust properties with $D_f = 1.77$ and $k_f = 8.1$ (with standard deviations of 2 and 10%, respectively) based on recent sampling and light scattering measurements.
2. Use of the Rayleigh-Debye-Gans scattering approximation, in conjuction with the fractal properties of polydisperse soot aggregates, yields predictions of soot scattering properties that agreed with measurements within experimental uncertainties. The evaluation considered soot from a variety of fuels over the following test ranges: x_p of 0.128–0.330, $\bar{N}$ of 31–467 and $\bar{N}^2/(\bar{N})^2$ of 1.82–3.23. However, additional evaluation of this methodology would be desirable, emphasizing small-angle (Guinier) behavior, larger values of primary particle diameters and aggregate sizes, and effects of departures of actual soot geometry and physical properties from spherical primary particles in point contact and having uniform refractive indices.
3. The performance of other widely used soot aggregate scattering theories generally was not satisfactory over the same range of conditions: Rayleigh scattering underestimated scattering levels and yielded extinction cross sections up to roughly less than half the correct values for large soot aggregates in the visible portion of the spectrum, while Mie scattering for an equivalent sphere gave inconsistent results because the optical properties of open-structured wispy aggregates having fractal dimensions less than 2 are not well represented by compact objects. Nevertheless, errors from use of the Rayleigh scattering approximation are small in comparison to other uncertainties in the infrared, where this approximation can be used for predictions of the radiative properties of soot and interpretation of extinction measurements to find soot volume fractions.
4. Use of Rayleigh-Debye-Gans polydisperse fractal aggregate theory provides a potential for nonintrusive measurements of both soot concentrations and structure in flame environments but several areas of uncertainty must be addressed to realize this potential, as follows: reliable solution algorithms for the inverse problem, to infer soot structure from scattering and extinction measurements, must be developed; and effects of fuel type and of flame conditions on soot refractive indices must be better understood. In particular, long-standing uncertainties and controversies concerning

the refractive indices of soot are the greatest impediment for reliable nonintrusive optical diagnostics to find soot properties in flame environments and resolving these difficulties merits high priority for future research.

ACKNOWLEDGEMENTS

The authors research on soot structure and optical properties is supported by the Building and Fire Research Laboratory of the National Institute of Standards and Technology. Grant No. 60NANB1D1175, with H. R. Baum serving as Scientific Officer; the Microgravity Science and Applications Division of NASA, Grant No. NAG3-1245, under the technical management of D. L. Urban of the Lewis Research Center and the Office of Naval Research. Grant No. N00014-93-1-0321 with G. D. Roy serving as program manager.

NOMENCLATURE

C	Optical cross section
d_p	Primary particle diameter
D_f	Mass fractal dimension
$E(m)$	Refractive index function, $\mathrm{Im}\,((m^2 - 1)/(m^2 + 2))$
$f(qR_g)$	Aggregate form factor, Equation (3)
f_s	Soot volume fraction
f_{VA}, f_{VR}	Soot volume fractions from aggregate and Rayleigh theories
$F(m)$	Refractive index function $|(m^2 - 1)/(m^2 + 2)|^2$
$g(kR_g, D_f)$	Aggregate total scattering factor
i	$(-1)^{1/2}$
k	Wave number, $2\pi/\lambda$
k_f	Fractal prefactor
m	Refractive index of soot, $n + ik$
n	Real part of refractive index of soot
N	Number of primary particles per aggregate
N_c	Aggregate size at onset of power-law regime, Equation (8)
$p(N)$	Probability density function of aggregate size
q	Modulus of scattering vector, $2k \sin (\theta/2)$
Q	Volumetric optical cross section
R_g	Radius of gyration of an aggregate
$\bar{R}_{gG}^2$	Mean square radius of gyration for the Guinier regime
$\bar{R}_{gL}^2$	Mean square radius of gyration for the power-law regime
x_p	Primary particle size parameter, $\pi d_p/\lambda$
θ	Angle of scattering from forward direction
k	Imaginary part of refractive index of soot
λ	Wavelength of radiation
ρ	Density
ρ_{sa}	Ratio of scattering to absorption cross section

Subscripts

a	absorption
d	differential

e	extinction
h	horizontal polarization
ij	incident (i) and scattered (j) polarization direction
s	total scattering
v	vertical polarization

Superscripts

a	aggregate property
p	primary particle property
$(^-)$	mean value over a polydisperse aggregate population

REFERENCES

Berry, M. V. and Percival, I. C. (1986) Optics of Fractal Clusters such as Smoke. *Optica Acta*, **33**, 577.
Bohren, C. F. and Huffman, D. R. (1983) *Absorption and Scattering of Light by Small Particles*, John Wiley and Sons, New York, 477–482.
Borghese, F., Denti, P., Saija, R., Toscano, G. and Sindoni, O. I. (1984) Multiple Electromagnetic Scattering from a Cluster of Spheres. I. Theory. *Aerosol Sci. Tech.*, **3**, 227.
Bowman, C. T. (1992) Control of Combustion-Generated Nitrogen Oxide Emissions Driven by Regulation. *Twenty-Fourth Symposium (International) on Combustion*. The Combustion Institute. Pittsburgh, 859–878.
Chang, H. and Charalampopoulos, T. T. (1990) Determination of the Wavelength Dependence of Refractive Indices of Flame Soot. *Proc. R. Soc. London*, **A430**, 577.
Charalampopoulos, T. T. (1992) Morphology and Dynamics of Agglomerated Particulates in Combustion Systems Using Light Scattering Techniques. *Prog. Energy Combust. Sci.*, **18**, 13.
Chen, H. Y., Iskander, M. F. and Penner, J. E. (1990) Light Scattering and Absorption by Fractal Agglomerates and Coagulations of Smoke Aerosols. *J. Modern Optics*, **2**, 171.
Dalzell, W. H. and Sarofim, A. F. (1969) Optical Constants of Soot and Their Application to Heat Flux Calculations. *J. Heat Trans.*, **91**, 100.
Dalzell, W. H., Williams, G. C. and Hottel, H. C. (1970) A Light Scattering Method for Soot Concentration Measurements. *Combust. Flame*, **14**, 161.
Dobbins, R. A. and Megaridis, C. M. (1987) Morphology of Flame-Generated Soot as Determined by Thermophoretic Sampling. *Langmuir*, **3**, 254.
Dobbins, R. A. and Megaridis, C. M. (1991) Absorption and Scattering of Light by Polydisperse Aggregates. *Appl. Optics*, **30**, 4747.
Dobbins, R. A., Santoro, R. J. and Semerjian, H. G. (1990) Analysis of Light Scattering From Soot Using Optical Cross Sections for Aggregates. *Twenty-Third Symposium (International) on Combustion*. The Combustion Institute, Pittsburgh, 1525–1532.
Dobbins, R. A., Mulholland, G. W. and Bryner, N. P. (1993) Comparison of a Fractal Smoke Optics Model with Light Extinction Measurements. *Atmospheric Environment*, in press.
Erickson, W. D., Williams, G. C. and Hottel, H. C. (1964) Light Scattering Measurements on Soot in a Benzene-Air Flame. *Combust. Flame*, **8**, 127.
Faeth, G. M., Gore, J. P., Chuech, S. G. and Jeng, S.-M. (1989) Radiation from Turbulent Diffusion Flames. *Annual Review of Numerical Fluid Mechanics and Heat Transfer*, **2**, 1.
Farias, T. and Carvalho, M. G. (1993) Personal Communication.
Felske, J. D., Charalampopoulos, T. T. and Hura, H. (1984) Determination of the Refractive Indices of Soot Particles From the Reflectivities of Compressed Soot Pellets. *Combust. Sci. Tech.*, **37**, 263.
Freltoft, T., Kjems, J. K. and Sinha, S. K. (1986) Power-Law Correlations and Finite Size Effects in Silica Particle Aggregates Studied by Small-Angle Neutron Scattering. *Physical Review*, **B33**, 269.
Iskander, M. F., Chen, H. Y. and Penner, J. E. (1989) Optical Scattering and Absorption by Branched Chains of Aerosols. *Appl. Optics*, **28**, 3083.
Jullien, R. and Botet, R. (1987) *Aggregation and Fractal Aggregates*, World Scientific Publishing Co., Singapore, 46–50.
Kerker, M. (1969) *The Scattering of Light*, Academic Press, New York, 414–486.
Köylü, Ü. Ö. and Faeth, G. M. (1991) Carbon Monoxide and Soot Emissions from Liquid-Fueled Buoyant Turbulent Diffusion Flames. *Combust. Flame*, **87**, 61.

Köylü, Ü. Ö. and Faeth, G. M. (1992) Structure of Overfire Soot in Buoyant Turbulent Diffusion Flames at Long Residence Times. *Combust. Flame*, **89**, 140.

Köylü, Ü. Ö. and Faeth, G. M. (1993a) Radiative Properties of Flame-Generates Soot. *J. Heat Trans.*, **115**, 409.

Köylü, Ü. Ö. and Faeth, G. M. (1993b) Optical Properties of Overfire Soot in Buoyant Turbulent Diffusion Flames at Long Residence Times. *J. Heat. Trans.*, **116**, 152.

Köylü, Ü. Ö. and Faeth, G. M. (1993c) Optical Properties of Soot in Buoyant Laminar Diffusion Flames, *J. Heat. Trans.*, **116**, 152.

Köylü, Ü. Ö., Sivathanu, Y. R. and Faeth, G. M. (1991) Carbon Monoxide Emissions from Buoyant Turbulent Diffusion Flames. *Third International Symposium on Fire Safety Science*, Elsevier, London, 625–634.

Lahaye, J. and Prado, G. (1981) Morphology and Internal Structure of Soot and Carbon Blacks, in Siegla, D. C. and Smith, G. W. (eds.), *Particulate Carbon*, Plenum Press, New York, 33–55.

Lee, S. C. and Tien, C. L. (1980) Optical Constants of Soot in Hydrocarbon Flames. *Eighteenth Symposium (International) on Combustion*, The Combustion Institute, Pittsburgh, 1159–1166.

Lin, M. Y., Lindsay, H. M., Weitz, D. A., Ball, R. C., Klein, R. and Meakin, P. (1989) Universality of Fractal Aggregates as Probed by Light Scattering. *Proc. Roy. Soc. London*, **A423**, 71.

Magnussen, B. F. (1974) An Investigation into the Behavior of Soot in a Turbulent Free Jet C_2H_2-Flame, *Fifteenth Symposium (International) on Combustion*, The Combustion Institute, Pittsburgh. 1415–1425.

Martin, J. E. and Hurd, A. J. (1987) Scattering from Fractals. *J. Appl. Cryst.*, **20**, 61.

Medalia, A. I. and Heckman, F. A. (1969) Morphology of Aggregates—II. Size and Shape Factors of Carbon Black Aggregates from Electron Microscopy. *Carbon*, **7**, 567.

Megaridis, C. M. and Dobbins, R. A. (1990) Morphological Description of Flame-Generated Materials. *Combust. Sci. Tech.*, **77**, 95.

Mountain, R. D. and Mulholland, G. W. (1988) Light Scattering from Simulated Smoke Agglomerates. *Langmuir*, **4**, 1321.

Nelson, J. (1989) Test of a Mean Field Theory for the Optical of Fractal Clusters. *J. Modern Optics*, **36**, 1031.

Newman, J. S. and Steciak, J. (1987) Characterization of Particulates from Diffusion Flames. *Combust. Flame*, **67**, 55.

Puri, R., Richardson, T. F., Santoro, R. J. and Dobbins, R. A. (1993) Aerosol Dynamic Processes of Soot Aggregates in a Laminar Ethene Diffusion Flame. *Combust. Flame*, **92**, 320.

Rossler, D. M. and Faxvog, F. R. (1980) Optical Properties of Agglomerated Acetylene Smoke Particles at 0.5145 μm and 10.6 μm Wavelengths. *J. Opt. Sci. Am.*, **70**, 230.

Rosner, D. E., Mackowski, D. W. and Garcia-Ybarra, P. (1991) Size- and Structure-Insensitivity of the Thermophoretic Transport of Aggregated 'Soot' Particles in Gases. *Combust. Sci. Tech.*, **80**, 87.

Samson, R. J., Mulholland, G. W. and Gentry, J. W. (1987) Structural Analysis of Soot Agglomerates, *Langmuir*, **3**, 272.

Santoro, R. J., Semerjian, H. B. and Dobbins, R. A. (1983) Soot Particle Measurements in Diffusion Flames. *Combust. Flame*, **51**, 203.

Tewarson, A. (1988) Generation of Heat and Chemical Compounds in Fires, in *SPFE Handbook on Fire Protection Engineering*, National Fire Protection Association, Quincy, MA, pp. 1–179.

Tien, C. L. and Lee, S. C. (1982) Flame Radiation. *Prog. Energy Combust. Sci.*, **8**, 41.

Vaglieco, B. M., Beretta, F. and D'Alessio, A. (1990) In Situ Evaluation of the Soot Refractive Index in the UV-Visible from the Measurements of Scattering and Extinction Coefficients in Rich Flames. *Combust. Flame*, **79**, 259.

van de Hulst, H. C. (1957) *Light Scattering by Small Particles*, Dover Publication, New York.

Viskanta, R. and Mengüc, M. P. (1987) Radiation Heat Transfer in Combustion Systems. *Prog. Energy Combust. Sci.*, **13**, 97.

Wersborg, B. L., Howard, J. B. and Williams, G. C. Physical Mechanism in Carbon Formation in Flames. *Fourteenth Symposium (International) on Combustion*, The Combustion Institute, Pittsburgh, 929–940 (1972).

Application of Soot Mass Growth Rates Determined in a Plug Flow Reactor on Soot Prediction in Turbulent Diffusion Flames

B.BARTENBACH, M. HUTH and M. LEUCKEL   *Engler-Bunte-Institute,*
University of Karlsruhe (TH), Combustion Research Section
Postfach 6980, 76128 Karlsruhe (Germany)

Abstract—This paper shows the results of investigations on soot formation or, more precisely, soot growth in a plug flow reactor as a function of well-defined conditions of temperature, stoichiometry and residence time. Measurements of soot concentrations (by gravimetry), particle sizes, temperatures and concentrations of stable organic and inorganic gas species are performed. The measurements are summarized in correlations describing the corresponding influences on soot growth rate and final soot yield.

For an application of those correlations from the plug flow reactor as a "calibrated" model for technical flames, extensive field measurements in free turbulent jet diffusion flames are performed. Variation of local temperatures, carbon concentrations (i. e. mixture fractions), gas species concentrations, soot concentrations, soot yields and growth rates along the axis of the flame are shown and compared for two flames with different exit gas velocities, i. e. exit Reynolds numbers, in order to exemplify possible influences on soot content and emission of this flame type.

Key Words: Soot formation, plug flow reactor, turbulent diffusion flames, gravimetry, calibrated model

INTRODUCTION

The motivation for investigating soot formation in flames is to be attributed to two well-known, sometimes contrary aspects of soot occurrence in technical hydrocarbon combustion:

1. increased radiative heat transfer when soot is present in a flame, which effects e.g. the thermal efficiency of technical combustion processes
2. soot emissions in case of uncomplete soot burn-out are regarded as unacceptable pollution

In principle both aspects can be treated by experimental investigations directly in applied or close-to-application systems of interest. As this would be expensive and difficult in most cases, experimental results from simplified but characteristic experimental combustion systems should be applicable to a universal prediction of soot content and emission of technical turbulent diffusion flames. To meet this aim, some type of correlation or calibrated model from measurements in simplified systems should be available. In this context it must be taken into account that with regard to soot formation related reactions a reliable detailed chemistry has not yet been established.

Therefore, the objective of our work has been the development of a "calibrated" model based on systematic experimental data collected in different model systems. We used a turbulent plug flow reactor with well defined conditions of temperature, flame gas stoichiometry and residence time for the soot formation process. To apply the results to a system with varying conditions and field gradients, extensive field measurements in turbulent jet diffusion flames have been performed in parallel to the plug flow reactor experiments.

Referring to the lack of knowledge in the field of soot formation chemistry, we tried to apply a more global kinetics only depending on temperature and mixture fraction which both are already computable by reliable models in various laboratory-scale and technical flames. Knowing these data the intended correlation should allow to compute soot concentrations in turbulent diffusion flames with a degree of accuracy sufficient for engineering applications.

Similar efforts with simplified kinetic correlations have been made by a number of authors in the past, e.g. Gore and Faeth (1986), Steward, Syed and Moss (1991), based on various experimental systems and fuels.

EXPERIMENTAL

A plug flow reactor (Fig.1) was used for the basic investigations on soot growth under well defined but variable conditions. As a feed stream for the reactor hot flue-gas is produced from natural gas in a precombustion chamber.

The hot gas flow leaving this combustion chamber enters the actual reactor tube at temperatures between $1000°C$ and $1600°C$, controlled by air preheaters and cooling rods and therefore independent of combustion stoichiometry in a certain range. The stoichiometry of the primary burner was varied in the range 0.8 to 2.0 of the air equivalence ratio during the experiments.

Through a multihole injection nozzle, placed in the lower part of the reactor tube, soot-forming hydrocarbons like methane, propane and acetylene have been injected perpendicularly into the main flow. The mixture of fuel and exhaust-gas produced in this way is quite similar to mixtures occurring in turbulent diffusion flames, due to internal reverse flow back-mixing and/or large turbulent eddies with high unmixedness. The initial molar fraction of the hydrocarbons was varied from 0 to 6% for methane, up to 3% in case of propane and up to 4% for acetylene. The corresponding overall stoichiometry after the injection covers a range of air equivalence ratios from 0.4 to 0.9. The temperature regime within the reactor tube can be extended up to $1700°C$.

The mixing zone (mixing time < 10 ms) is followed by the starting soot mass growth combined with characteristic gas phase reactions. In order to perform residence time resolved measurements of soot and gas species concentrations, soot particle size distributions and gas temperatures, various suction probes were introduced at different distances downstream of the injection nozzle. The range of residence time investigated was 10 to 100 ms.

Soot mass concentration is being measured by extracting gas samples through a suction probe with a sintered bronze filter fixed in a small metal tube and placed directly behind the tip of the probe, as shown in Figure 2, and subsequent soot mass measuring by gravimetry.

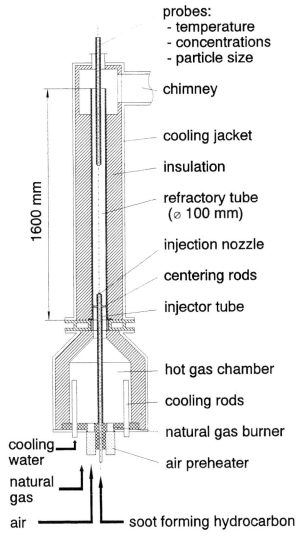

FIGURE 1   Plug flow reactor.

Particle size distributions are determined by measuring the mobility of charged particles in an electrical field using a differential mobility particle sizer (DMPS). Gas temperature is measured by suction pyrometry. The methods applied to determine gas species concentrations continuously are infrared absorption (CO, CO_2, (CH_4)), thermal conductivity (H_2) and paramagnetism (O_2). For discontinuous determination of hydrocarbons up to C_4 gaschromatography is being applied.

For the required measurements in free turbulent jet diffusion flames with fuels equal to those used in the plug flow reactor, the same measuring techniques are being applied, although extended to velocity measurements by LDV or Prandtl-probes to determine

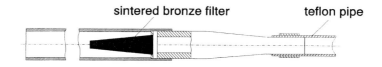

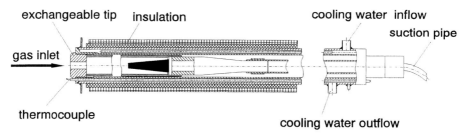

FIGURE 2   Soot sampling suction probe.

flow field structures and residence times along several trajectories, and simultaneous measurements of soot volume fraction and particle size by the extinction ratio technique based upon three wavelengths. Investigations have been performed recently (Lege *et al.* (1993)) to get informations about the comparability of gravimetrically determined soot mass concentrations and soot volume fractions measured with extinction techniques as frequently used in soot formation research. Good qualitative agreement was found and even quantitative agreement can be reached when using a suitable soot density for converting mass concentrations to volume fractions.

Up to now we used a 4 mm nozzle burner with hydrogen co-flow stabilization for our investigations of free turbulent jet diffusion flames. With propane as a fuel we covered Reynolds numbers, defined at the nozzle exit, from 28000 up to 70000. Thus we were able to investigate a certain range of residence times and mixing conditions. In addition, the effect of buoyancy still changes with Reynolds number variation in that regime, which is important in the soot burn-out zone and therefore interesting for later investigations of this process.

Soot concentration field measurements in enclosed turbulent diffusion flames are started to check the present results in systems with more complex flow field structures.

RESULTS

Plug Flow Reactor

Using results from the plug flow reactor, the influence of stoichiometry can be attributed to at least two independent parameters: the stoichiometric ratio of the incoming primary flue-gas (expressed as "primary air equivalence ratio λ_I"), and the stoichiometry after secondary fuel injection into the hot gas flow (expressed as "initial molar fraction of fuel" x_{fuel}^0, or "secondary air equivalence ratio" λ_{II}, or "secondary C/O-ratio" C/O_{II}). By investigating both factors in a wide range, further understanding of the various processes leading to soot formation and growth could be achieved. Thus progress in modelling of soot formation and growth due to similar mixing processes occurring in turbulent diffusion flames on one hand and in the plug flow reactor on the other, *i.e.* mixing of hot lean and cold rich gas regimes, could be expected.

Experimental Results

A large number of runs were performed in the plug flow reactor, especially with propane as a soot forming fuel. In Figure 3 residence time resolved soot concentration measurements are shown plotted versus residence time, with the characteristic reactor tube temperature as a parameter; an exemplary stoichiometry of $\lambda_I = 1.37$ as primary air equivalence ratio and $\lambda_{II} = 0.32$ as secondary air equivalence ratio was chosen for presentation. The secondary air equivalence ratio expresses the amount of free oxygen left in the incoming flue-gas related to the amount of oxygen required for a complete burn-out of the injected fuel. In the low temperature regime it is obvious that soot growth rates are approximately constant, i.e. independent of residence time in the observable range up to 90 ms (at about 1100°C) or up to 50 ms (at about 1350°C), respectively. The influence of temperature on soot growth rates is positive in this temperature regime. Looking at runs 3 and 4 where temperature exceeds 1350°C, deviations from a linear correlation between soot concentrations and residence time are observed at residence times lower than 40 ms. Soot growth shows an exponential decay with time, similar to the behaviour in laminar premixed flat flames (*e.g.* investigations by Baumgärtner *et al.* (1984), Bockhorn *et al.* (1984), Harris and Weiner (1983)), may be due to desactivation of soot surface as discussed by Haynes. Haynes and Wagner (1982) used these experimental data to propose an empirical model for residence time dependence of soot growth basing on a first order law. An extended

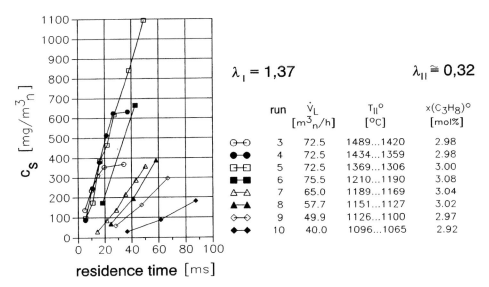

FIGURE 3   Soot mass concentration versus residence time.

application of this empirical model on results from the plug flow reactor will be discussed in the following.

In order to present an example for the influence of a "sub-divided stoichiometry", *i.e.* different results for a constant overall stoichiometry λ_G composed of varying primary and secondary stoichiometries, some additional results with propane as injected fuel are given in Figure 4. In every case of soot appearance in combustion, one important criterion is the critical value of stoichiometry for the beginning of detectable soot formation. The critical values for the secondary C/O-ratio, i.e. injected amount of carbon related to free oxygen, and the corresponding initial molar fraction of propane are plotted versus primary air equivalence ratio.

The theoretical value for the critical C/O-ratio from thermodynamic equilibrium calculation should be 1. We found much higher values near an air equivalence ratio λ_I of 1. On the other hand, values of only about 0.6 are found for very lean conditions of the primary flue-gas. The value of 0.6 is in quite good agreement with critical C/O-ratios determined in laminar hydrocarbon diffusion flames. Remarkably, that there is still a definite amount of fuel necessary for detectable soot formation even at rich pyrolytic primary conditions ($\lambda_I < 1$). This shows the important effect of primary combustion products, in particular H_2O, upon the soot formation limit.

In Figure 5 several runs are compared to illustrate the influence of secondary stoichiometry, expressed as the initial molar fraction of the injected propane, upon soot mass concentration and soot particle diameter, respectively. Parameter is the primary air equivalence ratio λ_I. The runs are representing a stepwise increase of injected propane into flue-gas of constant stoichiometry and temperature. As a consequence, the characteristic temperature after the mixing zone decreases with increasing fuel

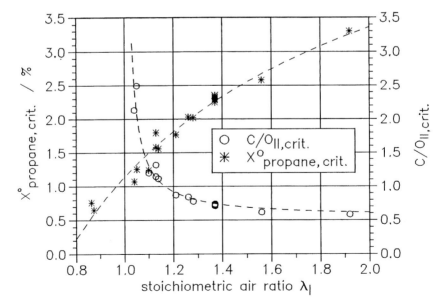

FIGURE 4 Critical C/O-ratio and critical initial molar fraction of propane versus primary air equivalence ratio.

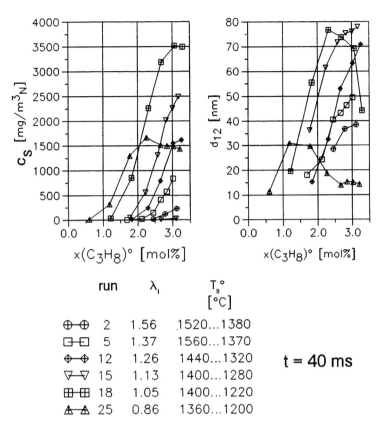

run	λ_I	T_{II}° [°C]
⊕–⊕ 2	1.56	1520...1380
☐–☐ 5	1.37	1560...1370
◆–◆ 12	1.26	1440...1320
▽–▽ 15	1.13	1400...1280
⊞–⊞ 18	1.05	1400...1220
▲–▲ 25	0.86	1360...1200

t = 40 ms

FIGURE 5   Soot mass concentration and particle size versus initial molar fraction of propane with primary stoichiometry as a parameter.

injection, *i.e.* no isothermal conditions for a run with constant primary stoichiometry could be realized. Nevertheless it can be stated that an increasing amount of injected propane leads to increasing soot yield as well as to increasing particle diameters in the range of very lean incoming flue-gas ($\lambda_I > 1.1$), *i.e.* the presence of free oxygen in the flue-gas. Under these oxygen-rich primary conditions, soot formation and growth reactions are accompanied by oxidation reactions based on free oxygen. For more or less pyrolytic conditions ($\lambda_I < 1.1$), *i.e.* no or only small amounts of oxygen left in the incoming flue-gas, the observed behaviour of soot concentration and particle size is changing. Soot concentrations are stagnating though more propane is injected, and mean particle sizes are even decreasing with increasing amount of fuel. Both effects are probably due to the comparatively cold reaction conditions in the extremely rich mixture: less soot is formed for decreasing temperatures. The low values of the mean particle diameter have been evaluated from a bimodal particle size distribution determined at runs 18 and 25, *i.e.* rich primary conditions, which differs from all the others with more lean conditions. It is suggested that the small-diameter fraction of these distributions ($d < 20$ nm) is composed of agglomarated condensates, *e.g.* soot precursors, which are gaseous under flame conditions. That means soot formation partially has been stopped at the step of precurser formation because of very low temperatures and therefore slow reactions.

Figure 6 shows the influence of the amount of injected hydrocarbon, fuel specification and temperature on soot concentration at constant residence times. An increasing soot yield, that means soot concentrations related to the molar fraction of injected fuel-bound carbon, is determined from methane to propane up to acetylene. Influence of temperature seems to be positive in the range of higher carbon concentrations. Critical values of carbon concentrations (soot formation threshold) do not differ significantly for different fuels and temperatures.

Correlations

Data obtained from the investigations on soot formation in the plug flow reactor were used to establish a soot growth correlation which relates soot mass concentration, soot yield and soot growth rates with temperature, initial fuel concentration and residence time. Equation (1) shows this correlation for soot mass concentration which represents all influences considered (Huth (1992)).

$$c_s = [c_\infty \rho_s \frac{T}{T^\theta} \exp[-\frac{1}{2}\left(\frac{T - C_{Tmax}}{C_{\sigma T}}\right)^2] \Delta x_{fuel}^n + c_{s\infty,min}]$$
$$\cdot [1 - \exp(-k_f(t - t^o))]$$

(1)

The expression in the first line of the equation stands for the final soot concentration, reached after long residence times. It depends on the excess molar fraction of the

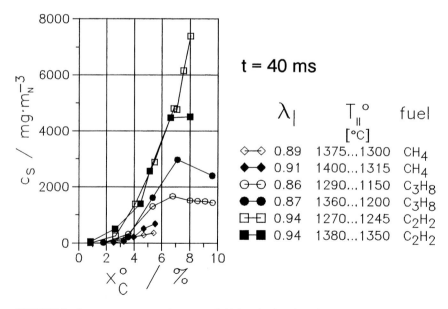

FIGURE 6   Soot mass concentration versus initial molar fraction of carbon from different fuels.

injected fuel multiplied by an exponential bell-shaped temperature function, implying a maximum of final soot concentration in the range of 1100–1250°C. The second line factor represents the time dependence, i.e. the global kinetics. Figure 7 represents equation (1) as a whole. It can be seen that residence time and, therefore, soot formation kinetics looses influence upon the amount of formed soot at high temperatures, because of the very fast kinetics in this temperature range, whereas in the lower temperature range kinetics is important.

Within the time factor expression, k_f is the rate constant determined by an arrhenius law (Fig.8). The activation energy resulting from the experiments is 231 kJ/mole. Basis for the stoichiometry and residence time part of this empirical correlation was the empirical model of Haynes and Wagner (1982), already mentioned above. Their correlation is based on a k_f-coefficient corresponding to the dotted line in the arrhenius plot in Figure 8, representing measurements in premixed laminar flat flames with an activation energy of 180 kJ/mole.

The parameters c_∞, $c_{T\,max}$, $c_{\sigma T}$, and n were chosen to fit the measurements.

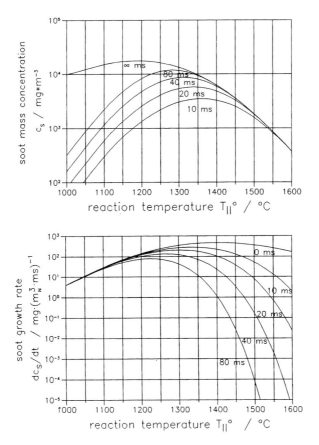

FIGURE 7 Soot mass concentration and growth rate versus reaction temperature (3% propane).

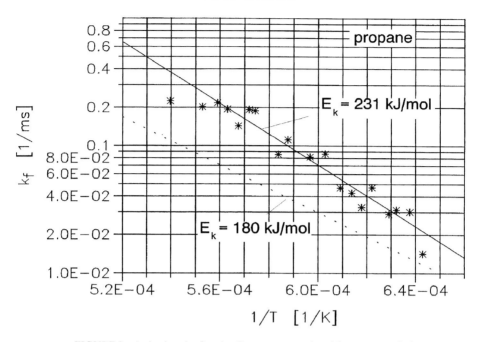

FIGURE 8   Arrhenius plot for plug flow reactor results with propane as fuel.

The standard deviation of predicted and measured soot concentrations for a set of constant values of $c_x = 3.39 \times 10^7 \text{mole}\%^{-n}$, $c_{T\max} = 1175°C$, $c_{\sigma T} = 146°C$, and $n = 1.44$, determined from all runs in the plug flow reactor, is 108mg/m^3. t is the residence time and t_0 the delay time of the soot growing process. Δx_{fuel} stands for the initial molar fraction of fuel minus its critical value (excess molar fraction). T is temperature and $T^{\ominus} = 273,15$ K stands for the reference temperature. This equation is able to describe the conditions in the plug flow reactor when using propane as a fuel, referring to the experimentally determined critical values of stoichiometry mentioned above. It has to be generalized by future experimental data to involve fuel specific parameters. The present measurements with methane and acetylene are not yet adequate for this.

The presented correlation does not take into account the effects of variable composition of mean overall stoichiometry corresponding to the primary flue-gas stoichiometry and the amount of secondary injected fuel. An application for a correlation considering those effects would be interesting for simulating the mixing influences on soot formation in turbulent combustion. This aspect will be another objective of further investigations.

Free Turbulent Jet Diffusion Flames

Experimental Results   The results presented here are based on two exemplary propane diffusion flames. The first one is defined by an exit velocity of 28 m/s and a nozzle diameter of 4 mm corresponding to an exit Reynolds number of 28000. The second one is defined by a exit velocity of 70 m/s and a Reynolds number of 70000.

Main objectives were measurements of local soot mass concentration, or soot volume fraction in case of extinction measurements, respectively, local temperature and local stoichiometry. Stoichiometry is detectable as total carbon concentration, which is directly equivalent to the mixture fraction. The determination of all stable species concentrations is requested for this.

In particular, the concentrations of small hydrocarbons like acetylene, ethylene and methane are interesting for their assumed participation in soot growth processes as carbon sources. Certainly, the decomposition of fuel components is of strong interest. Figure 9 shows the variation of propane (fuel), acetylene, ethylene and methane along the axis of the flame with Re = 70000. The strong formation of methane and ethylene at the beginning is due to the pyrolytic decomposition of propane shown in parallel. As ethylene is partially dehydrated to acetylene with increasing flame temperature, there is less ethylene than methane in the range of high acetylene concentrations. On the other hand, methane is a more stable species than ethylene in general. Since acetylene is regarded as being particularly important for soot surface growth, it is remarkable that detectable concentrations are found up to $x/d_0 = 240$, accompanied by somewhat smaller amounts of methane.

Variation of temperature and total carbon concentration (equivalent to mixture fraction) are plotted in Figure 10. The dilution occurring in the turbulent jet is represented by the decreasing values of local carbon concentration, independent of reaction. Temperature assumes a maximum of about 1300°C at $x/d_0 = 180$, the same range were the soot concentration shows its maximum value, as can be seen in Figure 11.

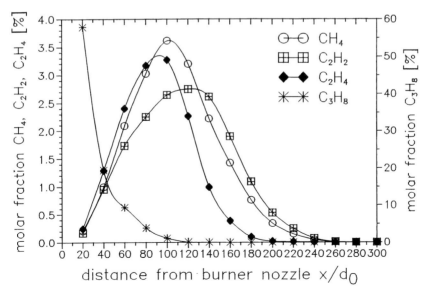

FIGURE 9   Molar fractions of fuel (propane) and intermediate hydrocarbon species versus burner distance.

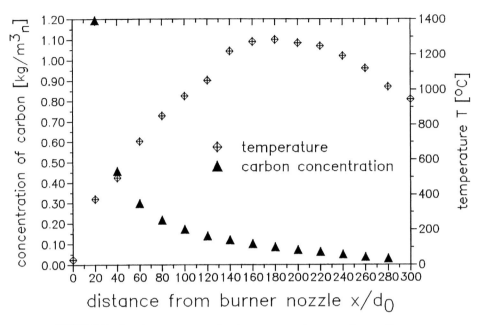

FIGURE 10   Local temperature and carbon concentration versus burner distance.

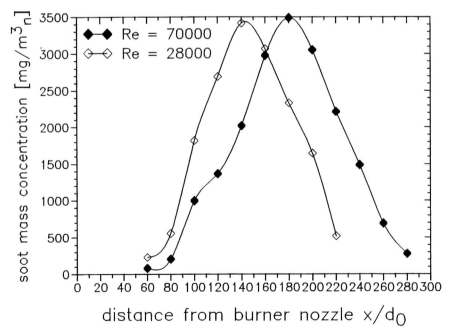

FIGURE 11   Comparison of soot mass concentration versus burner distance.

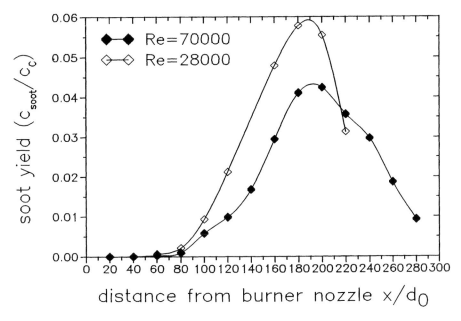

FIGURE 12   Comparison of soot yields for two different flames

To show the influence of residence time and, partially, mixing conditions, soot concentrations in two flames with Re = 70000 and Re = 28000, respectively, are compared in Figure 11, plotted along the axis of the flames. The maximum values are approximately equal, the location of the maxima is differing due to the differing residence times. Soot growth kinetics has to be taken into account, so that the observed delay for the flame with lower residence time (Re = 70000) is not unexpected.

Another type of comparison between the two flames discussed above is shown in Figure 12, where related soot concentrations or soot yields, i.e. related to the local concentration of total carbon, are plotted versus burner distance. It is remarkable that the flame with Re = 70000 shows a significantly lower maximum value of the soot yield, though the maximum values are nearly equal for the soot concentrations as shown in Figure 11. On the other hand, the location of the maxima of soot yields is identical, which is different in case of the comparison of soot concentrations

Numerical Modelling   For numerical modelling of soot concentration fields in turbulent flames the correlation shown in equation (1) is derived by time to calculate growth rates. The expression for fuel concentration Δx_{fuel} is converted into an expression of local mixture fraction to make it applicable for flow field calculations. Additionally the influence of residence time expressed by the arrhenius law was neglected for the weak effect on growth rates for typical flame temperatures $< 1350°C$ and residence times $< 100ms$, referring to Figure 7. For the flow field calculation a suitable turbulence model has to be used, which will be the k-$\in$ model. To avoid the implementation and solution of radiation and reaction/heat release submodels, where reliable results are difficult to get, the measured temperature field is being adapted for numerical imulation. A predicted soot concentration field, reliable enough to be ready for presentation, is not yet available. This is due to the lack of an adequate soot oxidation model and remaining uncertainties in flow field calculations. Both, the implementation of a correlation from literature and the development of a suitable model from additional plug flow reactor experiments are in progress and will be presented in future work.

CONCLUSIONS

Major Results from Plug Flow Reactor Experiments

– Soot mass growth rates turned out to be constant at low temperatures or residence times, respectively. For higher values of temperature and residence time, soot growth shows an exponential decay, corresponding to a first order law as already proposed in literature for premixed laminar flat flames.
– The general influence of the amount of injected fuel upon soot mass concentration and particle size is shown to be positive for comparatively oxygen-rich and hot reaction conditions. These tendencies are being modified for cold pyrolytic conditions.
   Mean particle radii where observed in the range of 10 to 80 nm.
– Increasing tendencies towards soot formation in the sequence methane/ propane/ acetylene were determined within the plug flow reactor.
– Correlations are developped describing soot concentrations and growth rates occurring in the plug flow reactor as a function of temperature, stoichiometry and residence time. Temperature dependence could be expressed as a bell-shaped function with a maximum occurring in the range of 1100 to 1250°C. The activation energy corresponding to the rate constant for soot growth from propane was found to be 231 kJ/mole.

Major Results from Investigations on Turbulent Diffusion Flames

– For the evaluation of extensive field measurements in free turbulent jet flames soot concentrations have been related to the local concentration of total carbon, in order to eliminate influence of dilution. Soot growth rates are expressed as temporal changes of the soot yield, correspondingly.
– A comparison of flames with varied exit gas velocities and corresponding Reynolds numbers in the range of 28000 to 70000 gives an idea of the influence of mixing conditions and residence times. A delay of soot formation for the "fast" flame is shown, i.e. the maximum value of soot concentration is situated further downstream, and the soot yield of the "fast" flame was measured to be lower.

FUTURE WORK

An application of the correlations resulting from the measurements in the plug flow reactor on soot prediction in free turbulent jet diffusion flames is in progress. For this purpose the influence of stoichiometry in equation (1) has to be changed from an initial secondary fuel concentration into a local flame stoichiometry, expressed by either the local air equivalence ratio or the local mixture fraction, which is detected by measuring local concentrations of total carbon in the flame. In addition, the expression for a definite soot concentration will be transformed into a function for soot growth rates. Thus, the influence of dilution in the free jet flame can be avoided.

Detailed investigations of soot formation from methane and acetylene in the plug flow reactor are in progress as well. In this context the addition of a soot oxidation rate correlation also based on measurements under well-defined plug flow conditions is essential to improve the predictability of soot content and emissions of turbulent diffusion flames.

In order to supply a reliable data base for further verification of the correlations proposed, additional field measurements in free-burning or confined turbulent jet flames of various fuels have been started.

ACKNOWLEDGEMENT

The investigations reported were performed within the joint research program of the 'Sonderforschungsbereich 167', financed by the **Deutsche Forschungsgemeinschaft**, Bonn, Germany. The authors would like to express their gratitude for this support.

NOMENCLATURE

c_c	local concentration of total carbon
c_S	soot mass concentration (in general)
c_{soot}	local soot mass concentration
c_x	pre-factor of bell-shaped temperature function
c_{Tmax}	maximum value of bell-shaped temperature function
$c_{\sigma T}$	deviation factor of bell-shaped temperature function
$c_{Sx,min}$	final soot concentration corresp. to detectable threshold
$C/O_{II,crit.}$	critical secondary molar relation carbon to oxygen
d_0	diameter of burner nozzle
d_{12}	mean diameter of soot particles
E_k	activation energy
k_f	soot growth rate constant
n	fuel-specific exponent
Re	Reynolds number
t	residence time
t^0	delay time of soot growing process
T	local temperature
$T^{\ominus}$	= 273.15 K, reference temperature
T_{II}^0	characteristic temperature in the reactor tube
V_L	volumetric air flow
x	distance from burner nozzle
$x(C_3H_8)^0$	initial molar fraction of propane
$x_{propane,crit}^0$	critical initial molar fraction of propane

x_{fuel}^0	inital molar fraction of fuel
$\Delta x_{fuel}{}^n$	initial molar fraction of fuel minus its critical values (excess molar fraction)
λ_I	primary air equivalence ratio
λ_{II}	secondary air equivalence ratio
λ_G	overall air equivalence ratio
ρ_s	soot density

REFERENCES

Bartenbach, B., Huth M. and Leuckel, W. Investigations on Soot Mass Growth in a Plug Flow Isothermal Reactor Combined with Soot Concentration Field Measurements in Turbulent Diffusion Flames, *Proceedings of the Anglo-German Combustion Symposium*, The British Section of The Combustion Institute, Cambridge, 491 (1993).

Bockhorn, H., Fetting, F., Heddrich, A. and Wannenmacher, G. (1984) *20th Symposium (Int.) on Combustion*, Investigations of the Surface Growth of Soot in Flat Low Pressure Hydrocarbon Oxygen Flames, The Combustion Institute, Pittsburgh, 979.

Böhm, H., Hesse, D., Jander, H., Luers, B., Pietscher, J., Wagner, H. Gg. and Weiss, M. *22nd Symposium (Int.) on Combustion*, The Influence of Pressure and Temperature on Soot Formation in Premixed Flames, The Combustion Institute, Pittsburgh, 403 (1988).

Gore, J. P. and Faeth, G. M. *21st Symposium (Int.) on Combustion*, Structure and Spectral Radiation Properties of Turbulent Ethylene/Air Diffusion Flames, The Combustion Institute, Pittsburgh, 1115 (1986).

Harris, S.J. and Weiner, A.M. (1983) Soot Particle Growth in Premixed Toluene/Ethylene Flames, *Comb. Sci. Tech.*, **38**, 75.

Haynes, B. S. and Wagner, H. Gg. (1982) The Surface Growth Phenomenon in Soot Formation, *Z. Phys. Chem. N.F.*, **133**, 201.

Huth, M. (1992) *Untersuchung der Russbildung bei partieller Brennstoffoxidation und Pyrolyse*, Ph.D. Thesis, Dept. of Chem. Eng., University of Karlsruhe (TH).

Kent J. H. and Wagner, H. Gg. Why Do Diffusion Flames Emit Smoke?, *Comb. Sci. Tech.*, **41**, 245.

Lege, R., Bartenbach, B., Müller, A. and Leuckel, W. (1993) Application of a Multiple Wavelength Extinction Ratio Technique for Soot Particle Determination in Turbulent Diffusion Flames, *Proceedings of the Anglo-German Combustion Symposium*, The British Section of The Combustion Institute, Cambridge, 483 (1984).

Stewart, C. D., Syed, K. J. and Moss, J. B. (1991) Modelling Soot Formation in Non-prmixed Kerosine-Air Flames, *Comb. Sci. Tech.*, **75**, 221.

Acoustic Pyrometry: A Correlation Between Temperature Distributions and the Operating Conditions in a Brown Coal Fired Boiler

W. DERICHS, F. HEß, K. MENZEL and E. REINARTZ   *RWE Energie AG Kruppstraße 5 45128 Essen - Germany*

Abstract—Acoustic pyrometry utilizes the physical relationship linking the speed of sound to the temperature of the medium through which the sound is propagated in order to derive the temperature and temperature distribution in combustion chambers.

During a several months lasting investigation, the temperature distribution in the combustion chamber of a brown coal fired 600 MW_e-boiler was measured via acoustic pyrometry under different boiler conditions. The effect of burner configuration, recirculated flue-gas, staged air and coal quality on temperature was investigated. The derived temperature distributions were correlated with boiler fouling behaviour and emission values.

Key Words: Acoustic pyrometry, brown coal boiler, operating conditions, temperature distributions, fouling emission

INTRODUCTION

Acoustic temperature measurement is a novel, non-interfering temperature measuring process, suitable for obtaining temperatures and their distributions over the cross-section of large furnaces "Recent work (Derichs and König, 1990; Derichs *et al.*, 1991; Breuer, Derichs and Heß, 1992)".

This method has undergone extensive testing at RWE Energie for the past several years. During this trial period and with the support of the manufacturer, RWE Energie has applied the experience gained from numerous measurement campaigns in power plants to make this technique more reliable with regard to operation. In order to meet the needs of the boiler operator, a new software program has been developed to better process and represent the measurement data for effective process control.

An investigation lasting several months was initiated to evaluate the operational usefullness of acoustic pyrometry. The tasks of this investigation were to determine

– what effects operational parameters
– mill/burner arrangement,
– mill/burner load,
– recirculated flue-gas,
– overfire air operation,
– load reduction,
– soot blower operation,   would have on furnace temperature and their distribution,
– whether temperatures and their distribution could be correlated with boiler fouling and
– whether temperature and their distribution could be correlated with NO_x-emissions.

DESCRIPTION OF FACILITIES AND MEASUREMENT SYSTEM

Boiler

The investigations were carried out on a lignite fired boiler with a steam rate of 1800 t/h and a capacity of 600 MW$_e$ (Fig. 1). Of the eight tangential register burners available, seven are typically in operation as the mill assigned to the eighth burner undergoes maintenance. With higher heating values, only 6 burners need be running. As NO$_x$-reduction measures

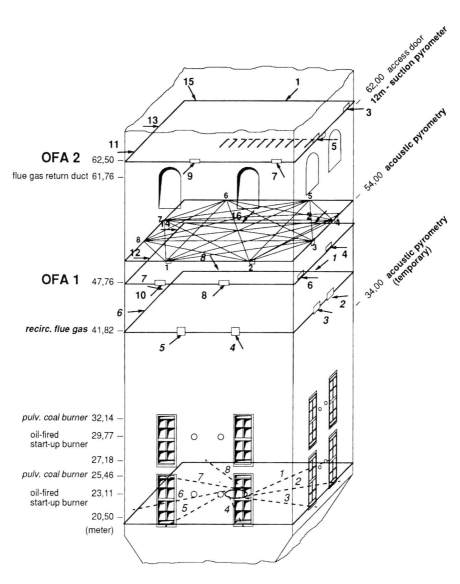

FIGURE 1   600 MW$_e$ boiler.

– Consolidation of burners to lower burner levels
– Air staging (burner air, overfire air, OFA levels 1 and 2; about 300 °C)
– Flue-gas recirculation (FGR, 160–190 °C)

have been applied to the furnace.

The recirculated flue-gas is tangentially blown into the furnace, similar to the fuel and combustion air. The overfire air enters the furnace at two levels, at each level perpendicular to the wall, but offset from the middle. Both the recirculated flue-gas and the overfire air support the upwards spiraling flame column.

At full load, the recirculated flue-gas flowrate is adjusted to approximately 180,000 m³/h (valve position 25 %). The overfire air control valves are adjusted to valve positions of 70 % (level 1, around 230,000 m³/h) and 100 % (level 2, around 240,000 m³/h). The recirculated flue-gas is not necessary to meet NO_x-emission standards, but lowers the furnace temperature and thereby reduces fouling.

Acoustic Pyrometry

Acoustic pyrometry is based on the temperature dependency of the speed of sound (c) in gas, according to the equation from La Place:

$$c = (\kappa\, R T/M)^{1/2}$$

Knowing the composition of the gas through which the sound is transmitted, the ratio of specific heats ($\kappa = C_p/C_v$) and the molecular weight (M) can be determined. The universal gas constant (R) is known.

The speed of sound is determined by measuring the propagating time of a sound wave along a known path lenght. In solving the La Place equation for *T*, the mean integral temperature along any given path lenght can be calculated with knowledge of *R*, κ and *M*. For the investigation, eight transceivers were mounted on the furnace periphery at a height of 54 m, resulting in a grid of 24 measurement paths (Fig. 1). For a short time, a second system was installed at a height of 34 m, yielding data from that level simultaneously.

Integral temperature values determined along these measurement paths can be used with the help of mathematical deconvolution techniques – similar to those used in computer tomography – to calculate a temperature distribution. A detailed description of the mathematics and physics behind acoustic temperature measurement, including a comprehensive error analysis, can be found in "Derichs and König, 1990". According to this analysis, the integral path temperatures can be determined with an uncertainty of ± 1%.

At any given time, one unit is switched to "transmitter mode", while the others function as receivers, such that with each impulse seven measurement paths are simultaneously analyzed. Per cycle, every unit functions once as a transmitter. A dense grid of individual measurement paths results, with each path being sounded from both directions. A falsification of integral temperature will occur with tangential-fired furnaces, namely, through a "pulling effect" by the flue gases flowing parallel to the measurement paths. The propagating time will either be extended or shortened – depending on the relative directions of signal and flue-gas. Since the measurement paths are sounded from both directions, this falsification of the integral temperature is compensated for.

One measurement period (7–10 minutes in duration) consists of 10–14 cycles, such that the integral temperature is obtained 20 times for each measurement path (10 times in any one direction). At the end of each measurement period, the installed software tests the integral temperatures for plausibility. This is done by taking 20 measurement values from each path, averaging them, and calculating a percental standard deviation. If this standard deviation is larger than 2 %, the outliers are removed from the group - until an average value with a standard deviation less than 2 % is obtained for each measurement path. These average values are then used to calculate an isothermal map.

In previous measurement campaigns, the calculated tomographic temperature distribution based on integral measurement values was verified with suction pyrometry. The confidence range of suction pyrometry is known to be $\pm 20\,^{\circ}C$ "Michelfelder and Thielen, 1992". Applying this data to statistically evaluate the pair of temperature variates, one arrives at a confidence range of $\pm 28\,^{\circ}C$ for the tomographically obtained temperature profile. Figure 2 illustrates a comparison of temperature profiles from boiler wall to the furnace middle obtained by acoustic pyrometry and suction pyrometry. In consideration of the confidence intervals, the temperature curves agree well with each other.

For purposes of ruling out any deviations generated by the deconvolution mathematics, the following results section will contain just one tomographic representation - the remaining temperature distributions will be represented in the form of integral path temperatures (i. e., direct measurement values).

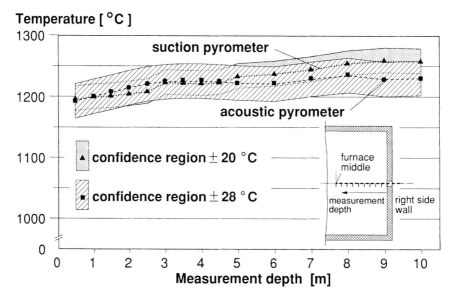

FIGURE 2   Comparison of the temperature progressions obtained by suction pyrometry (62 m) and acoustic pyrometry (54 m) 600 MW$_e$ boiler.

TEST DESCRIPTION AND DOCUMENTATION

The acoustical temperature measurement system was installed for a period of 16 months on a 600 MW_e boiler. The experiments took place from August to November of 1992. Certain boiler operating conditions were held constant for a days length to investigate the temperature effects of mill/burner configurations, flue-gas recirculation, and overfire air. The boiler temperature, fouling and emission behaviour were investigated under

– 8 different mill/burner combinations,
– varying fuel allocation to the mills,
– varying amounts of flue-gas recirculation (valve position: 5 % to 40 %).
– varying overfire air (level 1) flowrates with uniform nozzle distribution (valve positon: 0 % to 100 %),
– varying overfire air (level 1) flowrates with non-uniform nozzle distribution (valve positon: 35 % to 100 %),
– load reduction,
– soot blower operation.

As a qualitative indicator of boiler fouling, the condition of the measurement port openings was documented and evaluated in five degrees from no deposits at all to extremely heavy fouling/slagging. The mineral composition of the ash deposits was not analysed.

As a documentation of coal quality, the results of the proximate analysis of shift averaged coal samples (8 h - mean) and the classification according to the coal quality code (Tab. I) evaluating the ash, iron and potassium content were collected.

Emission data was processed daily via the central emissions monitoring system (ZEUS, 30 min-mean).

Temperature data was collected and represented as path temperature distributions (10 min-mean). As a degree for the eveness of the temperature distributions are the temperature differences between

– the maximum and the minimum path temperature $\Delta T_{min, max}$
– the maximum and the minimum arithmetical mean of the six path temperatures at a measurement port $\Delta T^*_{min, max}$
– the maximum and the minimum corner path temperature $\Delta T_{corner\ path}$
  stated in the figures.

RESULTS

In the results section, the effect of operational parameters (mill/burner, flue-gas recirculation, overfire air, load and soot blowing) on the temperature and its distribution will be discussed. Additionally the correlation between temperature and boiler fouling and NO_x-emissions will be reviewed.

TABLE I

Classification of Rhinish brown coal

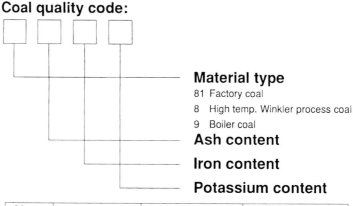

Coal quality code:

Material type
81 Factory coal
8   High temp. Winkler process coal
9   Boiler coal

Ash content

Iron content

Potassium content

Class	Ash (%)	Iron (ppm)	Potassium (ppm)
1	0 - 2,5	0 - 2000	0 - 400
2	2,5 - 7	2000 - 3000	
3	7 - 12	3000 - 4000	400 - 800
4	12 - 16	4000 - 5000	800 - 1000
5		5000 - 6000	1000 - 1200
6	16 - 24		
7		6000 - 8000	1200 - 1600
0		> 8000	> 1600

▓▓▓ Limit for mixed coal (class ≥ 4 $\overset{\wedge}{=}$ problem coal, P)

Influence of Operating Parameters on Temperature and its Distribution

The main parameters chosen for investigation were mill/burner configuration and loading, flue-gas recirculation, overfire air and load reduction.

Mill/burner operation   During the investigation period the temperature distribution at a height of 54 m was observed under various mill/burner arrangements and loads. Each mill/burner combination generated a characteristic temperature distribution at full load (FGR at 25 %, OFA 1 at 70 % and OFA 2 at 100 %). The temperature distribution of a given mill/burner combination varies slightly, e. g. due to heat value fluctuations. Notwithstanding the temperature distributions remain typical for its given mill/burner configurations.

   A characteristic for each combination is a low temperature zone situated across from the non-operating burner, as seen in Figure 3 for mill/burner 7 out of operation. Mill/burner 2 out of operation revealed a temperature distribution different in its characteristics from the other mill/burner combinations. Although there is still a zone

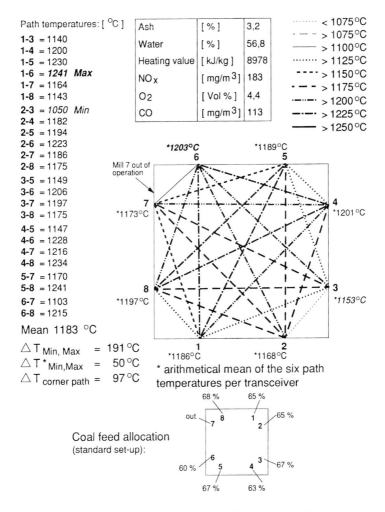

Path temperatures: [°C]

1-3 = 1140
1-4 = 1200
1-5 = 1230
1-6 = *1241 Max*
1-7 = 1164
1-8 = 1143

2-3 = *1050 Min*
2-4 = 1182
2-5 = 1194
2-6 = 1223
2-7 = 1186
2-8 = 1175

3-5 = 1149
3-6 = 1206
3-7 = 1197
3-8 = 1175

4-5 = 1147
4-6 = 1228
4-7 = 1216
4-8 = 1234

5-7 = 1170
5-8 = 1241

6-7 = 1103
6-8 = 1215

Mean 1183 °C

$\triangle T_{Min, Max}$ = 191 °C
$\triangle T^*_{Min,Max}$ = 50 °C
$\triangle T_{corner\ path}$ = 97 °C

Ash	[%]	3,2
Water	[%]	56,8
Heating value	[kJ/kg]	8978
NO_x	[mg/m^3]	183
O_2	[Vol %]	4,4
CO	[mg/m^3]	113

.... < 1075°C
– – – > 1075°C
——— > 1100°C
············ > 1125°C
- - - - > 1150°C
· — · — > 1175°C
·—··—· > 1200 °C
—··—· > 1225 °C
——— > 1250 °C

*1203°C   *1189°C

Mill 7 out of operation

*1173°C   *1201°C

*1197°C   *1153°C

*1186°C   *1168°C

* arithmetical mean of the six path temperatures per transceiver

Coal feed allocation
(standard set-up):

68 %   65 %

out

65 %

60 %

67 %   63 %

67 %

FIGURE 3   Path temperatures resulting from a standard fuel allocation set-up (% = conveyor belt speed).

with low temperatures opposite of the not operating mill/burner 2, a confinedly limited area of high temperature still opposite to mill/burner 2 was found. However the temperature distribution was even after long periods of time and for different coal qualities consistently recognised as typical for mill/burner 2 out of operation.

Besides the combination of the operating mills/burners is their fuel allocation, recorded as the coal feed allocation in %, of considerable influence on the temperature distribution. Starting from a standard set-up (Fig. 3), with the characteristic temperature distribution for mill/burner 7 not operating, the coal feed allocation was increased respectively decreased keeping the sum of the coal feed constant.

While changing the allocations for the different mills/burners the average furnace temperature, calculated from the 24 path temperatures, remained nearly constant. At first a bad trimming of the coal feed led to a very uneven temperature distribution. After a further change in the coal feed allocation the temperature distribution returned to its former evenness and after modification of the allocation in several steps a considerably more even temperature distribution was obtained (Fig. 4). At all different coal feed allocations the temperature distribution showed a more or less spread zone with low temperature opposite of the not operating burner remaining recognizable as typical for mill/burner 7 not operating.

During the tests of the trimming of the coal feed the trend of the wall temperatures, which are not shown in the figures with path temperature distributions, were recorded. The wall paths have only qualitative information value. The absolute temperature

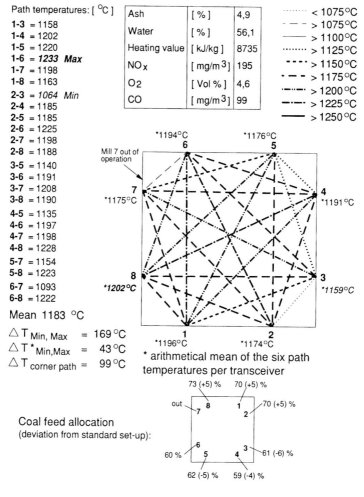

FIGURE 4   Path temperatures resulting from a changed fuel allocation.

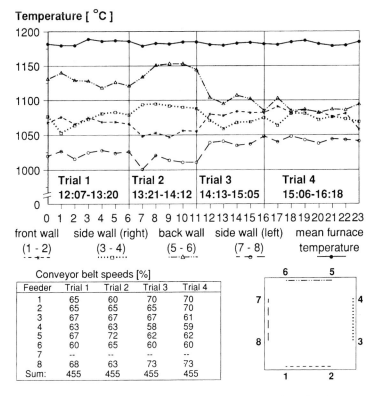

FIGURE 5   Wall temperatures resulting from different fuel allocations.

cannot be evaluated due to the steep temperature gradient near the walls causing Fermat bending, see "Derichs and König, 1990". But the trend of the wall path temperatures in Figure 5 still revealed, as Figures 3 and 4, that an ideal trimming of the coal feed allocations results in very even temperature distributions.

The Figure 6 presents two temperature distributions mathematically deconvoluted out of the 24 path temperatures. They were obtained simultaneously at the 34 m and 54 m level, the one at the 34 m level representing a higher mean and maximum temperature than the 54 m one. The differences between the distribution of the zones with high respectively low temperatures point out the more turbulent flame situation just downstream of the burner level. At the same time Figure 6 clearly unveils how the asymmetric distribution caused by the mill/burner combination upraises spiralwise with the fire column due to the tangential firing.

Recirculated flue-gas operation   Increasing the flowrate of the recirculated flue-gas lowers, as expected, inversely proportionally the mean furnace temperature, Figure 7. The temperature distribution remained qualitatively unchanged at all different test adjustments. The minimum and the maximum arithmetical mean of the six path

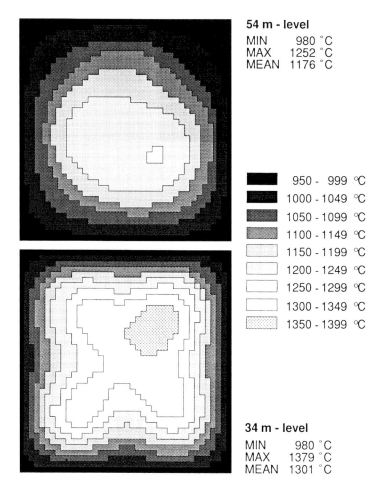

FIGURE 6   Simulations temperature distributions at the 34 m - and 54 m - level.

temperatures at the measurement ports, T^* in the figures, as well as the position of the minimum path temperatures kept the same position in all the temperatures distributions. Only the maximum path temperature changed its position slightly during the test with the control valve at 40 % from path 4–8 to path 5–8.

Over fire air operation   The task of the overfire air is to afterburn unburned carbon-monoxide (CO) and to keep the CO-emission small at simultaneously low NO_x-emissions. The temperature distribution in the measurement plane is directly influenced only by the lower level of the overfire air, OFA 1. Its 300 °C hot air is blown into the furnace 7 m upstream of the measurement plane. Nevertheless a different distribution of the OFA changes the air ratio in the burner level. Thus the modified air ratio in the burner level, and therefore indirectly the OFA 2, affects the temperature distribution in the measurement plane.

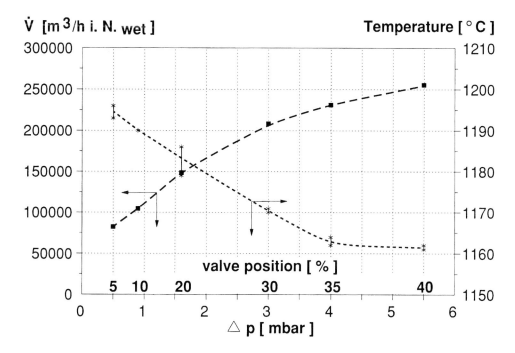

FIGURE 7 The dependence of recirculated flue gas flowrate and mean furnace temperature on the pressure difference across the flue gas fan control valve position.

OFA 1 With uniform distribution Figure 15 presents the trend of the mean furnace temperature in the measurement plane depending on the control valve position of the OFA 1. The NO_x-, CO- and CO_2- curves will be discussed in the chapter "correlation between temperature und NO_x-emission". The temperature decreased reproducibly to a minimum at a valve position of 70 % (1179 °C) and increased slightly at a valve position of 100 % (1190 °C). The reason for the increase of the temperature in the measurement plane at a 100 %-valve position of OFA 1 is presumable the after burning of unburned CO. Up to a valve position of 70 % the distribution of combustion air from the burner level toward the OFA 1 lowers the temperature in the measurement plane by retarding the combustion and spreading it over a longer section of the furnace. But at a 100 % -valve position the air ratio at the burner level is so low, that the much higher amount of unburnt CO leads to a considerable afterburning creating so much more heat than the furnace walls absorb that the temperature increases.

Concerning to the reference values for the evenness of the temperature distributions $(\Delta T_{min, max}; \Delta T^*_{min, max}; \Delta T_{corner\ path})$ the temperature distributions became up to the 70 %-valve position more even and again uneven at 100 %-valve position. During all different valve positions of the OFA 1 the temperature distribution remained typical for mill/burner 5 not operating with low temperatures in the area of the back wall and of the right side wall. Hence, uniformly distributed OFA does not decisively affect the qualitative temperature distribution.

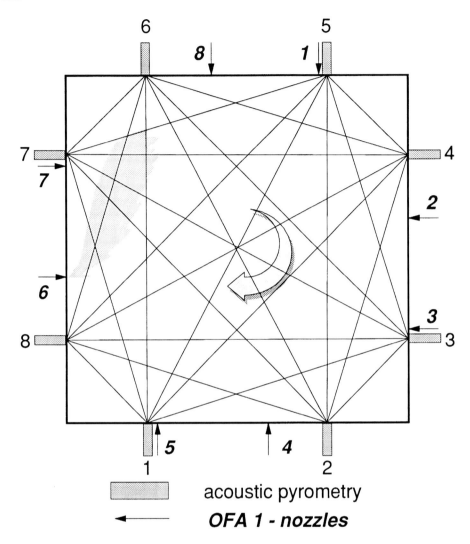

FIGURE 8   Arrangement of the overfire air (level 1) nozzles and the acoustic pyrometry transceivers.

OFA 1 with non-uniform distribution   To illustrate the influence of the overfire air on the temperature Figure 8 demonstrates the arrangement of the eight overfire air nozzles at level 47,76 m relative to the position of the eight transducers of the acoustic pyrometry at level 54 m and schematically the diversion of the overfire air vane. When blown into the furnace the overfire air is caught by the spiralwise upraising flame column and diverted clockwise. Hence, the overfire air influences in each case the temperature zone left of the nozzles, in Figure 8 e.g. OFA-nozzle 6 the zone in front of transceiver 7 and directly path 6–8. Temperature changes in these zones are indicated by a change in the arithematical mean of the six path temperatures at each transceiver and by a change of the directly affected measuring paths.

The Figures 9 and 10 demonstrate the different effects of single OFA-nozzles on the temperature in the measurement plane. The adjustments of the OFA control valves are stated in the figures.

In the considered mill/burner combination adding more "cold" OFA - 300 °C - at the nozzles 2 to 7 to the "hot" flue-gas leads to a decrease in temperature in the zone of the transceiver units left of the individual nozzles and in the path temperature of the paths directly influenced, e. g. see Figures 9 and 10 and Table II; nozzles No. 5/7 influencing transceiver 8/6 and paths 1–7/5–7.

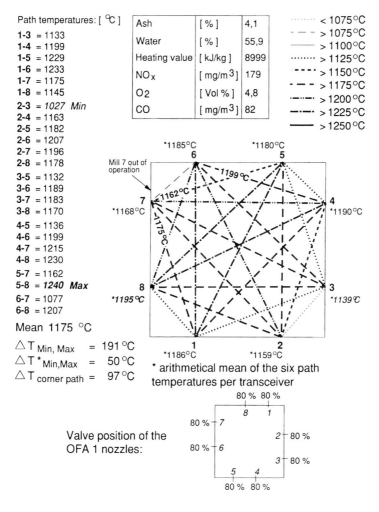

FIGURE 9   Path temperatures resulting from a uniform overfire air distribution (level 1).

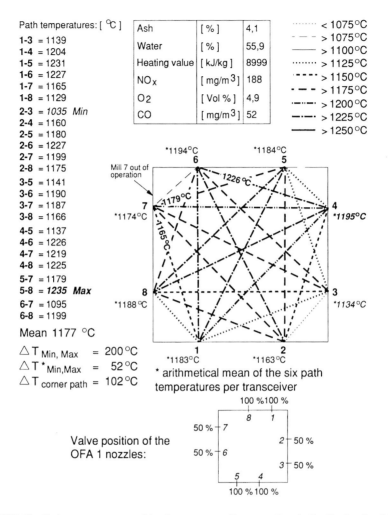

FIGURE 10   Path temperatures resulting from a non-uniform overfire air distribution (level 1).

Adding more OFA at the nozzles No. 1 and 8 on the other hand cause an increase in temperature in the zone of transceiver 5 respectively 4. The reason for this contrary behaviour is presumable a thread with a high amount of unburned matter in the area influenced by the OFA-nozzles No. 1 and 8. This assumption is confirmed by the CO-concentration, which is the lowest with highest OFA-flowrates at nozzle No. 1 and 8, Figure 10. In these areas the after burning produces such an amount of heat energy, that the flue-gas temperature increases although "cold" OFA is added. An afterburning happens in the near of the other OFA-nozzles (No. 2–7), too. But the heat production is not large enough to prevent the decrease of the flue-gas temperature when "cold" OFA is added to the "hot" flue-gas.

TABLE II

Different effects of overfire air on temperature

OFA-nozzle No.	Valve position %	transceivers No./T [°C]	path No./T [°C]
5	80	8/1195	1–7/1175
5	100	8/1188	1–7/1165
7	50	6/1194	5–7/1179
7	80	6/1185	5–7/1162
8	80	5/1180	4–6/1199
8	100	5/1184	4–6/1126

Load Reduction

A representative number of periods with load reduction were evaluated to determine the influence of load reduction on the furnace temperature. In Figure 11 the mean furnace temperature, determined as the arithemetical mean of the 24 path temperatures, is outlined as a function of the boiler load. The temperature values for full load were recorded just before and after the evaluated periods with reduced load. As expected, the mean furnace temperature decreases with load reduction. The large scatter of the measured values is partly traceable to the natural variations in the heating value of the lignite. However in Figure 11 a considerable number of the measured values are distinctly situated above the main collective. Fouling/slagging on the furnace walls and thus a bad heat transfer is the reason for the high temperatures of these measured values.

The temperature distributions found at reduced load can be more even as well as more uneven than the temperature distribution of preceeding respectively succeeding full load. The scatter of the evenness of the temperature distributions at reduced load is comparable to the one at full load.

From the investigation period typical temperature distributions for reduced load cannot be presented. As there are up to three mills/burners simultaneously not operating and as the reduced boiler load differs from 395 MW to 530 MW, there are more possibilities existing to combine the mills/burners as compared with relatively strictly fixed full load operation. Besides the temperature behaviour at reduced load was not systematically investigated. The boiler was operated according to the directions of the load dispatcher without planning to adjust the same reduced load situations.

Soot Blower operation

Soot blowers – using plain water in the radiation part and reheated steam in the convection part – are operated as required. The acoustic pyrometry is temporary affected only by the operation of the soot blowers using reheated steam, although these are situated further downstream in the convection part of the boiler. Operating these soot blowers increases the ground noise level considerably. The moment, at which the acoustic signal arrives at the receiver, is difficult to detect due to the higher ground noise level. Measuring paths affected by the temporary malfunctioning of the units show in general for a measuring cycle (around 10 min) distinctly recognizable measuring errors.

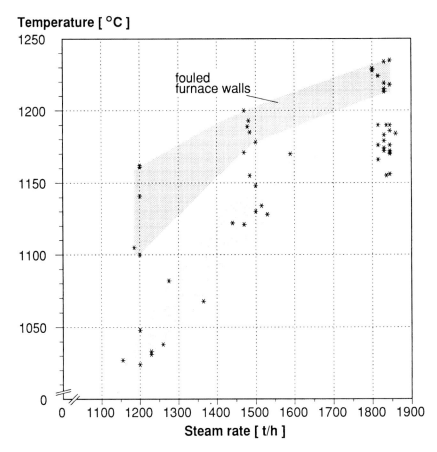

FIGURE 11   Mean furnace temperature as a function of boiler load.

The mean flue-gas temperature, determined of the 24 path temperatures, decreases after the use of the soot blowers in the radiation part up to 40 °C according to the cleaning effect of the soot blowers and the degree of the fouling/slagging at the furnace walls.

Correlation of Temperature and Fouling/Slagging

Fouling/slagging problems occured during the period of the investigations at different temperature levels and different coal qualities, as to be expected according to "Bals, *et al.*, 1981". The temperature influence is illustrated by the fouling/slagging found at the transition of the measuring ports to the furnace wall. Fouling/slagging (see + − symbols in Fig. 12) occurs mainly at the "hot" measuring ports, i. e. the ports with highest arithemetical means of the six path temperatures. Fouling/slagging is less frequent at low temperatures. Nevertheless fouling/slagging may occur even at low temperatures with coals of nonproblematic iron and/or potassium contents according to the coal quality code. A further decrease of temperature would possibly avoid

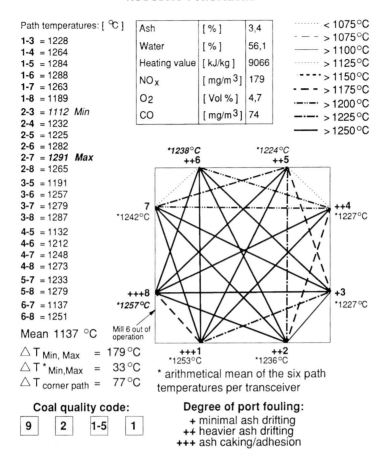

Path temperatures: [°C]

1-3 = 1228
1-4 = 1264
1-5 = 1284
1-6 = 1288
1-7 = 1263
1-8 = 1189
2-3 = 1112 *Min*
2-4 = 1232
2-5 = 1225
2-6 = 1282
2-7 = 1291 *Max*
2-8 = 1265
3-5 = 1191
3-6 = 1257
3-7 = 1279
3-8 = 1287
4-5 = 1132
4-6 = 1212
4-7 = 1248
4-8 = 1273
5-7 = 1233
5-8 = 1279
6-7 = 1137
6-8 = 1251

Mean 1137 °C

$\triangle T_{Min, Max}$ = 179 °C
$\triangle T^*_{Min, Max}$ = 33 °C
$\triangle T_{corner\ path}$ = 77 °C

Ash	[%]	3,4
Water	[%]	56,1
Heating value	[kJ/kg]	9066
NO_x	[mg/m^3]	179
O_2	[Vol %]	4,7
CO	[mg/m^3]	74

.... < 1075°C
- - - > 1075°C
——— > 1100°C
.... > 1125°C
····· > 1150°C
- · - > 1175°C
·—··— > 1200°C
—·—· > 1225°C
——— > 1250°C

*1238°C
++6

*1224°C
++5

7
*1242°C

++4
*1227°C

+++8
*1257°C

+3
*1227°C

Mill 6 out of operation

+++1
*1253°C

++2
*1236°C

* arithmetical mean of the six path temperatures per transceiver

Coal quality code:

| 9 | 2 | 1-5 | 1 |

Degree of port fouling:
+ minimal ash drifting
++ heavier ash drifting
+++ ash caking/adhesion

FIGURE 12   Path temperatures and fouling of the transceiver ports.

fouling/slagging. However the furnace temperature cannot be reduced at full load to such a low level that even with problematic coal qualities fouling/slagging does certainly not occur.

Therefore the leading parameter for fouling/slagging is for lignite, not the temperature, but the compositon of the fuel. Events, as grate surcharging as occured during the investigation period, cannot be foreseen using the acoustic pyrometry, although the increase of the mean furnace temperature due to bad heat transfer at slagged furnace walls amounts up to 50 °C.

With coal of more constant composition, e.g. bituminous coal, it is nevertheless likely that the acoustic pyrometry can be used to optimize soot blower operation and to avoid fouling/slagging.

Correlation of Temperature and NO_x-emission

The influence of changes in the temperature on the NO_x-emission is at the operating temperatures of a lignite fired boiler with temperature peaks up to 1300 °C small

NO$_x$ [mg/m^3]

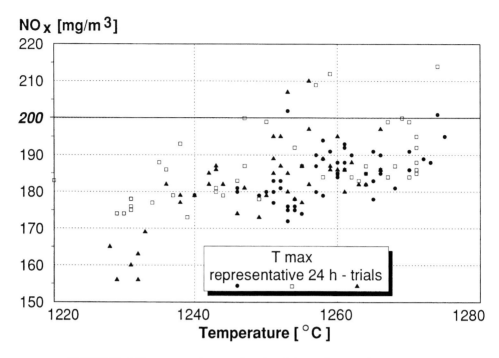

FIGURE 13   NO$_x$ concentrations as a function of the maximum path temperature.

compared to the effects of the overfire air and the coal compositon, "Recent work (Dürselen, H. J., 1992)". Nevertheless the evaluation points out, that NO$_x$-emissions are lower at low furnace temperatures and even temperature distributions.

Figure 13 presents the 1/2 h-mean NO$_x$-values as a function of the maximum path temperature of the temperature distributions determined simultaneously to the NO$_x$-emission.

The NO$_x$- and temperature values measured during three days while operating at full load were examplary evaluated. Although the NO$_x$-values scatter up to ± 15 mg/m^3, mainly an effect of the coal composition, they increase with temperature. Notwithstanding the NO$_x$-emission limit of 200 mg/m^3 can be met in general even at temperatures of 1300 °C. However it is noteworthy, that NO$_x$-emissions higher than 200 mg/m^3 occur only when the maximum path temperature is greater than 1250 °C.

In Figure 14 the above NO$_x$-values are plotted against the evenness of the simultaneously determined temperature distributions. As a measure of the evenness serves the difference of the minimum and the maximum path temperature related to the mean temperature. The NO$_x$-emission scattering around ± 15 mg/m^3 decreases with more even temperature distributions, i.e. smaller temperature differences.

The dependency of the NO$_x$-, CO-, O$_2$-emission and the mean furnace temperature on the OFA 1-valve position is illustrated in Figure 15. All parameters reach a minimum at a valve position of 70 %, the usual operating status. As described already in the chapter "Overfire air operation" it is the operating status with the most even temperature distribution.

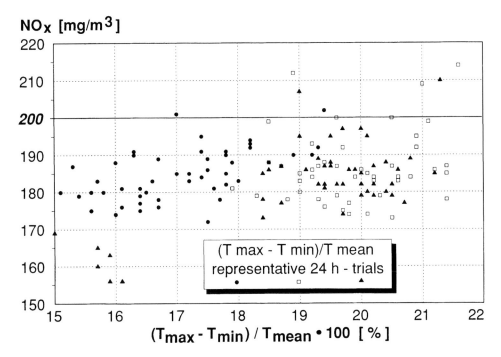

FIGURE 14　NO_x concentration as a function of temperature evenness.

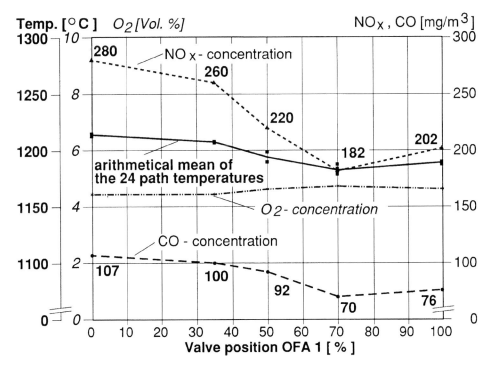

FIGURE 15　NO_x, CO, O_2 concentration and the arithmetical mean of the 24 path temperatures as a function of overfire air (level 1) value position.

CONCLUSION

During a several months lasting trial period the acoustic pyrometry was installed at a lignite-fired 600 MW_e-boiler to investigate the temperature and its distribution for different operating conditions. The investigations pointed out the influence of the operating mode of the mills/burners, the recirulated flue-gas, the overfire air, the load reduction and the soot blowers. Alterations in the operating mode of these parameters have a reproducible influence on the temperature distribution. The operating mode of the mills/burners, the recirculated flue-gas and the overfire air results – depending on their allocation – in distinctly more even temperature distributions in the measurement plane. The effects on temperature by alterations in the operating mode are recognized immediately by the acoustic pyrometry.

Fouling in particular slagging occurs mostly in the hot zones determined by the acoustic pyrometry. High temperatures promote, of course, fouling/slagging in the boiler, but the extremly varying composition of the Rhinish lignite is a main influencing parameter. Fouling/slagging cannot be avoided when burning coals with high iron and/or potassium content. To recognize boiler fouling early is not possible by acoustic pyrometry alone, when the fuel composition varies extremly. Nevertheless the mean furnace temperature increases distinctly – up to 50 °C - when fouling/slagging occurs. It is likely, that for constant coal compositions, as e. g. bituminous coal, the acoustic pyrometry could recognize boiler fouling early enough to prevent larger problems.

The overfire air operating mode and the coal composition are of substantial impact on NO_x-emissions. The influence of the temperature and its distribution on NO_x-emissions is small due to relatively low flue-gas temperatures with peaks of about 1300 °C. Nevertheless the investigations unveiled that the NO_x-emissions increases with the temperature. Where as NO_x-emissions decreases with the temperature distribution becoming more even.

ACKNOWLEDGEMENT

The investigations could not have been realised without the intense support of the engineering staff of the technical and the production department of the power station Neurath and the staff of the department for measurements at the head quarters.

REFERENCES

Bals, M., Glaser, W., Hein, K., Reimer, G. and Rüsenberg, D. (1981) Untersuchung der Eigenschaften Rheinischer Braunkohlen und deren Einfluß auf die Verschmutzungsneigung bei der Verfeuerung in Großkessel, In *Report RWE Energie.*
Breuer, N., Derichs, W. and Heß, F. (1992) Application of acoustic pyrometry to a lignite- fired utility boiler. In *VGB-Kraftwerktechnik*, **72**(1) (1992).
Derichs, W. and König, J. (1990) Die Schallpyrometrie - Ein Meßverfahren zur Bestimmung der Temperaturverteilungen in Kesselfeuerungen. In *DVV-Kolloquium*, Sept 1990.
Derichs, W., Dewenter, U. and König, J. (1991) Die Schallpyrometrie, Möglichkeiten und Grenzen eines Meßverfahrens zur Bestimmung der Temperaturverteilung in Kesselfeuerungen. In *VDI-Flammentag*, Sept. 1991.
Dürselen, H. J. (1992) Feuerungsseitige NO_x-Minderung bei Braunkohlestaubfeuerungen, Einfluß von brennstoff- und betriebsseitigen Parametern. Dissertation, Ruhr-University Bochum, July 1992.
Michelfelder, S. and Thielen, W. (1992) Verfahren zur Ermittlung von Flammentemperaturen (ed.) Technik der Messung hoher Temperaturen. In *VDI-Seminar* 36-19-09.

Evaluation of an Amplitude Sizing Anemometer and Application to a Pulverised Coal Burner

N. G. ORFANOUDAKIS and A. M. K. P. TAYLOR   *Imperial College of Science Technology and Medicine, Department of Mechanical Engineering, Exhibition Road, London SW7 2BX, United Kingdom*

Abstract—The accuracy and precision of a sizing anemometer, based on diffraction of light from a laser beam of 240 μm diameter and 514.5 nm wavelength at a collection angle of 3.82 with a central "pointer volume" formed by a laser Doppler test volume of 41 μm diameter and 488 nm wavelength and using non-spherical coal particles, have been evaluated experimentally. The required independent measurement of the coal partgicle size was based on an aerodynamically determined diameter which was related to the measured exit velocity of the coal particles at the exit of a rapidly converging nozzle. The accuracy and precision were found to be $+ 2, + 7, - 20$ μm and $\pm 2, \pm 7, \pm 10$ μm, respectively, for mean values of 15.30. 60 μm at a coal feeding of 0.0133 (g/min), corresponding to a number density of $7.1\ 10^{-10} (m^{-3})$. The poor accuracy for the largest size is due to saturation of the sizing photodetector. The accuracy and precision became worse by 10% and 50%, respectively, when the feeding rate increased by a factor of 35 and thus when the number density also increased from $7.1\ 10^{-10}$ to $2.5 \cdot 10^{-8}$. The application of the instruments illustrated by simultaneous measurements of size and velocity in a small scale, 11.3 kW, burner operated at a swirl number of 0.5 and piloted by a gas pilot flame supporting a 0.24g/min coal flame. The mean velocities showed that particle size classes characterised by mean diameters of 12 and 36 μm recirculated while those of 64 μm penetrated the recirculation zone. The arithmetic and Sauter mean diameters decreased with axial and radial distance, primarily due to the centrifuging effects of the swirl air.

Key Words: Particle sizing, laser doppler velocimetry, pulverised coal burners

INTRODUCTION

Knowledge of the location at which coal particles mix with recirculated hot gases is important because this controls the rate of volatile release and initiation of burnout. The amount of volatile release, particularly in the region of the internal recirculation zone (which shall be referred to as IRZ in the following text), is determined by the residence time in this region which is dependent on both the size and velocity of the pargicles, as well as on the aerodynamic field of the swirling gases. The recent predictions of, e.g., Costa *et al.* (1990) of species concentrations, temperatures, flow patterns and particle trajectories showed that an increase of swirl number by about 28% caused a reduction on the depth of penetration of the large (105 μm) particles by about 20%; consequently the reduction in penetration depth for higher swirl results in increased residence times in the IRZ. Additionally, it is important for NO_x emission reduction to maintain fuel rich conditions in the IRZ. Abbas *et al.* (1993) have used three particle size distributions to study the effect of particle size on NO formation in a laboratory scale coal-fired furnace and they found that small (25 μm) and large particle size segregations (121 μm) produced flames with lowr NO emissions. However, there are few papers which report measurements of either coal particle velocity or

1115

size. Exceptions include the work of Clausen *et al.* (1990) in a 500 kW furnace and Dugué and Weber (1992) who provided velocity information for the near burner area of a semi-industrial scale coal furnace. Clausen *et al.*, were able to measure velocity, but not size, and therefore assessed, theoretically, the error in mean gas velocity due to averaging over all coal particle sizes and they found it to be about 10% away from the IRZ but they pointed out the need for simultaneous size and velocity measurement. In the work of Dugué and Weber (1992) only primitive sizing information was available using the amplitude of the pedestal of the Doppler signal in order to discriminate, qualitatively, between the small and large particles. They found the particle slip is large in the fuel jet region close to the coal injector. Bonin and Queiroz (1991) have provided some of the few available size and velocity measurements, although their instrument was able to provide statistical, rather than instantaneous, correlations between these two variables and also measured far downstream of the burner area of an industrial furnace. There are thus few measurements of the fate of coal particles immediately after their injection in the secondary swirling air flow and none with simultaneous measurement of size and velocity, which are necessary for understanding of the phenomena related to volatile release in the IRZ.

The primary reason for this state of affairs is the difficulty of making the necessary measurements although there has been much development of instruments for simultaneous measurements of size and velocity such as, for example, the phase Doppler anemometer for spherical droplets. For non-spherical particles, such as pulverised coal, other principles must be used, including relating either the amplitude of light scattered from a laser beam (e.g. Maeda *et al.*, 1986), or the so-called visibility of the signal from a laser Doppler anemometer (e.g. Adrian and Orloff, 1977 and Negus and Drain, 1982), to the particle diameter. Both methods must remove the so-called trajectory ambiguity problem, one way, and that used in this work, is the pointer volume technique (e.g. Yeoman *et al.*, 1982). Comprehensive and up to date literature reviews are provided by Taylor (1992) and Jones (1993).

The advantage of the visibility, over the amplitude, technique is that it is, to a first approximation (see Kliafas *et al.*, 1990), independent of the incident laser power. However, the experimental execution of visibility measurement is more difficult than the amplitude technique, mainly because of the complicated electronic instrumentation involved. Additionally, visibility is a multivalued function of size and therefore the maximum size that can be measured is limited, sometimes unacceptably so. In this work the amplitude technique is investigated.

For the measurement of size of non-spherical particles by an instrument based on the above methods, the collected light should be overwhelmingly scattered by diffraction rather than reflection or refraction. Diffraction is insensitive to particle shape and refractive index, neither of which can be known accurately in applications such as the combustion of pulverised coal particles. Predominantly diffractively scattered light is collected by placing the receiving optics at a small angle to the direction of the laser radiation incident on the particle to be sized. The effect of uncertainty in refractive index on sizing accuracy has been established experimentally (e.g. Hong and Jones, 1976, 1978) but is readily found theoretically using Mie theory (e.g. Negus and Drain, 1982 and Holve and Self, 1979) and the error is of the order of 10%.

The effect of shape on sizing accuracy has been investigated recently by Or-fanoudakis and Taylor (1992) who concluded that the accuracy of size measurements of pulverised coal, over the range 10–90 μm, decreased from 19% to 9% when the mean collection angle was reduced from 2.5° to 1.4°. Although these results are some of the few to quantify the likely errors, their technique had some shortcomings because the experiments involved "static" mounting of particles on a glass flat. This implied simulation of the transient signal, no investigation of the potential effects of multiple occupancy, and the population size used for the evaluation was small (about 100 in total) because the mounting method was so time consuming.

The preceding paragraphs suggest that further tests on the optical sizing accuracy of the amplitude technique would be desirable, incorporating a pointer volume system and providing larger sample sizes on which to base statements of sizing tolerance. The required independent measurement of size was based on an aerodynamically deter-mined diameter (Marple and Rubow, 1976 and Wilson and Liu, 1980) using the flow nozzle reported by Hardalupas *et al.* (1988). Thus, the original contribution of this paper is to quantify the tolerance that can be expected of typical amplitude-sizing instruments for *irregular* particles: estimates of this tolerance are absent in the literature and thus statements of experimental accuracy cannot be made.

The method was based on introducing a small number of the particles to be sized on the centreline and at the upstream end of an abrupt nozzle with air flowing through it. As the air accelerated through the nozzle, the particles were dragged forward by the air, but moved with a slip velocity relative to the air because of their intertia. The slip velocity between the air and the coal particles increased with the aerodynamic diameter of the particles. Hence, a measurement of the *velocity* of a particle leaving the nozzle exit could be converted, by calculation or previous calibration of the nozzle, to the *aerodynamic diameter* of the particle. At the same time, measurements of the maximum amplitude of light scattered by the same particle provided, through previous calibra-tion, the *optical diameter* of the particle.

The purposes of this paper are: (a) to quantify the accuracy and precision of an instrument to measure the size of non-spherical particles based on amplitude and the pointer volume arrangement and (b) to report preliminary measurements using the instrument in a small coal burner and investigate the flow in the area near the burner quarl exit. The experimental method used is described in the following section and the results, together with discussion, are presented under the third section; the more important conclusions are summarised in the final section.

EXPERIMENTAL METHOD

Optical and Electronic Arrangement

Optical system   The instrument developed and used in this work was based on the maximum amplitude of the scattered light for the sizing and a conventional dual beam laser Doppler anemometer system for the velocity measurement. A two colour pointer volume was used and the instrument is shown in Figures 1a and 1b.

An Ar-ion laser (Spectra Physics 164) was operated in multi line mode to provide the two colours necessary for the pointer volume arrangement. Two Pellin-Brocca

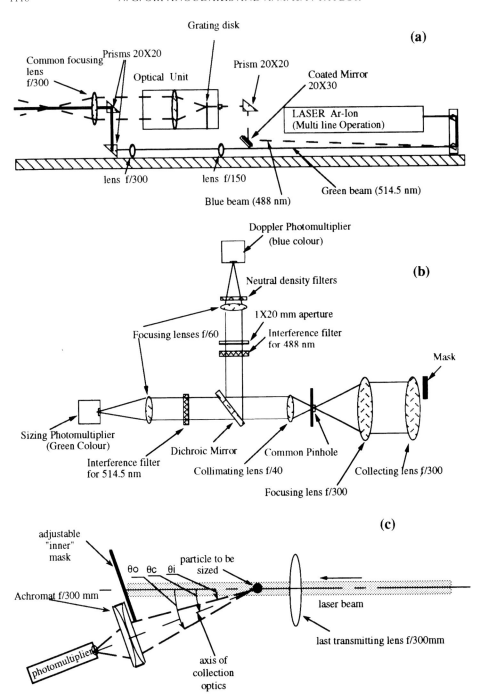

FIGURE 1 Schematic of the *sizing* anemometer. (a) Transmitting optics. (b) Receiving optics. (c) Layout of the optical system, with definition of the collection, θ_c, the inner masking, θ_i, and the outer masking, θ_0 angles.

dispersion prisms separated the colours of the laser beam: the blue line, at 488 nm, was passed through a radial diffraction-grating based optical unit to form the laser Doppler *pointer* volume with diameter $(1/e^2)$ of 41 μm using a 300 mm focal length lens, which was common to both colours. The green line, at 514.5 nm, formed the beam from which sizing information was derived and was guided through a pair of lenses (300 mm and 150 mm focal length) to provide, by axial translation of the lenses, adjsutment of the diameter of the sizing beam.

In order to overcome the trajectory ambiguity problem, due to the Gaussian intensity distribution in the sizing beam, it is necessary to make sure that particles selected for sizing are illuminated by as nearly constant and intensity region of sizing beam as possible. This is achieved, as is obvious, by increasing the ratio between the $1/e^2$ diameters of sizing and pointer volume. In Orfanoudakis (1994), three methods to measure – as opposed to calculate – the size of the probe volumes are given. The ratio of the diameters of the two probe volumes can be increased: (i) by adjusing the lenses in the sizing beam optical path and consequently increasing the sizing beam diameter, (ii) by decreasing the Doppler-photomultiplier voltage which reduces the pointer volume effective diameter, or (iii) by using neutral density filters in front of the Doppler-photodetector, which again is equivalent to the reduction of the pointer volume diameter. After a series of preliminary experiments, using a 25 μm precision pinhole as a scatterer which was passed through the pointer volume to give a valid Doppler signal, it was found that the signal amplitude produced by this pinhole did not change by more than 10% for ratios of the two measuring volume diameters greater than 5. Thus, the position of lenses for the sizing beam were adjusted in such a way that the magnified images of both probe volumes, using a microscope objective lens, resulted in a diameter ratio of about 6. The voltage of the power supply for both photomultipliers was set to 1200 V for most of the measurements reported below. Additionally, a combination of neutral density filters of 3.15% and 50% transmission was used with the glass beads measurements. These tests resulted in the choice of 240 μm for the diameter of the sizing beam which overlapped, and was concentric with, the doppler/pointer test volume. The details of the procedure for alignment, and precautions with laser mode operation, are fully described by Orfanoudakis (1994). It should be noted that trajectory ambiguity is removed at the expense of overall test volume dimensions which are larger than those of the anemometer and may therefore result in undesirably low limitations to the particle number density if multiple occupancy is also to be avoided.

Although Orfanoudakis and Taylor (1992) suggested the use of the smaller of the two mean collection angles, defined in Figure 1c, which they investigated, namely $\theta_c = 1.4°$, preliminary experiments performed with particles in the accelerating nozzle showed that this value resulted in frequent multiple occupancy of the test volume. The larger collection angle value of $\theta_c = 3.82°$, with $\theta_i = 1.4°$ as inner mask angle, resulted in fewer multiple particle occurrences and was therefore adopted for extensive tests.

Figure 1b shows the arrangement of the receiving optics in which an achromat pair, each of 300 mm focal length, focussed the scattered light of both colours onto a pinhole of 300 μm diameter, acting as a spatial filter. An achromat of 40 mm focal length subsequently collimated the light through a dichroic mirror which transmitted the green (514.5 nm) and reflected the blue (488 nm) light. Thereafter, interference filters reduced the "cross-talk" between the two colours and two 60 mm focal length achromats focussed the light onto pinholes at the front of the housings of photomultipliers (THORN-EMI 9817 B and 9815B). Measurement showed that the cross-talk from any of the photomultipliers to the other one was less than 6%. The resulting overall magnification of the receiving system was 1.5 and the optical characteristics are summarised in Table I. To improve the visibility of the Doppler signal, an aperture of 1×20 mm was used, see also Figure 1b.

It is worth emphasizing that *the procedure used for the adjustment of the receiving optics and electronics was as important for the accuracy and precision of the instrument as the choice of nominal values of, for example, the mean collection angle and pinhole size.* Full details are provided by Orfanoudakis (1994). The need for the maintenance of alignment *during* experiments using sizing anemometers based on the amplitude of scattered light cannot be over-emphasized and this was ensured by daily, and sometimes more frequent, checks on alignment and calibration which, although time-consuming, were feasible because the flow configurations were unconfined — in other words the beams did not have to pass through glass windows — and small scale. In applications where these conditions do not obtain, on-line monitoring arrangements, such as those described by Holve (1993), are mandatory.

TABLE I

Parameters of the optical system

Transmitting optics	
Laser type	Ar-Ion
Laser operation	Multi-line TEM$_{00}$
Total laser power (operation)	420mW
Exit diameter of beam (e^2):	1.5mm at 514.5 nm and 1.2 nm at 488 nm
Doppler volume characteristics	
Focal length/diameter of focusing lens on grating disk	80mm/5mm
Focal length/diameter of collimating lens	300mm/100mm
Focal length/diameter of last focusing lens	300mm/95mm
Beam separation	48.3mm
Beam crossing angle	9.3
Fringe spacing	3.04μm
Frigne number in e^2 diameter (calculated)	13
Frequency shift	1.21 and 5.7 MHz
Doppler volume diameter (calc.):	41.5μm
Doppler volume length (calc.):	500μm
Sizing volume characteristics	
Diameter of sizing volume (e^2):	240μm
Focal length/diameter of first lens (from laser side):	300mm/31.5mm
Focal length/diameter of second lens:	150mm/22.4mm
Focal length/diameter of focusing lens:	300mm/95mm
Receiving Optics	
Angle of collection	3.82
Focal length/diameter of receiving lens	300mm/52mm
Focal length/diameter of focusing lens	300mm/52mm
Focal length/diameter of collimating lens	40mm/18mm
Focal length of lenses focusing on photomultipliers:	60mm/25.4mm
Common pinhole diameter	300μm

Calibration Procedure

Calibration permitted conversion of the measured value of the maximum amplitude of light scattered by an irregular particle into an optically determined size. The size inferred from measurement of amplitude refers to the diameter of a calibration pinhole providing the same maximum amplitude as the particle in question. For the experiments reported here, a stationary pinhole together with a beam chopper was used, mainly because of its ease of use over alternative methods (see Orfanoudakis, 1994, for an extensive discussion). The calibration of the system and the measurements were conducted using the laser working in multi line operation and at a power of 420 mW which corresponded to 19 A current drawn through the tube. The accuracy of the power setting was limited through the setting of the current to ± 0.5 A and the laser mode was continuously monitored to ensure operation in TEM_{00}. The laser power used was determined as a compromise between various factors such as detectability of small particles, photodetector saturation limits and maintenance of the laser mode, and the value of 420 mW is specific to the system described here. Different arrangements, such as use of a laser with different beam output diameters, would require re-evaluation of the appropriate power. The laser power stability was checked using a 50μm precision pinhole as scatterer and observing the amplitude of the signal of the sizing photodetector. Continuous observation over half an hour showed that the value for the maximum amplitude did not change by more than 5%.

Electronic System

The electronic instruments used, together with the two colour system described above, are shown schematically in Figure 2. These consisted of appropriate high pass filters, a zero crossing counter and a transient recorder (a fast analogue to digital converter with 4 kbytes of memory) interfaced to, and controlled by, a PDP-11 computer. The Doppler frequency measurements were made by a 16 bit custom-built zero crossing counter described by Hardalupas (1989) and, in this configuration, can be expected to be accurate to within about 2% percent. The temporal variation of the amplitude of the

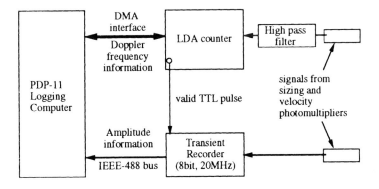

FIGURE 2   Schematic of the block diagram of the eletronic instruments' arrangement.

light scattered by the particle traversing the green beam was recorded using an eight bit transient recorder (Datalab DL912); the maximum sampling rate of this instrument was 20 MHz and it was operated with pre-tigger timebase and external trigger mode. The repeatability of the measured amplitude was 3%, the digitisation error 3% and the maximum uncertainty 6%. The repeatability is also connected to the drift of the photomultiplier which was examined by Orfanoudakis (1994) and was found to be less than 5%.

The measurement sequence was continguent on the *detection* of a Doppler signal by the counter: this instrument, in turn, controlled the start of the operation of the transient recorder. The pre-tigger time of the transient recorder, together with the sampling rate, were set as values appropriate to the measurement location.

The validation of Doppler frequency measurements from the zero crossing counter, was based on the well known 5/8 concept, implemented in software. The digitized signal from the sizing channel was examined in software to exclude multiple occupancy events by examining whether there was more than one peak on the main signal signature, or there were other peaks within a user-set time window on either side of the signal of interest.

Flow Configurations

Accelerating nozzle   The accelerating nozzle is shown in Figure 3a. Nozzle, or "co-flowing", air was introduced through four inlets and passed through 3 perforated plates with holes of 5 mm in diameter and 10 mm centre-to-centre pitch to improve flow

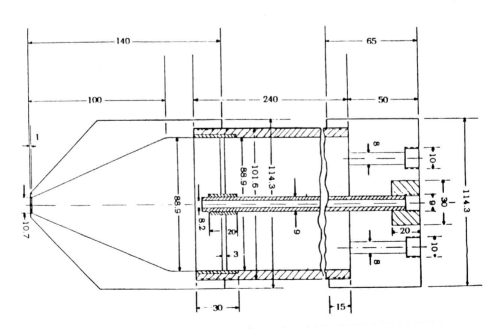

FIGURE 3(a)   Schematic of the accelerating nozzle.

uniformity. The co-flowing air passed through flow straighteners upstream of the inlet of the nozzle. The air was accelerated by passing through the nozzle area ratio of 67.7:1 and, for ease of manufacture, the nozzle was machined as a cone rather than with a contoured shape. The particles to be sized were introduced at the upstream end of the nozzle by being pneumatically transported down a central pipe made of stainless steel with 8.2 mm (ID) and 9.55 mm (OD) diameter and, for most of the experiments, of 299 mm length. The material was chosen to avoid static electricity because of the smooth internal bore. The coal particles were from the same coal seam (Bentinck) as that for which particle shapes have been extensively reported by Orfanoudakis and Taylor (1992), although the size distribution used here corresponds to that reported by Abbas et al. (1993).

The air flow rate was metered upstream of the inlets using two rotameters; one for the co-flowing air and one for the transporting air. The co-flowing air flow rate was 245 (1/min) and that of the transporting air was 6.5 (1/min) and these values corresponded to mean co-flowing air nozzle exit velocity of 47.1 (m/s) at the nozzle exit and 0.66 (m/s) at the nozzle inlet, whereas the particle velocity at the exit of the central pipe was 2.05 (m/s).

The velocity, measured by the Doppler anemometer at the exit and on the centreline of the accelerating nozzle using only micrometer-sized seeding particles, was 45.7 (m/s) and the turbulence intensity was 4.8%. The discrepancy between the calculated and measured air exit velocity was of the order of 3%.

To avoid multiple particle occupancy of the sizing test volume, and to ensure that the velocity of the air in the nozzle was not disturbed by the introduction of the particles, the coal was fed at low flow rates using a vibrating plate type feeder with variable speed.

Small Coal Swirl Stabilized Burner

The small coal burner is shown schematically in Figure 3b. This laboratory scale burner involved axial injection of coal particles through a central pipe of 3 mm internal diameter and radial injection of natural-gas (94% methane) through six 1.0 mm holes, into a coaxial swirling stream and a diffusing quarl. A contraction with area ratio of about 19 upstream of the burner exit contributed to the establishment of symmetrical flow at the entrance of the quarl. The quarl ($L/D_{\text{throat}} = 2$ and $\phi = 20°$) could be easily replaced by other quarls with different geometry or even removed in order to measure the boundary conditions at the exit of the contraction. This burner configuration, which is smilar to those found in industrial furnaces, provided symmetrical and attached flames for a wide range of Reynolds and swirl numbers, as for example $Re = 30500$ (the calculation was based on the hydraulic diameter of the annulus, the bulk velocity, U_0, and the kinematic viscosity of air at room temperature) and $S_w = 0.5$, values which were used for the experiments reported here (see also Tab. II). It should be noted that the small scale of the experiment results in little or no effects of beam misalignment and defocussing due to gradients of refractive index in the flame and this is important for maintaining the accuracy of the measurements with this design of sizing anemometer.

The conditions used for the experiments presented here are shown in Table II and the flow rates were 9.51/min and 3001/min for the primary and secondary air through

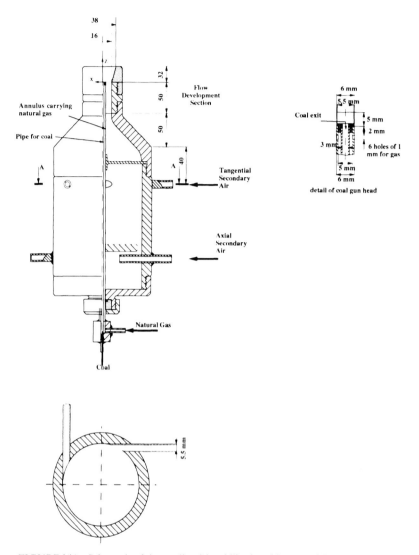

FIGURE 3(b)   Schematic of the small swirl stabilised coal burner with gas support arrangement.

the burner correspondingly. These values resulted in air bulk velocity, U_0, of 29 m/s at the entrance of the quarl and particle exit velocity, U_p, of 22.4 m/s at the exit of the coal gun; both values were calculated using the flow rates and the burner geometry shown in Figure 3b. The coal batch used in the burner had the same size distribution as that referred to as $d_{mean} = 121$ μm in Abbas *et al.* (1993) and Orfanoudakis *et al.* (1993). The momentum ratio, R_m, was defined as the ratio of the ratio of the primary air and coal momentum to the secondary air momentum. For these measurements the coal equivalence ratio. Φ_c, was deliberately kept to a low level, see Table II, (a) to minimize the possibility of multiple occupancy in the sizing instrument probe volume, (b) to aid

TABLE II

Burner flow characteristics

Quantity	
Primary-transporting air (1/min)	9.5
Secondary axial air (1/min)	100
Secondary swirl air (1/min)	200
Bulk velocity at quarl inlet U_0, (m/s)	29
Particle exit velocity U_p, (m/s)	22.4
Coal feeding rate (g/min)	0.24
Natural-gas flow rate (1/min)	30
Fuel Φ_c equivalence ratio (coal)	0.007
Fuel Φ_g equivalence ratio (gas)	0.85
Potential total heat release (kW)	11.3
Swirl number S_w	0.5
Momentum ratio (R_m)	1/40
Geometry of the burner quarl used	$L/D_{throat} = 2$ and $\phi = 20°$
Reynolds number	30500

evaluation of computer codes by considering the coal to be a passive scalar without the complication introduced by coal combustion and (c) because in any case the trend in the particle velocities and trajectories would be the same as at higher Φ_c. The gas fuel equivalence ratio, Φ_g, was 0.85 and was kept well above the extinction limits of the flame. Given the feeding rate of the coal and the flow rate of the gas, the potential total heat release was about 11.3 kW, to which the coal contributed less than 5%.

Calculated Correlation Between the Aerodynamic Diameter and Velocity at Nozzle Exit

The evaluation of the instrument was performed by the comparison between the optical size measured by the instrument and an independent measurement (*aerodynamic*) size of particles derived from the particle exit velocity from the accelerating nozzle, measured by the laser Doppler anemometer. Orfanoudakis (1994) describes the calculation, based on the low Reynolds number drag coefficient, to relate the exit velocity of a particle from the nozzle to its aerodynamic diameter, in μm, given the air velocity variation along the nozzle centreline which is known from continuity. The calculations encompassed particle densities between 1000 (kg/m³) and 2950 (kg/m³): the density of coal particles usually lies between 1000 and 1400 (kg/m³). Tests, to be explained below, involved the use of 40 μm and 80 μm glass beads and the densities for these were 2410 and 2950 (kg/m³) respectively. The effect of uncertainty in the coal particle density decreases in importance if the nozzle air flow rate and, hence, the particle Reynolds number is reduced (e.g. Wilson and Liu, 1980); however, reduction of this flow rate also results in reduction in the sizing resolution (i.e. μm/(ms⁻¹)) of the measurements for small particles. Thus, the flow rate must be a compromise between these two conflicting effects. Orfanoudakis (1994) shows that for the typical uncertainty of coal particle density, namely from 1000 to 1500 (kg/m³), the denser prarticles of, for example, 10 μm diameter can have 10% lower velocity than that of the lighter ones at the exit of the accelerating nozzle. Although it is likely that there are particle-to-particle differences in coal density, these calculations show that the magnitude of these differences should have little effect on their nozzle exit velocity, at least for particles larger than about 20 μm.

RESULTS AND DISCUSSION

Instrument Evaluation with Particles (Glass Beads) of Known Size

The sizing instrument was tested with commerically available, quasi-monodisperse glass beads of 40 μm and 80 μm nominal diameter. The results are presented here for the larger diameter, in Figure 4, in the form of a scatter plot between the measured maximum amplitude, which is equivalent to *optical* size *via* the calibration with the precision pinholes, and the measured velocity, which is equivalent to *aerodynamic* size *via* calculation. For convenience, this Figure and the next retain the primitive measurements, namely voltage and velocity, to provide an indication of the magnitudes of the quantities measured in this experiment and which can be readily compared with their experimental errors. The results, show that the rms scatter in the plotted data, for both maximum amplitude and velocity, is of the order of 5% of the respective means. This could increase up to 25% if the system was even slightly misaligned and the rms was strongly dependent on the kind and degree of misalignment.

Measurements with Coal Particles

Inert flow in the accelerating nozzle The total number of measurements taken at the exit of the accelerating nozzle was limited to 1200, because the combination of the

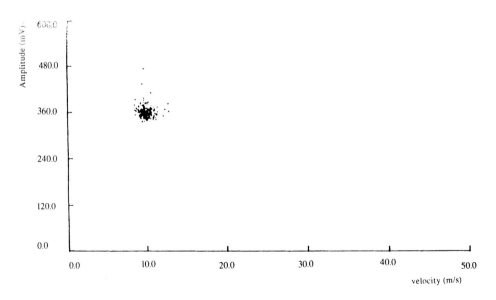

FIGURE 4   Measurements of optical (amplitude) and aerodynamic diameter (velocity) with particles of known size: 80 μm glass beads.

logging computer and the transient recorder, which provided the amplitude information, was about 3 s per measurement in this system, which was designed primarily for
flexibility to aid instrumentation development. Faster systems can certainly be constructed but would be based on custon-built electronics, such as for example those used
for the signal processing of the so-called shadow Doppler anemometer (Hardalupas *et
al.*, 1994). In addition, an amplitude validation (see Orfanoudakis, 1994) rate of only
25% was achieved, so these measurements took a total of about 4 to 5 hours. The
statistical error on the velocity measurement for each velocity bin, assuming that the
number of measurements in each velocity bin was roughly 1/10 of the total, on the mean
velocity was about 3%, according to Yanta (1973). Note that the rms velocity was not
relevant to this experiment.

Figure 5a shows the scatter plot of measured amplitude and velocity of coal particles,
displaying the expected trend of negative correlation between the two variables. The
correlation is not expected to be perfect, mainly because of the non-spherical shape and
so, at any given value of nozzle exit velocity, there is a finite spread of maximum
amplitudes and the implications for the accuracy and precision of measurement will be
quantified below.

Seeding particles, Al_2O_3 in these experiments, were also introduced into the nozzle
and the resulting scatter plots are superposed in Figure 5a which, because of the small
size of the seeding particles, follow the air flow with high fidelity and give smaller
amplitudes than the coal particles.

The data of Figure 5a was further processed to provide quantitative statements
about accuracy and precision. Thus, the total velocity range was divided into ten
equally spaced windows and, in each velocity window, the mean and rms values of the
measured maximum amplitudes were calculated. The choice of the number of windows

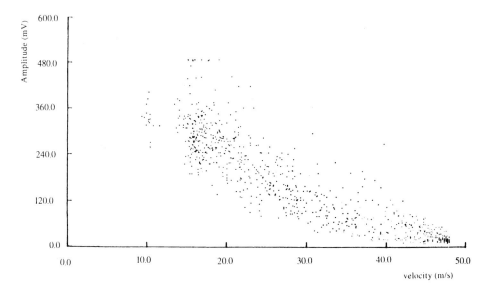

FIGURE 5(a)   Measurements of optical (amplitude) and aerodynamic diameter (velocity) with coal particles.

was a compromise between minimising the variation of mean amplitude across each window, which gives rise to artifical broadening of the rms values, and having a statistically significant number of measurements in each window. These ten experimentally-derived values, with measured nozzle exit velocity as independent variable, were compared with the response curve for five particle diameters of particle density 1000 (kg/m^3), using the *calculation* prescribed in Orfanoudakis (1994). The amplitudes associated with each of the five particles were inferred from the calibration curve, which is also provided in Orfanoudakis (1994).

Figure 5b shows the results for a coal feeding rate of 0.0133 (g/min), which corresponds to a mass fraction of 41.10^{-6}, calculated using the total air mass flow rate through the nozzle: the retention of signal amplitude as the ordinate, with the corresponding diameters inserted as a subsidiary ordinate on the right hand side of the graph, serves as a reminder that the size resolution–determined by the tolerance on the measurement of amplitude–is greater at lower diameters. An alternative way to present the data shown in Figure 5b is to convert the measured amplitude to size (i.e optical diameter, d_{opt}) using a calibration curve, and the measured exit velocity to size (aerodynamic diameter, d_{aero}) *via* a calculation (see Orfanoudakis, 1994). The outcome is shown in Figure 5c, where the optical diameter is plotted against the aerodynamic diameter. To assist the comparison, the "ideal line" of $d_{aero} = d_{opt}$ is also plotted on this graph. In this Figure, the rms values of *amplitude* were also converted to *size* and shown as error bars. From the comparison between the optical-aerodynamic diameter line and the ideal line, the accuracy and precision can be inferred. Note that measurements obtained with seeding particles, about 300 in number, are included as well. In this Figure, the accuracy – *defined here as the difference in the mean values between the two curves* of the instrument was estimated to be about $+2\,\mu m$, $+7\,\mu m$ and $-20\,\mu m$ at central values of 15,30 and 60 µm respectively. Thus there is a systematic over- or

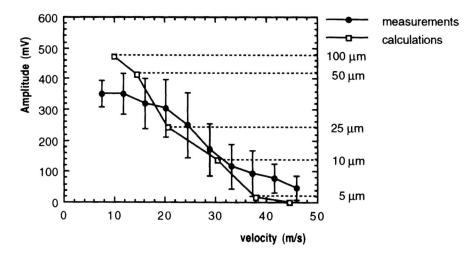

FIGURE 5(b)   Evaluation of accuracy and precision of the sizing anemometer at coal feeding rate of 0.0133(g/min): *optical* (µm) versus *aerodynamic* (µm) diameter obtained from the measured amplitude velocity plot via *calibration* curve (Orfanoudakis, 1994) and *calculation* of the particle exit velocity of a given particle size, respectively.

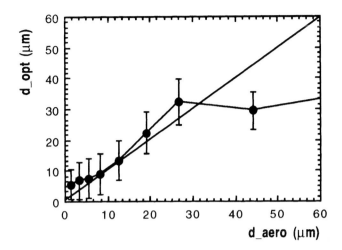

FIGURE 5(c)   *Optical* (μm) versus *aerodynamic* (μm) diameter obtained from the measured amplitu velocity plot via *calibration* curve (Orfanoudakis, 1994) and *calculation* of the particle exit velocity of a given particle size (Orfanoudakis, 1994), respectively.

under-estimation of the size for different size ranges, although it should be recalled that the number of samples of the largest sizes is small and the confidence level on the value of $-20\,\mu m$ is smaller than for the other two values. The corresponding values for precision – *defined here as the rms spread of values about the mean* – are $\pm 2\mu m$, $\pm 7\,\mu m$ and $\pm 10\,\mu m$ and an examination of Figure 5b shows that the best resolution is of the order of 4 μm and occurs at about 10 μm. Although the differences in mean amplitude between calculations and experiment at a given velocity might be caused by the *gradual* saturation of the sizing photodetector as particles approach the *known* saturation border associated with 80 μm diameters (for this experiment; see Orfanoudakis, 1994), and, indeed, probably lower as the results below suggest, the impression is also likely e.g., flake-like particles which could give rise to a large optical, but small aerodynamic, diameter. A further reason could also be the presence of porous particles giving rise to large optical, but small aerodynamic, diameters.

The values presented here are comparable with those quoted in Orfanoudakis and Taylor (1992) for accuracy and precision (namely, 1 μm and 5 μm, respectively) for the small to medium particle sizes, which also constituted the larger part of the measurements. Generally, the accuracy and precision are, especially for the samll and medium particles of the examined range, comparable to, but somewhat worse than, those obtained by static calibration experiments of Orfanoudakis and Taylor. This was partly due to multiple particle occupancy, even at the low feeding rate of 0.0133 g/min used here. The large inaccuracy at a particle size of 60 μm is a result *not* expected from the effects of the irregular shape of the coal particles. It is highly likely that this is the result of *gradual* saturation of the photodetector at this diameter and larger. In principle, the effects of saturation must be avoided by using two detectors, with dynamic ranges adjusted through setting of the high voltage supply, to avoid saturation effects up to diameters in excess of 100 μm. In practice, the aerodynamic behaviour of particles above 60 μm is not expected to be a function of the diameter and hence it is arguable that precise sizing is not required.

The main restriction on the use of the instrument is the requirement to avoid multiple occupancy as the particle density increases. To check the attenuation of the sizing beam due to the density of the flow, the following experiment was conducted: a 50 μm precision pinhole was used as scatterer and at the time of measurements coal particles were dropped "upbeam" of the pinhole at the maximum rate of the coal feeder. The signal amplitude again did not change by more than 5% and thus the effects of attenuation of the laser beam as it passed through the particle field, of the type discussed by Kliafas et al. (1987, 1990), was negligible in this experiment. In general, however, the extent and likely effects of beam attenuation from this source must be carefully assessed in each application and the latter two references provide a basis for this.

The effects of increasing the quantity of coal fed up to 0.473(g/min) into the nozzle were investigated which corresponded to a mass fraction (kg of coal/kg of coal-air mixture) of betwen $41 \cdot 10^{-6}$ and $1.45 \cdot 10^{-3}$, and to a number density (number of particles/μm³) of between $7.1 \cdot 10^{-10}$ and $2.5 \cdot 10^{-8}$, respectively. The results in Orfanoudakis (1994) show that the increase in the coal feeding rate and, hence, in the number density of about a factor of 35 reduces the accuracy and precision up to about 10% and 50%, respectively, possibly because of multiple occupancy.

Reacting Flow in the Small Coal Burner

The coal particle velocity characteristics are presented below as averages over three size classes designated as small for $d_p \leqslant 24$ μm, medium for $24 < d_p < 48$ μm and large for $d_p \geqslant 48$ μm. In the following text these three size classes will be referred to as 12, 36 and 64 μm particles, from the mean-size bin diameters of the three classes respectively. The total number of measurements taken at each point was 900, for reasons already mentiond, and the error on the velocity measurement was estimated separately for each size class. For small particles and for the smallest number of measurements the error on the mean and rms velocity was of the order of 2% and about 9% (Yanta, 1973). For the medium particles the numbers were of the order of 10% and 45% and for the large particles about 8% and 10% for the mean and rms velocity correspondingly. These numbers are quoted for the worst case for each size class but account for distributions in the mean and particularly the rms distributions not being smooth. The relatively large error for the medium size class is due to the few particles that were measured at each point and which in turn is due to the particular coal size distribution used for these experiments being relatively deficient in this size class (28.2% under 75 μm and 9.9% under 25 μm, $d_{10} = 48$ μm). Radial profiles were taken only on one side of the geometrical axis of symmetry owing to the time-consuming nature of the experiments. However, measurements (Orfanoudakis, 1994) have shown that the flow in this burner is not far from being symmetric.

Figure 6a shows the variation of the mean axial velocity U/U_0 along the centreline and Figure 6b the radial profile of U/U_0 at $z/D_{throat} = 2.13$ downstram of the quarl exit for conditions indicated in Table II. The cartesian axes z and x are defined in Figure 3b. The 12 μm and 36 μm particles recirculate, apparently following the air motion,

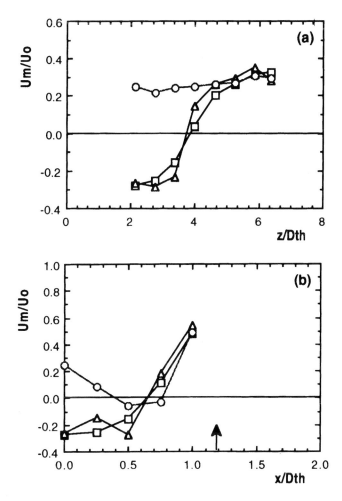

FIGURE 6   Measurements of mean axial velocity: (a) centreline and (b) radial profiles at $z/D_{th} = 2.13$. Symbols: *open squares* small particles with $d_b \leqslant 24$ μm, *open triangles* medium particles with 24 μm $< d_p \leqslant 48$ μm and *open circles* large particles with $d_p \geqslant 48$ μm. Vertical arrow on abscissa denotes the location of the inside at the exit of the quarl.

whereas the 64 μm particles, due to their momentum acquired from the primary air stream, maintain positive velocities everywhere on the centreline. Table III shows that the mean Stokes number, defined in the footnote to the table, for 12 μm particles is of order of 10, providing the expectation that the particles that belong to this size class should follow the air mean velocity with high fidelity, which in turn means that the length of the gas recirculation zone is similar to that indicated by the 12 μm size class. The 36 μm particles have a mean Stokes number of the order of about unity, which also suggests fairly good response to the gas phase flow and indicates that results for the particles obtained from the sizing anemometer are plausible. Figure 6b shows that the widths over which the 12 μm and 36 μm particles recirculate extended to about

TABLE III

Stokes numbers

Category of particles	τ_{relax} (ms)[a]	St_m[b]	St_t[b]
Small ($d_p \approx 12\,\mu$m)	0.15	9	6
Medium ($d_p \approx 36\,\mu$m)	1.4	0.94	0.62
Large ($d_p \approx 64\,\mu$m)	4.5	0.29	0.19

[a]The relaxation time is defined as:

$$\tau_{relax} = \frac{\rho_p \cdot d_p^2}{18 \cdot \mu g}$$

where ρ_p, d_p, are the particle density and diameter respectively and μ_g is the gas phase dynamic viscosity. The effect of the irregular shape of coal particles may be accounted for by reducing the above value by a form factor of 1.7, see also Clausen et al. (1990).

[b]The mean and turbulent Stokes numbers can be defined correspondingly as:

$$St_m = \frac{T_m}{\tau_{relax}} \text{ and } St_t = \frac{T_t}{\tau_{relax}}$$

where T_m and T_t are mean and turbulent flow time scales, here defined as the ratio of the quarl *exit* diameter to the bulk air velocity at the quarl *inlet*

$$T_m = D_{qexit}/U_0 = 1.31 \text{ ms}$$

with U_0 taken from Table II and, following Hardalupas et al., (1990), we use the approximation $v' \approx 0.15\,U_0$ in the formula

$$T_t \approx 0.3\,D_{qexit}/(3v') \approx 0.87 \text{ ms}$$

$x\,D_{throat} = 0.65$ for $z/D_{throat} = 2.13$. The 64 μm particles had a very narrow recirculation region with values of x/D_{throat} between 0.4 and 0.8 and this result, together with the absence of a recirculation zone on the centreline, is also expected from the magnitude of the mean Stokes number being of the order of 0.1.

In Figure 7 the centreline and radial profile measurements of u'/U_0, for the same conditions as Figure 6, are shown. In Figure 7a, it can be observed that the 64 μm particles had higher rms velocities than the 36 μm and 12 μm. This, of course, is a surprising result because the turbulent Stokes numbers, St_t, as shown in Table III, might be taken to imply that the 12 μm particles would be expected to respond better to the air turbulence. A comparable result has been reported by Hardalupas et al. (1990) in particle laden jets who explained the phenomenon as being generated by the superposition of particle trajectories from regions of different mean axial velocity and has been confirmed by the calculations of Mastorakos et al. (1990). Thus the magnitude of the rms velocity was not related to the transfer of turbulent energy from the air to the particles and can explain the rms velocities results obtained here. Figure 7b shows that this trend is maintained in the radial direction in the central region up to $0.3\,D_{throat}$.

The arithmetic and Sauter mean diameters (abbreviated to AMD or d_{10} and SMD or d_{32} respectively in the following text) are defined as:

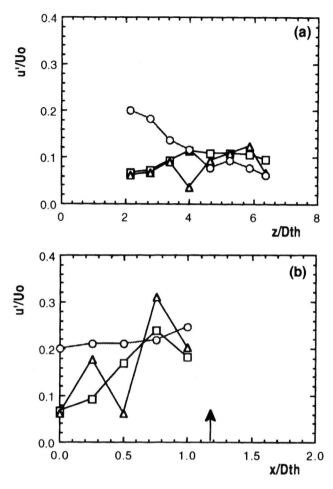

FIGURE 7 Measurements of rms axial velocity: (a) centreline and (b) radial profiles at $z/D_{th} = 2.13$. Symbols *open squares* small particles with $d_p \leqslant 24\,\mu$m, *open triangles* medium particles with $24\,\mu$m $< d_p \leqslant 48\,\mu$m and *open circles* large particles with $d_p \geqslant 48\,\mu$m. Vertical arrow on abscissa denotes the location of the inside surface at the exit of the quarl.

arithmetic mean diameter

$$d_{10} = \frac{\displaystyle\sum_{i=1}^{k} n_i \cdot d_i}{\displaystyle\sum_{i=1}^{k} n_i}$$

and Sauter mean diameter

$$d_{32} = \frac{\displaystyle\sum_{i=1}^{k} n_i \cdot d_i^3}{\displaystyle\sum_{i=1}^{k} n_i \cdot d_i^2}$$

where n_i is the number density of the coal particles and d_i is the diameter in size class "i" and summation is over all "k" size classes. The number density for the coal particles was corrected using the residence time in the pointer-Doppler volume as in Hardalupas (1989). The accuracy with which the number density can be measured is a question which also arises in the context of phase Doppler anemometry and is still the subject of intense debate (see Taylor, 1994). Here, we make the plausible assumption that it is comparable to that available to phase Doppler anemometry, given that the same technique is used, namely systematic and random errors of the order of 20% and 10% (Hardalupas *et al.*, 1994).

In Figure 8, the centreline and radial profile, the latter at $z/D_{throat} = 2.13$ (i.e., at the quarl exit), measurements of AMD and SMD are included to show that the instrument

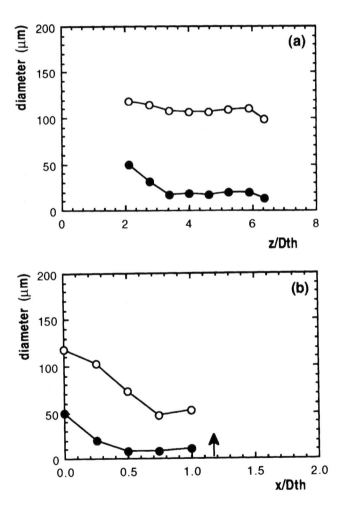

FIGURE 8   Measurements of AMD and SMD (a) centreline and (b) radial profiles at $z/D_{th} = 2.13$. Symbols: *blocked circles* arithmetic and *open circles* Sauter mean diameter. Vertical arrow on abscissa denotes the location of the inside surface at the exit of the quarl.

successfully detected the expected decrease in the values of both AMD and SMD with axial and radial distance due to radial dispersion of the particles generated mainly by centrifuging due to swirl. Figure 8b shows the radial profiles of AMD and SMD.

These measurements have been extended to investigate and quantify the effect of various parameters such as swirl number, burner quarl geometry and momentum ratio on the particle dispersion and the potential volatile release in the IRZ. The results, together with the velocity boundary conditions and initial particle size distribution in the detail required for evaluation of CFD predictions, are reported by Orfanoudakis (1994).

CONCLUSIONS

Experiments have been performed to evaluate the sizing accuracy and precision of a coal particle sizing instrument developed specially for sizing non-spherical particles. The work described in Orfanoudakis and Taylor (1992) for assessments of accuracy and precision was based on a relatively small sample size and was improved to a larger sample size and under flow conditions. The measurements compare the optical and aerodynamic size of coal particles and also of glass beads of known size. In addition, the instrument was used to take simultaneous measurements of size and velocity in an unconfined small scale coal burner to examine the characteristics of the flow in the near burner area. The main findings can be summarised as follows.

1. The reliable operation of the pointer volume technique imposed the selection of a value of 5.8 for the ratio of the $1/e^2$ diameters of the sizing and Doppler measuring volumes.
2. A computer program was used to calculate the exit velocity for certain scatters' size and, thus, provide an independent method of sizing for the evaluation of the sizing instrument. The use of this program also provided the means to quantify the effect of density uncerainty and it was found that for typical values, namely from $1000-1500(\text{kg/m}^3)$, the velocity and, hence, the aerodynamic diameter could change by about 10%.
3. The accuracy and precision of the instrument for coal particle sizing measurements was found to be $+2, +7, -20\,\mu m$ and $\pm 2, \pm 7, \pm 10\,\mu m$, respectively, for mean values of 15, 30, 60 μm at the lowest coal feeding of 0.0133(g/min). The accuracy and precision for this coal feeding rate, corresponding to a number density of $7.1 \cdot 10^{-10}(\text{m}^{-3})$, were comparable to those for the *static* calibration experiments reported in Orfanoudakis and Taylor (1992).
4. Saturation of the sizing photodetector certainly occurred for particles larger than about 80 μm, limiting the sizing range of the instrument to this value for the set-up used in these experiments, and is also likely to be responsible for the poor accuracy for the 60 μm size range.
5. The best resolution that could be achieved, for a collection angle of $\theta_c = 3.82°$, was 4 μm at a diamter of around 10 μm.
6. The accuracy and precision became worse by 10% and 50%, respectively, when the feeding rate increased by a factor of 35 and thus when the number density also increased from $7.1 \cdot 10^{-10}$ to $2.5 \cdot 10^{-8}$.

7. The alignment procedure proved to be very important for proper correlation between the optical and aerodynamic size; the procedure needed for good alignment is given in Orfanoudakis (1994).

8. The instrument has been used to size pulverised coal particles in a burner and provide coal particle velocities for three size classes in the near burner region. The smallest two classes recirculated but the largest did not, showing the value of size-discriminated velocity measurements. In addition, the instrument can provide measurements of arithmetic and Sauter mean diameter, which in this flow provide evidence of the effect of centrifuging on particle dispersion.

ACKNOWLEDGEMENTS

The authors are glad to acknowledge financial support from the Commission of the European Communities, by contract JOULE-0042-GR. The authors are indebted to Professor J. H. Whitelaw for the criticism of the results and his guidance. The help and advice of Dr. Y. Hardalupas of Imperial College and of Dr. K. Hishida, of Keio University. Yokohama and Dr. G. Wigley, of AVL, Graz, during the early stages of the development of the sizing instrument are gratefully acknowledged. Dr. E. Mastorakos wrote the first version of the software for the aerodynamic diameter calibration calculations. The continuous and meticulous support of Mr. P. Trowell, Mr. G. Rasmussen and the late Mr. O. Vis, in modifying the experimental apparatus, is gratefully appreciated.

REFERENCES

Abbas, T., Costen, P., Lockwood, F. C. and Romo-Millares, C. A. (1993) The Effect of Particle Size on NO Formation in a Large-Scale Pulverized Coal-Fired Laboratory-Furnace: Measurements and Modelling, *Combustion and Flame*, **93**, 316–326.

Adrian, R. J. and Orloff, K. L. (1977) Laser Anemometer Signals: Visibility Characteristics and Application to Particle Sizing, *Appl. Opt.*, **16**(3), 677–684.

Bonin, M. P. and Queiroz, M. (1991) Local Particle Velocity, Size and Concentration Measurements in an Industrial-Scale Pulverized Coal-Fired Boiler. *Combustion and Flame*, **85**, 121–133.

Clausen, S., Jensen, P. A. and Rathmann, O. (1990) LDA Velocity Measurement in a 500 kw Pulverized Coal Flame, (*Proceedings of 5th International Symposium of Applications Laser-Doppler Anemometry to Fluid Mechanics* Eds., Adrian R. J *et al.*, Lisbon, paper 35.1.

Costa, M., Costen, P., Lockwood, F. C. and Mahmud, T. (1990) Detailed Measurements and Modelling of An Industry-Type Pulverised-Coal Flame. *23rd Symposium International on Combustion. The Combustion Institute*, pp. 973–980.

Dugue, J. and Weber, R. (1992) Laser Velocimetry in Semi-Industrial Natural-Gas, Oil and Coal Flames by Means of a Water-Cooled LDV Probe. *Proceedings of 6th International Symposium of Laser-Doppler Anemometry to Fluid Mechanics*, Eds., Adrian R. J. *et al.*, Lisbon, paper 22.3.

Hardalupas, Y., Hishida, K., Maeda, M., Morikita, H., Taylor, A. M. K. P. and Whitelaw, J. H. (1994) Shadow Doppler Technique for Sizing Particles of Arbitrary Shape. *Applied Optics*, **33**, 8417–8426.

Hardalupas, Y., Taylor, A. M. K. P. and Whitelaw, J. H. (1988) Measurements in Heavily-laden Dusty Jets with Phase-Doppler Anemometry In *Transport Phenomena in Turbulent Flows:Theory, Experiment and Numerical Simulation*, Eds., M. Hirata and N. Kasagi, pp. 821–835. Hemisphere, washington, DC.

Hardalupas, Y., Taylor, A. M. K. P. and Whitelaw, J. H. (1989) Velocity and Particle-flux Characteristics of Turbulent Particle-Laden Jets. *Proc. R. Soc. Land.*, **A426**, 31–78.

Hardalupas, Y. (1989) Experiments with Isothermal Two Phase Flows. *Ph.D. Thesis*, University of London.

Hardalupas, Y., Taylor, A. M. K. P. and Whitelaw, J. H. (1990) Velocity and Size Characteristics of Liquid-Fuelled Flames Stabilised by a Swirl Burner, *Proc. R. Soc. Land.*, **A428**, 129–155.

Hardalupas, Y., Taylor, A. M. K. P. and Whitelaw, J. H. (1994) Liquid Fuel Mass Flux, Fraction and Concentration Distributions in a Swirl Stabilised Flame. *Int. J. Multiphase Flow*, 20 (suppl), pp. 233–259.

Holve, D. J. (1983) A Single Particle Counting Diagnostic System for Measuring Fine Particles at High Number Densities in Research and Industrial Applications. *Sandia National Laboratories*, Report SAND83–8246.

Holve, D. J. and Self, S. A. (1979) Optical Particle Sizing for In situ Measurements; Part 1 and 2. *Appl. Opt.*, **18**(10), 1632–1645 and 1646–1652.

Hong, N. S. and Jones, A. R. (1976) A Light Scattering Technique for Particle Sizing Based on Laser Fringe Anemometry. *J. Phys. D: Appl. Phys.*, **9**, 1839–1848.

Hong, N. S. and Jones, A. R. (1978) Some Aspects of Light Scattering in Laser Fringe Anemometers. *J. Phys. D: Appl. Phys.*, **11**, 1963–1967.

Jones, A. R. (1993) Light Scattering for Particle Characterization, in *Instrumentation for Flows with Combustion* (Ed. Taylor, A. M. K. P.). Academic Press London, pp. 323–404.

Kliafas, Y., Taylor, A. M. K. P. and Whitelaw, J. H. (1987) Errors in Particle Sizing by LDA Due to Turbidity in the Incident Laser Beams, *Exp. Fluids*, **5**, 159–176.

Kliafas, Y., Taylor, A. M. K. P. and Whitelaw, J. H. (1990) Errors Due to Turbidity in Particle Sizing Using Laser-Doppler Velocimetry. *Trans. ASME. J. Fluids Eng.*, **112**, 142–149.

Maeda, M., Hishida, K., Sekine, M. and Watanabe, N. Measurements on Spray Jet Using LDV System with Particle Size Discrimination. *3rd International Symposium of Applications of Laser-Doppler Anemometry to Fluid Mechanics* Eds.: (1986) Adrian, R. J., *et al.*, Lisbon, paper 20.3.

Marple, V. A. and Rubow, K. L. (1976) Aerodynamic Particle Size Calibration of Optical Particle Counters. *J. Aerosol Sci.*, **7**, 425–433.

Mastorakos, E., McGuirk, J. J. and Taylor, A. M. K. P. (1990) The Origin of Turbulence Acquired by Heavy Particles in a Round, Turbulent Jet, *Part. Part. Syst. Characterisation*, **7**, 203–208.

Negus, C. R. and Drain, L. E. (1982) Mie Calculations of the Scattered Light From a Spherical Particle Traversing a Fringe Pattern Produced by Two Intersecting Laser Beams. *J. Phys. D: Applied Physics*, **15**, 375–402.

Orfanoudakis, N. G. Measurements of Size and Velocity of Burning Coal. *Ph.D. Thesis*, (1994) University of London.

Orfanoudakis, N. G. and Taylor, A. M. K. P. (1992) The Effect of Particle Shape on the Amplitude of Scatered Light for a Sizing Instrument. *Part. Part. Syst. Characterisation*, **9**, 223–230.

Orfanoudakis, N. G., Taylor, A. M. K. P. and Whitelaw, J. H. (1993) Measurements of Size and Velocity in a Swirl Stabilized Coal Burner with Gas Support. Report TF/93/10. Thermofluids Sections, *Department of Mechanical Engineering*, Imperical College, London.

Taylor, A. M. K. P. (1992) Optically-Based Measurement Techniques for Dispersed Two Phase Flows, in *Combusting Flow Diagnostics*, pp. 233–289. (Eds. Durao, D. F. G. *et al.*). Kluwer Academic Publishers, Dordrecht.

Taylor, A. M. K. P. (1994) Two Phase Flow Measurements in *Optical Diagnostics for Flow Processes*, (Eds. Lading, L. *et al.*, Plenum Press, New York, pp. 205–228.

Wilson, J. C. and Liu, B. Y. H. (1980) Aerodynamic Particle Size Measurement by Laser-Dopplker Velocimetry, *J. Aerosol Sci.*, **11**, 139–150.

Yanta, W. J. (1973) Turbulence Measurements with a Laser Doppler Velocimeter. Naval Ordnance Laboratory, White Oak, Silver Spring, Maryland, NOLTR 73–94.

Yeoman, M. L., Azzopardi, B. J., White H. J., Bates, C. J. and Roberts, P. J. Optical Development and Application of a Two Colour LDA System for the Simultaneous Measurement of Particle Size and Particle Velocity, *Engineering Applications of Laser Velocimetry*, Winter Annual Meeting, (1982) *ASME, Phoenix, Arizona*, 14–19 November.

Flame Visualization in Power Stations

H. J. M. HULSHOF, A. W. THUS and A. J. L. VERHAGE
*KEMA - Power Generation Utrechtseweg 310 6800
ET Arnhem, The Netherlands*

Abstract—The shapes and temperature of flames in power stations, fired with pulverized coal and gas, have been measured optically. Spectral information in the visible and near infrared is used.

Coal flames are visualized in the blue part of the spectrum, natural gas flames are viewed in the spectral region of CH-emission.

Temperatures of flames are derived from the best fit of the Planck-curve to the thermal radiation spectrum of coal and char, or to that of soot in the case of gas flames.

A measuring method is presented for the velocity distribution inside a gas flame, employing pulsed alkali salt injection. It has been tested on a 100 kW natural gas flame.

Key Words: Flame visualization, radiation temperature, pulverized coal flame, gas flame, power station, pulsed salt injection.

INTRODUCTION

Operators of power stations want to know about the state of each of the burners, but usually they are only supplied by overall parameters: the concentration of NO_x in the flue-gas, the percentage of burn-out, the rate of steam production, etc. From these overall figures it is difficult to conclude which of the burners performs below average. Optical inspection of the flames of the burners reveals possible differences in flame shape and radiation intensity. This information may then be used to tune the individual burners, using the overall parameters to check the efficacy of the tuning operation. In this way, flame visualization can lead to a cleaner environment. Burners are often tunable with respect to the swirl of the combustion air while also the ratio fuel to air can be varied. In most cases the rating between the burners can be changed. Inhomogenities caused by fouling or corrosion can be detected optically, enabling a well-concerted effort during a stop, reducing down time.

Flame shapes are viewed by scanning the flame in the light of a preselected part of the spectrum. In coal-fired power stations the spectrum is found to be rather featureless. Only the sodium- and potassium doublets are clearly distinguishable. Imaging of coal flames is done in the blue part of the spectrum where background emission is smallest. Gas flames have a number of resolved emission bands, such as those of OH, CH, C_2 and H_2O. In gas-fired power stations the CH-emission around 430 nm is used as it delineates the flame fronts well while background radiation is relatively low.

If the radiating medium such as char, fly-ash, soot, is considered as a grey body, the temperature can be derived by fitting a Planck curve to the spectrum. The emissivity of the medium for the different wavelengths is taken as a constant. The temperature can also be derived from the ratio of the line intensities of alkali doublets. For gas flames this means that alkali salt must be injected into the flame, preferably a cesium salt (Braam *et al.*, 1991). If salt is injected in short pulses it is possible to follow its distribution in time. In this way a map of the velocity profile of a gas flame can be obtained. The last part of this paper gives the results of experiments with pulsed salt injection applied to a 100 kW gas burner. The preceding sections contain measurements on the flames of powder coal- and gas burners in power stations.

MEASUREMENTS OF PULVERIZED COAL FLAMES

Optical measurements in power stations require the insertion of a viewing element through a porthole in the wall of the furnace. In our case, this is a 1 m water-cooled endoscope, with a viewing angle of 60°. There are two objectives, one looking forward in an axial direction, the other at a 70° angle with that axial line. The endoscope has a continuously variable diaphragm. Window protection is achieved by means of a flow of compressed air.

The image of the endoscope falls on a slit of the spectrograph. The slice of the image is dispersed in wavelength and a part of the spectrum falls on the chip of a silicon CCD-camera with 512 × 512 pixels and a sensitivity of 10 photons per count at 700 nm. A full image is formed by scanning the slit of the spectrograph through the image plane of the endoscope in maximally 256 steps. The exposure time of the chip can be varied from nominally 6.3 ms to tens of seconds by means of a mechanical shutter. The intensity scale has a range of 18 bits. The image of the chip is stored as a 1 Mbyte file in computer memory or on hard disk. Images are processed with a 486-computer. The general set-up is depicted in Figure 1, where the endoscope views in the forward direction.

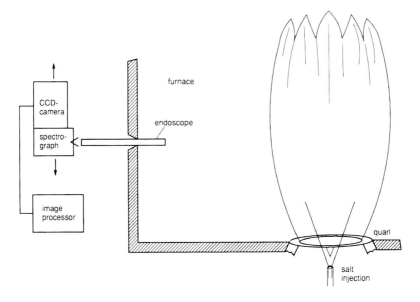

FIGURE 1   Schematic view of the flame visualization equipment. The entrance of the spectrograph can move through the image plane of the endoscope.

In the visited coal-fired power station 36 burners, each with a thermal rating of 45 MW at full load, are arranged in 3 rows of 6 in either of two opposing walls (Fig. 2). The flames have been observed from the positions A, B and C. The intensity contours of the flames of two burners are shown in Figure 3. The flame shapes are observed to be different for these two burners. With help of these pictures the swirl of some of the burners has recently been adjusted and the slag hoods have been removed. The fuel distribution has also been modified in favour of the outer burners. With help of iso-intensity plots the so-called flame parameters (Kurihara *et al.*, 1986) have been derived from the pictures. With these the shape, and therefore the swirl, can be expressed in numbers.

Non-intrusive measurements of the temperature have been made in two ways. The first method, using thermal radiation, has served to determine the temperature at a distance of 1 m from the mouth of the pulverized coal burner. During these measurements the power station operated at full load (630 MW_e), burning low ash, low sulphur coal, ground down to 70 μm particles. The slice of the flame at 1 m from the burner (nr. 35 in Fig. 2) was monitored through port C above the burner. The spectrum was recorded in steps of 2 nm between 400 and 1000 nm. The data were deconvoluted with help of a calibrated tungsten strip lamp at 1927 K and also with a calibrated oven at 1373 K. As a function of wavelength, the resulting intensity was fitted to a Planck curve by varying the temperature as a parameter. The emissivity of the coal and char was fixed at a value of 0.9 (see e.g. Best *et al.*, 1986). A good fit is obtained at T = 1785

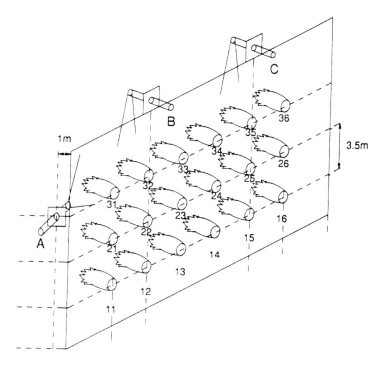

FIGURE 2   Lay-out of one of the two furnace walls that harbours 18 pulverized coal burners. The measuring locations are indicated with the letters A, B, C.

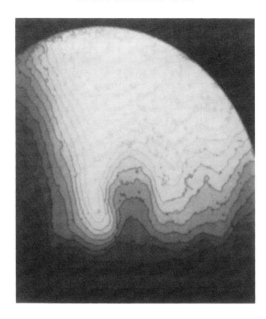

FIGURE 3   Intensity distribution of the coal flames of the burners 22 and 23 (see Figure 2) viewed from position B.

K with a standard deviation of about 20 K (Fig. 4). Each of the 300 points represents of the average value over a spectral band of 5 nm, integrated during 5 s. In spite of this averaging large fluctuations due to flame movements occur, leading to deviations as large as 40% from the Planck curve. Therefore, two colour pyrometry is likely to be less accurate.

The other way to determine the temperature in the pulverized coal flame utilizes salt injection. A solution of 0.6 mole CsCl in water was injected through the central oil lance of the burner at a rate of 5 cm^3/s. The salt solution passed through a nozzle at the tip of the lance, pressurized with 300 kPa air (Fig. 1). Both lines of the Cs-doublet (852 and 894 nm) were observed but their intensity was only a few percent of that of the fluctuating background level. The error in the ratio of the two intensities was too large to serve as a check for the other method. Potassium from the coal shows up as a stronger emitter than cesium introduced from outside, but in this case the errors in the temperature are also large due to the small difference in wavelength between the lines of the doublet.

The optical depth of the flames was checked in the following way:

In measuring position A (Fig. 2) the emission of Li, injected as LiCl into burner 31, was measured at 670 nm. The signal was a few percent of the background value, but it disappeared when the injection was shifted to burner 32. Apparently, the emission at 670 nm could not pass the flame of burner 31. Therefore, it is concluded that it is virtually impossible to observe a large coal flame through another similar flame at optical wavelengths.

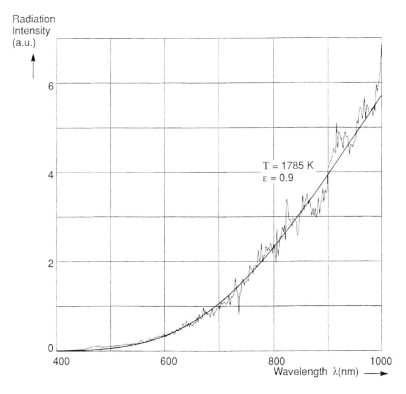

FIGURE 4   The temperature of a pulverized coal flame is derived from the Planck curve taking $\varepsilon = 0.9$. It is measured in a slice of the flame positioned at 1 m in front of burner 35 (position C in Figure 2).

MEASUREMENTS IN A GAS-FIRED POWER STATION

The same equipment was used to perform measurements in a gas-fired power station with 18 burners rated at 88 MW_{th} each. They are mounted in two opposite walls, one of which is shown in Figure 5. Unlike pulverized coal flames the gas flames are transparent at optical and near infrared wavelengths: burners in the background can be seen through the nearest flame. The spectrum of the gas flames exhibits the CH-band around 430 nm and the C_2-bands at 470 and 515 nm, especially near the mouth of the burners. The flame of burner 31 imaged using the emission of both CH (430 nm) and C_2 (515 nm) is depicted in Figure 6. The H_2O-bands at 720, 820, 900–980, and 1120–1160 nm are present everywhere inside the furnace. To our surprise, emission of the Na-doublet at 589 nm and the K-doublet (766.5, 769.9 nm) was also observed. Natural gas does not contain alkali metals but traces may have entered with the combustion air. The CH-radical occurs where combustion of CH_4 takes place so its emission outlines the flame front. Images of CH-emission, taken at 430 nm, with help of background subtraction, demonstrate the difference in flame shape between burner 21 and 31. The contours of equal intensity show that burner 31 has more swirl than 21 (Fig. 7). The burners further away are also visible in Figure 7. The flame shapes of the more distant

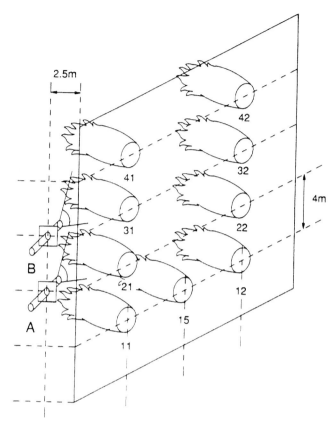

FIGURE 5   Lay-out of one of the two furnace walls of a gas-fired power station. Viewing locations are marked A and B.

burners are also different from one another. It is possible to assign flame parameters that are characteristic for the swirl.

Injection of alkali-salt, in this case NaCl, has been attempted through the oil lance, similar to the one in the pulverized coal burner. Apparently, this time the local temperature was much higher and the solution rapidly turned into steam, blocking the supply from the container. The NaCl injected for short periods raised the intensity of the doublet at 589 nm by more than an order of magnitude. In this way, one flame is made to stand out amidst the others. From the intensity distribution a relative temperature can be derived. It requires an injection of cesium salt to pinpoint the absolute value of the temperature with the two line method. Using a solution of 0.5 mole CsCl in water dispersed at a rate of 3 cm^3/s the temperature profile of a gas flame of a 88 MW$_{th}$-burner could be determined, showing temperatures up to 2200 K. The results are very recent and will be published[1].

---

[1] Note to the editor: This has been performed succesfully in May 1994 in a 88 MW$_{th}$-gas flame.

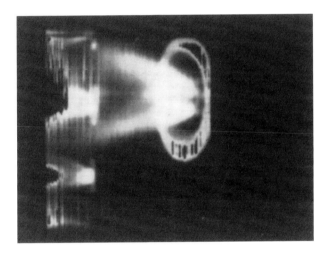

FIGURE 6   Images of burner 31 (as marked in figure 5) formed by CH emission at 430 nm (upper picture) and C_2 emission at 515 nm.

Thermal radiation emanating from soot has been fitted to a Planck-curve, keeping the emissivity fixed at a value of 0.9. This method relies on the assumption that molecular radiation of a gas flame between 400 and 880 nm is weak compared to the thermal radiation of soot particles. Each soot particle is considered to radiate as an almost black body. There are so many soot particles in a large flame that sufficient radiation is produced to record a thermal spectrum. At the other hand, the density is such that the flame is optically thin and background radiation is also detected. The temperature of the walls is too low, however, to emit substantially in the band between 400 and 880 nm. As shown in Figure 8 the temperature of the flame at a distance of

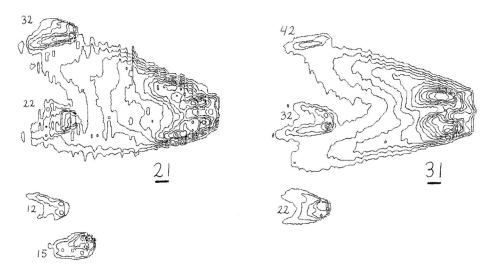

FIGURE 7   Contour plot of lines with equal intensity of radiation for the burners 21 and 31, as depicted Fin Figure 5. Note the difference in burner adjustment.

1.4 m from the burner mouth is close to 2000 K. The Planck-curve does not fit the data above 880 nm. Our explanation is that the strong emission of H_2O between 900 and 1000 nm dominates the emission by soot in this region. Water vapour must be present everywhere at a concentration of 18 volume percent, while there is relatively little soot judged from the transparency of the flames. Since the radiation intensity depends strongly on the temperature the recorded value of the temperature is likely to be associated with the hottest region of the observed slice of the flame. A flaw of the method is the possible dissociation of soot in regions of high temperature. This leaves salt injection as the most reliable approach to the passive measurement of temperatures in flames.

PULSED SALT INJECTION

If salt is injected into the fuel or the air, during a short period, the band of salt vapour will be carried by the gas and emit light in regions of high temperature. The journey of the band of salt can be witnessed with a camera taking pictures at a moment that is delayed with respect to the time of injection. By varying the delay time the velocity field of the flame can be mapped.

The experimental set-up is depicted in Figure 9. The valve in the nozzle is opened electromagnetically during 2 ms, and a small volume of salt solution is nebulized into the primary air of the burner. Aerodynamic friction will quickly reduce the excess speed of the droplets to that of the carrier air. The nozzle is situated at a distance of 12 cm

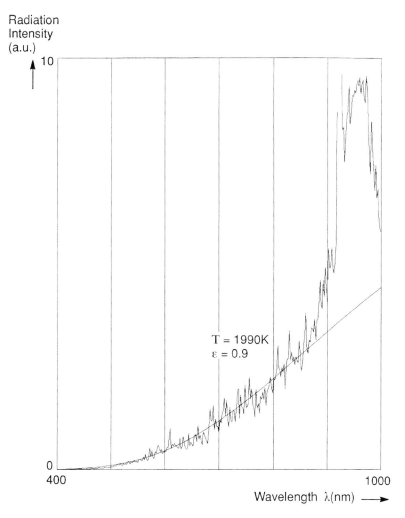

FIGURE 8   The gas temperature derived form a Planck curve fitted to the emission from a slice of the flame at 1.4 m in front of burner 31 (position B in Figure 5).

behind the mouth of the burner. After its release from the nozzle, it takes 12 ms before the salt vapour reaches the burner mouth, so the average speed is 10 m/s (Fig. 10).

Three ms after its entry into the combustion chamber the front of the salt emission has advanced over 1.5 burner diameters, i.e. 14 cm (Fig. 10c). The average droplet speed has increased to 47 m/s, due to rapid heating and expansion of the burning gas.

Figure 10d shows the salt at 17 ms after its front emerged from the burner mouth. The front has then covered a distance of 2.5 burner diameters (23 cm) corresponding to 14 m/s. This indicates a decrease of the forward gas velocity when the flame expands into the furnace. In the depicted case no swirl was present. In experiments with high swirl the flames show a stronger expansion.

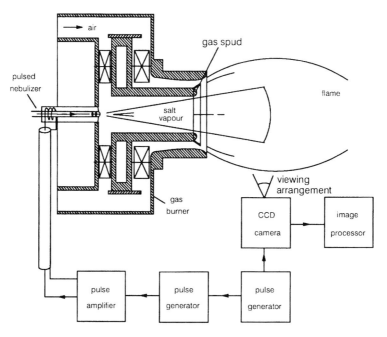

FIGURE 9   Schematic set-up of the experiment for pulsed salt injection into a 100 kW gas flame.

The relatively long camera shutter time of 8 ms plays an important role when the data are interpreted. The horizontal size of the clouds of salt in Figure 10 is, for a greater part, determined by the 8 ms opening-closing cycle of the shutter; the salt travels 11 cm during 8 ms if the gas speed is 14 m/s, whereas the measured length in Figure 10d is 16 cm. Another factor that plays a role in the cloud size is the injection time of about 2 ms.

For practical applications in gas-fired power stations much shorter shutter times are necessary. The speed of air and gas is there about 50 m/s and the desired spatial resolution is 10 cm or better. Therefore, the total shutter cycle should not exceed 1 ms, while the pulsed injection must be performed within 1 ms also. In principle, these requirements can be met if the design of the injector is improved.

CONCLUSION

Flame visualization provides useful information about flame shape and flame intensity of individual burners in a power station. This information can be used to adjust the burners whereafter the results can be checked by an other optical inspection.

Flame temperatures with relatively low errors are obtained from a fit of the Planck-curve to the thermal emission spectrum between 400 and 1000 nm both for coal- and gas flames in power systems. In the case of gas flames, temperatures from cesium

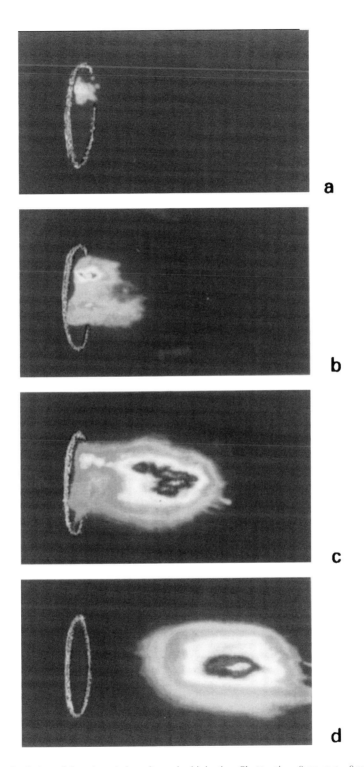

FIGURE 10   Evolution of the salt emission after pulsed injection. Shutter time: 8 ms. a: $t = 0$ ms, the first emission is observed: salt enters the oven b: $t = 1$ ms: further advance of the salt vapour into the oven. c: $t = 3$ ms: integrated view of the salt emission, $t < t_{shutter}$; d: $t = 17$ms: integrated view of the salt emission, $t > t_{shutter}$ The thermal power of the flame is 100 kW, no swirl.

emission are basically more reliable but data are still scarce. Salt injection into gas flames offers interesting prospects for   velocity mapping provided salt is administered in short pulses. Imaging times should be correspondingly brief.

ACKNOWLEDGEMENT

This work was carried out by order of and in close cooperation with the Dutch electricity producing companies, notably EPON and EPZ.

REFERENCES

Braam, A. L. H., Hulshof, H. J. M., de Jongh, W. and Thus A.W. (1991) An optical diagnostic method for pulverized coal fired power stations, First *Int. Conf. on Combustion Technologies for a clean environment*, **1**, 20.3.
Kurihara, N., Nishikawa, M., Watanabe, A., Satoh, Y. and Ohtsuka, K. (1986) A combustion diagnosis method for pulverized coal boilers using flame image recognition technology, *IEEE Transactions on Energy Conversion*, 99–103.
Best, P. E., Carangelo, R. M., Markham, J. R. and Solomon, P. R. (1986) Extension of Emission-Transmission Technique to Particulate Samples using FT-IR, *Combustion and Flame*, **66**, 47–66.

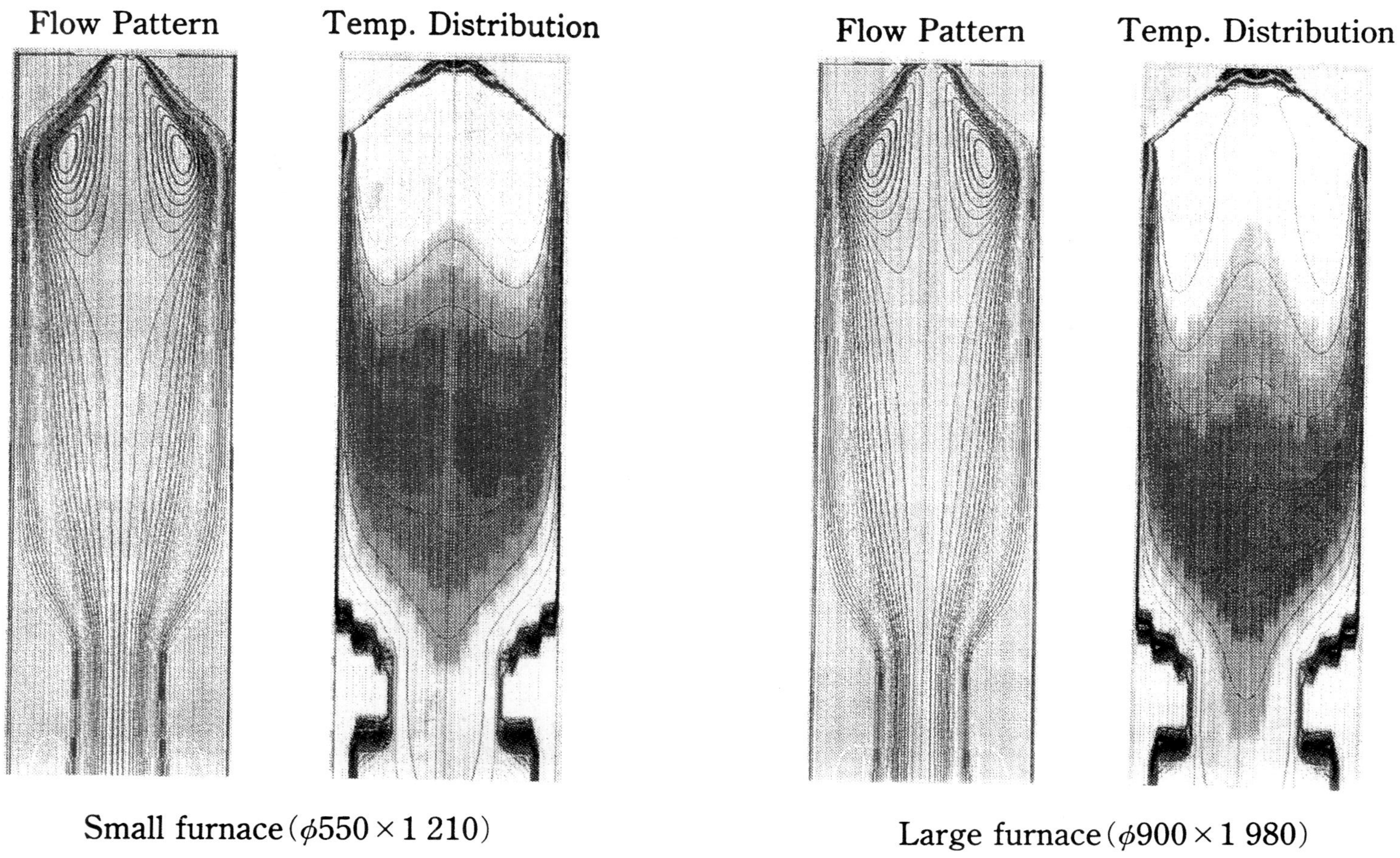

Temp. Distribution

Flow Pattern

Large furnace ($\phi 900 \times 1\,980$)

Temp. Distribution

Flow Pattern

Small furnace ($\phi 550 \times 1\,210$)

COLOR PLATE I. *See* T. Suzuki *et al.*, Figure 10. page 705.

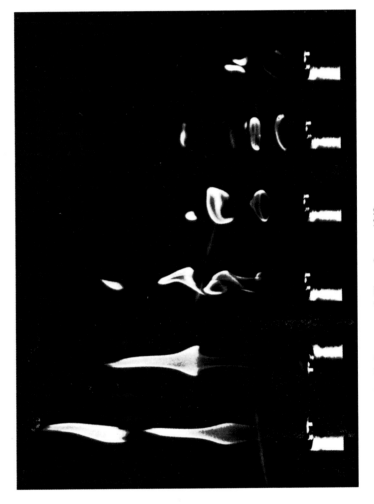

COLOR PLATE II. *See* D. Proctor *et al.*, Figure 2. page 1048.